EXPERIMENTAL CHAOS

Related Titles from AIP Conference Proceedings

617 Bayesian Inference and Maximum Entropy Methods in Science and Engineering: 21st International Workshop on Bayesian Inference and Maximum Entropy Methods in Science and Engineering
Edited by Robert L. Fry, May 2002, 0-7354-0063-6

597 Nonequilibrium and Nonlinear Dynamics in Nuclear and Other Finite Systems: International Conference
Edited by Zhuxia Li, Ke Wu, Xizhen Wu, enguang Zhao, and Fumihiko Sakata, November 2001, 0-7354-0041-5

574 Modeling Complex Systems: Sixth Granada Lectures on Computational Physics
Edited by Pedro L. Garrido and Joaquín Marro, July 2001, 0-7354-0013-X

568 Bayesian Inference and Maximum Entropy Methods in Science and Engineering: 20th International Workshop
Edited by Ali Mohammad-Djafari, June 2001, 0-7354-0004-0

567 Bayesian Inference and Maximum Entropy Methods in Science and Engineering: 19th International Workshop
Edited by Joshua T. Rychert, Gary J. Erickson, and C. Ray Smith, May 2001, 0-7354-0003-2

560 Nanoscale Linear and Nonlinear Optics: International School on Quantum Electronics
Edited by Mario Bertolotti, Charles Bowden, and Concita Sibilia, April 2001, 1-56396-993-9

553 Disordered and Complex Systems
Edited by Peter Sollich, A. C. C. Coolen, L. P. Hughston, and R. F. Streater, February 2001, 1-56396-983-1

548 Fundamental Issues of Nonlinear Laser Dynamics
Edited by Bernd Krauskopf and Daan Lenstra, December 2000, 1-56396-977-7

468 Nonlinear and Collective Phenomena in Beam Physics — 1998 Workshop: International Committee on Future Accelerators
Edited by Swapan Chattopadhyay, Max Cornacchia, and Claudio Pellegrini, April 1999, 1-56396-862-2

411 Applied Nonlinear Dynamics and Stochastic Systems Near the Millennium
Edited by James B. Kadtke and Adi Bulsara, November 1997, 1-56396-736-7

To learn more about these titles, or the AIP Conference Proceedings Series, please visit the webpage **http://proceedings.aip.org**

EXPERIMENTAL CHAOS

6th Experimental Chaos Conference

Potsdam, Germany 22–26 July 2001

EDITORS

Stefano Boccaletti
Istituto Nazionale di Ottica Applicata, Florence, Italy

Bruce J. Gluckman
George Mason University, Fairfax, VA

Jürgen Kurths
Universität Potsdam, Potsdam, Germany

Louis M. Pecora
Naval Research Laboratory, Washington, DC

Mark L. Spano
University of Missouri at St. Louis, St. Louis, MO

SPONSORING ORGANIZATION
Office of Naval Research

AMERICAN INSTITUTE OF PHYSICS

Melville, New York, 2002
AIP CONFERENCE PROCEEDINGS ■ VOLUME 622

Editors:

Stefano Boccaletti
Istituto Nazionale di Ottica Applicata
Largo E. Fermi, 6
I 50125 Florence
ITALY

E-mail: stefano@ino.it

Louis M. Pecora
Naval Research Laboratory
Code 6340
Washington, DC 20375
USA

E-mail: louis.m.pecora@nrl.navy.mil

Bruce J. Gluckman
Neural Dynamics Laboratory
The Krasnow Institute for Advanced Study
George Mason University
Fairfax, VA 22030
USA

E-mail: bgluckma@gmu.edu

Mark L. Spano
University of Missouri at St. Louis
Center for Neurodynamics
8001 Natural Bridge Road
St. Louis, MO 63121
USA

E-mail: mark.spano@ece.gatech.edu

Jürgen Kurths
Institut für Physik
Universität Potsdam
Am Neuen Palais 10
D-14469 Potsdam
GERMANY

E-mail: juergen@agnld.uni-potsdam.de

L.C. Catalog Card No. 2002106317
ISBN 0-7354-0071-7
ISSN 0094-243X
Printed in the United States of America

CONTENTS

DATA ANALYSIS

APPLICATIONS

TRAFFIC

CONDENSED MATTER

SYNCHRONY

DYNAMO

ECOLOGY

Preface

The 6th Experimental Chaos Conference was perhaps the most international of all the conferences thus far. This was the second Experimental Chaos Conference (ECC) held outside the US (the 3rd ECC was held on Edinburgh, Scotland in 1995). The location in Germany proved to be a good one drawing a wide-ranging group of scientists. The large, diverse crowd drove the topics into a broad spectrum of areas many of which were linked to the conference only by the use of nonlinear dynamics concepts. But that is what made so many seemingly disparate talks and posters interesting to everyone.

In the realm of chemistry I. Kiss and J. Hudson show in corrosion experiments that not only can corrosion reactions be chaotic, but they can be synchronized. Electronics design takes a step forward into the hyperchaotic realm with the talk by K. Murali et. al which shows that one can begin to find rules to guide the design and construction of hyperchaotic circuits. Often circuits are connected to communication devices and techniques and in that connection A. Chenys et. al show that hyperchaotic circuits might be good candidates for broadband signals.

In Geophysics and Astrophysics the behavior of atmospheres, both the earth's and the sun's appear to be well characterized by multifractal geometry. In the case of the solar wind that geometry is related to the chaotic nature of the wind's motion.

Neuroscience was the topic of several talks and papers for this conference. Emerging themes appear to be the control of neuronal activity and the construction of hybrid neuron-circuit systems. The former comes out in B. Gluckman's talk in which seizing clusters of neurons are controlled to a non-seizing state by the application of a static, external electric field. The hybrid systems were introduced in Selvertson's talk in which circuits modeled on neurons were substituted for real neurons in the lobster. Such hybrid circuits actually appear to function properly. Another neuroscience theme that surfaced at the conference was that of using nonlinear data analysis techniques for the characterization of physiological pathologies. For example, the talk by N. Wessel et. al shows how these approaches can be applied to cardiac arrhythmias and that of W. Klonowski et. al to the detection of problems in EEG signals.

Data analysis talks focus a lot on the deduction of dynamics from data. The H.U. Voss et. al and I. Tokuda et. al emphasized this. Surrogates remain a concern in data nonlinear data analysis and the talk by K. Dolan emphasizes this with the introduction of a new surrogate producing method.

The dynamics of combustion which appeared in early Experimental Chaos Conferences was present here in a very interesting set of experimental papers by DeGroote (on Criticality on Flame Spreading), Daw (on Controlling Cyclic Combustion Variations), and Gorman (on Dynamics of Pulsating and Cellular Flames using a High-speed, High-sensitivity Camera). In this area it is obvious that experiments lead theory by many years. The phenomology is quite varied.

Optical sciences remained a large component of this Experimental Chaos Conference as it has in the past. The emphasis on several lasers or on spatiotemporal laser systems has begun with the interesting "optical turbulence" talk of P.L. Ramazza et. al and the study of coupled lasers by B. Forsmann et. al. Several applications of nonlinear lasing behavior were presented all in the realm of communication from Chaotic On-off Keying by Uchida to Optical Data Encryption by Paul and to the still interesting topic of synchronization by Wedekind.

Condensed matter systems typically show a wide variety of dynamical behavior. This year an emphasis appeared to be on stochastic forcing of condensed matter systems. Such driving produced On-off intermittency (U. Behn) and unexpected stochastic resonances in Ferroelectric systems (Diestelhorst).

Synchronization of chaos which showed up in optical systems was also presented in a session of its own. For this conference the topic has broadened to include different kinds of synchronization (e.g. Generalized synchronization in lasers by Tang) and synchronization under extremely noisy conditions using a two-time-scale transmitter/receiver system.

Dynamo experiments, which are traditionally difficult and dangerous, appear to be moving toward full-scale success with the work of Sisan on Liquid Sodium held in a spherical configuration and with water models of the Sodium experiment by Marie et. al. Such systems now produce reliable data and make it possible to understand the electrodynamics of liquid conductors (e.g. sodium) in rotational systems. The rotation makes the experiments highly relevant to stellar and planetary internal dynamics.

New topics that had not been presented before at the previous Experimental Chaos Conferences were Ecology and Traffic dynamics. Despite their apparent distance from physics, chemistry and biology, the main themes of previous conferences, the new topics found favorable reception in the general audience. One could see heads nodding in recognition of patterns as D. Helbing and S. Eubank showed their simulation of crowd and traffic flow. Similarly, ecological dynamics showed that state space concepts can also be applied to populations and insect networks as in L. Stone and E. Ben-Jacob's talks.

The number of talks at this conference was greater than at any previous conference and the breadth of topics much larger than ever before. We hope we can continue the trend of broad topic ranges in the next conferences. This is more difficult than it sounds since much attention must be paid to keeping an audience of wide interests curious about topics from their field. But we feel it is worth the effort since the cross fertilization from previous Experimental Chaos Conferences has shown that seeds sown in other fields can flower abundantly.

The Editors,

Stefano Boccaletti, Bruce Gluckman, Juergen Kurths, Lou Pecora, and Mark Spano

CHEMISTRY

Forcing, Coupling and Feedback of Chaotic Electrochemical Oscillators

István Z. Kiss, Wen Wang, and J.L. Hudson

Department of Chemical Engineering, 102 Engineers' Way, University of Virginia, Charlottesville, Virginia 22904-474

Abstract. Experiments on the dynamic behavior of nickel electrodes in sulfuric acid are discussed. We first describe the effect of periodic forcing of a single chaotic oscillator and show the occurrence of phase synchronization and suppression of chaos through intermittency. The collective dynamics of arrays of chaotic oscillators are then considered. The applications of global coupling, of periodic forcing, and of feedback to the array are discussed and the effects of the three are compared. Synchronization and both intermittent and stable chaotic cluster configurations are seen in all three cases; other details differ, however. The feedback and periodic forcing produce configurations of more than two stable periodic clusters; up to four clusters occur with forcing.

INTRODUCTION

Synchronization of chaotic oscillators has received considerable attention in the past few years. Theoretical investigations have shown the interesting dynamic behavior of small sets (usually two) of chaotic oscillators [1-3] and the conditions for complete synchronization of identical chaotic systems have been analytically determined [1,4-6]. Phase synchronization during the periodic forcing of a single chaotic oscillator, that is, in the case of unidirectional coupling, has been investigated [7,8]. A few experiments have been carried out on synchronization of small sets of chemical oscillators [9,10].

The rich dynamics of larger sets of coupled chaotic systems, including some behavior not possible in smaller sets, has been investigated using maps [11-14] and ordinary differential equations [15-17]. Turbulent, partially ordered, ordered, and coherent states can occur [18]. Phase synchronization of sets of coupled differential equations by periodic forcing has been seen in simulations [8]. Suppression of spatiotemporal chaos have also been shown by forcing [19] and feedback [20].

In this paper we review some of our experimental results on coupled chaotic oscillators. We use as our system an electrochemical reaction, the electrodissolution of nickel in sulfuric acid. We first describe experiments on phase synchronization with forcing of a single oscillator [21] and then consider the collective dynamics of larger arrays. The effects of global coupling [22], feedback [23], and forcing [24] on a set of chaotic oscillators are investigated. Stable and intermittent clustering and synchronization occurring in the three types of systems are compared [25,26].

CP622, *Experimental Chaos: 6th Experimental Chaos Conference*, edited by S. Boccaletti et al.

EXPERIMENTS

A standard electrochemical cell consisting of a nickel working electrode, a $Hg/Hg_2SO_4/K_2SO_4$ reference electrode, and a platinum mesh counter electrode was used. Experiments were carried out in 4.5 M H_2SO_4 solution at a temperature of 11 ^{0}C. A schematic of the apparatus is shown in Fig. 1.

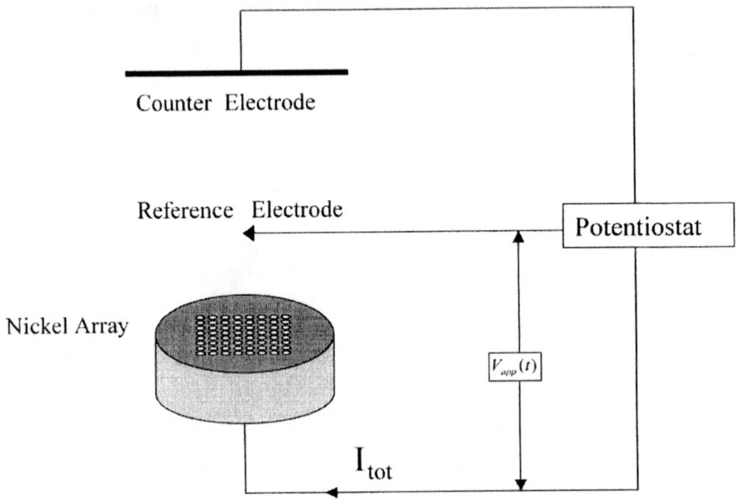

FIGURE 1. Schematic of the apparatus.

The 64-electrode array in an 8x8 geometry is shown; for experiments with a single oscillator the array is replaced with one electrode. The working electrodes are embedded in epoxy and reaction takes place only at the ends. The diameter of the electrode in the experiments on forcing of one oscillator is 2 mm and those of the 64 elements in the array are 1 mm. The currents of the electrodes are measured independently at a sampling rate of 100 Hz and thus the rate of reaction as a function of position and time is obtained. Additional details on the experiments can be found in the original papers [22,27].

In the forcing experiments of both a single electrode and the array a sinusoidal signal with frequency Ω and amplitude A is superimposed on the applied potential:

$$V_{app}(t) = V_0 + A\sin(2\pi\Omega t) \tag{1}$$

Global coupling is added to the array with the use of resistors. Since the electrolyte concentration is high, the external resistance is larger than that of the electrolyte and the added global coupling dominates over the non-global coupling inherent in the system. The total external resistance is held constant while the fraction dedicated to individual currents is varied. The external resistors are added individually (R_{ind}) or collectively (R_{coll}). A total resistance (R_{tot}) is then defined as

$$R_{tot} = R_{coll} + \frac{R_{ind}}{n} \quad (2)$$

where n is the number of electrodes. In these experiments R_{tot} = 14.2ohms. The collective resistor couples the electrodes globally: the current on one electrode affects the dynamics of the other electrodes since current through any given electrode influences the potential drop on all electrodes equally. The parameter (ε) is the ratio of collective to total resistance:

$$\varepsilon = \frac{R_{coll}}{R_{tot}} \quad (3)$$

This value of ε takes on values from zero to one as the global coupling increases.

In the feedback experiments a feedback signal is generated with gain (K) and delay (τ). The perturbation of output voltage is proportional to the difference of the total current ($I_{tot}(t - \tau)$) and the pre-calculated mean total current I_{tot}, mean; the applied potential is then:

$$V_{app}(t) = V_o + K(I_{tot}(t-\tau) - I_{tot,mean}) \quad (4)$$

Since the feedback is a small amplitude perturbation of the applied potential, we assume that the mean total current does not change significantly after the feedback is imposed. The mean total current ($I_{tot,mean}$) is used as the offset of the feedback signal to maintain the applied potential in the chaotic region.

RESULTS

Periodic forcing of single chaotic oscillator

An attractor reconstructed by the use of time delays from a time series of a chaotic state of a single (unforced) oscillator is shown in Fig. 2a; this chaotic state is reached via a period-doubling bifurcation sequence as the applied potential is changed.

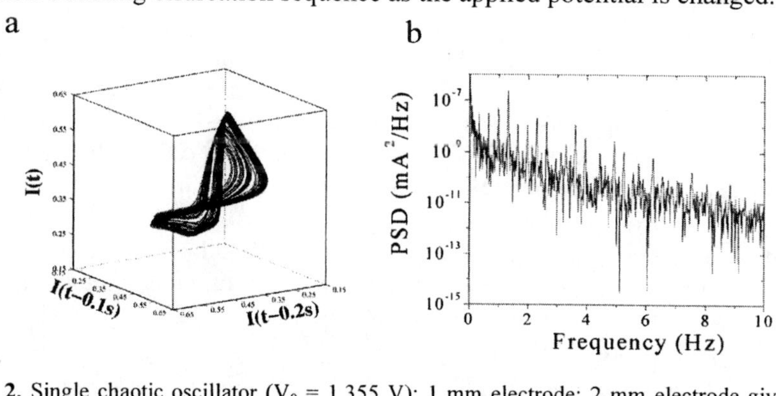

FIGURE 2. Single chaotic oscillator (V_0 = 1.355 V); 1 mm electrode; 2 mm electrode gives similar behavior. a. Chaotic attractor reconstructed from the current of a single chaotic oscillator using time-delay coordinates. b. Power spectrum.

The information dimension of the attractor is 2.2. A power spectrum made from the time series is shown in Fig. 2b. The frequency of the chaotic oscillation is $\omega_0 = 1.325$ Hz.; this was found by use of the Hilbert transform to obtain the instantaneous phase and amplitude of the time series [8,28].

Periodic forcing was applied to the single chaotic oscillator. The phase-locked region in forcing frequency/forcing amplitude parameter space is shown in Fig. 3.

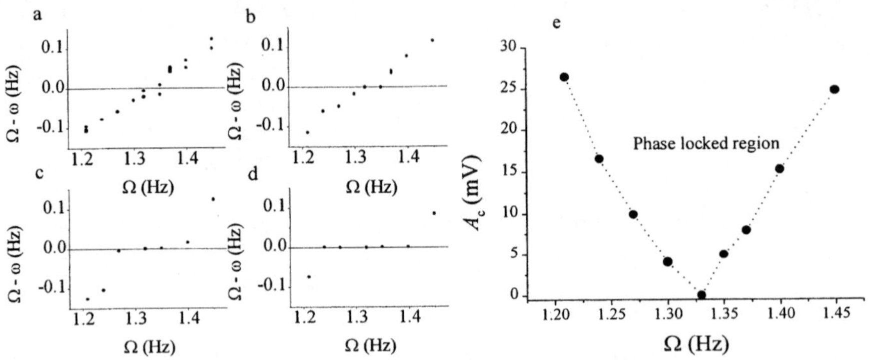

FIGURE 3. Single forced chaotic oscillator. The frequency difference (Ω-ω) as a function of the forcing frequency (Ω) for a) A=0 mV, b) A=6.6 mV, c) A=13.2 mV, d) A=16.5 mV. e) The phase locked region in A-Ω space.

For forcing frequency Ω close to the unforced chaotic frequency ω_0 the phase locking results in a chaotic state, that is, the chaotic signal takes on the frequency of the forcing (ω = Ω) and the oscillations have an approximately fixed, nonzero phase difference.

Forcing was carried out over a range of frequencies between 1.21 Hz and 1.45 Hz. The frequency differences Ω - ω vs. forcing frequency Ω are presented in Figure 3 for various amplitudes. At A=0 (without forcing, Figure 3a) the points lie approximately on a line with a slope of unity; deviations are due to the small variations of ω_0 . For a small amplitude (A = 6.6 mV, Figure 3b) phase synchronization occurs only for frequencies close to ω_0. As the amplitude is made larger, the phase synchronized frequency region increases as can be seen in Figs. 3c and 3d. At the smallest (Ω =1.21Hz) and the largest (Ω = 1.45 Hz) frequencies phase synchronization occurs at larger values of forcing amplitude ($A \approx$ 25mV). Figure 3e shows the critical forcing amplitude A_c at which phase synchronization is observed. The shape of the experimentally determined synchronization tongue is approximately symmetric around ω_0.

For larger forcing frequencies (Ω ≥ 1.40Hz) the transition into the phase-locked region does not occur via a chaotic phase synchronization but rather there is an intermittent transition from chaos to a periodic state; the periodic state is P2 for Ω =1.45 Hz, and P4 for Ω =1.40Hz.

Global coupling of arrays of chaotic oscillators

We now turn to an array of 64 electrodes. The behavior of the weakly coupled array of 64 oscillators (without added coupling, feedback, or forcing) is shown in Fig. 4a.

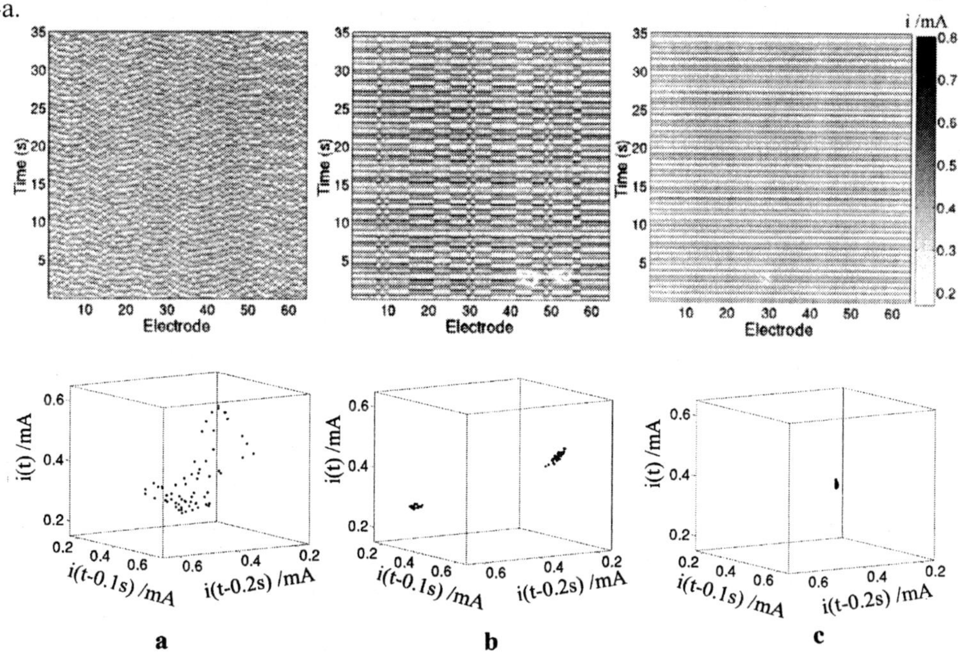

FIGURE 4. An array of 64 Ni electrodes at different global coupling strengths. a) ε=0, uncoupled oscillators. b) ε =0.725, chaotic clustering. c) ε =1, synchronized chaotic oscillators. Top: space/time plots of the currents of the electrodes, dark corresponds to high, light to low current. Bottom: snapshots in state space of the 64 elements.

A snapshot showing the positions of the 64 elements in state space is seen above. Note that the individual positions are distributed over the attractor. The currents from all 64 electrodes are shown below in a space/time plot; dark corresponds to high and light to low current.

As the global coupling strength is increased the oscillators undergo the sequence of behavior: Very weakly coupled chaos → Intermittent chaotic clusters → Stable chaotic clusters → Intermittent chaotic clusters → Chaotic synchronized state [22]. These transitions are depicted schematically in the middle row of Fig. 5. Space/time plots and snapshots in state space are shown in Fig. 4b and 4c for a stable chaotic cluster and a synchronized state, respectively and representative cluster configurations are shown in Fig. 5.

In the 64-electrode system many stable cluster configurations are possible. We have observed clusters with as few as 18 elements so that the observed range is (18,46) to (32,32). A finite range of stable two-cluster configurations has also been observed in theoretical studies with coupled maps [12,13]. Two of the cluster configurations obtained in the nickel experiments are shown in Fig. 5; one of these is a fairly balanced configuration (30,34) whereas in the other there are 19 and 45 elements in each cluster, respectively.

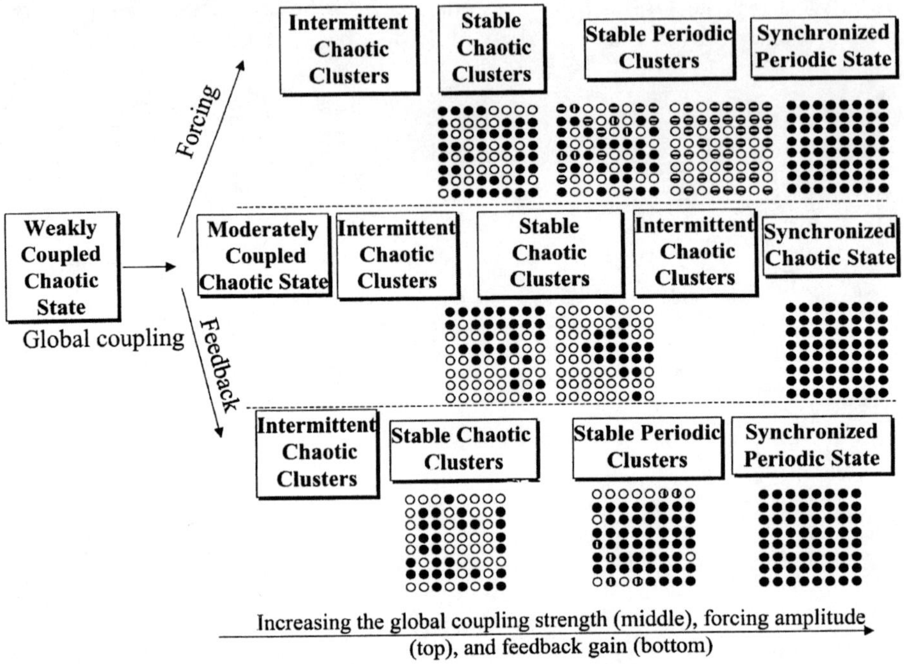

FIGURE 5. The collective behavior of an array of 64 Ni electrodes with global coupling, forcing, and feedback. Top: with increasing the amplitude of the forcing. Middle: with increasing global coupling strength. Bottom: with increasing the gain of the feedback. In the stable clustering region some representative cluster configurations are shown.

Note in both configurations of Fig. 5 that the edge elements have a tendency to be in the same cluster. As mentioned above, there is a relatively small (but not zero) coupling that is inherent in the system and that exists even without the added coupling; coupling among the edge regions is somewhat stronger than it is among elements in the interior of the array. Although the stable cluster configurations are unaffected by small perturbations, a sufficiently large disturbance can move an element or elements form one cluster to another.

In the intermittent cluster regions on either side of the stable cluster region, clusters from and break up. Sometimes two clusters form and at other almost complete synchronization occurs. However, neither the cluster formation nor the synchronization is stable. Both break up after a short time.

Periodic forcing of arrays of chaotic oscillators

The forcing is applied to a base state in which some global coupling is present, that is, to a state with a global coupling strength of $\varepsilon = 0.56$. We use this base state, rather than the very weakly coupled state because the forcing is not strong enough to synchronize the turbulent state ($\varepsilon = 0.0$) under the conditions of these experiments and because we wish to compare the transitions into and through the cluster states of forcing (and the feedback in the next section) with the globally coupled case; we thus start the sets of experiments at a state just below the stable cluster region. The forcing frequency is chosen to be near the dominant frequency of the unforced chaotic signal, ω_0 = 1.3 Hz. With variations in the forcing amplitude over the range 0 to 50 mV the system, as shown in Fig. 5, goes through the following sequence: Unsynchronized chaos → Intermittent chaotic clusters → Stable chaotic clusters → Stable periodic clusters → Periodic synchronized state [24].

At 5 mV forcing amplitude the tendency of clustering and synchronization can be already seen but the clustered states are not stable and exist for short times. As the forcing amplitude is increased to 25 mV intermittent clustering can be more clearly seen; clusters form more often and remain intact for longer times before breaking up. The clusters are still unstable but can hold together for times up to approximately 15 oscillation cycles. The clusters became stable as the as the forcing amplitude was increased to 30 mV. In the stable chaotic cluster region two clusters are always seen. The arrangement seen in Fig. 5 is (26,38), i.e., the two clusters contain 26 and 38 elements, respectively.

As the amplitude is increased further from the stable chaotic cluster region, stable periodic clusters form. Two to four stable clusters are seen, depending on conditions. Additional increase to a forcing amplitude of 50 mV results in a period-one, synchronized state. Representative cluster arrangements are shown in Fig. 5. The four-cluster periodic configuration obtained at A = 35 mV is (23,25,11,5). As the forcing amplitude was further increased to 45 mV two periodic clusters were observed. For higher forcing amplitudes, 50 mV, a synchronized periodic oscillation was obtained.

Two, three, and four cluster states have been observed. The sizes of the groups differ. When four clusters are formed there is a tendency for the clusters to group, that is, for two of the clusters to have similar time series but with a phase lag; the other two clusters then have a different time series. Thus two of the clusters are on the same (periodic) attractor and the other two have a different cycle. When only two clusters exist, the limit cycles of the two clusters differ and thus the two do not just differ by a phase lag. The four-cluster state gives way to a two-cluster state as the forcing amplitude is increased. Although we do not have a sufficiently high resolution in the parameter (amplitude of forcing), it appears that the transition from a state with four clusters to one with two clusters may occur by a mechanism in which the sets of clusters with the same time series (but different phases) merge through a loss of the phase difference.

Feedback on arrays of chaotic oscillators

Feedback was added to the chaotic oscillators using the same base state as that used in the forcing studies described in the section above [23]. The behavior was studied as a function of the feedback gain. The results presented here will all be for a zero time delay (τ = 0 s in Eqn. 4). The dependence on feedback gain is also shown in Fig. 5, i.e., with increasing feedback gain we see: Unsynchronized chaos → Intermittent chaotic clusters → Stable chaotic clusters → Stable periodic clusters → Periodic synchronized state → Stable steady state.

Representative cluster arrangements are shown in Fig. 5 for three values of the feedback gain. In the chaotic region two clusters are always obtained and the numbers of elements in the two clusters are approximately evenly balanced. The distribution ranged from (32,32) to (28,36); one of the (29,35) configurations is shown in Fig 5. As the feedback gain is increased and the behavior becomes periodic, two trends are noticed in the configurations: an imbalance in cluster sizes occurs and a third cluster arises. As the feedback gain is further increased, one of the clusters begins to dominate as seen in Fig. 5 and with further increase this cluster encompasses the entire array and synchronization is obtained.

CONCLUDING REMARKS

We have reviewed some experimental studies on the dynamics of coupled chaotic oscillators using as a model system the electrodissolution of nickel in sulfuric acid solution.

Phase synchronization during periodic forcing of a single oscillator was first discussed. Unsynchronized, intermittently synchronized (with phase slips), and phase synchronized states were observed with increases in the forcing amplitude. In the phase synchronized region the phases are locked with non-zero phase difference. The bifurcations responsible for phase synchronization of chaotic oscillators have been described theoretically [29]; the role of phase locking of the unstable periodic orbits embedded in the chaotic attractor was emphasized. In our experiments when the forcing frequency is sufficiently removed from ω_0 an intermittent transition took place resulting in a periodic phase-synchronized state.

We also investigated populations of oscillators using electrode arrays to which global coupling, periodic forcing, and feedback were applied. The three types of experiments all produce intermittent and stable chaotic cluster states and synchronization with sufficiently large values of global coupling, feedback gain, and forcing amplitude, respectively. The three cases differ, however, in other details. With global coupling the stable chaotic cluster region is bounded on both sides by less ordered regions of intermittent clusters. With feedback, increases in gain past the stable chaotic cluster region yield periodic clusters, first two, and then three. As the gain becomes larger one of the three clusters grows and the array is synchronized. With the imposition of periodic forcing as many as four periodic stable clusters emerge.

The imposed coupling differs in the three types of experiments. Global coupling produces direct interactions among the elements. The external forcing has a different role. There is some very weak inherent local and long-range coupling (through the electrolyte) and also moderate global coupling (controlled by system electronics) in the system to which the forcing is applied. This intrinsic coupling was too weak to produce clustering or synchronization but it did furnish interactions among the elements and contributed to the collective dynamics of the system. The application of the periodic forcing changes the dynamics of the individual elements and clustering and synchronization become possible with the same strength of intrinsic coupling.

The feedback influences the dynamics of the individual oscillators and adds an additional global coupling.

Although many theoretical studies of collective dynamics have been carried out with examples in biology, physics, and chemistry, far fewer experimental studies are known. The electrochemical arrays are ideal for such experiments since the dynamics can be measured at each site and the coupling can be carefully controlled.

ACKNOWLEDGMENTS

We acknowledge with thanks the support of the Office of Naval Research and the National Science Foundation.

REFERENCES

1. Fujisaka, H. and Yamada, T., *Prog. Theor. Phys.* **69,** 32 (1983).
2. Pecora, L. M. and Carroll, T. L., *Phys. Rev. Lett.* **64,** 821-824 (1990).
3. Pikovsky, A. S. and Grassberger, P., *J. Phys. A-Math. Gen.* **24,** 4587-4597 (1991).
4. Pecora, L. M., Carroll, T. L., Johnson, G. A., Mar, D. J., and Heagy, J. F., *Chaos* **7,** 520-543 (1997).
5. Pecora, L. M., *Phys. Rev. E* **58,** 347-360 (1998).
6. Heagy, J. F., Carroll, T. L., and Pecora, L. M., *Phys. Rev. E* **50,** 1874-1885 (1994).
7. Stone, E. F., *Phys. Lett. A* **163,** 367-374 (1992).
8. Pikovsky, A. S., Rosenblum, M. G., Osipov, G. V., and Kurths, J., *Physica D* **104,** 219-238 (1997).
9. Zeyer, K. P., Holz, R., and Schneider, F. W., *Ber. Bunsen-Ges. Phys. Chem.* **97,** 1112-1119 (1993).
10. Kiss, I. Z., Wang, W., and Hudson, J. L., *PCCP Phys. Chem. Chem. Phys.* **2,** 3847-3854 (2000).
11. Kaneko, K., *Physica D* **37,** 60-82 (1989).
12. Kaneko, K., *Physica D* **41,** 137-172 (1990).
13. Kaneko, K., *Physica D* **54,** 5-19 (1991).
14. Xie, F. and Cerdeira, H. A., *Phys. Rev. E* **54,** 3235-3238 (1996).
15. Zanette, D. H. and Mikhailov, A. S., *Phys. Rev. E* **57,** 276-281 (1998).
16. Osipov, G. V., Pikovsky, A. S., Rosenblum, M. G., and Kurths, J., *Phys. Rev. E* **55,** 2353-2361 (1997).
17. Zanette, D. H. and Mikhailov, A. S., *Phys. Rev. E* **62,** R7571-R7574 (2000).
18. Kaneko, K., *Phys. Rev. Lett.* **63,** 219-223 (1989).
19. Wu, S., He, H., and Huang, Z., *Phys. Lett. A* **260,** 345-351 (1999).
20. Battogtokh, D. and Mikhailov, A., *Physica D* **90,** 84-95 (1996).
21. Kiss, I. Z. and Hudson, J. L., *Phys. Rev. E* in press (2001).
22. Wang, W., Kiss, I. Z., and Hudson, J. L., *Chaos* **10,** 248-256 (2000).
23. Wang, W., Kiss, I. Z., and Hudson, J. L., *Ind. Eng. Chem. Res.* in press (2002).
24. Wang, W., Green, B. J., and Hudson, J. L., *J. Phys. Chem. B* in press (2001).
25. Kiss, I. Z., Wang, W., and Hudson, J. L., *Chaos* submitted (2001).
26. Wang, W., Kiss, I. Z., and Hudson, J. L., *Phys. Rev. Lett.* **86,** 4954-4957 (2001).
27. Kiss, I. Z., Wang, W., and Hudson, J. L., *J. Phys. Chem. B* **103,** 11433-11444 (1999).
28. Gábor, D., *J. IEE London* **93,** 429 (1946).
29. Pikovsky, A., Zaks, M., Rosenblum, M., Osipov, G., and Kurths, J., *Chaos* **7,** 680 (1997).

ELECTRONICS

Design Principles of Hyperchaotic Circuits

K. Murali*, E. Lindberg+, H. Leung*

*Department of Electrical and Computer Engineering, University of Calgary, 2500 University Drive N.W., Calgary, AB T2N 1N4, Canada

+Orsted.DTU Department, 348 Technical University of Denmark, DK-2800, Kgs.Lyngby, Denmark

Abstract. An introduction to the design methodology of simple hyperchaotic circuits with stable and unstable oscillators is considered. A number of configurations, which may serve as the physical mechanisms behind hyperchaotic behaviour are studied. The basic design principle is demonstrated by means of different simple hyperchaotic circuits with core RC sinusoidal oscillator and diode as the single nonlinear element. Further, a simple mixed mode hyperchaotic circuit is introduced for chaos based digital communication applications.

INTRODUCTION

The design of chaotic oscillator circuits has been the subject of active interest due to the usage of these circuits in various engineering applications and in particular chaos based communications [1-3]. The performance of these systems mainly depends on the individual characteristic nature of the inherent chaos circuits. Therefore for an appropriate application, a specific chaotic circuit with desirable properties has to be used. Especially for signal encryption and communication applications, increasing interest has been developed to use hyperchaotic circuits characterized by multiple positive Lyapunov exponents (LE) [1-3]. In order to obtain hyperchaotic oscillations from an electronic circuit it should be at least a fourth-order one [4-21]. However, among the different types of chaotic circuits proposed in the literature, majority of the circuits are often modeled by a third-order system of nonlinear differential equations and few circuits are described by fourth-order system of equations [4-6,13,14,20,21]. Also with the available fourth-order chaotic circuits, only a handful of them have exhibited hyperchaos whose chaotic attractor is characterized by more than one positive Lyapunov exponent. Therefore increasing attention has been paid recently towards systematic methods for designing higher-order chaotic circuits [13-19]. In particular, active linear building blocks and passive nonlinearities are preferable because different chaotic circuits can be synthesized once a specific design rule is arrived [13,22].

Hyperchaos was first reported from computer simulations of hypothetical ordinary differential equations in [23]. The first observation of hyperchaos from a real experimental system, a fourth-order electrical circuit, was reported in [4]. During recent times, different types of hyperchaotic oscillators and circuits have been proposed[5,6,14-18]. These circuits include the nonlinearly coupled 2nd order linear oscillators[6,14,16,17], arrays of coupled nonlinear chaotic oscillators[2,24], infinite

CP622, *Experimental Chaos: 6th Experimental Chaos Conference*, edited by S. Boccaletti et al.

dimensional delay line based oscillators[25], etc. Interestingly, a common feature of the circuits reported in [1,3,13-15] is the nonlinear interaction between different harmonic oscillators. Thus almost all the hyperchaotic circuits proposed so far contain at least one unstable or active oscillator in the circuitry.

In this study, we propose a systematic design principle to construct simple hyperchaotic circuits. In the core of the present hyperchaotic circuits, there exists harmonic oscillators (stable or unstable) and how they are nonlinearly coupled by simple passive or active nonlinear elements made up of diodes. Further, we demonstrate here that almost all the circuits introduced here can be physically decomposed either to a resonator or to another classic sinusoidal oscillator through passive only nonlinear coupling. It is confirmed that hyperchaotic phenomena are observed in all the present circuits by circuit experiments, computer simulations of the governing differential equations and PSPICE circuit simulations. In Section 2, we discuss the present architecture of the system models. Section 3 deals with the circuit configurations of different hyper-chaotic circuits with active oscillators and resonators. In Section 4, we introduce the new circuit realizations based on core classic sinusoidal oscillators. Section 5 reports the dynamics of simple mixed-mode hyperchaotic circuit. Finally, Section 5 focuses on the conclusion.

ARCHITECTURE OF SYSTEM MODEL

We use the following system models shown in Fig. 1 for our design methodology, which consists of an active oscillator, a resonator and nonlinear coupling element. Basically for an active oscillator structure, a negative resistance oscillator with (LC-R) structure is used, in which a negative impedance converter (NIC) is parallely coupled with an LC resonator. The NIC is constructed with an op-amp and three linear resistors [20,21,26]. For the case of resonator, a passive LC oscillator is used. The nonlinear elements are constructed with either passive or active circuit elements. Mainly general purpose diodes are used to realize the nonlinear elements. The characteristic curves of the different nonlinear elements are depicted in Figs. 2(a-c). In the structures represented in Fig. 1, in order to admit hyperchaos, the following minimal conditions are being satisfied: (i) at least one of the blocks is active (ii) at least four energy storage elements exist and (iii) atleast one nonlinear element exists. Here for simplicity the piecewise-linear model of the nonlinear elements are considered. The characteristic curves of Fig. 2(a-c) are modelled as [20,26]

$$\text{(i)}\quad I_N = f(V) = 0.5G_b(|V-B_P|+V-B_P), \tag{1a}$$

$$\text{(ii)}\quad I_N = f(V) = G_bV+0.5(0-G_b)(|V+B_P|-|V-B_P|), \tag{1b}$$

$$\text{(iii)}\quad I_N = f(V) = G_aV+ 0.5G_b(|V-B_P|+V-B_P), \tag{1c}$$

corresponding to Fig. 2(a), 2(b) and 2(c) respectively. Here G_a, G_b and B_p are the negative slope, positive slope and the break-point voltage respectively. In the present study passive nonlinear coupling is used in the architectures of Figs. 1(a) and 1(b). For Fig. 1(c), active nonlinear coupling is employed.

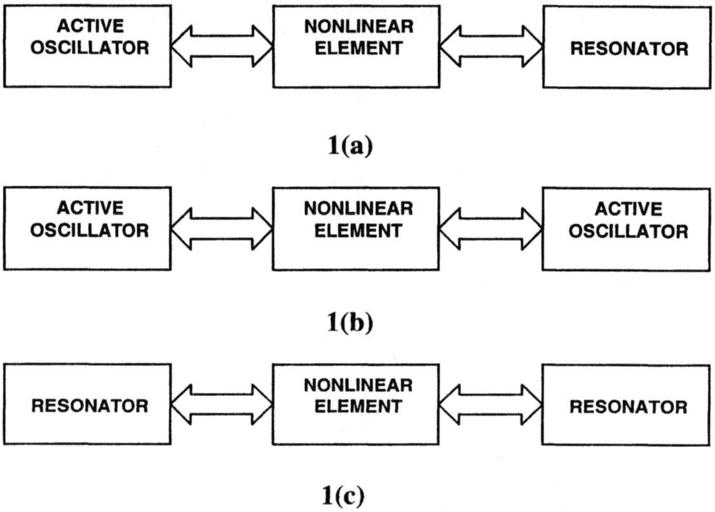

FIGURE 1. General architecture of autonomous hyper-chaotic oscillators: Here active oscillator is the classic LC-R oscillator with frequency f_a, the resonator is a classic passive LC tank circuit with frequency f_p and the nonlinear element is made-up of either active or passive devices. $f_a \approx \pi f_p$(irrational frequency ratio).

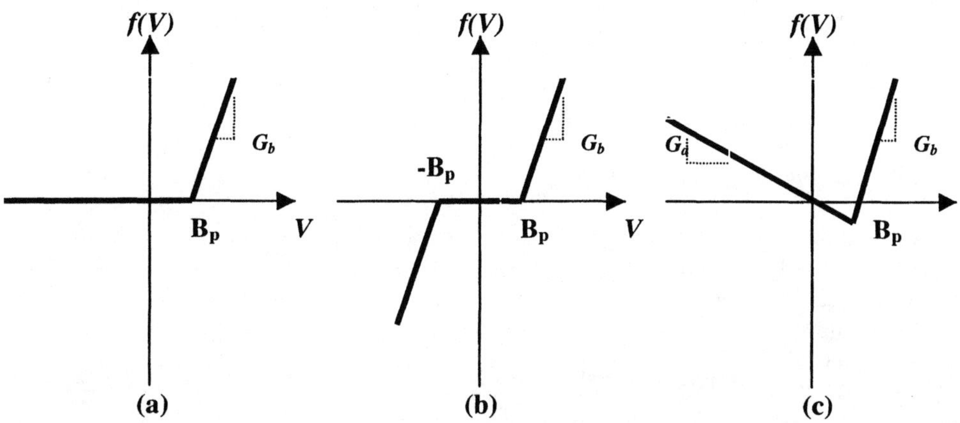

FIGURE 2. The characteristic curves of different nonlinear elements (NLEs). (a) and (b) correspond to passive nonlinear elements and (c) corresponds to active nonlinear element.

HYPER-CHAOTIC CIRCUITS WITH ACTIVE OSCILLATORS AND RESONATORS

In this section, we have introduced the various hyper-chaotic circuit configurations shown in Fig. 3(a-d) belonging to anyone of the architectures of Fig. 1 with either passive or active nonlinear coupling element. The dynamics of circuit in Fig. 3(b) are described by Eq. (2) as [16,17]

$$C_1\frac{dV_1}{dt}=-\frac{V_1}{R_1}-I_{L1}-I_N,\ L_1\frac{dI_{L1}}{dt}=V_1,\ C_2\frac{dV_2}{dt}=-\frac{V_2}{R_2}-I_{L1}+I_N,\ L_2\frac{dI_{L2}}{dt}=V_2. \quad (2)$$

Equation (2) is used as the generalized equations to represent the dynamics of circuits of Fig.3(a-d). By introducing the notations

$$x_1=\frac{V_1}{B_p},\ x_2=\rho\frac{I_{L1}}{B_p},\ x_3=\frac{V_2}{B_p},\ x_4=\rho\frac{I_{L2}}{B_p},\ \rho=\sqrt{L_1/C_1},\ \tau=\sqrt{L_1C_1},\ \varepsilon=\frac{C_2}{C_1},$$

$$\mu=\frac{L_2}{L_1},\ G_b=\frac{1}{r_d+R_s},\ b=G_b\rho,\ \theta=\frac{t}{\tau},\ \dot{u}=\frac{du}{d6},\ \alpha=\frac{\rho}{|R_1|},\ \beta=\frac{\rho}{|R_2|},\ k=\frac{\rho}{R_c},$$

with $y = x_1 - x_3$, we obtain the following set of normalized circuit equations:

$$\dot{x}_1=\alpha x_1-x_2-G(y),\ \dot{x}_2=x_1,\ \varepsilon\dot{x}_3=\beta x_3-x_4+G(y),\ \mu\dot{x}_4=x_3, \quad (3)$$

where

$$G(y)=0.5b(|y-1|+y-1), \quad (4a)$$

$$G(y)=b(y)+0.5(0-b)(|y+1|-(|y-1|), \quad (4b)$$

$$G(y)=a(y)+0.5b(|y-d|+y-d). \quad (4c)$$

are the rescaled mathematical representations of Eqs. (1a),(1b) and (1c) respectively. For the case of Fig. 3(d), $G_a = 1/(R + R_s)$, $G_b = 1/(R_p + R_s)$, $R_p = (R.r_d)/(R + r_d)$,

TABLE 1(a). The circuit parameters used for PSPICE simulations and circuit experiments of Fig.3(a-d)[6,16-18].

Fig.No.	R_1[Ω]	C_1[nF]	L_1[mH]	DIODE	L_2[mH]	C_2[nF]	R_2[Ω]	R_s[Ω]
3(a)	1800	220	256	1N914	84	68	∞	0
3(b)	1800	220	256	1N914	84	68	4500	550
3(c)	1800	220	256	1N914	25.6	68	4500	0
3(d)[1]	∞	220	256	1N4148	84	68	∞	155

[1]R_{L1}=15 Ω, R_{L2}=7.5 Ω and R=-1900 Ω.

TABLE 1(b). The rescaled parameters used for numerical simulations of Eqs.(3) for the circuits of Fig.3(a-d).

Fig.No.	α	β	ε	μ	a	b	D
3(a)	0.55	0	0.31	0.33	0	40.0	1.0
3(b)	0.6	0.24	0.31	0.33	0	1.88	1.0
3(c)	0.6	0.24	0.31	0.1	0	34.24	1.0
3(d)	0	0	0.31	0.33	-0.62	6.0	1.0

$a = G_a\rho$, $b = G_b\rho$, $d = 1\text{-}R_s/|R|$ [18]. The different circuit parameters used for present study for the circuits of Fig. 3 are tabulated in Table. 1(a,b).

Analysis

Figure 4 shows respective hyper-chaotic attractors observed by PSPICE simulation results (a), circuit experimental results (b) and numerical simulation results (c) of rescaled circuit equations (3) for the circuit models of Fig. 3(a-d). As the control parameters are varied typically these circuits exhibit the transition from Hopf bifurcation from periodic orbit to quasiperiodic orbit then to the hyper-chaotic attractors after torus-breakdown. The parameters used are tabulated in Tables (1a) and

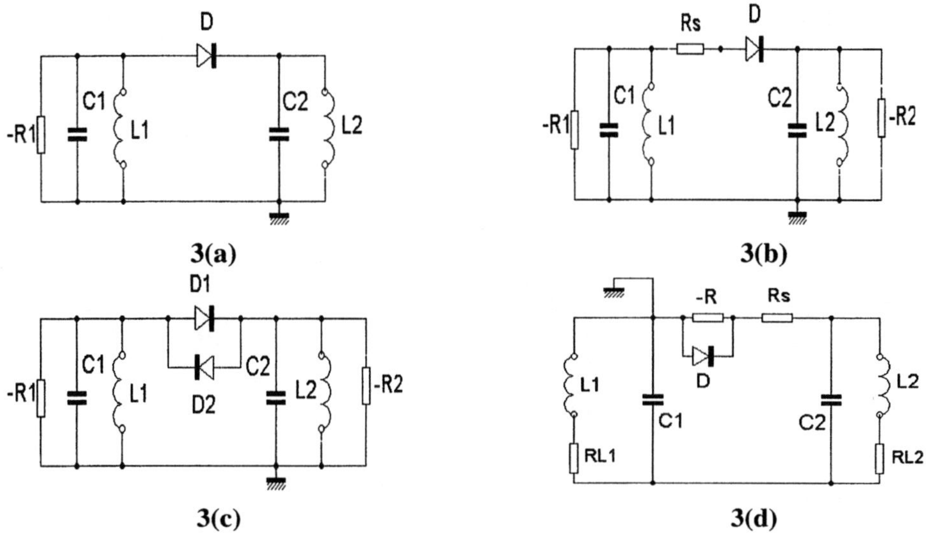

FIGURE 3. Fourth-order hyperchaotic circuits. (a) Active oscillator coupled to a resonator [6]. The characteristic curve of the nonlinear coupling element is shown in Fig. 2(a). (b) Two active oscillators are nonlinearly coupled [16,17]. The characteristic curve of the nonlinear coupling element is shown in Fig. 2(a). (c) Two active oscillators are nonlinearly coupled [22]. The characteristic curve of the nonlinear coupling element is shown in Fig. 2(b). (d) Two resonators are nonlinearly coupled [18]. The characteristic curve of the nonlinear coupling element is shown in Fig. 2(c). The reference node is switched because the negative resistor is a three terminal element [18].

(1b). For actual realization of nonlinear elements only general purpose diodes are used. For hardware implementation, circuits 3(a)-(c) are constructed with only passive nonlinear elements(diodes and linear resistors) and circuit of 3(d) uses an active nonlinear element made up of negative resistor, a general purpose diode and a linear resistor. For the circuit 3(d), the reference node is switched because the connected negative resistor is a three terminal element[18]. However for numerical simulation of Eqs. (3) and (4c) with rescaled parameters correspond to Fig. 3(d), a two-terminal nonlinear element is assumed without switching the reference point[18].

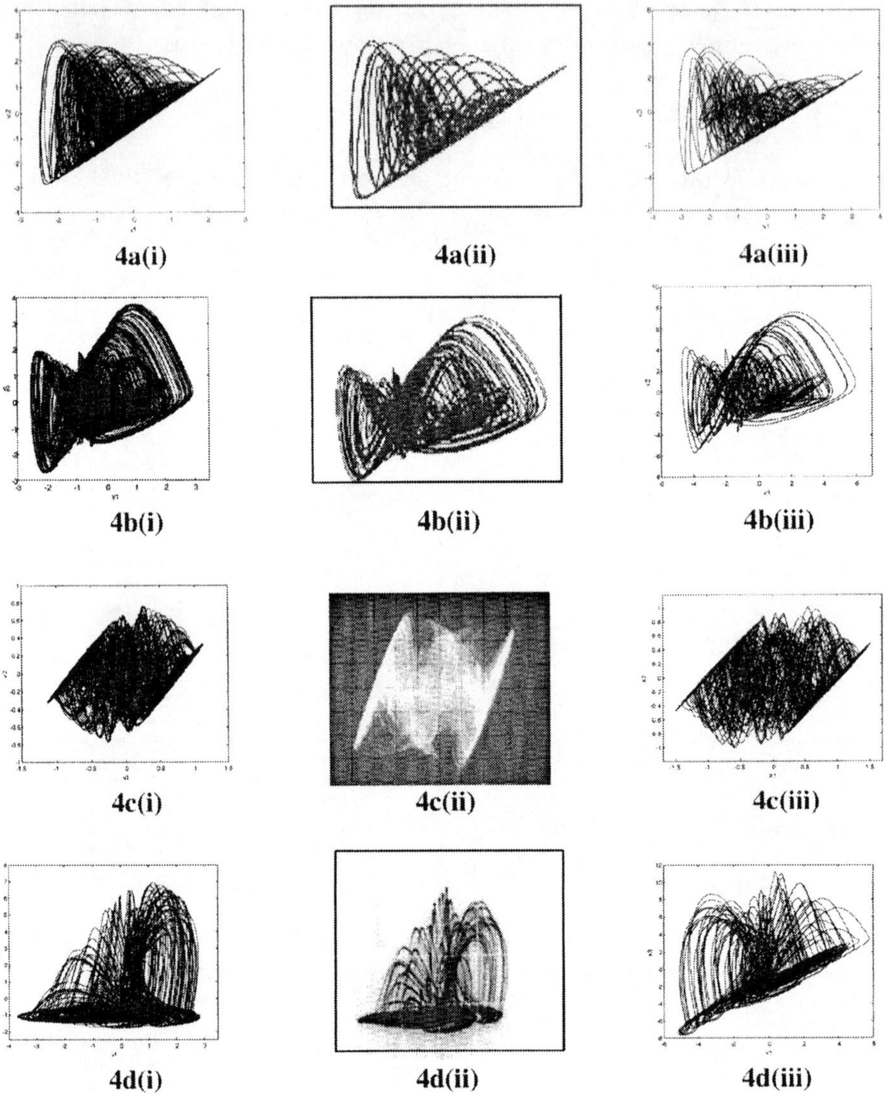

FIGURE 4. Phase-portraits of hyper-chaotic attractors corresponding circuits of Fig. 3(a), (b), (c) and (d) respectively. (i) PSPICE simulation result in $(V_1\text{-}V_2)$ plane. (ii) Experimental result in $(V_1\text{-}V_2)$ plane. (iii) Numerical simulation results of the normalized circuit equations in $(x_1 - x_3)$ plane.

HYPERCHAOTIC CIRCUITS BASED ON SINUSOIDAL OSCILLATOR

Recently, the applicability of RC sinusoidal oscillators in the design of different types of chaotic circuits has attracted considerable interest [7-13]. These circuit designing principles are motivated mainly because at least one chaotic oscillator can be obtained from any sinusoidal oscillator [10-13]. The basic idea here is that one can replace the active oscillator circuit, i.e., the LC-R configuration with that of classic RC sinusoidal oscillator (Wien-bridge, Twin-T, 2R-2C type, etc.). Accordingly, in the following we show the possibility of synthesizing different simple configurations of hyperchaotic circuits based with at least one RC sinusoidal oscillator namely, the Wien-bridge oscillator. In these cases any of the active oscillator (negative resistance oscillator LC-R) of Fig. 3 is substituted with Wien-bridge oscillator. The new circuit implementations are depicted in Fig. 5(a-d). The Wien-bridge oscillator is constructed with conventional op-amp (AD712 or μ A741). The gain is controlled by the ratio of the resistors R_A and R_B as $K = 1 + R_B/R_A$ [7-13]. The Wien-bridge sinusoidal oscillator is ideally described in dimensionless form by:

$$\dot{x}_1 = (K-2)x_1 - x_2, \qquad \dot{x}_2 = (K-1)x_1 - x_2, \tag{5}$$

when equal-R equal-C design ($R_1 = R_2 = R_0$, $C_1 = C_2 = C_0$) is considered. Here the theoretical gain required to start oscillations is $K = 3$ and the frequency of oscillation $\omega_0 = 1/R_0C_0$. Practically, one can write $K = 3 + \Delta K$, where ΔK is a control parameter of the circuit which is mainly used to compensate for any losses and to guarantee that oscillations start [13]. By setting $\tau = t/R_0C_0$, $x_1 = \frac{V_1}{B_p}$, $x_2 = \frac{V_2}{B_p}$, $x_3 = \frac{V_3}{B_p}$, $x_4 = \frac{RI_{L1}}{B_p}$, $\beta = \frac{R_0^2 C_0}{L}$, in the following the dynamics of circuits in Fig. 5 are considered as

(a) the hyper-chaotic circuit of Fig. 5(a) is represented by the following dimensionless set of equations:

$$\dot{x}_1 = (K-2)x_1 - x_2 - G(y),\ \dot{x}_2 = (K-1)x_1 - x_2,\ \dot{x}_3 = -x_4 + G(y),\ \dot{x}_4 = \beta x_3, \tag{6}$$

where $y = x_1 - x_3$ and $G(y)$ is represented by (4a). The actual circuit parameters are fixed for PSPICE and circuit experiments as $R_B = 2800\Omega$, $R_A = 1000\Omega$, $C_1 = C_2 = C_3 = C_0 = 68nF$, $R_1 = R_2 = R_0 = 1800\Omega$, $L = 18mH$ and the diode is 1N4148. For numerical simulations of (6), the parameters are fixed as $K = 3.85$, $b = 88$ and $\beta = 18$. The simulation results are shown in Fig. 6a(i-iii).

(b) The hyper-chaotic circuit of Fig. 5(b) is represented by the following dimensionless set of equations as

$$\dot{x}_1 = (K-2)x_1 - x_2 - G(y),\ \dot{x}_2 = (K-1)x_1 - x_2,\ \dot{x}_3 = \mu x_3 - x_4 + G(y),\ \dot{x}_4 = \beta x_3, \tag{7}$$

where $\mu = \frac{R_0}{R}$ and $G(y)$ is represented by (4a). The actual circuit parameters are fixed for PSPICE and circuit experiments as $R_B = 2800\Omega$, $R_A = 1000\Omega$, $C_1 = C_2 = C_3 = C_0 = 68nF$, $R_1 = R_2 = R_0 = 2390\Omega$, $L = 18mH$, $Rs = 500\Omega$, $R = 8000\Omega$ and the diode is 1N4148. For the present case for the numerical simulations of (7), the parameters are fixed as K = 3.6, b = 4.56, $\mu = 0.38$ and β = 21.6. The simulation results are shown in Fig. 6b(i-iii).

(c) the hyper-chaotic circuit of Fig. 5(c) is represented by the following dimensionless set of equations as (7) but with $G(y)$ is represented by (4b). The actual circuit parameters are fixed for the PSPICE and circuit experiments *as* $R_B = 3200\,\Omega$, $R_A = 1000\,\Omega$, $C_1 = C_2 = C_3 = C_0 = 68nF$, $R_1 = R_2 = R_0 = 2000\,\Omega$, $L = 18mH$, $R = 10000\,\Omega$ and the diode is 1N4148. For numerical simulations of (7) with (4b), the parameters are fixed as K = 4.2, b = 80, $\mu = 0.2$ and β = 15.2. The simulation results are shown in Fig. 6c(i-iii).

(d) The hyper-chaotic circuit of Fig. 5(d) is represented by the following dimensionless set of equations as

$$\dot{x}_1 = (K_1 - 2)x_1 - x_2 - G(y), \qquad \dot{x}_2 = (K_1 - 1)x_1 - x_2,$$

$$\varepsilon\dot{x}_3 = \mu\{(K_2 - 2)x_3 - x_4\} + G(y), \; \varepsilon\dot{x}_4 = \mu\{(K_2 - 1)x_3 - x_4\}, \tag{8}$$

where $\mu = \dfrac{R_1}{R_2}, \varepsilon = \dfrac{C_2}{C_1}$ and $G(y)$ is represented by (4b). The actual circuit

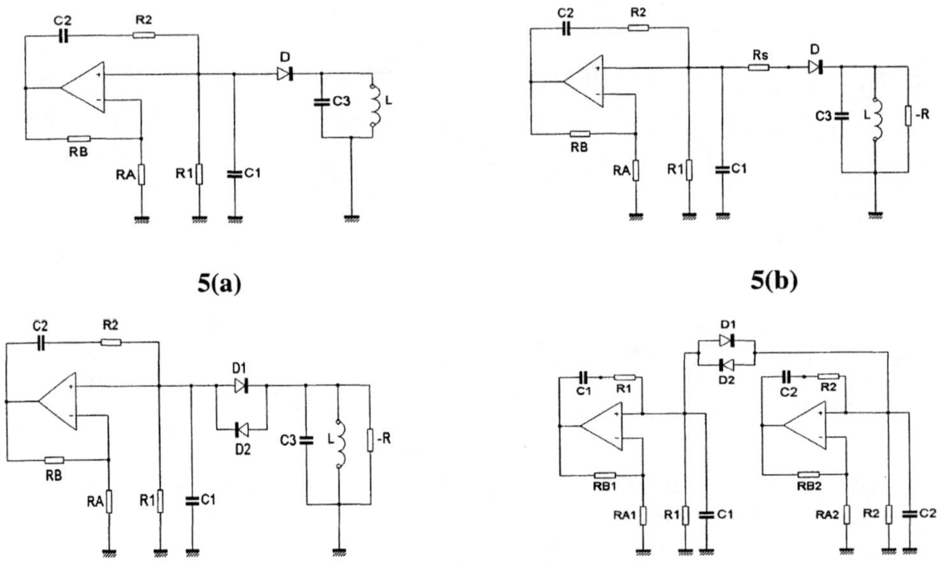

FIGURE 5. Classic Wien-bridge oscillator (WBO) based hyper-chaotic circuits. (a) WBO coupled to a resonator. The characteristic curve of the nonlinear coupling element is shown in Fig. 2(a). (b) WBO coupled to an active oscillator. The characteristic curve of the nonlinear coupling element is shown in Fig. 2(a). (c) WBO nonlinearly coupled to an active oscillator. The characteristic curve of the nonlinear coupling element is shown in Fig. 2(b). (d) Two WBOs are nonlinearly coupled. The characteristic curve of the nonlinear coupling element is shown in Fig. 2(b).

parameters are fixed for the PSPICE and circuit experiments as $RB_1 = 2700\Omega$, $RA_1 = 1000\Omega$, $C_1 = 68nF$, $R_1 = 1930\Omega$, $RB_2 = 2940\Omega$, $RA_2 = 1000\Omega$, $C_2 = 22nF$, $R_2 = 1100\Omega$ and the diode is 1N4148. For numerical simulations of (8), the parameters are fixed as $K_1 = 3.7$, $K_2 = 3.94$, $b = 76.4$, $\mu = 1.74$ and $\varepsilon = 0.324$. The simulation results are shown in Fig. 6d (i-iii).

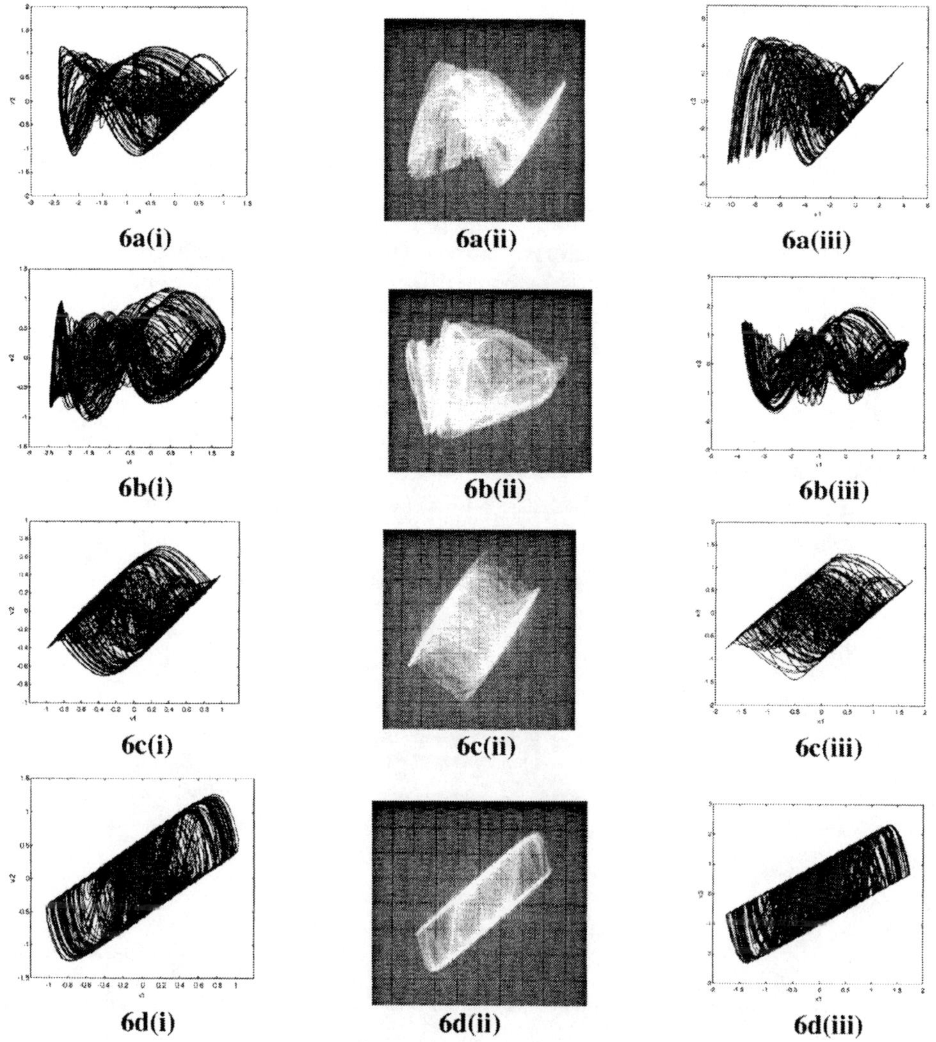

FIGURE 6. Phase-portraits of hyper-chaotic attractors corresponding to Fig. 5(a), (b), (c) and (d) respectively. (i) PSPICE simulation result in (V_1-V_2) plane. (ii) Experimental result in (V_1-V_2) plane. Horizontal axis is V_1 [0.5V/div]. Vertical axis is V_2 [0.5V/div]. (iii) Numerical simulation results of the normalized circuit equations in (x_1-x_3) plane. The parameters are fixed as in text.

Thus it is now clear that by using suitable sinusoidal oscillators as an active block, one can able to design suitable hyperchaotic circuits by using only passive nonlinear coupling elements. Also interestingly, by considering two generic sinusoidal oscillators coupled mutually by two passive diodes only, one can arrive at a simple inductorless hyper-chaotic circuit configuration. We have tested this design approach with other generic sinusoidal oscillators using 2R-2C type, Twin-T, etc. and observed similar type of hyperchaotic phenomena exhibited by the circuit of Fig. 5(d).

MIXED-MODE HYPER-CHAOTIC CIRCUIT

An increasing interest in studying the behaviour of nonlinear mixed-mode [27] chaotic circuits has recently developed. This is stimulated by their potential application in chaos based communications. Thus it is important to design suitable hyper-chaotic circuits whose structure is able to provide greater reliability in the form of a wide range of circuit parameter variations and easy implementability of certain chaos based secure communication schemes, namely, chaos-shift-keying (CSK) or chaos-on-off-keying (COOK) [28]. Presently, a novel mixed- mode hyper-chaotic circuit is considered with two switches as shown in Fig. 7(a). In this circuit, the common dynamics of the two hyper-chaotic circuits of Figs. 3(a) and 3(b) were combined using a switching method. The switching is controlled by square-wave or pulse signal, this mixed-mode hyper-chaotic circuit exhibits both the chaotic dynamical behaviour of circuits of Figs. 3(a) and (b). In this circuit if the states of the switches are set as S1-ON (Q=1) and S2-OFF ($\overline{Q}$ =0) then the dynamics of circuit Fig. 3(a) can be studied. When the states of switches are set as S1-OFF (Q=0) and S2-ON($\overline{Q}$ =1), then the dynamics of circuit Fig. 3(b) can be studied.

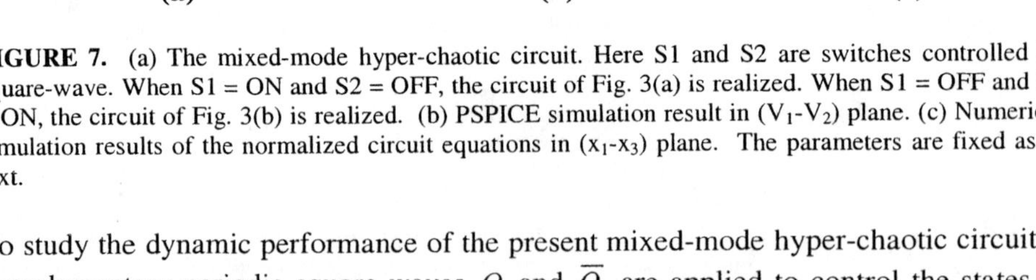

FIGURE 7. (a) The mixed-mode hyper-chaotic circuit. Here S1 and S2 are switches controlled by square-wave. When S1 = ON and S2 = OFF, the circuit of Fig. 3(a) is realized. When S1 = OFF and S2 = ON, the circuit of Fig. 3(b) is realized. (b) PSPICE simulation result in (V_1-V_2) plane. (c) Numerical simulation results of the normalized circuit equations in (x_1-x_3) plane. The parameters are fixed as in text.

To study the dynamic performance of the present mixed-mode hyper-chaotic circuit, a complementary periodic square waves Q and $\overline{Q}$ are applied to control the states of switches S1 and S2 in order to have continuous switching operation. The PSPICE and numerical simulation results of the governing dynamical equations are depicted in Figs. 7(b) and (c) respectively. For PSPICE simulations, a periodic pulse waveform with pulse period(T_p) = 1ms and pulse width(T_w) = 0.5ms is used. For numerical simulations, a periodic pulse waveform with pulse period = 50s and pulse width = 25s is used.

In Table-2, we have tabulated the calculated Lyapunov exponents for all the above circuits by using the TISEAN program [29]. The Lyapunov exponents are calculated from the circuit simulation data. For all cases, the maximal two Lyapunov exponents (λ_1, λ_2) are positive thus confirming the reported circuits are exhibiting hyper-chaotic phenomena. The third LE approaches zero and the fourth LE is deeply negative.

TABLE 2. The calculated two positive Lyapunov exponents from the circuit data[29].

Circuit No.	λ_1	λ_2
Fig.3(a)	0.14	0.07
Fig.3(b)	0.12	0.05
Fig.3(c)	0.11	0.03
Fig.3(d)	0.13	0.04
Fig.5(a)	0.13	0.05
Fig.5(b)	0.14	0.03
Fig.5(c)	0.07	0.028
Fig.5(d)	0.06	0.03
Fig.7(a)*	0.11	0.055

*When switches S1 and S2 are controlled by a square-wave.

CONCLUSIONS

In this work we have reported two classes of hyper chaotic circuits based on harmonic oscillators. In the first part we have considered the cases with active LC-R oscillators, resonators and nonlinear coupling element based hyper chaotic circuits. In the second part, we have substituted the active LC-R oscillators with that of Wien-bridge oscillator by maintaining the same circuit structure configurations and used only passive nonlinear circuit elements. We have also shown that the possibility of realizing novel hyper-chaotic circuit consisting only two classical sinusoidal oscillators coupled mutually by two passive diodes only. Thus a large number of realizations of this particular type of circuit can be designed by simply using any standard sinusoidal oscillator. Finally, we have introduced a kind of mixed-mode hyper-chaotic circuit, whose dynamics can be studied by properly controlling the two switches with suitable square-wave or pulse signal, so that the dynamics of two different hyper-chaotic circuits can be studied with a single mixed-mode circuit. Particularly, this kind of circuit is designed for potential ensuing chaos based digital communication applications.

ACKNOWLEDGMENTS

The authors wish to thank Professor Arunas Tamasevicius for his helpful discussions.

REFERENCES

1. Kocarev, L., and Parlitz, U., *Phys. Rev. Lett.*, **74**, 5028-2031 (1995).
2. Kocarev, L., Parlitz, U., and Stojanovski, T., *Phys. Lett. A* **217**, 280-284 (1996).
3. Brucoli, M., Cafagna, D., Carnimeo, L., and Grassi, G., *Int. J. Bifurcation & Chaos* **8**, 2031-2040 (1998).
4. Matsumoto, T., Chua, L.O., and Kobayashi, K., *IEEE Trans. Circuits Syst.* **33**, 1143-1147, (1986).
5. Suzuki, T., and Saito, T., *IEEE Trans. Circuits Syst.* **41**, 876 – 884 (1994).
6. Tamasevicius, A., Cenys, A, Mykolaitis, G., Namajunas, A., and Lindberg, E., *Electron. Lett.* **33**, 542-544 (1997).
7. Tamasevicius, A., Mykolaitis, G., and Cenys, A., *Electron. Lett.* **34**, 606 – 608 (1998).
8. Morgul, O., *Electron.Lett.* **31**, 2058 - 2059 (1995).
9. Morgul, O., *Electron.Lett.* **32**, 1403 – 1404 (1995).
10. Elwakil, A.S., and Kennedy, M.P., *Electron. Lett.* **34**, 1161-1162 (1998).
11. Elwakil, A.S., and Kennedy, M.P., *Int. J. Bifurcation & Chaos* **10**, 1981-1985 (2000).
12. Elwakil, A.S., and Kennedy, M.P., *Int.J. Circuit Theory Appl.* **28**, 319-334 (2000).
13. Elwakil, A.S., and Kennedy, M.P. *IEEE Trans. Circuits Syst.-I* **48**, 289 - 307 (2001).
14. Tamasevicius, A., and Cenys, A., *Chaos, Solitons & Fractals* **9**, 115-119 (1998).
15. Tamasevicius, A., Cenys, A., Namajunas, A., and Mykolaitis, G., *Chaos, Solitons & Fractals* **9**, 1403-1408 (1998).
16. Murali, K., Tamasevicius, A., Mykolaitis, G., Namajunas, A., and Lindberg, E., "Unstable Oscillators Based Hyperchaotic Circuit", *Proceedings of 7th Int. Workshop on Nonlinear Dynamics of Electronic Systems* NDES'99, Ronne, Denmark, 1999, pp. 241-244.
17. Murali, K., Tamasevicius, A., Mykolaitis, G., Namajunas, A., and Lindberg, E., *Nonlinear Phenomena in Complex Systems* **3**, 7-10 (2000).
18. Lindberg, E., Murali, K., and Tamasevicius, A., "Hyperchaotic Circuit with Damped Harmonic Oscillators", *Proceedings of the IEEE Int. Symposium on Circuits & Systems*, ISCAS'2001 **III**, Sydney, Australia, 2001, pp. 759-762.
19. Hosokawa, Y., Nishio, Y., and Ushida, A., "A Design Method of Chaotic Circuits Using Oscillator and a Resonator", *Proceedings of the IEEE International Symposium on Circuits & Systems*, ISCAS'2001 **III**, Sydney, Australia, 2001, pp. 373-376.
20. Lakshmanan, M., and Murali, K., *Chaos in Nonlinear Oscillators: Controlling and Synchronization*, World-Scientific, Singapore, 1996.
21. Ogorzalek, M. J., *Chaos and Complexity in Nonlinear Electronic Circuits*, World-Scientific, Singapore, 1997.
22. Lindberg, E., "Chaotic Oscillator Design", *Seminar Notes*, Seminar at Competence Centre for Circuit Design, Dept.Applied Electronics, Lund University, Ole Romers vag 3, P.O.Box 118, SE-221 00 Lund, Sweden, (2000).
23. Rossler, O.E., *Phys.Lett.A* **71**, 155-157 (1983).
24. Anishchenko, V. S., Vadivasova, T. E., Astakhov, V. V., Sosnovtseva, O.V., Wu, C.W., and Chua, L.O., *Int. J. Bifurcation & Chaos* **5**, 1677-1699 (1995).
25. Namajunas, A., Pyragas, K., and Tamasevicius, A., *Phys. Lett.A* **201**, 42-46(1995).
26. Chua, L.O., Desoer, C.A.,and Kuh,E.S., *Linear and Nonlinear Circuits*, McGraw-Hill, Singapore, 1987.
27. Kilic, R., Alci, M., and Tokmaci, M., *Electron. Lett.* **36**, 103 - 104 (2000).
28. Kennedy, M. P., Rovatti, R., and Setti, G., *Chaotic Electronics in Telecommunications,* CRC Press, Boca Raton, 2000.
29. *TISEAN: NonlinearTime Series Analysis*,website: **http://mpipks-dresden.mpg.de/~tisean**

Chaos-Based Signal Processing

Maciej J. Ogorzałek

Department of Electrical Engineering, University of Mining and Metallurgy, 30-059 Kraków, Poland

Abstract. Nonlinear systems exhibiting chaotic behavior can be considered as a source of a great variety of signals. Given a time series measured from a known or an unknown dynamical system we address a series of problems, such as section-wise approximation of the measured signal by pieces of trajectories from a chosen nonlinear dynamical system (model); signal restoration when the measured signal has been corrupted e.g. by quantization; signal coding and compression. The key to attack these problems is estimation of the initial conditions for a dynamical system which is used as the generator of approximating waveforms.

INTRODUCTION

Processing of signals and data measured from real physical objects or processes become of growing importancein various applications [1, 9, 12, 17]. Observations in such disciplines as meteorology, astronomy, seismology, physiology, medical observations and measurements, electronics and telecommunications typically are sensed by electronic means, convereted to numerical form, stored and possibly further processed. Signal processing is understood here in a very wide way encompassing storage, compression, feature extraction, interpretation, any kind of filtering, shaping etc.

Researchers from different backgrounds have different approaches to signal processing tasks - some concentrate on just "number crunching" and effective algorithms to solve the task while others try to rely on the knowledge of the underlying phenomena. Physicists' and system-level researchers' approach is to find first mathematical models of such processes (in terms of equations describing their time evolution). Despite the effort furnished into development of such models they were so far of little use for signal processing [9].

On the other hand there is a definite need for advanced tools for nonlinear signal processing which could exploit *a priori* phenomenological knowledge. Nonlinear models could offer a very powerful tool when combined with other efficient methods. We would like to show that a nonlinear (possibly chaotic) deterministic model together with nonlinear toolkit can be used to perform a variety of signal processing tasks such as:

• Section-wise approximation of the measured signal.

Several approaches of this kind have been widely used in signal processing – section-wise approximation of waveforms using various functions and their combinations is a commonly accepted technique (Bessel, Chebyshev, Hermite and other polynomial approximations, splines, radial-basis functions, various versions of Gaussian functions and many others). Nonlinear dynamical systems and systems generating chaos in particular

CP622, *Experimental Chaos: 6th Experimental Chaos Conference*, edited by S. Boccaletti et al.

could be of particular interest here considering choices of models/sources of approximating waveforms because of specific features of chaotic signals. Chaotic systems produce signals over a broad band of frequencies; trajectories show sensitive dependence on initial conditions - which means that changing the starting point of the process we can expect completely uncorrelated output waveforms (although bounded in space and confined by the same geometrical object - an attractor). This richness in dynamic behavior could offer new possibilities for signal processing as we will see in the following sections.

In experiments we confirm that some of the Generators of Approximating Waveforms (GAW) having "rich dynamics" in many cases give very good results of approximation.

• Signal restoration. In the continuous-time case the approximation task is performed in such a way that we always obtain a trajectory which is best in the least squares sense. Continuous-time approach has intrinsically the capability of smoothing out the quantization effects.

• Signal coding and compression. By nature of the two above described tasks we obtain a specific type of coding of sections of the measured signal by a series of initial condition points in the state space. Thus the signal can be compressed - instead of storing the whole waveform we can just store the model equations and a set of initial conditions which when applied to the model can regenerate (approximately) the measured trajectory.

Suppose we have a complicated time-waveform measured from some real system e.g. ECG, EEG, speech or any other. Storage of such signals is a difficult problem as to maintain the accuracy the signal must be sampled at a high frequency and in a digital form requires large word-length. Suppose we have some signal measured during some time interval T. Let us divide the interval T in an arbitrary way into sub-intervals. In this way we define a sequence of points. The idea now is to store/code the waveform using these chosen points only. Such kind of coding can be effective only in the case when the decoder could reproduce the pieces of waveform between the samples.

We propose the following solution to this problem. Let us assume we have a generator producing chaotic waveforms over a wide band of frequencies and sufficient dynamic range. We would like to approximate our measured waveform between the samples i and $i+1$ by a piece of a suitably chosen chaotic trajectory. This means that on the basis of one measured waveform ("output" from some process) we need to find initial condition (state) of the chaotic system which will give rise to a trajectory and corresponding output sufficiently close to the goal one — the measured waveform to be stored/coded.

In the nonlinear dynamics literature closely related problems known as **shadowing** and **noise removal** have also been considered [1, 6, 7, 12].

CONTINUOUS–TIME APPROACH

Let us assume that the model of dynamics of the system under consideration is in the Lur'e form [5]:

$$\begin{array}{rcl} \dot{\mathbf{x}}(t) & = & \mathbf{A}\mathbf{x}(t)+\mathbf{B}u(t), \\ y(t) & = & \mathbf{C}^T\mathbf{x}(t), \\ u(t) & = & f[y(t)] \end{array} \quad , \tag{1}$$

where $\mathbf{x}(t)$, $\mathbf{B}$, $\mathbf{C} \in \mathbf{R}^n$, $y(t)$, $u(t) \in R$, $\mathbf{A}$ - $n \times n$ real matrix, $f : \mathbf{R} \to \mathbf{R}$ is a locally Lipschitz function for $t \geq 0$.

Now suppose that we observe a transient signal of finite length τ_f which is the output of the system $\mathcal{S}$ in a time interval $[t_0,\ t_0+\tau_f]$. Suppose that we do not have access to the state variables of $\mathcal{S}$ at any time. Let us suppose further that the trajectory associated with the signal $\{y(t)\}$ begins with an initial state $\mathbf{x}_0 = \mathbf{x}(t_0)$ and ends at a final state $\mathbf{x}_f = \mathbf{x}(t_0+\tau_f)$. Let us call $\mathcal{T}$ the orbit associated with the trajectory. Our problem is to find with maximum precision the initial conditions of the state variables of $\mathcal{S}$ - $\mathbf{x}_0$ on the basis of the observation of $\{y(t)\ for\ t_0 \leq t \leq t_0+\tau_f\}$ only, in order to be able to reproduce accurately the signal $y(t)$ observed during the interval $[t_0,\ t_0+\tau_f]$. In the control literature this problem is called *exact observability* [14].

Solution – State Observer

For the Lure' type of systems it is possible to write explicitly the equation for the output signal:

$$y(t) = \mathbf{C}^T e^{\mathbf{A}t}\mathbf{x}_0 + \mathbf{C}^T \int_{t_0}^{t} e^{\mathbf{A}(t-\tau)}\mathbf{B}f[y(\tau)]d\tau. \tag{2}$$

in a simplified form we can write:

$$\mathbf{C}^T e^{\mathbf{A}t}\mathbf{x}_0 = y(t) - (g*f)(t), \tag{3}$$

where g is the impulse response of the linear part of the system and $*$ denotes the convolution.

Iff $y - g*f$ belongs to the range of the observability map

$$\mathbf{P} : R^n \ni \mathbf{x}_0 \longrightarrow \mathbf{P}\mathbf{x}_0 \in L^2(t_0,t_k),\ (\mathbf{P}\mathbf{x}_0)(t) = \mathbf{C}^T e^{\mathbf{A}t}\mathbf{x}_0. \tag{4}$$

then equation (7) will lead to Then using the orthogonal projection theorem one gets

$$\mathbf{G}\mathbf{x}_0 = \mathbf{q}, \tag{5}$$

where $\mathbf{G}$ be the Gram matrix of this systems and q is the vector of the scalar products (in $L^2(t_0,t_k)$) of this system with the right-hand side of (3). This equation can be solved giving the best least-square approximation of the initial point $\mathbf{x}_0$ [14] (in control engineering terms we build an *observer* which reproduces the initial condition on the basis of the output signal). Using the pseudo-inverse concept we obtain the following least-squares approximate formula for the initial condition [5]:

$$\mathbf{x}_0 = (\mathbf{G}^T\mathbf{G})^{-1}\mathbf{G}^T\mathbf{z}. \tag{6}$$

The procedure implemented in Matlab we introduced also more realistic effects such as signal quantization, non-ideality of elements and noise.

There exist other solutions for finding the initial condition. One of the approaches is to use adaptive procedures, In [18] we used the procedure originally proposed by Maybhate and Amritkar [16] for signal coding. Such procedures often fails as there is no guarantee of convergence nor finding the proper solution.

DISCRETE-TIME APPROACH

Let us consider now that we have a sampled version of the measured time series $y(i)$, $i = 0 \ldots N$.

Let us suppose next that we have a discrete-time dynamical system (map)

$$z(i+1) = F[z(i)],\ z(0) = z_0. \tag{7}$$

We would like to find a map F and an initial condition z_0 which would reproduce a sequence $z(i)$ in such a way that $z(i) - y(i) \leq \varepsilon$ ($z(i) = y(i)$ in an ideal case) where ε is an assumed acceptable error bound.

Universal Nonlinear Approach to Coding

The main result can be formulated as the following **Proposition:**

Arbitrary time-series of any length can be coded using single initial condition for a suitably chosen one-dimensional map [19].

The reasoning is the following:
- The amplitudes of the measured signal can be rescaled to the unit interval,
- One can assign to the time series its symbolic representation accordingly with the chosen coder (map)
- thus instead of the original time series we can further consider a series of bits (0s and 1s)
- Let us consider simple Bernoulli shift map - it is a well established fact that such a map can reproduce any sequence of symbols. Thus finding a suitable initial point one can reproduce the considered trajectory with given symbolic sequence.

Comments:

There exists a trade-off between the length of the data set and the precision of the initial condition - the longer the time series the better precision (number of bits) is required for storing the initial condition. This seems to be a paradox but the chaotic system is itself reconstructing the lost information! There exist a number of approaches for finding initial conditions for a chaotic map given an output time series [13, 21]. In the experiments described below we used the simplest case of a Bernoulli shift map and the extended halving method [21] or the backward iteration approach as proposed in [2].

Estimation of initial condition

For our application we will use algorithm based on the general approach proposed by C. Ling *et al.* [2]. We will assume that our we have a version of the measured signal which is scaled to the interval $[0;1]$ and we will denote it by $y(n) = x(n) + w(n)$, where $x(n)$ will be assumed to be generated by a chaotic map

$$x(n+1) = F[x(n)], \tag{8}$$

and $w(n)$ is the measurement noise, $F : D \longrightarrow D$ where D is the phase space of the map (in our examples always equal to the unit interval). We will introduce a partition of the phase space into a number of disjoint sets $E = \{E_0, E_1, ..., E_{q-1}\}$ in such a way that $\cup_{i=0}^{q-1} E_i = D$, $E_i \cap_{i \neq j} E_j = \emptyset$. Assuming that at time n the point $x(n) \in E_i$ we assign to it a symbol $s(n) = i$ $(i = 1, ...q-1)$. In this way any orbit of our map can be assigned a string of symbols $S = \{s(0), s(1)..., s(n)...\}$ called a symbolic sequence. We have thus a one-to-one mapping Θ assigning to an initial condition $x(0)$ the sequence S

$$\Theta(x(0)) = S, \tag{9}$$

such that $F^n x(0) \in E_{s(n)} \iff x(0) \in F^{-n} E_{s(n)}$ for $n \geq 0$, where F^n denotes the n–fold composition of the map F, and $F^{-n} E_{s(n)}$ is the set of points that are in the set $E_{s(n)}$ after n iterations. A symbolic sequence is called admissible if there exists a $x(0)$ such that $\Theta(x(0)) = S$. The partition E is called generating partition if for each admissible sequence the set $\cap_{n=-\infty}^{\infty} E_{s(n)}$ contains only one point.

Let us denote by F_i the restriction of F to E_I. Let us assume that F_i is invertible. Suppose now that we have measured time series $\{x(0), x(1), \ldots x(N-1)\}$and corresponding symbolic sequence $\{s(0), s(1), ...s(N-1)\}$. Using this symbolic sequence we can locate the initial point $x(0)$ to within the set $\cap_{i=0}^{N-1} F^{-i} E_{s(i)}$. The estimation becomes better with growing N - for generating partitions $x(0)$ is determined uniquely for $N \longrightarrow \infty$. Ling Cong *et al.* [2] proposed the following formula to estimate the initial condition $x(0)$:

$$\tilde{x}(0|N) = F_{s(0)}^{-1} \circ F_{s(1)}^{-1} \circ \ldots \circ F_{s(N-1)}^{-1}(\xi), \tag{10}$$

where ξ is a point in the domain of $F_{s(N-1)}^{-1}$. In [2] a proof of convergence is given for the case of E being a generating partition and each F_i being an onto map. For maps not being "onto" there is no proof of convergence! - although the algorithm works quite well in many tested cases – for details see [2].

For coding purposes it is sufficient to use simple cases of Bernoulli map or tent map which satisfy the conditions of the Cong *et al.* algorithm. The generating partition consists of two subsets $E_1 = [0,0.5)$, $E_2 = [0.5,1]$, symbols are defined according to:

$$s(i) = \begin{cases} 0 & x(i) < 0.5 \\ 1 & x(i) \geq 0.5. \end{cases} \tag{11}$$

The inverse maps for the case of Bernoulli map are of the form

$$F_s^{-1}(x) = \frac{x+s}{2}. \tag{12}$$

For the tent map

$$F_s^{-1}(x) = s + \frac{(1-2s)x}{2}. \tag{13}$$

The algorithm has been implemented in Matlab. One can either use fixed length of the segment of time series or adaptively adjust the length according to imposed allowable approximation error.

TRADE-OFFS IN TRAJECTORY RECONSTRUCTION

To see the tradeoffs in the coding approach we will use experimental data generated by Chua's circuit [15] – we stored just one variable x_1 — voltage across the C_1 capacitor in Chua's circuit. In the experiments we used a sampled sequence of 5000 data points.

Effect of precision of the initial condition

First let us show the tradeoffs existing when changing the precision for storing the initial condition calculated using the continuous-time approach I. Figure 1 shows results of reconstruction (left column) and the reconstruction error (right column) when the "time window" is 5000 data points with the precision on the initial condition is varying from 24 bit down to 12 bit. Comparing this error of reconstruction in the considered cases one can see that for 24 bit precision the reconstruction is excellent – to within 1% accuracy. With diminishing precision of initial condition one observe first that the quality of reconstruction becomes worse towards the end of the reconstruction interval as in the case of 16 bit precision. For 12 bit we see serious discrepancy of reconstructed trajectory due to the sensitive dependence to initial condition.

Our simulations show that there exists a tradeoff between the length of the trajectory and the precision of the initial conditions. For a given precision in initial condition one cannot extend the length of the time series which provides an acceptable error of reconstruction beyond a certain limit (due to the effect of the sensitivity to initial condition). Increasing the precision of initial condition beyond a certain limit will not decrease however the error variance because there is an inherent lack of precision of initial state estimation due to accumulation of the calculation errors which results in eventual separation of the trajectory starting at this initial condition from the original one.

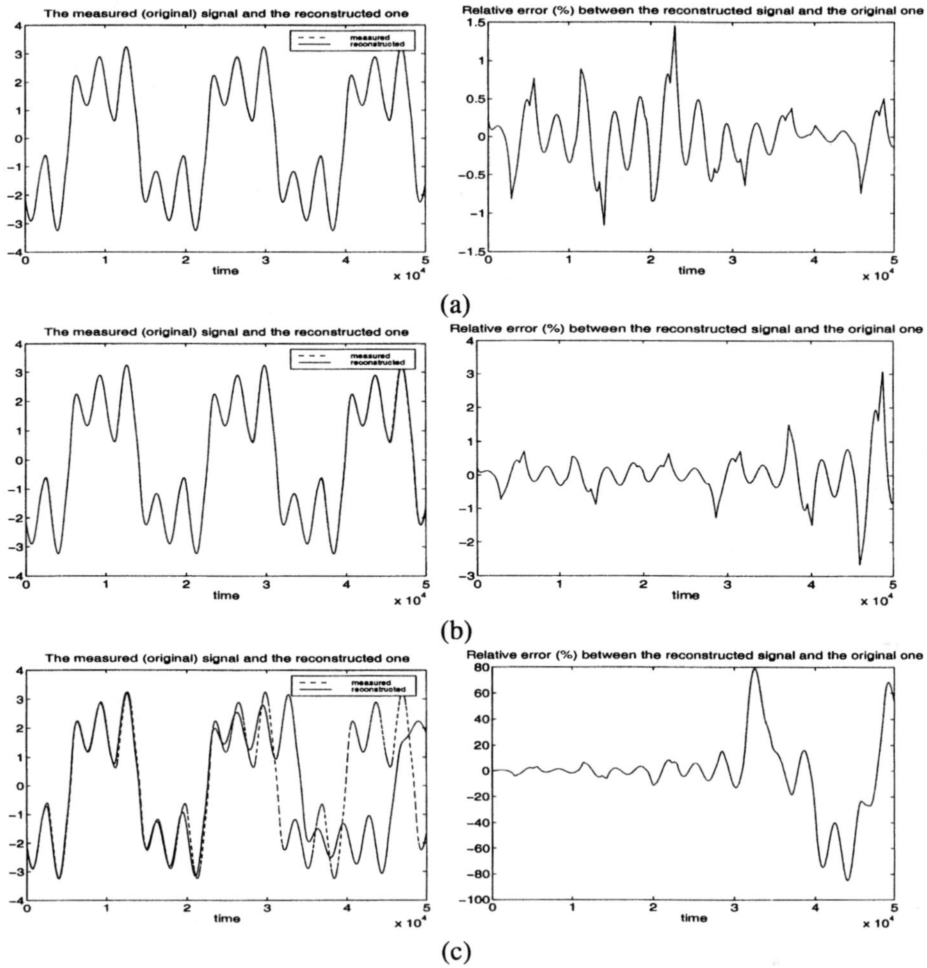

FIGURE 1. Relative reconstruction error for a sequence of 5000 samples but when the precision on the initial condition changes from 24 bit (a) through 16 bit (b) down to 12 bit (c). Figures in the first column show comparison between the original and the reconstructed signal and in the second column the relative error.

Immunity to Quantization Errors

We investigated also the ability of the reconstruction method to deal with the quantization effects introduced by the measurement channel – this means that the measured time-series is discretized in value.

The continuous-time algorithm in this case show that it is possible to find a true system trajectory close to the original one according to some least squares criteria. For a reasonable horizon we can see that the method tends to locate an approximate orbit which smoothes out the quantization effects. Figure 2 shows the performances of the reconstruction process for four different quantization levels of $y(t)$, i.e., 16, 8, and 4 bits. One can see that even for the modest level of quantization of 8 bit the reconstruction process is efficient.

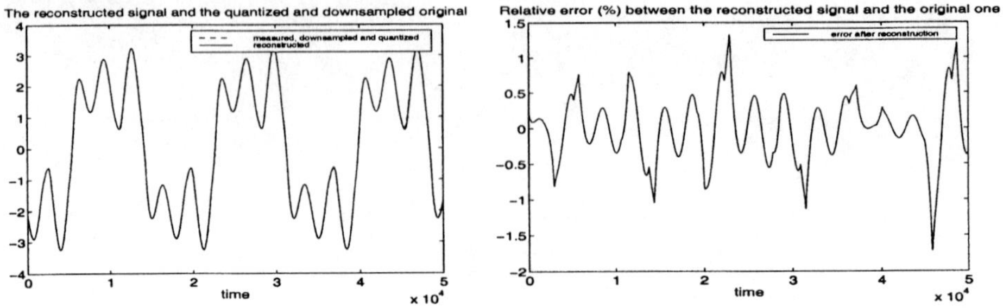

(a)

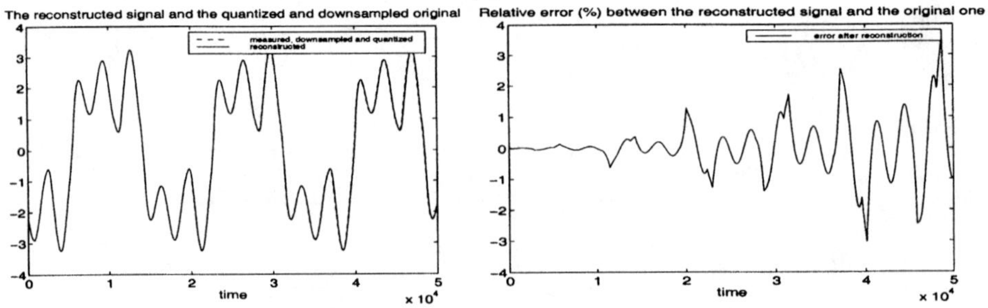

(b)

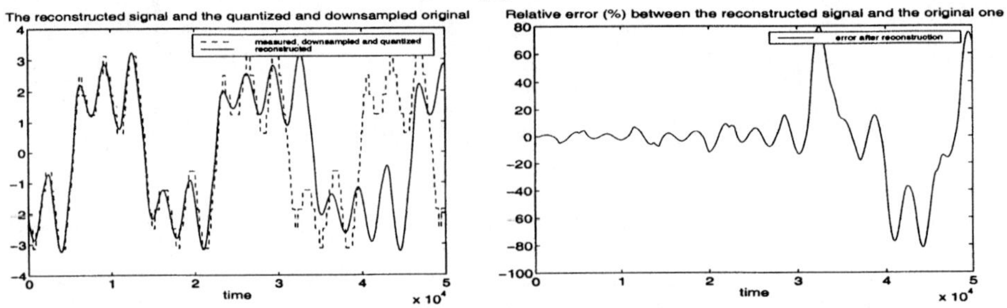

(c)

FIGURE 2. Results of reconstruction for the sequence of 5000 samples. Successive rows of the figure show the reconstruction behavior for quantization levels of $y(t)$, at 16 bit (a), 8 bit (b), 4 bit (c).

PROCESSING OF SIGNALS WHEN NO MODEL OF DYNAMICS IS KNOWN

In the experiments described below we try to use an arbitrary model i.e. we approximate the measured signal interval by interval by pieces of trajectories of an arbitrarily chosen (chaotic) oscillator. In the experiments we used both continuous-time and discrete-time approximation methods based on arbitrary GAW as described before but we introduced in the algorithm an upper limit for error introduced by approximation in the considered interval. We used an adaptive scheme for changing the approximation interval length in order not to exceed the given error bounds. In this way we obtained a sequence of initial conditions for each of the maximum length intervals and ensured that the approximation error satisfied the assumed upper bounds.

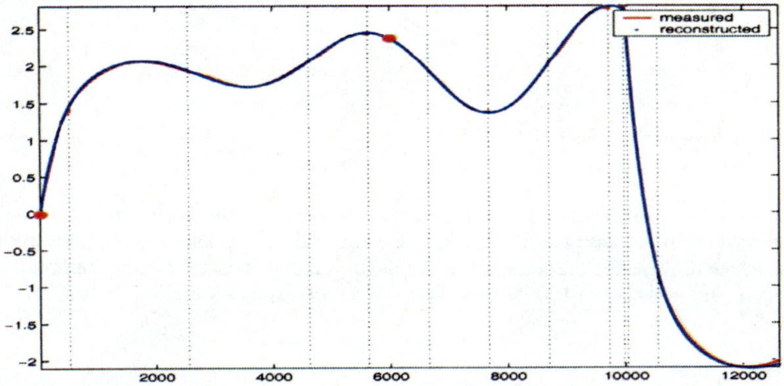

FIGURE 3. Artificially generated signal and the 12 pieces reconstructed by the model (Chua's circuit) – blue dots almost coincide with the original measured waveform. In the discrete-time case the approximation is very good - we need just two points (red circles) with floating-point accuracy to reproduce the sequence of 600 points on the trajectory. The maximum error has been fixed to 2.5% and precision of the initial conditions for each segment to 16 bit.

Approximation and Coding of a Synthetic Signal

To make the first test we generated artificially 40000 samples of a special kind of signal by a Matlab program. The signal was composed of square wave of amplitude 2.0 and frequency $f_0 = 500$Hz which was added with a sinusoidal signal of frequency $f_1 = 2400$Hz modulated by a saw-tooth waveform of frequency f_0. The combined signal was low-pass filtered by a first-order filter with a cut-off frequency of 4000Hz. The obtained waveform can be seen in the Fig.3.

To generate approximating pieces of trajectories we used the standard Chua's circuit operating in the chaotic mode. We obtained an excellent approximation of a signal which does not belong to the attractor of Chua's circuit!. The maximum error has been fixed to 2.5% and precision of the initial conditions for each segment to 16 bit. Maximum interval length was set to 2048 iterates. The whole waveform could be stored and reconstructed using just 12 initial conditions! We had to store 36 numbers representing initial conditions for 12 intervals.

We tested also the performance of the Bernoulli shift approximator using floating point coding of the initial condition. The reproduction is excellent – for a series of 1200 points (down-sampled by factor of 10 version of the measured signal) we just need two initial points to be stored!

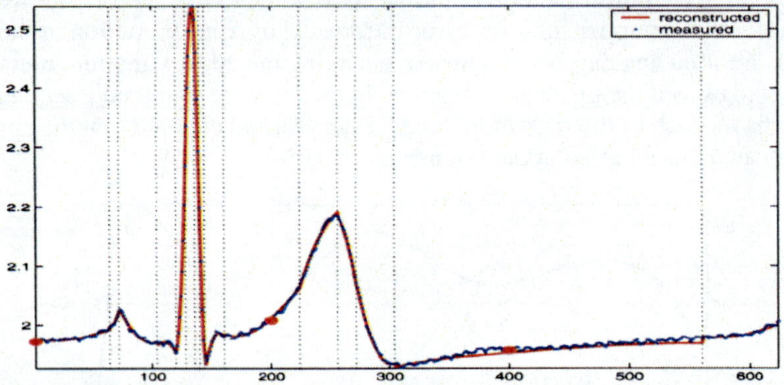

FIGURE 4. The ECG signal is approximated by 15 pieces of chaotic trajectories of Chua's circuit. The compression ratio in this case is of 8.76. Using Bernoulli shift map and floating point coding for initial condition (three red dots represent the initial points for sections recovered using Bernoulli shift map) the compression ratio is far above 100! with accuracy of reproduction within 2%.

Approximation and Coding of an ECG Signal

In our earlier papers exploiting the continuous -time approach we used one interval of the ECG waveform as our benchmark. Using continuous-time approximation, using 16 bit precision for data and 2.5% allowable error, the compression ratio was 8.76. We need to store 45 parameters at 16 bit each (initial conditions for 15 intervals) + 15 times the lengths of the intervals coded at 9 bit.

Truly amazing result is obtained by the Bernoulli shift approximator - we are able to reproduce 500 points on the trajectory using just one or two (depending on the precision required for reproduction) initial points !! Thus reducing the number of stored points by a factor of 20-50 but requiring high precision for storing the initial condition! This is a compression ratio far above the reach of any existing technique! In the case shown in the figure three initial points (scalar!) were sufficient to reproduce the waveform with 2.5% precision.

Comparing the results obtained in terms of ECG compression ratios it should be underlined that typically state-of-the-art ECG compression methods give results in the range of compression ratios 12-15 (at the 500Hz sampling rate) while using Bernoulli shift map and floating point coding for initial condition the compression ratio is far above 100!

TABLE 1. Performance comparison of various ECG compression methods

Method	CR	f_s Hz	PRD (%)	Comments
AZTEC	10.0	500	28.0	Poor *P* and *T* fidelity
TP	2.0	200	5.3	Sensitive to Sampl. rate
CORTES	4.8	200	7.0	Poor *P* fidelity
SAPA	3.0	250	4.0	High fidelity
Entropy Coding	2.8	250	-	
DPCM Lin Pred	2.5	250	-	High fidelity
DPCM Lin Pred + Entropy cod.	7.8	500	3.5	High fidelity
Karhunen-Loeve	12.0	250	-	
Adapt. Hermite Functions	11.6	360	-	info on ectopic beat

For comparing this method's performance, see Table 1 (after [11]). Of course, our technique needs to be optimized for use in compression.

CONCLUSIONS

Nonlinear models with rich, possibly chaotic dynamics are potentially a very powerful source of segments of waveforms which can be used for signal approximation/reconstruction and also coding and compression.

Proposed approach for finding initial conditions can be performed recursively giving an excellent approximation of a given (measured) signal by sections of trajectories of a chosen model (waveform generator). Astonishingly this approximation/reconstruction approach can be also used when no good model of system dynamics is known. Taking a "sufficiently rich" generator of approximating waveforms (GAW) (e.g. a suitable chaotic oscillator or map) we have shown that an arbitrary waveform could be approximated/reconstructed. Furthermore calculation of initial conditions for each interval offers an interesting possibility for signal coding and compression. The measured signal is automatically coded by a set of points in the state space At no additional computational cost we obtain compression of the measured signal - instead of storing the whole waveform one can just store the "code". Knowing the generating model and initial conditions one can easily regenerate the original signal with excellent accuracy. With the discrete time approach in many cases we obtain the compression ratios of 100 and more!

ACKNOWLEDGMENT

This research has been supported in part by the the University of Mining and Metallurgy, Krakow, Poland – Grant No.10.120.184. The author would like to acknowledge the hospitality of the of the Instituto de Microelectronica de Sevilla, Centro Nacional de Microelectronica, Sevilla Spain during the course of preparation of this manuscript. Stay at CNM was made possible thanks to "Ano sabatico" Award from the Ministerio de Education y Cultura of Spain.

REFERENCES

1. H. D. I. Abarbanel, *Analysis of Observed Chaotic Data*, Springer Verlag, 1996.
2. C. Ling, X. Wu and S. Sun, "A general efficient method for chaotic signal estimation", *IEEE Trans. Sign. Proc.*, vol.47, No.5, pp.1424-1428, 1999.
3. S. M. Cox and A. J. Roberts, "Initial conditions for models of dynamical systems", *Physica D*, vol. D85, pp.126-141, 1995.
4. H. Dedieu and M. J. Ogorzałek, "Signal Coding and Compression Based on Chaos Control Techniques", *Proc. 1995 IEEE International Symposium on Circuits and Systems*, Seattle, vol. 2, pp.1191-1194, 1995.
5. H. Dedieu and M. J. Ogorzałek, "Chaos-based signal processing", *Int. J. Bifurcation and Chaos*, vol.10, No.4, pp.737-748, 2000.
6. J. D. Farmer and J. J. Sidorowich, "Optimal Shadowing and Noise Reduction", *Physica D*, vol. D47, pp.373-392, 1991.
7. P. Grassberger, R. Hegger, H. Kantz, C. Schaffrath and T. Schreiber, "On noise Reduction Methods for Chaotic Data", *Chaos*, vol. 3, pp.127-141, 1993.
8. J. Guckenheimer, P.Holmes: *Nonlinear Oscillations, Dynamical Systems, and Bifurcations of Vector Fields*, Springer Verlag, 1983.
9. S. Haykin and J. Principe, J. "Making sense of a complex world", *IEEE Signal processing Magazine*, vol. 15, no. 3, pp.66-81, 1999.
10. K. Judd and A. Mees, "On Selecting models for Nonlinear Time Series", *Physica D*, vol. D82, pp.426-444, 1995.
11. S. M. S. Jalaleddine, C. G. Hutchens, R. D. Strattan, W. A. Coberly, "ECG Data Compression Techniques - A Unified Approach", *IEEE Trans. on Biomedical Eng.*, vol. 37, no. 4, April 1990.
12. H. Kantz and T. Schreiber,*Nonlinear Time Series Analysis*, Cambridge University Press 1999.
13. S. Key and V. Nagesha, "Methods for Chaotic Signal Estimation", *IEEE Trans. Sign. Proc.*, vol.43, no.8, pp.2013-2016, 1995.
14. W.S. Levine, (ed.), *The Control Handbook*, Chapter 37: Observers, CRC Press and IEEE Press 1996.
15. R. Madan, (Ed.), *Chua's circuit: A paradigm for chaos*, World Scientific Series, 1993.
16. A. Maybhate and R. E. Amritkar, "Estimation of initial conditions from a scalar time series", (private communication) 2000.
17. M. J. Ogorzałek, "Taming Chaos: Part I - Synchronisation", *IEEE Trans. Circuits Systems*, vol. CAS-40, no. 10, pp.693-699, 1993. "Part II - Control", *IEEE Trans. Circuits Systems*, vol. CAS-40, no. 10, pp.700-706, 1993.
18. M. J. Ogorzałek, "Piecewise-chaotic approximation of arbitrary time-series", Proc. NOLTA'2000, Dresden, 2000.
19. M. J. Ogorzałek, "Nonlinear-dynamics-based approximation and compression of arbitrary time-series", Proc. ISCAS'2001, Sydney, 2001.
20. G. Wade, *Coding techniques: An introduction to compression and error control*, Palgrave Publishers, 2000.
21. S. Wang, P. C. Yip, H. Leung, "Estimating initial conditions of noisy chaotic signals generated by Piece-Wise Linear Markov maps using itineraries", *IEEE Trans. Sign. Proc.*, vol. 47, no.12, pp. 3289-3302, 1999.

Broadband Hyperchaotic Oscillator With Delay Line

Antanas Čenys*, Erik Lindberg+, Antonis N. Anagnostopoulos¶, Gytis Mykolaitis*, Skaidra Bumelienė*, and Arūnas Tamaševičius*

*Semiconductor Physics Institute, A. Goštauto 11, LT-2600 Vilnius, Lithuania

+Ørsted•DTU Department, Technical University of Denmark, DK-2800 Lyngby, Denmark

¶Department of Physics, Aristotle University of Thessaloniki, GR-54006 Thessaloniki, Greece

Abstract. Dynamical systems with time delay can be employed as high dimensional hyperchaotic oscillators with multiple positive Lyapunov exponents. We describe an electronic circuit composed of a 3-stage amplifier and a delay line in the feedback loop. The 1st stage of the amplifier is a nonlinear one while the 2nd and the 3rd stages are linear ones. Microwave transistors having the threshold frequency 5 to 9GHz are used in the experimental circuits. The oscillators generate hyperchaotic signals in the frequency range from about 1MHz to more than 1000MHz with unevenness of the spectral density less than 20dB. Mathematical models are presented. The oscillators are described either by a scalar nonlinear DDE or by a set combined of one nonlinear DDE and two linear ODEs.

INTRODUCTION

Dynamical systems with time delay can exhibit very complex high-dimensional hyperchaotic behavior with multiple positive Lyapunov exponents thus can be exploited as noise like signal generators. An example of such a system provides the Mackey-Glass (MG) mathematical model presented by a delay differential equation (DDE) [1]. The MG model has been studied in details by Farmer [2]. To our best knowledge an electronic circuit imitating the MG system has been first described by Dmitriev and co-authors [3]. An acoustic device has been employed for the delay element. Later an analog computer design has been used by Losson and co-authors [4] to simulate the dynamics of a related to the MG model – the piecewise constant version (rectangular nonlinear function) proposed by an der Heiden and Mackey. The delay unit in the analog computer has been implemented by means of a sample-and-hold (S&H) IC, so-called "bucket brigade device" used along with a clock generator. Consequently the analog computer is a hybrid one containing time continuous and time discrete units. A tent type nonlinear function as well as a rectangular one have been considered by Schwarz and Mögel [5]. Moreover, the authors of the paper [5] suggested to replace the S&H device by a transmission line thus leading to fully time continuous system. Another version of a time continuous delay system has been proposed by Namajūnas *et al.* [6]. Two complementary junction field effect transistors

CP622, *Experimental Chaos: 6th Experimental Chaos Conference*, edited by S. Boccaletti et al.

(JFET) coupled in a special way [7] were used for the nonlinear device while the delay unit was implemented by means of a high order LCL low-pass filter. Fortunately, the nonlinear device composed of the complementary pair of JFETs is described by a function very similar in form to that introduced in the MG model. Therefore, this analog circuit [6] was employed to investigate various features of the hyperchaotic MG system, including control of (hyper)chaos [8,9] and synchronization of (hyper)chaos [10-12]. In addition, Pyragas [13] has considered the possibility to employ the MG hyperchaotic generators for secure communication.

We note that most experiments with the MG or related type electronic circuits have been performed at relatively low (kilohertz) frequencies [3,4,6,8-11]. High frequency delay line based chaotic oscillator has been considered both numerically and experimentally by Mögel and co-authors [14]. However, only a short note that a practical prototype generates noise like signals in a spectral range from DC to over 100MHz is presented. For chaos based communications and other applications chaotic and hyperchaotic oscillators operating at high megahertz to gigahertz frequencies are needed [15]. In the present paper we describe in details an experimental circuit generating broadband chaotic signals to over 1000MHz.

EXPERIMENTAL CIRCUIT

Circuit Description

The delay line based oscillator is depicted schematically in Fig. 1 and Fig. 2.

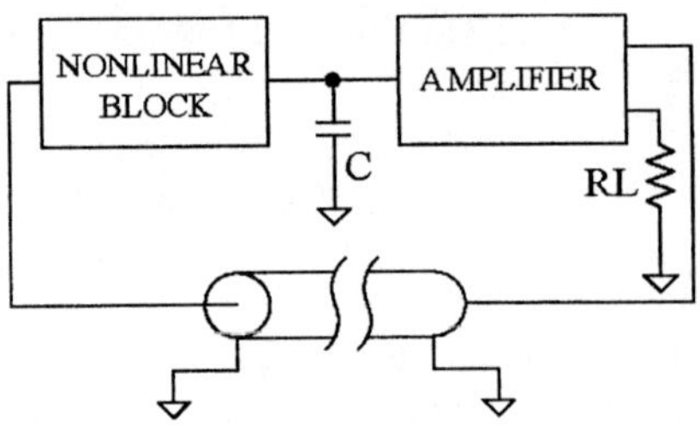

FIGURE 1. Block diagram of the oscillator.

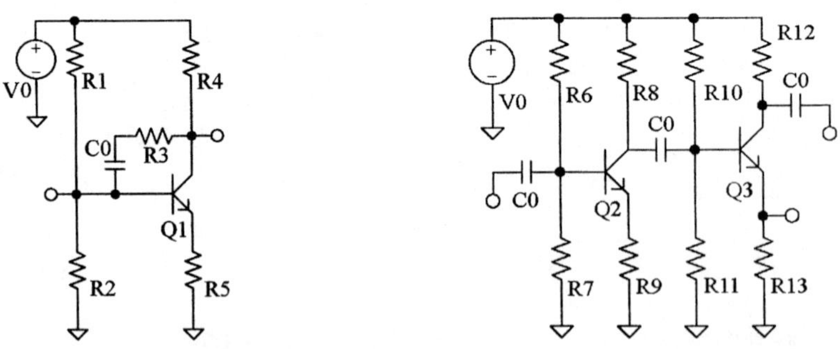

FIGURE 2. Circuit diagrams of non-linear stage (left) and linear amplifier (right).

The oscillator consists of a nonlinear unit, an integrating RC chain (the role of R plays the output resistance of the nonlinear unit), a linear amplifier, and a segment of a broadband transmission line in the feedback loop (Fig. 1).

The nonlinear unit actually is a single-transistor amplifier with a relatively small DC bias of the Q1 (Fig. 2, left). The by-pass resistor R3 together with the transistor Q1 switched between the ON and OFF states produce a nonlinear tent type response function. The slopes of the tent function can be controlled by the resistors R3...R5. The linear amplifier is a two stage non-inverting circuit with sufficiently large DC bias of the transistors Q2 and Q3 (Fig. 2, right). The load resistor RL (50Ω) is coupled in parallel to the emitter resistor R13. In the both sub-circuits C0 are the coupling capacitors (C_0=47nF).

Three different versions of the oscillator were built. Surface mount devices (SMD) were employed in all the circuits. The segments of length L of coaxial transmission cable RG58U (50Ω) were used for the delay lines.

The element parameters of the 1st oscillator were as follows: Q1...Q3 – BFR93A bipolar junction transistors (BJT) with the threshold frequency f_T=5GHz, L=64cm ($T_{del}\approx$3.3ns), V_0=8V, and the SMD parameters: C=22pF, R_1=10kΩ, R_2=1.2kΩ, R_3=R_4=68Ω, R_5=20Ω, R_6=R_{10}=3.3kΩ, R_7=R_{11}=510Ω, R_8=R_{12}=51Ω, R_9=R_{13}=16Ω.

The 2nd oscillator was composed of the following elements: Q1...Q3 – BFG520 type BJTs (f_T=9GHz), L=13cm ($T_{del}\approx$680ps), V_0=11V, and the SMD parameters: C=2.7pF, R_1=15kΩ, R_2=1.2kΩ, R_3=R_4=68Ω, R_5=20Ω, R_6=R_{10}=5.6kΩ, R_7=R_{11}=510Ω, R_8=R_{12}=51Ω, R_9=R_{13}=16Ω.

Finally, the 3rd oscillator is almost the same as the 2nd one , but with the capacitor C removed (C=0).

Experimental Results

The 1st oscillator is characterized by broadband continuous power spectrum in the frequency range up to 400MHz (at –20dB).

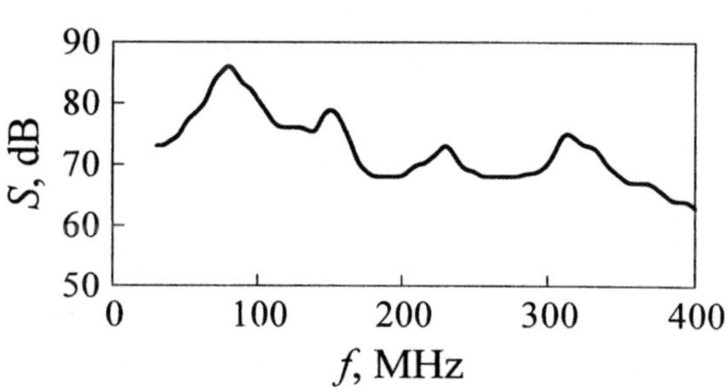

FIGURE 3. Power spectrum of the 1st oscillator. Q1...Q3 – BFR93A, T_{del}=3.3ns, C=22pF, V_0=8V. Spectral resolution is 120kHz. The level of 0dB corresponds to 1μV across the 50Ω load resistance.

Meanwhile, the 2nd and the 3rd oscillators exhibit extremely wide frequency band from about 1MHz to, apparently, over 1000MHz (Fig. 4). Unfortunately, we were not able to measure the spectra at f>1000MHz. Simulations performed with the "Electronic Workbench" (PSPICE based software) indicate high spectral densities up to 1500MHz. We note, that though the spectral density is about 15dB lower in the case of the 3rd oscillator, however, the overall spectrum is relatively more flat.

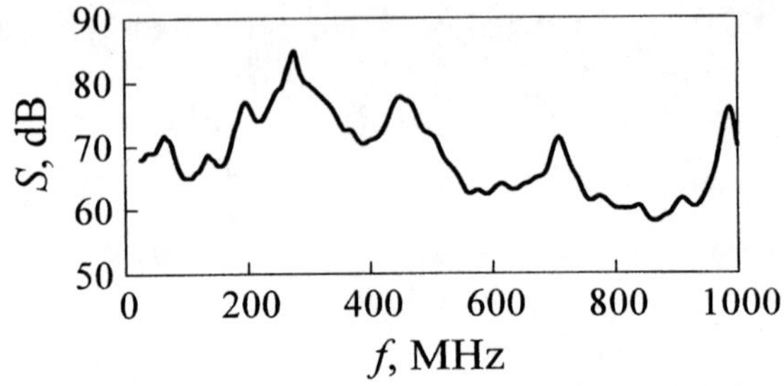

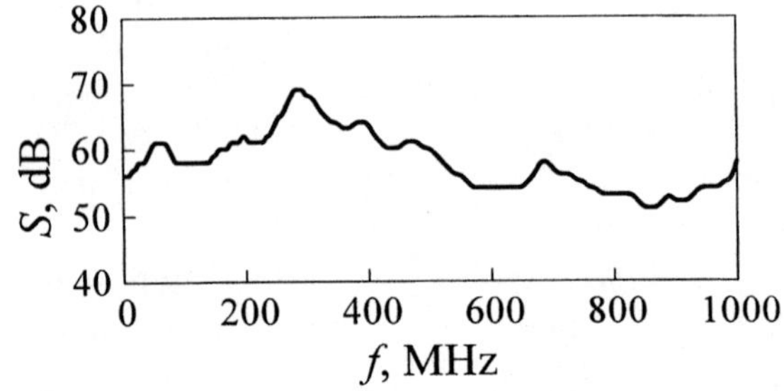

FIGURE 4. Power spectra of the 2nd (top, C=2.7pF) and the 3rd (bottom, C=0) oscillators. Q1...Q3 – BFG520, T_{del}=680ps, V_0=11V. Spectral resolution is 120kHz. The level of 0dB corresponds to 1μV across the 50Ω load resistance.

MATHEMATICAL MODEL

Equations

It is convenient to introduce the following dimensionless variables and parameters:

$$x = \frac{U_C}{U^*},\quad \theta = \frac{t}{RC},\quad \tau = \frac{T_{del}}{RC},\quad a = \frac{R_4}{R_3 + R_4 + \rho},\quad b = \frac{R_3}{R_3 + \rho},$$
$$c = \frac{\rho}{R_3 + \rho},\quad d = \frac{R_2 V_0}{(R_1 + R_2)U^*},\quad e = \frac{R}{R_5 + r},\quad R = \frac{R_4(R_3 + \rho)}{R_3 + R_4 + \rho}. \tag{1}$$

Here U_C is an AC component of the voltage across the capacitor C, ρ is the characteristic resistance of the transmission line, U^* is the break-point voltage and r is the small signal ON-resistance of the base-emitter junction. In (1) it is assumed for simplicity that the common-emitter forward current gain $\beta_F >> 1$, yielding high input resistance of the amplifying stages.

Then the 1st and the 2nd oscillators are described by the following DDE:

$$\frac{dx}{d\theta} = -x(\theta) + akx(\theta - \tau) - F[x(\theta), x(\theta - \tau)]. \tag{2}$$

Here k is the gain of the linear amplifier. The second and the third terms in the right hand part compose the nonlinear tent type function. Note, that in contrast to the MG model the nonlinear function depends on two variables, $x(\theta)$ and $x(\theta-\tau)$:

$$F = e\begin{cases} bkx(\theta-\tau) + cx(\theta), & bkx(\theta-\tau) + cx(\theta) > 1-d, \\ 1-d, & bkx(\theta-\tau) + cx(\theta) \le 1-d. \end{cases} \tag{3}$$

In the case of the 3rd oscillator (the external capacitor C removed, i.e. C=0) the parasitic capacitances $C_1 \ldots C_3$ of the transistor collector nodes should be taken into account. Then the dynamics of the oscillator is given by a set of a nonlinear DDE and two linear ODEs:

$$\begin{aligned} \frac{dx}{d\theta} &= -x(\theta) + akz(\theta-\tau) - F[x(\theta), z(\theta-\tau)], \\ \varepsilon_2 \frac{dy}{d\theta} &= x - y, \\ \varepsilon_3 \frac{dz}{d\theta} &= y - z. \end{aligned} \tag{4}$$

Here function F is similar to one given by (3), but $x(\theta-\tau)$ is replaced with $z(\theta-\tau)$:

$$F = e\begin{cases} bkz(\theta-\tau) + cx(\theta), & bkz(\theta-\tau) + cx(\theta) > 1-d, \\ 1-d, & bkz(\theta-\tau) + cx(\theta) \le 1-d. \end{cases} \tag{5}$$

In (4) and (5) the following notations have been introduced:

$$\begin{aligned} &x = \frac{U_{C1}}{U^*}, \quad y = \frac{U_{C2}}{k_2 U^*}, \quad z = \frac{U_{C3}}{k_2 k_3 U^*}, \quad k = k_2 k_3, \\ &\theta = \frac{t}{RC_1}, \quad \tau = \frac{T_{del}}{RC_1}, \quad \varepsilon_2 = \frac{R_2 C_2}{RC_1}, \quad \varepsilon_3 = \frac{R_3 C_3}{RC_1}, \end{aligned} \tag{6}$$

while the other parameters are the same as in (1). In (6) $U_{C1} \ldots U_{C3}$ are the AC components of the collector voltages of Q1…Q3. The k_2, k_3, and k are the gains of the Q2-stage, the Q3-stage, and of the overall linear amplifier, respectively.

Numerical Results

Chaotic attractors from the eqn. (2) and eqn. (4) are presented in Fig. 5. Both of them resemble the attractors of the MG model [2,6,8].

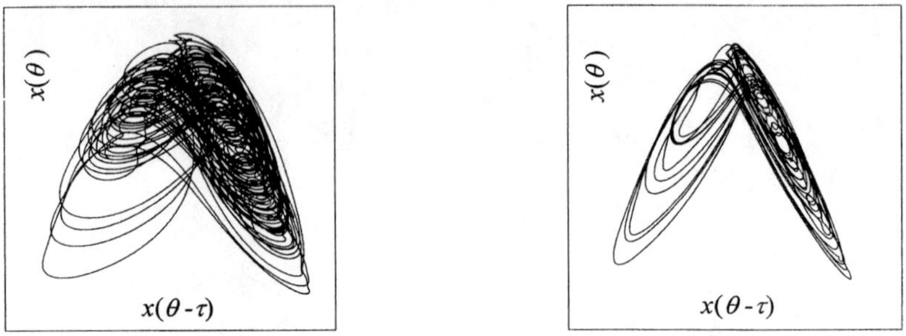

FIGURE 5. Phase portraits, $x(\theta)$ *vs.* $x(\theta-\tau)$. Left: from eqn.(2) with a=0.37, b=0.58, c=0.42, d=1.2, e=1.9, k=6, τ=3.3. Right: from eqn.(4) with a, b, c, d, e, k the same as in the left plot, $\varepsilon_2=\varepsilon_3=1.2$, τ=7.2.

To characterize the oscillators quantitatively the Lyapunov exponents (LE) have been calculated from the model. Even for relatively low delays (τ>2) the attractor has more than one positive LE (Fig. 6). Thus, the system exhibits *hyperchaotic* behavior.

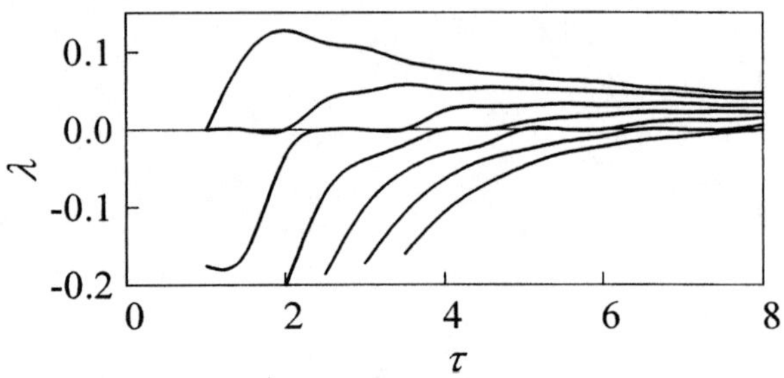

FIGURE 6. Seven maximal Lyapunov exponents λ *vs.* dimensionless delay time τ from eqn. (2). All the parameters are the same as in Fig. 5.

CONCLUSIONS

We have designed, built and investigated three versions of a delay line based oscillator. SMD, the passive ones along with microwave transistors have been used in the experimental circuits. Fast nonlinear unit has been implemented by means of a linear by-pass resistor and single-transistor low-biased common-emitter inverting amplifier operating in a nonlinear mode. The oscillators generate broadband noise like signals in the frequency range from about 1MHz to over 1000MHz. Multiple positive Lyapunov exponents calculated from the mathematical model confirm hyperchaotic behavior of the circuits.

ACKNOWLEDGMENTS

A. Čenys, G. Mykolaitis, S. Bumelienė and A. Tamaševičius were supported in part by USA European Office of Aerospace Research and Development under the contract F61775-01-WE065.

REFERENCES

1. Mackey, M. C. and Glass, L., *Science* **197**, 287-289 (1977).
2. Farmer, J. D., *Physica D* **4**, 366-393 (1982).
3. Dmitriev, A. S., Orlov, Yu. N., and Starkov, S. O., *Radiotekhnika i elektronika* **34**, 1980-1983 (1989), in Russian.
4. Losson, J., Mackey, M.C., and Longtin, A., *Chaos* **3**, 167-176 (1993).
5. Schwarz, W. and Mögel, A., "Chaos Generators with Transmission Line" in *Proc. 2nd Int. Workshop on Nonlinear Dynamics of Electronic Systems NDES'94*, Krakow, Poland, 1994, pp. 239-244.
6. Namajūnas, A., Pyragas, K., and Tamaševičius, A., *Phys. Lett. A* **201**, 42-46 (1995).
7. Chua, L. O., Yu, J., and Yu., Y., *IEEE Trans. Circuits and Systems* **32**, 46-61 (1985).
8. Namajūnas, A., Pyragas, K., and Tamaševičius, A., *Phys. Lett. A* **204**, 255-262 (1995).
9. Namajūnas, A., Pyragas, K., and Tamaševičius, A., *Int. J. Bifurcation and Chaos* **7**, 957-962 (1997).
10. Tamaševičius, A., Čenys, A., Mykolaitis, G., Namajūnas, A., and Lindberg, E., *Electron. Lett.* **33**, 2025-2026 (1997).
11. Tamaševičius, A., Čenys, A., Namajūnas, A., Mykolaitis, G., and Lindberg, E., "Synchronizing Hyperchaotic Circuits" in *Applied Nonlinear Dynamics and Stochastic Systems near the Millenium*, edited by J. B. Kadtke and A. Bulsara, AIP Conference Proceedings 411, New York: American Institute of Physics, 1997, pp. 81-86.
12. Tamaševičius, A., Čenys, A., Namajūnas, A., and Mykolaitis, G., *Chaos, Solitons & Fractals* **9**, 1403-1408 (1998).
13. Pyragas, K., *Int. J. Bifurcation and Chaos* **8**, 1839-1842 (1998).
14. Mögel, A., Schwarz, W., Lukin, K. A., and Zemlyanoy, O. V., "Chaotic Wide Band Oscillator with Delay Line" in *Proc. 3rd Int. Workshop on Nonlinear Dynamics of Electronic Systems NDES'95*, Dublin, Ireland, 1995, pp. 259-262.
15. *Toward a new digital communication technology based on nonlinear dynamics and chaos*, Strategic Assessment Report, U.S. Army Research Office, 1996.

Experimental behavior of a dissipative Toda-Rayleigh ring

E. del Río*, V. A. Makarov†, M. G. Velarde† and W. Ebeling**

*E.T.S.I. Aeronauticos, Universidad Politecnica de Madrid, Plaza Cardenal Cisneros, 3, Madrid 28040, Spain
†Instituto Pluridisciplinar, Universidad Complutense, Paseo Juan XXIII, 1, Madrid 28040, Spain
**Institute of Physics, Humboldt-Universität Berlin, Invalidenstrasse, 110, Berlin 10178, Germany

Abstract. Using a analog electronic circuit for a Toda-Rayleigh ring we study its oscillatory modes and we show with a simple and robust method how switching between them occurs.

DISSIPATIVE TODA-RAYLEIGH LATTICE

Toda [1] provided exact solutions, including solitons, for a Hamiltonian lattice system with exponential interaction. Subsequently, other (conservative) soliton bearing systems were proposed for nonlinear signal processing and transmission. However, observation of waves in nature and engineering devices shows that dissipation is unavoidable and may, on occasion, drastically alter the expected phenomena. Experiments with analog electrical circuit implementations of the Toda lattice and numerical tests have shown that, in practical terms, unavoidable dissipative effects lead to energy losses and indeed spoil the conservative nature of the Toda lattice [2, 3, 4, 5, 6]. Thus the interest arises of studying the role of dissipation and controlled energy pumping e.g. in the stationary wave propagation along the Toda lattice here taken with Rayleih's dissipation, as the concept of dissipative solitons recently, put forward by Velarde and collaborators [7] have proven quite useful in understanding dissipative waves in fluids.

Analog circuit implementation of the model

Singer and Oppenheim in Ref. [3] have developed a circuit for modeling the original, conservative, Toda system. Consequently, all observed waves come from chosen initial conditions imposed by a current generator. For a dissipative system like in ours the initial conditions loose significance and a well defined finite number of different solutions survive. We may chose between them by playing with an external voltage [9]. The circuit scheme for the Toda chain including Rayleigh's dissipation is shown in Fig. 1. According to the I–V relation of the double capacitor (DC), we have [3, 9].

$$\frac{d^2V_n}{dt^2} = \omega_v^2 R_{\rm dc} I_{\rm dc} = \omega_v^2 R_{\rm dc}\left(I_n - I_{n+1} - I_{\rm nr} - I_R\right), \quad \omega_v^2 = \frac{1}{C_1R_1C_2R_2} \tag{1}$$

CP622, *Experimental Chaos: 6th Experimental Chaos Conference*, edited by S. Boccaletti et al.

where I_R and $I_{\rm nr}$ are currents through the corresponding linear, R, and nonlinear, NR, resistors (Fig. 1). The current through the non-linear resistor (block NR in Fig. 1), $I_{\rm nr}$, is a nonlinear function of the voltage applied to its terminals, $\Delta V = V_n - V_a$ (Fig. 1).

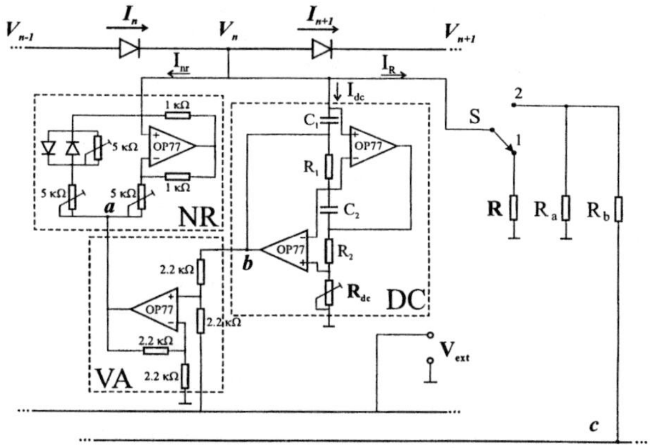

FIGURE 1. Scheme of the Toda-Rayleigh lattice including the three main blocks: DC is a double capacitor, NR accounts for the nonlinear dissipative function I_{nr} and VA provides variable shift of the nonlinear function I_{nr} to compensate voltage drift. **R** is an additional stabilizing high value resistor (R=120 kΩ). The diodes are 1n4148. For $\mathbf{R}_a$ and $\mathbf{R}_b$ see main text.

We approximate the voltage-current characteristics $I_{\rm nr}(\Delta V)$, with a cubic polynomial

$$I_{\rm nr}(\Delta V) = \frac{3\sqrt{3}I_{\rm max}}{2\widetilde{\Delta V}^3}(\widetilde{V}^2 - \Delta V^2)\Delta V \tag{2}$$

where $I_{\rm max}$ and $\widetilde{V}$ are, respectively, the maximum and zero of the function $I_{\rm nr}$. In our experiments we use $\widetilde{V} \approx 1.45\,{\rm V}$ and $I_{\rm max} \approx 45\,\mu{\rm A}$. Note that for ΔV we have [9]

$$\Delta V = \frac{1}{\omega_v^2 R_{\rm dc} C_1}\frac{dV_n}{dt} - V_{ext} \tag{3}$$

hence

$$\frac{d^2V_n}{dt^2} = \omega_v{}^2 R_{\rm dc}\left\{I_s\left[e^{\frac{V_{n-1}-V_n}{V_t}} - e^{\frac{V_n-V_{n+1}}{V_t}}\right] + I_{\rm nr}\left(\frac{dV_n}{dt}, V_{\rm ext}\right) - \frac{1}{R}V_n\right\} \tag{4}$$

Let us note that in general an excitation passing through a unit in the chain increases (or decreases) its voltage by a value depending on the type of excitation [3]. Thus a wave rotating in the ring continuously increases (respectively, decreases for a wave running in the opposite direction) the voltage in the units and, eventually, saturates the circuit. To avoid this problem, we have an additional electric block VA in Fig. 1. The term I_R in the r.h.s. of (4) comes from the large (120 kΩ) value of R (Fig. 1). This resistor compensates unavoidable differences between units and prevents saturation of the circuit due to small disturbances.

Oscillations in the ring chain

The system possesses $(N+1)$ different stable states [9]: $(N-1)$ oscillatory modes and 2 non-oscillatory, rotatory modes. The amplitude of oscillations depends on the mode. The modes $m = (N-1), (N-2), \ldots$ have the same amplitude as the modes with $m = 1, 2, \ldots$. They correspond to waves with the same space scales like waves with $m = 1, 2, \ldots$, but propagating in opposite directions. We can denote them as modes having negative mode number, $m = -1, -2, \ldots$. For even N there is a "special" (*optical*) mode with $m = \pm N/2$ (both signs define the same wave form when nearest units oscillate in anti-phase). Two longest, propagating in opposite directions, modes with $m = \pm 1$ have the largest amplitude. The amplitude decreases with the increase of mode number. In our case $N = 6$ (6 units) and hence we have $N - 1 = 5$ different waves ($m = \pm 3$ correspond to one and the same optical mode). Depending on initial conditions, after a transient time one of the seven (5 oscillatory + 2 non-oscillatory) modes appears in the ring.

Figure 2 shows experimental traces for 3 different modes taken for the same external voltage. The $m = 1$ wave has the largest amplitude while waves with $m = 2, 3$ have about the same amplitude, as theoretically predicted.

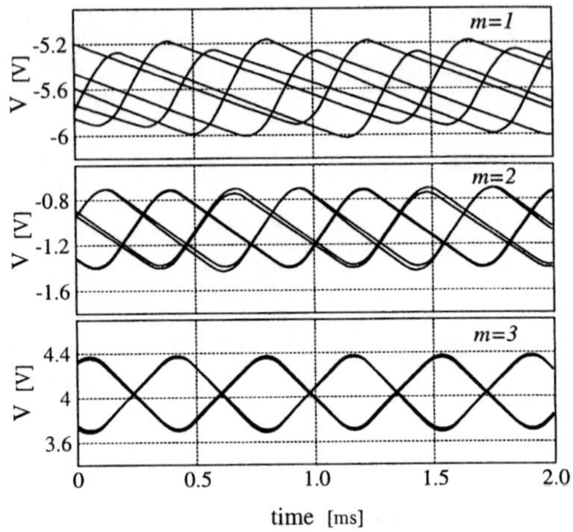

FIGURE 2. Oscilloscope traces of the voltages taken from all 6 units of the ring for 3 different modes. The switch S 1 is in position 1.

MODE COMPETITION AND SWITCHING

We used V_{ext} as a parameter to study the wave competition. We start from the non-oscillatory mode and a high, negative value of V_{ext} (-1.8 V). Then we slowly (quasi-statically) increase V_{ext} to high positive values and follow the mode dynamics. Figure 3(a) shows the domains of stability and jumps that occur between modes. Several modes can overlap for the same value of the external voltage. Thus as V_{ext} increases beyond a critical point the initial mode becomes unstable while two other modes are stable. Hence one of them can win in the competition. For example, for $V_{ext} \approx -0.85\,V$ the $m = -0$

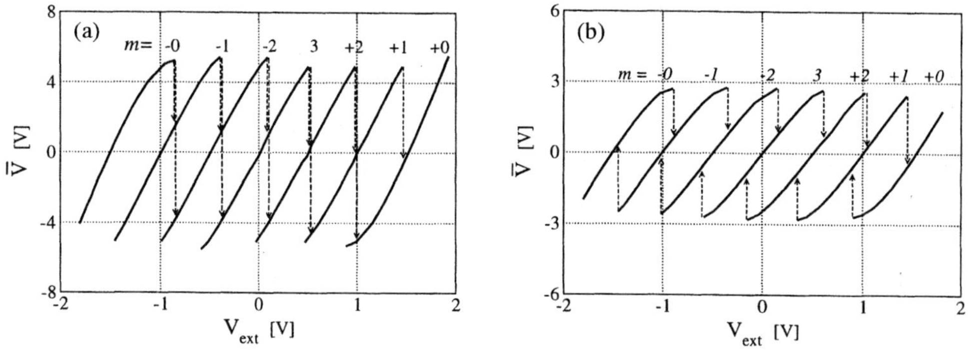

FIGURE 3. Mean voltage value on units for different modes and direction of jumps between them *vs* the external voltage, V_{ext}, in the ring of 6 units. Each curve corresponds to a certain mode with mode number marked m. Parameter values are the same as for Fig. 1 and for **a) R**=120 kΩ and for **b) R**=60 kΩ.

("anticlockwise" non-oscillatory mode with) becomes unstable but the two others with $m = -1, -2$ are stable. Then either $m = -1$ or $m = -2$ can win and be established in the ring. In our experiments we observed both possibilities in different runs. Further increasing V_{ext} leads to new jumps between modes as shown in Fig. 3(a). Analogous jumps are observed while decreasing the external voltage from 1.8 V down to -1.8 V (not shown on Fig. 3(a).

In practical terms, in contrast to what we have seen in Fig. 3(a), the absence of uncertainty in mode formation and jumps between them is usually preferable. We experimentally found that decreasing the value R to a half of it (from 120 kΩ to 60 kΩ), jumps only occur between the two nearest modes and there are no jumps to the next nearest neighbors. Figure 3(b) shows the sequence of all jumps between modes. There is a hysteretic behavior between any two nearest modes. Changing the value of the external voltage we can get the desired mode. Comparing Figs 3(a) and 3(b) we can see that their shapes are similar, but the mean value in Fig. 3(b) is half of the corresponding value in Fig. 3(a). Indeed, the mean value of the voltage disappears in all terms of Eq. (4) but not in the last one.

We can connect the resistor **R** to a common node **c**, and this to a voltage source giving a fixed voltage V_c. Its global effect is just the addition of a new constant in the last term of the Eq. (4),

$$\frac{d^2V_n}{dt^2} = \omega_v{}^2 R_{\text{dc}} \left\{ I_s \left[e^{\frac{V_{n-1}-V_n}{V_t}} - e^{\frac{V_n - V_{n+1}}{V_t}} \right] + I_{\text{nr}} \left(\frac{dV_n}{dt}, V_{\text{ext}} \right) - \frac{1}{R} (V_n - V_p) \right\} \qquad (5)$$

hence shifting the mean voltage value from what we have in the case of ground connection of the resistors **R**, to the actual one. Accordingly, $\overline{V}_{ng}$ them $\overline{V}_{ng} = \overline{V} + V_c$ and the shape of the figs. 3(a) and 3(b) do not change, there is only a shift along the vertical axis by the V_c. By switching **S**(Fig. 1) to position 2, we have replaced **R** by two resistors R_a and R_b: R_a is grounded and all R_b are connected to a common node **c** rising a voltage V_c. Note that according to Thévenin's theorem, we exchange in Eq. (5), **R** and V_c by the corresponding Thévenin equivalent [11]

$$V_c \to \frac{R_a}{R_a+R_b}V_c, \quad R \to \frac{R_aR_b}{R_a+R_b} \tag{6}$$

V_c is driven by the system. For this voltage we have the average over all units. Setting $V_n = \overline{V}_{ng} + u_n$, where u_n is the oscillatory part of V_n with $\overline{u}_n = 0$ we have

$$V_c = \frac{1}{N}\sum_{n=1}^{N} V_n = \overline{V}_{ng} + \frac{1}{N}\sum_{n=1}^{N} u_n \tag{7}$$

Note that in our case the mean voltage value $\overline{V}_{ng}$ is equal for all units. We experimentally found that the last term in Eq. (7) is small relative to $\overline{V}_{ng}$ so we can neglect it and from Eqs.(6) and $\overline{V}_{ng} = \overline{V} + V_c$ we have

$$\overline{V}_{ng} = \frac{R_a+R_b}{R_b}\overline{V} \tag{8}$$

which is equivalent to changing **R**. We have chosen $R_a = R_b = 120k\Omega$ and from the Thévenin equivalent, Eq. (6), we have $R = 60k\Omega$ like in Fig. 3(b). The effect of no connection of the node **c** to any voltage source, just described by Eq. (8) gives us a factor 2 in $\overline{V}$ whose behavior showed in Fig. 3(a). From Eq. (8), we get an unstable system by removing R_a .

We can use the variable V_c to drive V_{ext} by switching off V_{ext} from the external voltage source and making a direct connection between V_{ext} and the node **c**. By connecting a capacitor **C** between the node **c** and ground, the equation describing the actual behavior of V_{ext} is

$$\frac{dV_{ext}}{dt} = \frac{N}{CR_b}\left(\frac{1}{N}\sum_{n=1}^{N} V_n - V_{ext}\right) \tag{9}$$

The capacitor **C** produces a long time dynamical behavior of V_{ext} (relative to the oscillatory period of the modes) determined by the total time constant $\frac{R_bC}{N}$ which in our case, with 6 units, $R_a = R_b = 120k\Omega$ and $C = 530\mu F$, is approximately 5.3s.

As earlier said, the bracket in the r.h.s. of Eq. (9) is about $(\overline{V}_{ng} - V_{ext})$. From this equation we can see that $\overline{V}_{ng}$ tries to approach to V_{ext}. This means that the system must follow the line corresponding to its actual mode, increasing o decreasing V_{ext} according to the actual value of $\overline{V}_{ng}$: if $(\overline{V}_{ng} - V_{ext}) > 0$ then V_{ext} must increase, otherwise, if $(\overline{V}_{ng} - V_{ext}) < 0$ then V_{ext} must decrease.

Since V_{ext} is not too large, the system will approximately follow the lines of Fig. 3(b) with a minor shift, up or down, according to the sign of V_{ext} as it is shown in Fig.(4.a).

In Fig. 4(a), the system is running clockwise and can cross the equilibrium bisector line through several points, while it switches between modes. Since the equilibrium line is unstable, the system keeps running for ever through all oscillatory modes. At every jump between modes through the bisector line in Fig. 4(a), the system changes its direction according with to the jump: Jumping up to down, the system exchanges right to left in the V_{ext} evolution and vice versa. These jumps correspond with maxima or

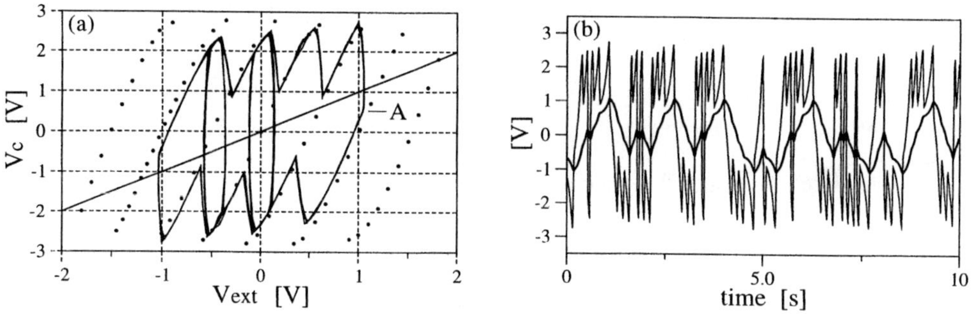

FIGURE 4. **a**) Same data as in fig.3(b) (dots) together the bisector line and the path followed by the system. **b**) Time realization of V_{ext} (small amplitude) and $\overline{V}_{ng}$ large amplitude) corresponding to **a** showing random jumps from one mode to other one.

minima in the time series of V_{ext}, as it is shown in fig.(4.b). Since the point **A** lies below the bisector line, the system skips the monotonic mode m=+0. In order to include this mode in the path followed by the system in Fig. 4(a), we need to lower the slope of the original bisector line, getting a new line passing below **A**. Connecting a voltage-divisor between the node **c** and V_{ext} entry we easily get this new line.

In contrast to what we see in Fig. 3(b), the system can randomly jump either between two nearest modes or to next nearest neighbors, due to conditions not being the same as in Fig.(3.b) even with $V_{ext} \approx 0$. Fig. 4(a) shows that the system never reaches the stationary mode, which is the only state for Fig. 3(b).

Research supported by Grants PB96-599 and PB 97-0574-C04-01 (Ministerio de Ciencia y Tecnologia, Spain) and by the Humboldt-Mutis Foundation.

REFERENCES

1. See e.g. M. Toda, *Theory of Nonlinear Lattices* (Springer, Berlin, 1981).
2. R. Hirota and K. Suzuki, Procs. IEEE **61**, 1483 (1973).
3. A. C. Singer and A. V. Oppenheim, Int. J. Bifurcation Chaos **9**, 571 (1999).
4. N. Islam, J. P. Singh and K. Steiglitz, J. Appl. Phys. **62**, 689 (1987).
5. Y. Okada, S. Watanabe and H. Tanaka, J. Phys. Soc. Jpn. **59**, 2647 (1990).
6. T. Kuusela and J. Hietarinta, Phys. Rev. Lett. **62**, 700 (1989).
7. M. G. Velarde, A. A. Nepomnyashchy and M. Hennenberg, Adv. Appl. Mech. **37**, 167 (2000).
8. V. A. Makarov, W. Ebeling and M. G. Velarde, Int. J. Bifurcation Chaos **10**, 1075 (2000).
9. V. A. Makarov, E. del Río, W. Ebeling, and M. G. Velarde Phys. Rev. E (2001) (to be appear).
10. Lord Rayleigh, *The Theory of Sound* (Dover, NY, 1945) Vol. 1., ch. III, p. 81.
11. P. Horowitz and W. Hill, *The Art of Electronics* (Cambridge University Press, NY, 1987).

Motor Maps for Nonlinear Control

Paolo Arena, Maide Bucolo, Luigi Fortuna, Mattia Frasca

Dipartimento Elettrico, Elettronico e Sistemistico
Universit degli Studi di Catania
Viale A. Doria 6, 95125 - Catania — Italy
Email: lfortuna@dees.unict.it

Abstract. In this paper the design of a motor map to control a chaotic system is presented. A feedback entrainment scheme is adopted: a system with different parameters is used to generate the reference trajectory for the chaotic system to be controlled, while the motor map provides the appropriate gain value of the feedback signal. As input of the motor map the state of the system to be controlled is considered. The motor map based adaptive controller offers high performances, specially in the case when the reference trajectory is switched into another one. In this case, a specialization of the neurons constituting the motor map is observed: while a group of neurons learns the appropriate control law for a reference trajectory, another group specializes itself to control the system when the other trajectory is used as reference. Moreover, a discrete components hardware implementation of the motor map has been realized.

Introduction

The importance of topology-preserving maps in the brain relies both in the representation of sensory input signals and in the ability of performing an action in response to a given stimulus. Indeed, neurons in the brain are organized in local assemblies able to perform a given task such as for example sending appropriate signals to muscles. These neural assemblies constitute two-dimensional layers in which the locations of the excitation are mapped into movements.

Topology-preserving structures mapping the input signals inspired the paradigm of Kohonen's Neural Networks [2]. In these maps neighboring neurons are excited by similar inputs. An extension of these neural structures is represented by the motor maps [3]. These are networks able to react to localized excitation by triggering a movement (as in the brain are the *motor cortex* or the *superior colliculus*). To this aim, unlike the Kohonen's networks, they should include the storage of an output specific to each neuron site. This is achieved by considering two layers: one devoted to the storage of input weights and one devoted to the output weights.

In this paper the self-organizing features of the motor map are exploited to perform the control of chaotic systems [1] and to build-up an autonomous system able to learn by itself the appropriate control law is presented.

Formally, a motor map can be defined as an array of neurons mapping the space V of the input patterns into the space U of the output actions:

$$\Phi : V \rightarrow U$$

It is worth remarking that the same plastic char cteristics of the input layer should be preserved also in the assignment of output values. In other words, the learning

CP622, *Experimental Chaos: 6th Experimental Chaos Conference*, edited by S. Boccaletti et al.

phase deals with both the input weights and the output ones updating. This allows the map to perform tasks such as for example a motor control task. Both supervised and unsupervised learning can be applied. However, in order to investigate about an autonomous self-organizing system for nonlinear control, unsupervised learning has been considered. In this case there is not *a priori* information on the appropriate control and no teacher is available, thus the algorithm should find by itself the correct control action. The *reward function* is the only source of information and provides a parameter indicating how well the control is being performed.

The learning algorithm is the key to obtain a spatial arrangement of either the input or output weight values of the map. This is achieved by considering an extension of the winner-take-all algorithm [3]. At each learning step, when a pattern is given as input, the winner neuron is individuated: this is the neuron that best matches the input pattern, i.e. whose output weight closely resembles the input pattern. Then, a neighborhood of the winner neuron is considered and an updating involving both the input weights and output ones for neurons belonging to this neighborhood is performed according to:

$$\begin{aligned} w_{i,in}(t+1) &= w_{i,in}(t) + \eta_{in} \cdot \xi_i \cdot (v(t) - w_{i,in}(t)) \\ w_{i,out}(t+1) &= w_{i,out}(t) + \eta_{out} \cdot \xi_i \cdot (s(t) - w_{i,out}(t)) \end{aligned} \tag{1}$$

where η, ξ_i, v, w_{in} and w_{out} are the learning rate, a function indicating the shape and the extension of the neighborhood, the input pattern, the input weights and the output ones, respectively: the index i takes into account the neighborhood of the winner neuron. $s(t)$ is the output of the map and in the case of *supervised learning* is given as input of the learning phase, while in the case of *unsupervised* learning this is varied as follows:

$$s(t) = w_{winner,out} + a_s \cdot \lambda$$

where a_s is a positive parameter analogous to the learning rate in (1), and λ is a gaussian variable with zero-mean and unit variance. The updating of the neuron weights is performed according to the value of the reward function: the weights are updated, only when the increase in the reward function exceeds the average increase gained at the lattice site of the winner neuron.

The Motor Map Control Scheme

For the chaos control purpose a feedback entrainment scheme is adopted. The purpose of this control is to entrain a nonlinear system to be controlled to a given trajectory. The control system is illustrated in Fig. 1. The system to be controlled is a chaotic system and the reference system is as the controlled one, but with a different parameter set such that for example a stable cycle limit exists. However, as discussed in the following, other reference trajectories, not properly entrainment trajectories, can be considered. The motor map controller acts on the system to be controlled by setting the value of the feedback gain. This gain modulates the error signal between the reference trajectory and the actual one and is fed back into the system to be controlled.

Even if a huge number of experiments have been carried out on different systems to be controlled, here the focus is on the *double scroll Chua attractor*. The equations of the Chua's circuit [4] are the following:

$$\begin{aligned} \dot{x} &= \alpha \cdot (y - h(x)) \\ \dot{y} &= x - y + z \\ \dot{z} &= -\beta \cdot y \\ h(x) &= m_1 x + 0.5 \cdot (m_0 - m_1) \cdot (|x+1| - |x-1|) \end{aligned} \tag{2}$$

and can be rewritten to include the feedback signals as follows:

$$\begin{aligned} \dot{x} &= \alpha \cdot (y - h(x)) + k_x \cdot (x_{ref} - x) \\ \dot{y} &= x - y + z + k_y \cdot (y_{ref} - y) \\ \dot{z} &= -\beta \cdot y + k_z \cdot (z_{ref} - z) \\ h(x) &= m_1 x + 0.5 \cdot (m_0 - m_1) \cdot (|x+1| - |x-1|) \end{aligned} \tag{3}$$

Instead of controlling the whole state space, only two variables have been fed back, i.e. in our experiments k_z was set to zero and not allowed to vary. The possibility of achieving good performances with this scheme seems to be not a general conclusion, but it holds for this specific chaotic attractor. However, the motor map control scheme is general and can include the feedback of the whole state space.

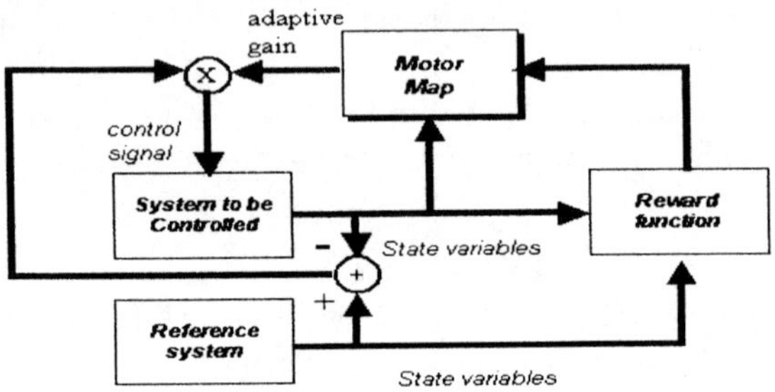

FIGURE 1. The motor map control scheme.

As above discussed, the motor map learning is unsupervised. To this aim a reward function should be introduced. This reward function takes into account the error between the state of the controlled system and the one of the reference system and thus constitutes a performance parameter to be minimized by the adaptive controller. It has been chosen as follows:

$$R = -((x_{ref} - x)^2 + (y_{ref} - y)^2) \tag{4}$$

Two schemes have been used to perform the control: in the former one a generic case is assumed, a motor map is built for each variable to be controlled; in the second scheme a unique motor map acts on the system controlling the variable having the largest error. In the first scheme, the two motor maps act independently onto the two state variables, i.e. x and y, by storing in the input layer the values of the corresponding input pattern and in the output one the values of the corresponding feedback gain k_x (first map) or k_y (second map). Instead, in the second scheme only a motor map is devoted to control both the two variables, by choosing at each step to control the variable having the largest error.

In both the two cases the winner-take-all law establishes which neuron has the most suitable gain value. Moreover, only the weights of the winner neuron are updated. The learning rate is constant, thus the learning of the motor map is always active and the motor map is able to deal with changes in the reference trajectory. Moreover to avoid ever increasing values of the gain k, a performance goal on the error is introduced: namely, when the error is under a given bound, the learning is stopped; if the error grows up and overcomes the threshold, the learning starts again.

The simplifications, here discussed, of the original learning algorithm have been introduced with the aim of building a discrete components hardware realization implementing the learning/working phase in the simplest way, compatibly with the quality of the control. To reduce the hardware requirements, by exploiting the noise-like behavior and the broad-band spectrum of a chaotic attractor, it can be assumed that the *stochastic term* λ is implemented by a chaotic signal from an analog circuit. In particular, it can be chosen a zero-mean variable as for example the y variable of another Chua s circuit having time constants smaller than the ones of the circuit to be controlled.

Results

In this Section some results are presented. Both the two control schemes described in the previous Section are considered. The system to be controlled is, in all the simulations here reported, the *double scroll Chua attractor,* described by equation (2) with the following parameter values:

$$\alpha = 9; \beta = 14.286; m_0 = -\frac{1}{7}; m_1 = \frac{2}{7} \tag{5}$$

First, the scheme with two motor maps is taken into account. As a first example a limit cycle has been chosen as entrainment trajectory for the system (3). This limit cycle is generated by choosing the parameters of the system (2) as follows:

$$\alpha = 7; \beta = 12; m_0 = -\frac{1}{7}; m_1 = \frac{2}{7} \tag{6}$$

Two maps, each having 4 neurons, have been trained for 390 epochs to achieve the performance specifications EB_x=0.08 (error bound for the x variable) and EB_y=0.04 (error bound for the y variable). The performance goal was met with settling time t_{sx}=517 epochs and t_{sy}=590 epochs. Fig. 2 refers to this first example.

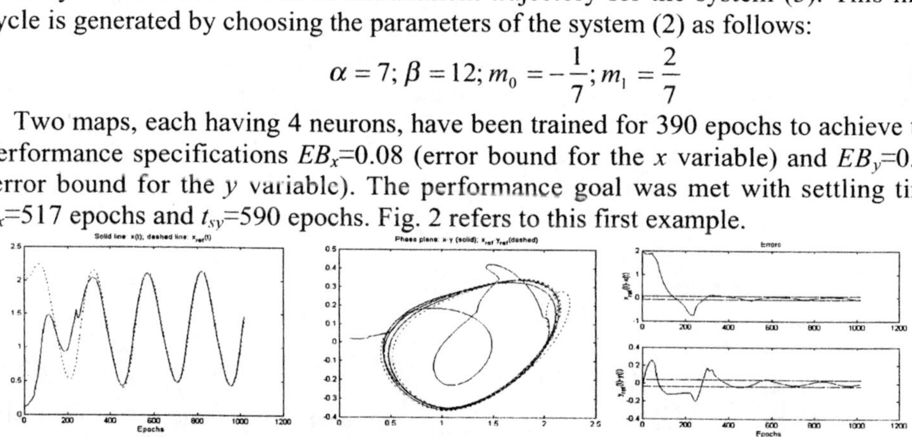

FIGURE 2. (a) Trends of the controlled variable x (solid line) and of the reference x_{ref} (dashed line). (b) limit cycle of the controlled system and of the reference system. (c) errors $e=x_{ref}-x$ and $e=y_{ref}-y$.

From these results it derives that the motor map efficiently learns by itself the control action to be performed on the system to be controlled. A further example illustrates how the self-organization is able to deal with different reference trajectories. In particular the case in which a double scroll Chua attractor with parameters as in (5) alternates with a limit cycle as in (6) to entrain the trajectory of the system is treated. This example is illustrated in Fig. 3: the controller is constituted by two motor maps (each having 9 neurons) trained for 10000 epochs. The reference system, initially a double scroll Chua attractor, is switched to a limit cycle at t=3800 epochs and comes back to a double scroll Chua attractor at t=5500 epochs. In Fig. 3(a) the trend of the variable x of the controlled system is shown also with the reference signal x_{ref}. As it can be noticed the motor map learns the right control action within the first 1000 epochs and the two signals are indistinguishable until the switching of the reference. When the limit cycle is assumed as reference trajectory, the error increases as shown in Fig. 3(b), but remains within low bounds. Of course this takes place since the map has to completely reorganize itself to follow the new trajectory.

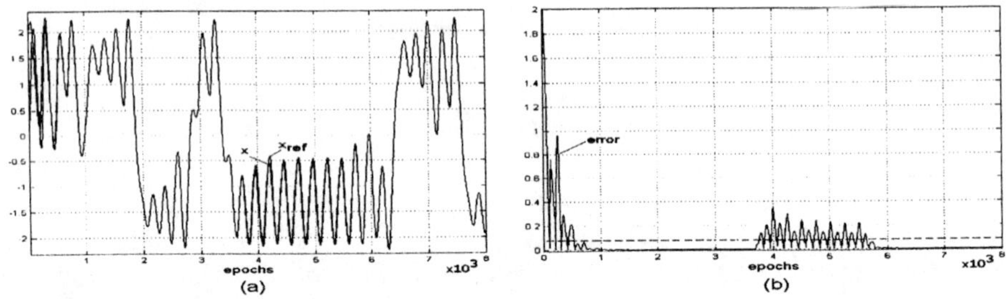

FIGURE 3. Results of the motor map control when the trajectory is switched from a double scroll to a limit cycle and again to a double scroll: (a) waveforms of the controlled state variable x and of the reference x_{ref}; (b) absolute value of the error $e=x_{ref}-x$.

In the case of the scheme with a unique motor map, as intuition suggests, the achievement of the performance specifications is more difficult. In this case the controller chooses to control the variable having the largest error from the reference one. Once completed the learning phase, a cluster of neurons will be devoted to the control of a variable, while another cluster will control the other variable.

In Fig. 4 some results obtained adopting such a scheme are shown. In Fig. 4(a) the reference is a single scroll with the following parameters:

$$\alpha = 8.5; \beta = 14.286; m_0 = -\frac{1}{7}; m_1 = \frac{2}{7} \quad (7)$$

In Fig. 4(b) another chaotic attractor generated by a Chua's circuit with the following parameters:

$$\alpha = -6.69191; \beta = -1.52061; m_0 = -\frac{1}{7}; m_1 = \frac{2}{7} \quad (8)$$

is assumed as reference system. Finally, in Fig. 4(c) an example of the control of another chaotic system (the Lorenz system [5]) is shown: the reference system is generated by the limit cycle of the Chua's circuit with parameters as in (6). In all the examples reported in Fig. 4, the reference trajectory is generated by a system with very different parameters, or by another system. The control performance are yet good, but the control signals have higher amplitudes than in the previously considered cases.

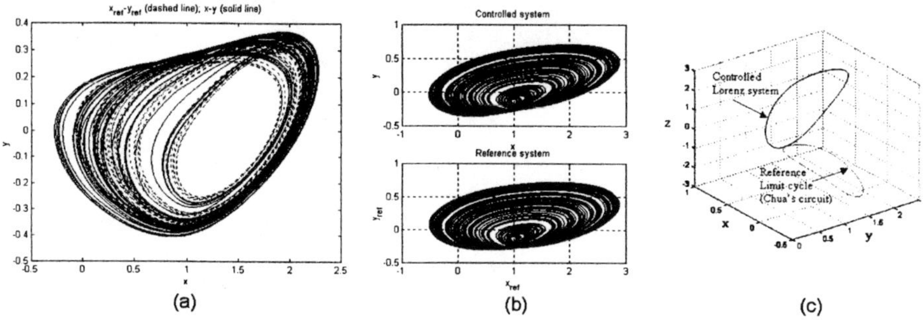

FIGURE 4. Motor Map based chaos control: (a) the reference is a *single scroll;* (b) the reference is a *chaotic attractor;* (c) controlling a Lorenz system with a limit cycle reference generated by a Chua s circuit.

Conclusions

In this paper a motor map based adaptive controller has been introduced to entrain a chaotic system to a trajectory generated by another attractor (both a limit cycle or other strange attractors have been analyzed as references).

A huge number of experiments have been performed. In most of them, the system to be controlled was a Chua's circuit and the feedback of only two state variables instead of the whole state space was found to be enough to guarantee good performance of the control system. The feasibility of a discrete components hardware realization is guaranteed by consideration that a small number of neurons is required to solve the task of chaos control and by several assumptions considered when numerical simulations have been carried out. To the aim of an hardware realization, it has been investigated the behavior of the map when only the winner neuron is updated instead of a neighborhood of this neuron, this assumption makes sense since the effect of cooperation in a small assembly of neurons is weaker than the one in a large layer of neurons. The hardware realization of the motor map controller shows results that agree with the numerical ones.

References

1. Chen G. and Dong X., *From Chaos to Order: Methodologies, Perspective and Apllications*, World Scientific Series on Nonlinear Science, Series A, vol. 24, 1998.
2. Kohonen T., Self-Organized Formation of Topologically Correct Feature Maps , *Bio. Cybern.* , 43, pp. 59-69, 1972.
3. Ritter H., Martinetz T. and Schulten K., *Neural Computation and Self-Organizing Maps*, Addison Wesley, 1992.
4. Madan R. N., *Chua's circuit: a paradigm for chaos*, World Scientific Series on Nonlinear Sciences, Series B, Vol. 1 (World Scientific, Singapore), 1993.
5. Strogatz S. H., *Nonlinear Dynamics and Chaos*, Perseus Book, 1994.
6. Chen G. and Dong X., *From Chaos to Order — Methodologies, Perspectives and Applications*, Worldscientific Series on Nonlinear Science, Series A, Vol. 24, 1998.

GEOPHYSICS AND ASTROPHYSICS

Chaos in the Music of the Spheres

J. Robert Buchler*, Zoltan Kolláth† and Robert Cadmus**

*University of Florida, USA
†Konkoly Observatory, HUNGARY
**Grinnell College, USA

Abstract. The light curves (time series of the radiated energy) of most large amplitude, pulsating stars such as the well known Cepheid stars are regular. However, a smaller group of variable stars that are located next to them in the Hertzsprung- Russell diagram undergoes irregular light variations and exhibits irregular radial velocities as well. The mechanism behind this irregular behavior was a long standing mystery. A flow reconstruction technique based on the observed lightcurves of six separate stars shows that their underlying dynamics is chaotic and low dimensional ($d = 4$). Furthermore, we present evidence that the physical mechanism behind the behavior is the nonlinear interaction of just two pulsation eigenmodes. In a generalized Shil'nikov scenario, the pulsation energy alternates continuously, but irregularly between a lower frequency mode that is linearly unstable and thus growing, and a stable overtone that gets entrained through a low order resonance (2:1), but that wants to decay. The flow reconstruction from the stellar light curve thus yields interesting physical insight into the pulsation mechanism.

INTRODUCTION

Almost every type of star becomes unstable to self-excited vibrations at least once during its lifetime. Typically, the stars that exhibit large amplitude variations in light are undergoing radial pulsations, *i.e.* pulsations in which the star remains spherically symmetric at all times. Typically, only the fundamental pulsation mode or the first two overtones are involved in the pulsation. In contrast, most small amplitude variable stars such as the Sun or white dwarfs undergo nonradial pulsations, but with many modes participating in the motion.

The classical Cepheid variables and RR Lyrae variables are the best known and best studied variable stars of the radial pulsator category. They owe this interest at least in part to their role as primary cosmological distance indicators; they are almost perfectly periodic and obey a period-luminosity relation that can be used for measuring their distances.

This paper concerns itself with variable stars that lie next to and slightly below the classical Cepheids in a long and broad curved strip in the Hertzsprung - Russell diagram (luminosity vs. temperature diagram) [1]. For historical reasons the stars at the lower luminosity end of the strip are called W Virginis type stars, further up RV Tauri type stars, then Semiregular stars and finally Mira variables, even though the transition between the different classes appears to be gradual. For convenience we shall refer to all of them here as semiregular stars, *largo sensu*. The W Vir stars that lie at the low luminosity end of the strip with concomitantly low period ($P \lesssim 20d$) have regular light curves, but for the slightly more luminous stars stars the observations show alternations

CP622, *Experimental Chaos: 6th Experimental Chaos Conference*, edited by S. Boccaletti et al.

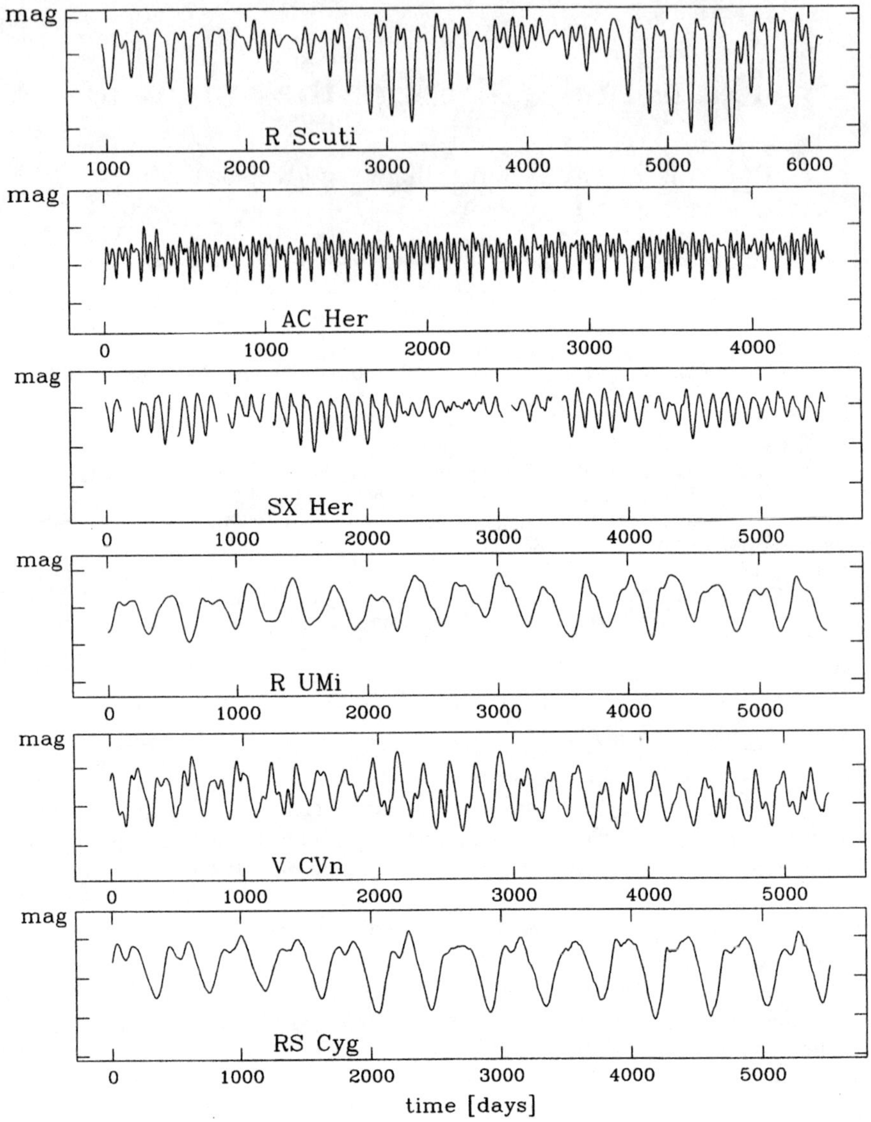

FIGURE 1. Smoothed observed light curves

in the pulsation cycles that appear to be due to period doubling [2]. In fact, numerical hydrodynamical models display cascades of period doublings [3], [4], [5] as well as tangent bifurcations [6], [7]. Observations also show that stars with higher luminosity have increasingly irregular pulsations [8].

Traditional astronomers like to think of stars are being multi-periodic (meaning that they consider the pulsation to be a superposition of a number of pulsation frequencies with steady amplitudes). It is true a phenomenologically motivated multiperiodic fit will always be 'successful' (as an interpolation), but on physical grounds it is possible to rule out such a description [9]. In a nutshell:

(1) The frequency peaks in the Fourier spectrum of the light curve are numerous, but in this type of star, and in the frequency range of interest, there do not exist enough radial pulsation eigenmodes, nor even low order nonradial pulsation modes that could be observable and excitable to such large amplitudes. No multi-periodic fits are possible that can reproduce the light curves within the observational accuracy while using modal frequencies that are acceptable on a *physical basis.*

(2) The frequencies in the Fourier spectra appear randomly variable from one section of the light curve data to another (for R Sct *cf.* [10]). If a multi-periodic star were evolving slowly, the structure of the peaks should evolve slowly as well. The absence of such correlations eliminates any explanation of the lightcurve as that of slowly evolving multi-periodic stars.

(3) One recent paper ([11]) 'explains' the irregular behavior of R Sct as the superposition of stochastically excited linear oscillators. While such a stochastic 'explanation' is perhaps mathematically valid, it is not meaningful on physical grounds because no mechanism is proposed (nor can be found) that could excite damped modes to such large pulsation amplitudes (up to factors of 40 in the light curve of R Sct).

Unfortunately the cycling times of these variable stars vary from as low as 20 days for the lower luminosity W Vir stars to as much as a year for the Semiregular variables. It is almost needless to say that very few stars have been observed with a coverage that lends itself to modern nonlinear analyses such as a flow reconstruction. Furthermore, for obvious reasons, the observations are generally not spaced with equal time intervals, so that an interpolation with concomitant introduction of noise is unavoidable.

A few years ago we were fortunate to obtain access to and analyze the data sets of two irregular variable stars, called R Scuti [12] [13] and AC Herculis [14], that have a sufficiently good coverage and exhibit a large enough number of pulsation cycles (cf. Fig. 1) to be representative of their long term behavior.

Our analysis was based on a flow reconstruction with a multivariate polynomial map $\mathcal{M}$ (or o.d.e.) in an embedding space of dimension d_e

$$X_{n+1} = \mathcal{M} \cdot X_n$$

where the state vector $X \in \mathrm{R}^{d_e}$ is constructed from the scalar time-series $\{x_i\}$,

$$X_n = (x_i, x_{i-\tau}, x_{i-2\tau}, \dots x_{i-(d_e-1)\tau})$$

where τ is a delay parameter that should be long enough so that noise does not kill the reconstruction, but small enough so that the map does not become too nonlinear [15]. We briefly summarize here the results that were obtained.

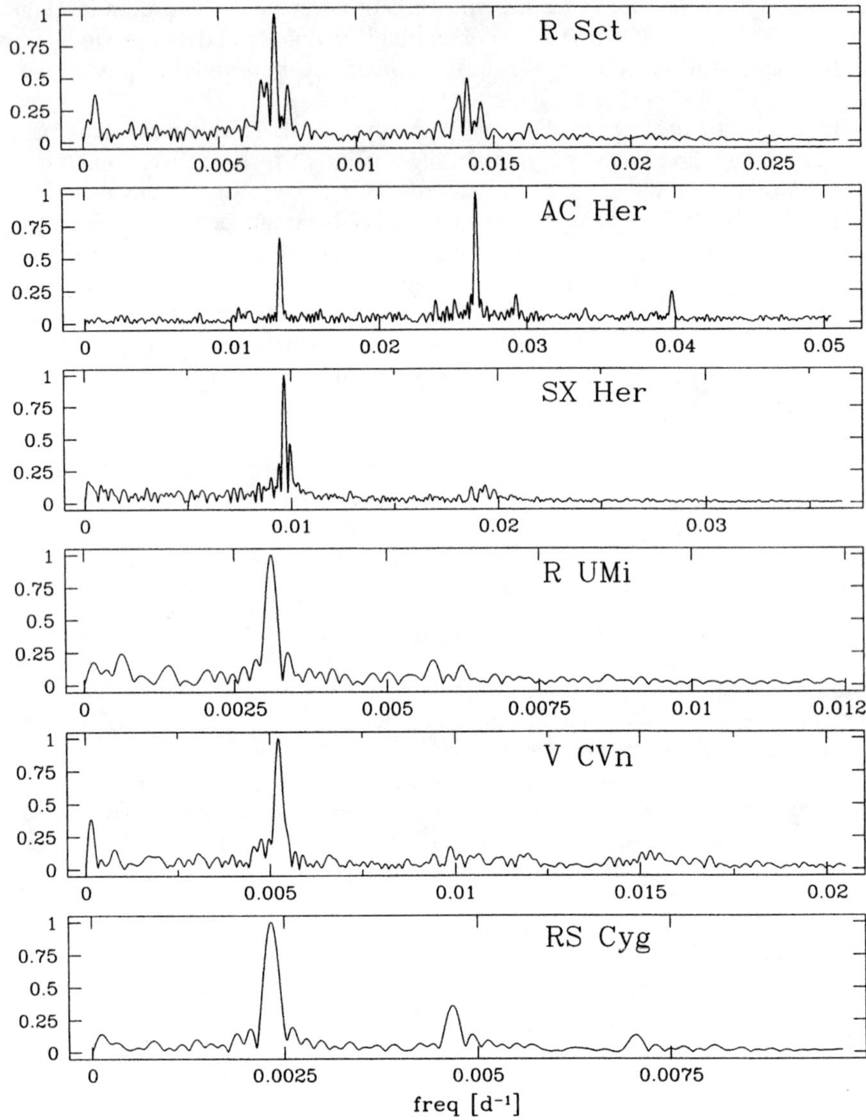

FIGURE 2. Amplitude Fourier spectra (relative amplitude scale)

The stars R Scuti and AC Herculis

The light curve data of R Sct were shown to be generated by a 4-dimensional dynamics (d=4) [12] [13]. This conclusion was based on several facts:
(a) the minimum embedding dimension is $d_e = 4$: first, synthetic light curves generated

in 3D bear no resemblance to the observations; second, the error of the fit levels off at $d_e = 4$; third, the nearest neighbor method indicates 4 as the minimum dimension. However, the strongest argument comes from a comparison of the Fourier spectra of the synthetic light curves (generated through an iteration of the map) with those of that observational data, and of a comparison of the respective Broomhead-King projections (onto the eigenvectors of the correlation matrix).

(b) the fractal (Lyapunov) dimensions [16] derived from the Lyapunov exponents of the synthetic signals fell in the range 3.1 – 3.2 for R Sct, *independently of the embedding dimension*. This is therefore a lucky situation where the bounds $d_L \sim 3.15 < d \leq d_e = 4$ uniquely determine the physical dimension to be $d = 4$.

(c) There is no *a priori* guarantee that a map or flow should capture the underlying dynamics. The fact that one of the Lyapunov exponents is always close to zero very strongly suggests that we have been successful in the flow reconstruction. Indeed, a map with a short time-step should be close to a flow for which we know that the corresponding Lyapunov exponent is exactly zero. Actually we have also reconstructed a true flow (system of o.d.e.'s) for R Sct, although we find that the reconstruction of the flow is a little less robust than that of the map.

(d) Perhaps the most interesting physical result comes from the linearization of the successful maps around their fixed points: Two spiral roots $\pm i\omega + \xi$ are found with the following properties: $\omega_2 \approx 2\omega_1$, with $\xi_2 > 0$, and $\xi_1 < 0$, and $|\xi_2| > \xi_1$. This is of course reminiscent of the Shil'nikov criterion [17].

These results allow us to give a physical interpretation of the motion. We already mentioned that the regular, classical Cepheid variables undergo pulsations in one of the two lowest modes of pulsation. It is seen that the semiregular stars to the same: the complex amplitudes of vibrational modes are the 'natural' coordinates for the phase space, so that $d = 4$ implicates the involvement of two vibrational modes. Furthermore, the low frequency mode, of frequency ω_1 is linearly unstable (self-excited) and grows in amplitude. In the nonlinear regime it interacts with a second mode because of a low order resonance condition ($\omega_2 = 2\omega_1$). However, this entrained mode is linearly stable and wants to decay. A chaotic motion of alternating growth and decay ensues.

Our analysis of AC Herculis [14] was a little less conclusive. The constructed maps and flows were less robust. By that we mean that for many seed values the iteration blew up much more rapidly. We believe that this lack of robustness may be due to the lower signal to noise ratio (Fig. 1) for AC Her. Despite these short-comings of the reconstruction, however, everything points at the minimal embedding dimension again being 4. The fractal dimension d_L turned out to be lower $d_L \sim 2.3$ than for R Sct.

As Fig. 1 shows the light curve of AC Her has a lot smaller amplitude swings, and hardly any phases of low amplitude oscillations. The linear part of the map which describes the vicinity of the fixed point is therefore less well determined, and so are the stability roots. We were therefore unable to verify whether the same resonant entrainment as in R Sct is operative in AC Her.

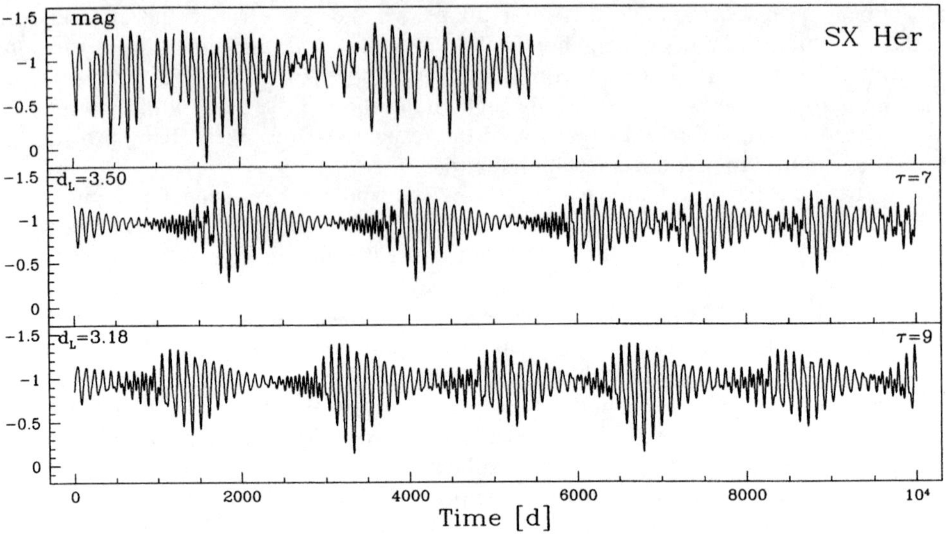

FIGURE 3. Flow reconstructions with the light curve magnitude of SX Her.

NEW RESULTS

In this paper we present an analysis of observational data of four additional stars, taken by one of us (RC) at Grinnell College over a 15 year period (cf. Fig. 1). These data are much higher quality than the amateur astronomer data previously analyzed, but they suffer from the drawback of lesser coverage. The stars go under the names of SX Herculis, R Ursae Minoris, RS Cygni and V Canum Venaticorum, and astronomers classify them as Semiregular variables (or Mira type for V CVn). The light curves of these stars are shown together with those of R Sct and AC Her which we analyzed earlier. The light curves of Fig. 1 represent cubic spline fits to the observational data.

Our cubic spline fits depend on the amount of smoothing which is controlled by the parameter σ ([18]), and by the size of the gaps that we allow. Interpolation is necessary because our analysis requires a time series with equal time intervals. Typically we use a value of $\sigma = 0.02$.

In Fig. 2 we also display the amplitude Fourier spectra of the six stars. Of the four stars RS Cyg has the most harmonic power whereas SX Her has almost none. But, as is well known, we do not learn too much about the dynamics from linear analyses, such as Fourier, MEM, ARMA or time-frequency [20].

The star SX Herculis

The smoothed 5500 day long observational light curve (with 1 day sampling) of SX Her is shown on top of Fig. 3 for reference. In the flow reconstruction we have

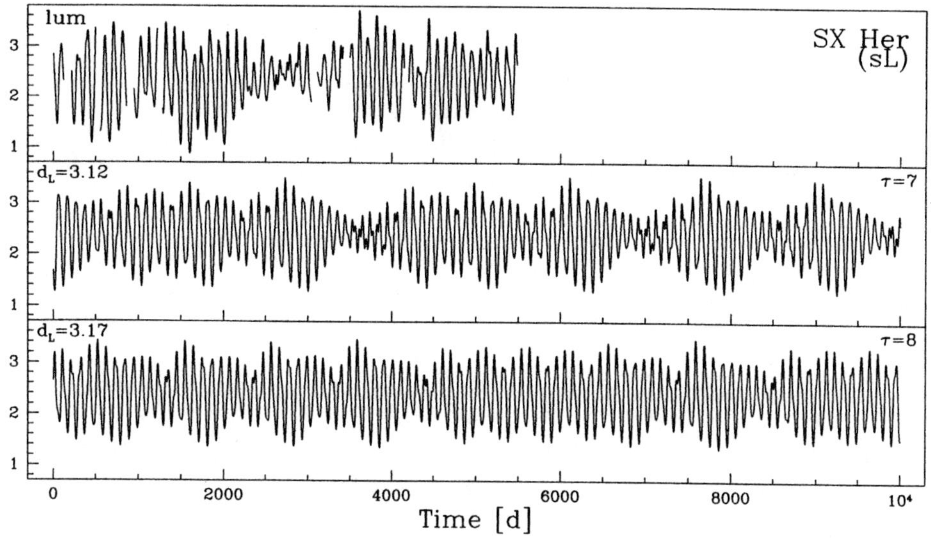

FIGURE 4. Flow reconstructions with the luminosity of SX Her.

used astronomical magnitudes, $m \equiv -2.5 \mathrm{Log} L$, rather than the more physical luminosity (L, energy radiated per unit time). Underneath, we display two of the best results of our global flow reconstructions in the form of segments of synthetic light curves that were generated with the help of 4D maps and two different values of τ, indicated in the upper right corner of the graphs. The synthetic light curves have been obtained through a 10500 fold iteration of the maps with different initial seeds, thus generating 10000 d long signals. (We have discarded the first 500 iterations to avoid transients).

We note however that for a given map not all seeds lead to this type of signal, some go to fixed points, some blow up. We attribute these failures at least partially to the shortness of the observational data set that was used to train the map. After all, the observational signal samples only a small part of phase space, and not very densely for that matter. As the map is iterated the trajectory eventually ends up in a part of phase space where the map is not well determined because of the training signal had no or few points there, and it runs off to infinity. For other parameter values the opposite can also occur, namely that the map is too stable leading to limit cycles because of a paucity of phase space points around this limit cycle. Stable synthetic signals resembling the observations could only be found for some values of τ (7 and 9), indicating that this reconstruction lacks robustness. Despite these shortcomings, however, the synthetic light curves display many of the characteristic features of the observed light curves, such as asymmetric bursts, but they are somewhat more regular.

In the upper left corners we show the Lyapunov dimensions of the synthetic signal. They have been computed with relatively short time-series (10000 points) and are thus not very precise. Nevertheless the maps that successfully produce synthetic signals with the characteristics of the observations all have a fractal dimension d_L in the range from 3.1 – 3.7.

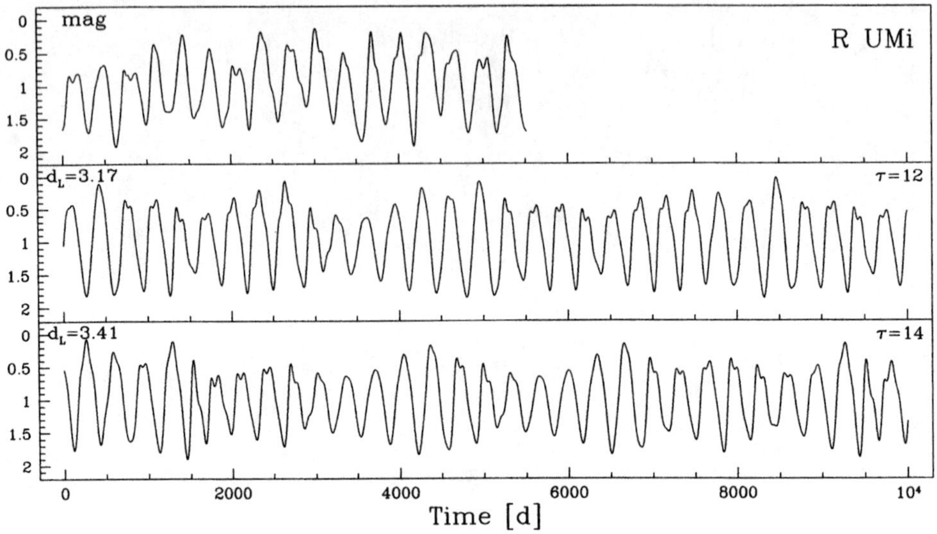

FIGURE 5. Flow reconstructions with light curve magnitude of R UMi.

Reconstructions in 3D are not successful at all. In 5D they are successful, but not very robust, most likely because of the short number of modulation cycles. However, the few synthetic signals that we were able to reconstruct had a fractal dimension less than 4, despite the fact that the embedding space was 5. This is of importance because it suggests again, as for R Sct that the physical dimension is $d = 4$.

When we linearize our 4D maps around their fixed point, we find that for the good reconstructions the fixed points of the maps are of a doubly spiral nature, as in the case of R Sct. (There can be more than one fixed point, but generally only one is located at the 'center' of the motion). The ratio of the two frequencies is found to be 2.1 – 2.2, and the stable growth rate of the higher frequency root (real part of the linear spiral eigenvalues) is always larger than the unstable one, again a generalized Shil'nikov scenario for chaos [17]. The linearization of the map is of course is closely related to the study of the linear vibrational properties of the equilibrium star. Thus for SX Her, as for R Sct earlier on, we arrive at a physical picture of the irregular pulsation mechanism, namely one of continual and alternating exchange of energy between a growing pulsation mode of frequency f_0 and a decaying resonant mode of frequency $\sim 2f_0$.

We have also made flow reconstructions with the luminosities (obtained from transforming the smoothed magnitude data) which are shown in Fig. 4. This reconstruction is seen to lead to better synthetic signals, although again only for two values of $\tau = 7$ and 8. However, physically, the results are not that different. One finds fractal Lyapunov dimensions 3.2 – 3.4, *i.e.* again larger than 3, but less than 4, and period ratios again close to and slightly greater than 2. (Here it has not been possible to construct robust 5D maps).

Although the quality of the reconstructions for SX Her are hampered by the shortness of the data set (not many modulation cycles), which shows up among other things in a relatively poor stability of the maps and in a narrow range of values of τ over which the reconstruction is possible, one can reasonably conclude that that the embedding dimension is 4 and that the fractal dimension is between 3.1 and $3.8 < 4$.

The star R Ursae Minoris

In Figure 5 on top we display the smoothed 5500 d long R UMi observational signal that we have used to train our maps. Underneath are two segments of synthetic signals that were generated from the iteration of four different reconstructed 4D maps, for values of τ going from 4 to 10 (days). Good synthetic signals are seen to be obtainable from a range of τ values, even though there are some intermediate values for which no good reconstruction has been obtained. The best 3D maps that we can construct are incapable of producing synthetic signals that bear a resemblance to the observational light curve, suggesting an embedding dimension $d_e = 4$.

The maps for R UMi are not as robust as those for R Sct, even though the observational accuracy is much higher. One of the probable reasons is that there are fewer cycles in the data. Indeed, for a successful flow reconstruction to be possible the data has to sample sufficiently well the dominant features of the dynamics in phase space. If some regions are only covered very lightly, then the maps can have 'leaks' in such regions, and the lack of robustness manifests itself in trajectories that blow up quickly. The synthetic signals that are constructed from the maps are therefore relatively short. It is also hard to obtain accurate Lyapunov exponents and fractal dimensions.

As for AC Her, the R UMi signal does not explore very well the linear neighborhood of the fixed point. Consequently it is not possible to derive any information about a possible resonance.

We find that the fractal dimensions for good synthetic signals hover around $d_L = 3.25 \pm 0.15$. Reassuringly this dimension is found to be stable in the sense that reconstructions in 5D also yield d_L in the same range, and importantly that they are less than 4, *i.e.* independent of the dimension of the embedding space. The conclusion that imposes itself, at least tentatively, is that the physical dimension which is sandwiched between $d_L \sim 3.25$ and $d_e = 4$, is $d = 4$ for this star as well.

The star V Canum Venaticorum

We have not been able to make a reconstruction with the smoothed magnitude data themselves, even with various values of smoothing, although the light curve data span some 27 cycles. The reason why our reconstruction is not successful could be that over the 15 years of observations the light curve does not explore its whole potential, *i.e.* it does not explore enough of the physical phase space to allow us to make a successful reconstruction. We cannot exclude the possibility that a higher order dynamics might be at work, but then again we would need more observational information to lay it bare. Perhaps the star undergoes intermittent mass loss which causes obscuration and leads

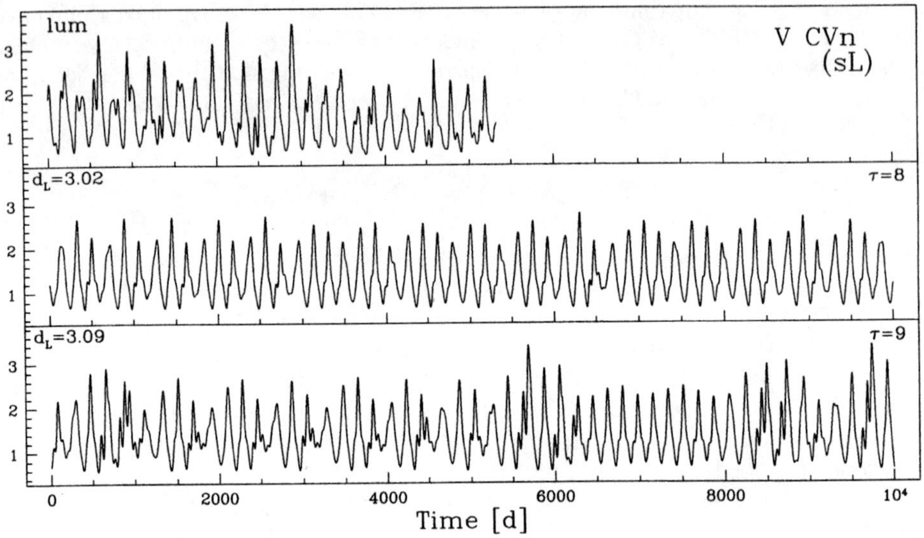

FIGURE 6. Flow reconstructions with the stellar luminosity of V CVn.

to variations in the overall luminosity. We think the first of these former may be the problem, because, after a prior conversion of the same magnitude data to luminosities we arrive at a reasonably good reconstruction (with two values of τ, 8 and 9). The smoothed observational luminosity data that are used in the reconstruction are shown in Fig. 6 together with two typical synthetic luminosity curves.

In conclusion, the data for V CVn are just barely sufficient for a reconstruction which therefore remains fragile.

The star RS Cygni

The smoothed observed light curve of RS Cyg is displayed in Fig. 7, followed by a couple of 4D synthetic light curves. The flow reconstructions are reasonably robust (with a range of τ from 11 to 17). The synthetic light curves appear good, but have some difficulties capturing the moving small RV Tau-like feature, and they have added wiggles that are nonexistent in the data. These difficulties are not astonishing because of the shortness of the data set, *viz.* only about 13 cycles – we should not expect to get more out the data than is in them. A comparison of the amplitude Fourier spectrum of the synthetic light curves with that of the data also shows that the reconstruction is somewhat lacking. The Lyapunov dimensions have a wide range of values 2.1 – 3.1. A 2:1 resonance condition is approximately satisfied for τ = 11, 12 and 15.

We find that flow reconstructions in 3D are not satisfactory. As to 5D there are too few data points to attempt a reconstruction, but our results, in particular Fig. 7, suggest that for RS Cyg an embedding dimension $d_e = 4$ is sufficient. The physical phase space

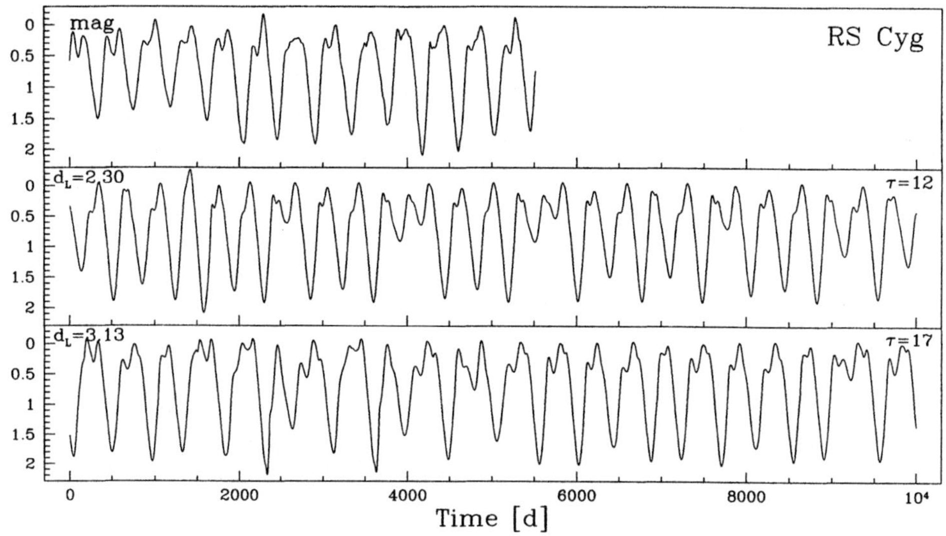

FIGURE 7. Flow reconstructions with light curve magnitude of RS Cyg.

dimension d is probably equal to 4 in order to accommodate two complex vibrational modes (strictly speaking, if $d_L < 3$ a value of 3 cannot be ruled out from $d_L < d \leq d_e$). Possibly there is again a 2:1 resonance condition between the two modes in this star. The analysis of RS Cyg is thus promising, but for a reliable flow reconstruction it will be necessary to gather more information about the light curve of this star.

DISCUSSION AND CONCLUSIONS

The reader may have been left wondering why the classical Cepheids undergo periodic pulsations while their neighbors in the Hertzsprung-Russell diagram pulsate irregularly. We want to address the difference between these two classes of stars briefly. They have the same luminosity range, but the irregular stars have lower masses, by about a factor of 10. As a result the coupling between the heat flow and the acoustic oscillation is strongly enhanced. This can best be seen with a linear stability analysis of the equilibrium stellar models. Let us call ω the eigenvalues of the pulsation modes (for an assumed $\exp(i\omega t)$ dependence). Then for the modes that are relevant here, *i.e.* the lowest frequency modes, we find that the relative growth rates of these modes $|\text{Im}\,\omega/\text{Re}\,\omega|$ are of the order of a few percent for the classical Cepheids, but they are of order unity for the semiregular stars. The mathematical consequence is that there exists a center manifold for the classical Cepheids, and their behavior can be captured with amplitude equations (normal forms) [21]; in particular, the periodic pulsations of the classical Cepheids owe their existence to the proximity of a broad Hopf bifurcation.

Clearly, a large value of the relative growth rates is a prerequisite for the occurrence of chaos (appreciable amplitude changes must be possible over a period). For the W Virginis type stars at the low luminosity end the growth rates are still small, and periodic pulsations (limit cycles) prevail, just as for the classical Cepheids. For the more luminous ones, such as the stars analyzed in this paper, the growth rates are of order unity, thus fulfilling the necessary condition for chaotic behavior.

We have applied the global flow reconstruction technique to the high quality observational data of SX Her, R UMi, V CVn and RS Cyg. In view of the relatively small number of observed cycles and the complexity of the light curves our conclusions must be taken with some caution. The flow reconstructions lack the robustness of those obtained earlier from the large amplitude variable star R Sct. Consequently, it is not possible to produce very long synthetic light curves by iterating the map, and the Lyapunov exponents and the fractal dimension are therefore beset with large fluctuations from one synthetic signal to another. Our flow reconstructions are best for SX Her and R UMi. For RS Cyg the time span of the observed light curve is clearly too short. Our poor success with V CVn is perhaps related to the complexity of its light curve.

We find that the light curves of six different large amplitude, irregularly pulsating stars all indicate a minimum embedding dimension of 4, although the only reconstructions that have some robustness are those for R Sct and SX Her. For both of these the fractal Lyapunov dimensions fall in the range 3.1 – 3.5, and one can conclude that the dimension of the physical phase space of the dynamics is 4. This in turn suggests that the 'natural' generalized coordinates in this phase space are the complex amplitudes of two vibrational modes.

Our analysis of both R Sct and SX Her suggests that the irregular behavior has the same physical mechanism. The irregular behavior arises through the nonlinear interaction between two vibrational modes, one of lower frequency that is unstable and one with higher frequency that is stable. Furthermore, the ratio of the two frequencies is close to 2:1 in both stars. We thus arrive at the same physical picture of the underlying mechanism for the irregular light curve. A lower frequency mode is self-excited and entrains an otherwise linearly stable overtone through a low order (2:1) resonance. The irregular pulsation occurs as a result of continual exchange of energy between the two resonant modes. The minimum embedding dimensions of the other four stars corroborate this finding although no information about a resonance could be gotten from the linearization of the map.

The fact that the underlying dynamics of these types of stars are low dimensional and chaotic is not surprising, and it was in fact predicted by numerical hydrodynamical simulations of W Vir type stars [15] [22]. Furthermore a topological analysis of the attractor [23] corroborated these conclusions.

The flow reconstruction technique thus has shed new light on the old mystery of the nature of the irregular pulsations of a large class of large amplitude pulsating variable stars..

ACKNOWLEDGMENTS

Work supported by the National Science Foundation and by MTA-OTKA (T-026031).

REFERENCES

1. Cox, J. P. *Theory of Stellar Pulsation*, Princeton Series in Astrophysics (1980).
2. Arp, H. C., *Astron. J.* 60, 1–11 (1955).
3. Buchler, J.R. & Kovács, G., *Astrophys. J. Lett.* 320, L57–62 (1987).
4. Kovacs G. & Buchler J.R., *Astrophys. J.* 334, 971–994 (1988).
5. Aikawa, T., *Astrophys. & Space Sci.* 164, 295–308 (1990).
6. Buchler, J. R., Goupil, M. J. & Kovács, G., *Phys. Lett*, A126, 177–180 (1987).
7. Aikawa, T., *Astrophys. & Space Sci.*, 139, 281–293 (1987).
8. Pollard, K. et al., "MACHO Observations of Type II Cepheids and RV Tauri Stars in the LMC", in *Variable Stars and the Astrophysical Returns of Microlensing Surveys*, Eds. R. Ferlet, J.P. Maillard & B. Raban, Editions Frontières, pp. 219–223, (1996).
9. Buchler, J. R. & Kolláth, Z., "Nonlinear Analysis of Irregular Variables", in *Nonlinear Studies of Stellar Pulsation*, Eds. M. Takeuti & D.D. Sasselov, Astrophy. & Space Sci. Lib. Ser. 257, pp. 185–213 (2000) [http://xxx.lanl.gov/abs/astro-ph/0003341].
10. Kolláth Z., *Month. Not. Roy. Astron. Soc.*, 247, 377–386 (1990).
11. Konig, M., Paunzen, E. & Timmer,J., *Month. Not. Roy. Astron. Soc.* 330, 297–301 (1999).
12. Buchler, J. R., Serre, T., Kolláth, Z. & Mattei, J. *Phys. Rev. Lett.* 74, 842–845 (1995).
13. Buchler, J. R., Kolláth, Z., Serre, T. & Mattei, J., *Astrophys. J.* 462, 489–504 (1996).
14. Kolláth, Z., Buchler, J. R., Serre, T. & Mattei, J., *Astron. & Astrophys.* 329, 147–155 (1998).
15. Serre, T., Kolláth, Z. & Buchler, J. R., *Astron. & Astrophys.*, 311, 833–844 (1996).
16. Ott, E., *Chaos in Dynamical Systems*, Cambridge Univ. Press (1993).
17. Glendenning, P. & Tresser, C., *J. Physique Lett.* 46, L347–352 (1985).
18. Reinsch, C. H., *Numerische Mathematik*, 10, 177–201 (1967).
19. Buchler, J.R., "Search for Low-Dimensional Chaos in Observational Data", International School of Physics "Enrico Fermi", Course CXXXIII, *Past and Present Variability of the Solar-Terrestrial System: Measurement, Data Analysis and Theoretical Models*, Eds. G. Cini Castagnoli & A. Provenzale, pp. 275–288, Società Italiana de Fisica, Bologna, Italy (1996). [http://xxx.lanl.gov/abs/chao-dyn/9707012]
20. Weigend, A.S & Gershenfeld, N. A., *Time Series Prediction* (Addison-Wesley: Reading) (1994).
21. Buchler, J.R., "A Dynamical Systems Approach to Nonlinear Stellar Pulsations", in *Nonlinear Phenomena in Stellar Variability*, Eds. M. Takeuti & J.R. Buchler, Dordrecht: Kluwer Publishers, reprinted from Astrophys. & Space Sci, 210, pp. 1–31 (1993).
22. Serre, T., Kolláth, Z. & Buchler, J. R., *Astron. & Astrophys.*, 311, 845–851 (1996).
23. Letellier, C., Gouesbet, G., Soufi, F., Buchler, J.R.& Kolláth, Z. *Chaos*, 6 (3), 466–476 (1996).

Multifractality and chaos in the solar wind

Wiesław M. Macek

Faculty of Mathematics and Natural Sciences, School of Exact Sciences, Cardinal Stefan Wyszyński Uniwersity in Warsaw, Dewajtis 5, 00-815 Warszawa; Space Research Centre, Polish Academy of Sciences, Bartycka 18 A, 00-716 Warszawa, Poland

Abstract. We analyze a time series of velocities of the solar wind plasma stream including Alfvénic fluctuations measured *in situ* by the Helios spacecraft in the inner heliosphere, which is the region of space dominated by the solar wind flow. We calculate the generalized dimensions directly from the cleaned experimental signal. The resulting spectrum of dimensions shows multifractal structure of the solar wind in the inner heliosphere. The obtained multifractal spectrum is consistent with that for the multifractal measure on the self-similar weighted Cantor set.

The generalized dimensions of attractors are important characteristics of *complex* dynamical systems [1]. Since these dimensions are related to frequencies with which typical orbits in phase space visit different regions of the attractors, they provide information about dynamics of the systems [2, 3]. More precisely, one may distinguish a probability measure from its geometrical support, which may or may not have fractal geometry. Then, if the measure has different fractal dimensions on different parts of the support, the measure is multifractal [4, 5]. The modern technique of nonlinear time series analysis allows to estimate the multifractal measure directly from a single time series.

Following space physics applications, e.g., [6, 7], we consider the inner heliosphere. Since the 1960s we have known that besides electromagnetic radiation, the Sun also radiates charged particles forming a plasma blowing nearly radially outward from the Sun. The solar wind plasma flowing supersonically away from the Sun is quite well modeled within the framework of the hydromagnetic theory. This continuous flow has two forms: slow (≈ 300 km s^{-1}) and fast (≈ 900 km s^{-1}) [8]. The fast wind is associated with coronal holes and is relatively uniform and stable, while the slow wind is quite variable in terms of velocities. We limit our study to the low-speed stream. Indication for a chaotic attractor in the slow solar wind has recently been given in [9, 10, 11, 12, 13]. In particular, Macek [9, 10] has calculated the correlation dimension of the reconstructed attractor and has provided tests for *nonlinearity* in the solar wind data, including a powerful method of singular-value decomposition [14] and statistical surrogate data tests [15]. Further, Macek and Redaelli [13] have shown that the Kolmogorov entropy of the attractor is *positive* and finite, as it holds for a *chaotic* system. The entropy is plausibly constrained by a *positive* local Lyapunov exponent that would exhibit sensitive dependence on initial conditions of the system.

In the present paper we have extended our previous results on the dimensional time series analysis [9]. Namely, we apply the technique that allows a realistic calculation of the generalized dimensions of the solar wind flow, directly from the cleaned experimental signal by using the Grassberger and Procaccia method [16]. The resulting spectrum

CP622, *Experimental Chaos: 6th Experimental Chaos Conference*, edited by S. Boccaletti et al.

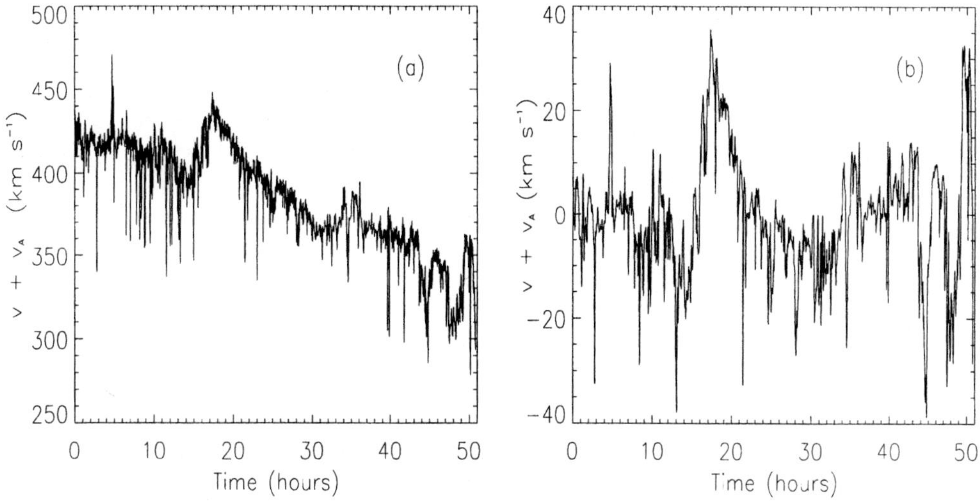

FIGURE 1. The flow velocity with included Alfvénic velocity, $v + v_A$ (Elssäser variable) observed by the Helios 1 spacecraft in 1975 from 67:08:20.5 to 69:11:07 (day:h:min) at distances 0.32 AU from the Sun, (a) the raw data, (b) the detrended and smoothed data.

of dimensions shows multifractal structure of the solar wind in the inner heliosphere. The obtained multifractal spectrum is consistent with that for the multifractal measure on the self-similar weighted Cantor set, e.g., [5].

These results provide supporting evidence that the *complex* solar wind is likely to have multifractal structure, cf. [7]. In this way, we have further supported our previous conjecture that trajectories describing the system in the inertial manifold of phase space asymptotically approach the attractor of low-dimension [9, 10, 13]. One can expect that the attractor in the low-speed solar wind plasma should contain information about the dynamic variations of the coronal streamers. It is also possible that it represents a structure of the time sequence of near-Sun coronal fine-stream tubes.

We analyze the Helios data using plasma parameters measured *in situ* in the heliosphere near the Sun, at 0.3 AU, Ref. [8]. The radial velocity component of the plasma flow, v, has been investigated in [9, 10, 13]. However, it is known that various disturbances are superimposed on the overall structure of the solar wind, including mainly Alfvén waves, which move away from the Sun. Therefore, in this paper we take also into account Alfvénic fluctuations of the flow. Namely, we analyze one of the so-called Elssäser variables $x = v + v_A$, where $v_A = B/(\mu_o \rho)^{1/2}$ is the Alfvénic velocity calculated from the experimental data: the radial component of the magnetic field of the plasma B and the mass density ρ (μ_o is the permeability of free space). These raw data of $N = 4,513$ points, with sampling time of $\Delta t = 40.5$ s, are shown in Figure 1 (a).

As in [9, 13], slow trends ($420.06 - 15.4\ t - 82.31\ t^2$, with t being a fraction of total sample) were subtracted from the original data $x_i = v(t_i) + v_A(t_i)$, in km s^{-1}, $i = 1,\ldots,N$ and the data with the initial several percent noise level, were (8-fold) smoothed (replacing each data point with the average of itself and its two nearest

TABLE 1. Characteristics of the solar wind, $v + v_A$, filtered data.

	Smoothed	Shuffled data	Shuffled phases
Skewness, κ_3	0.26	0.26	-0.26
Kurtosis, κ_4	0.88	0.88	0.26
Relative complexity	0.17	1.0	0.23
Autocorrelation time, t_a	3.5×10^3 s	40 s	3.3×10^3 s
Capacity dimension, D_0	4.4	5.8	4.6
Information dimension, D_1	4.0	5.0	4.5
Correlation dimension, D_2*	3.5	5.2	4.6

* The average slope for $6 \leq m \leq 8$ is taken as D_2.

neighbors). The detrended and smoothed data shown in Figure 1 (b). Certainly, this moving average filter removes considerable amount of noise, leaving only about 1%. The nonlinear filtering, which allows calculation of the entropy, has been discussed in [13]. It has been shown that after the nonlinear Schreiber filter the calculated dimension has been somewhat reduced. In this paper, we focus on the calculations of dimensions. Therefore, we prefer to use only moving average filtering, cf. [9]. It is worth noting that our solar wind data in Figure 1 (b) resemble somewhat the time series in Figure 2 of Ref. [4], which is a simulation of a sample from a multifractal measure.

Table 1 summarizes selected calculated characteristics of the detrended data cleaned by using the moving average filter. The probability distributions are clearly non-Gaussian, cf. Ref [13]. We have a large skewness of ~ 0.26 (as compared with its normal standard deviation 0.06) and a very large kurtosis of 0.88 (the latter was small for the analysis with no magnetic field), cf. Ref [9]. We have also estimated Lempel-Ziv measure of *complexity*, relative to white noise [17]. The calculated value ~ 0.17 is even smaller than in [9] (≈ 0.20); maximal complexity, or randomness, would have a value of 1.0, while a value of zero denotes perfect deterministic nonlinear predictability. We choose a time delay $\tau = 150\ \Delta t$, where the autocorrelation function has the first minimum and its value is very small. This value is smaller than $t_0 = 376\ \Delta t$, the first zero of the autocorrelation function, $(\langle x(t)x(t+t_0)\rangle - \langle x(t)\rangle^2)/\sigma^2 = 0$ with average velocity $\langle x\rangle = -0.36$ km s^{-1} and standard deviation $\sigma = 8.54$ km s^{-1}, cf. also t_a in Table 1 when the autocorrelation function decreases to $1/e$.

Using our time series of equally spaced, detrended and cleaned data, we construct a large number of vectors $\mathbf{X}(t_i) = [x(t_i), x(t_i+\tau), \ldots, x(t_i+(m-1)\tau)]$ in the embedding phase space of dimension m, where $i = 1, \ldots, n$ with $n = N-(m-1)\tau$. Then, we divide this space into a large number $M(r)$ of equal hypercubes of size r which cover the presumed attractor. If p_j is the probability measure that a point from a time series falls in a typical j-th hypercube, using the q-order function $I_q(r) = \Sigma(p_j)^q, j = 1, \ldots, M$, the q-order generalized dimension is given by [2, 3]

$$D_q = \frac{1}{q-1} \lim_{r \to 0} \frac{\ln I_q(r)}{\ln r}, \tag{1}$$

We see from Eq. (1) that the larger q is, the more strongly are the higher probability cubes (visited more frequently by a trajectory) weighted in the sum for $I_q(r)$. Only if $q = 0$, all the cubes are counted equally, $I_o = M$, and we recover the box-counting dimension, D_0.

Writing $I_q(r) = \sum p_j(p_j)^{q-1}$ as a weighted average $\langle (p_j)^{q-1} \rangle$, one can associate bulk with the generalized average probability per hypercube $\mu = \sqrt[q-1]{\langle (p_j)^{q-1} \rangle}$, and identify D_q as a scaling of bulk with size, $\mu \propto r^{D_q}$. Since the data cannot constrain well the capacity dimension D_0, we look for higher order dimensions, which quantify the multifractality of the probability measure on the attractor. For example, the limit $q \to 1$ leads to a geometrical average (the information dimension). For $q = 2$ the generalized average is the ordinary arithmetic average (the standard correlation dimension), and for $q = 3$ it is a root-mean-square average. In practice, for a given m and r,

$$p_j \simeq \frac{1}{n - 2n_c - 1} \sum_{i=n_c+1}^{n} \theta(r - \mid \mathbf{X}(t_i) - \mathbf{X}(t_j) \mid) \tag{2}$$

with $\theta(x)$ being the unit step function, and $n_c = 2$ is the Theiler's correction [19]. Finally, $I_q(r)$ is taken to be equal to the generalized q-point correlation sum [16]

$$C_q(m,r) = \frac{1}{n_{\text{ref}}} \sum_{j=1}^{n_{\text{ref}}} (p_j)^{q-1}, \tag{3}$$

where $n_{\text{ref}} = 500$ is the number of reference vectors. For large dimensions m and small distances r in the scaling region it can be argued that $C_q(m,r) \propto r^{(q-1)D_q}$, where D_q is an approximation of the ideal limit $r \to 0$ in Eq. (1) for a given q, Ref. [16].

First, we calculate the natural logarithm of the standard ($q = 2$) correlation sum $C_m(r) = C_2(m,r)$ versus $\ln r$ (normalized) for various embedding dimensions: $m = 4$ (dotted curve), $m = 5$ (diamonds), 6 (triangles), 7 (squares), 8 (crosses), 9 (plus), and 10 (stars) signs. The results obtained using the moving average filter are presented in Figure 2, while those obtained using the singular-value decomposition and nonlinear Schreiber filters have been discussed in [9, 10, 13]. The slopes $D_{2,m}(r) = \mathrm{d}[\ln C_m(r)]/\mathrm{d}(\ln r)$ in the scaling region of r should provide the correlated dimension. Since the correlation sum is simply an arithmetic average over the numbers of neighbors, this can yield meaningful results for the dimension even when the number of neighbors available for some reference points is limited in most real dynamical systems. If the D-dimensional attractor exists, we expect a plateau of the slopes for $m \geq D$ and in the worst case for $m > 2D$. For m large enough an average slope in the scaling region indicates a proper correlation dimension D_2. We have a clear plateau which appears already for $m = 4$ (dotted curve) and $m = 5$. For higher dimensions, $m \geq 8$, the plateau is still present but more smeared out by the statistical fluctuations at small r. In our case the slope of the calculated correlation sum saturates for $m > 5$, with an average for $6 \leq m \leq 8$ of $D_2 = 3.5 \pm 0.1$, cf. Ref. [9]; this is consistent with the attractor of the low-dimension.

Second, the generalized dimensions D_q in Eq. (1) as a function of q are shown in Figure 3. The error bars are purely statistical here. The spectrum of dimensions shows multifractal structure of the solar wind in the inner heliosphere. For comparison, the values calculated numerically for the standard Lorenz equations, using 2000 points at intervals of $\Delta t = 0.05$, $\tau = 5\,\Delta t$, are shown by plus signs. An extremely simple example of the multifractal system is the weighted Cantor set, where the probability of visiting one segment is p (say $p \leq 1/2$) and the probability of visiting the other segment is $1 - p$.

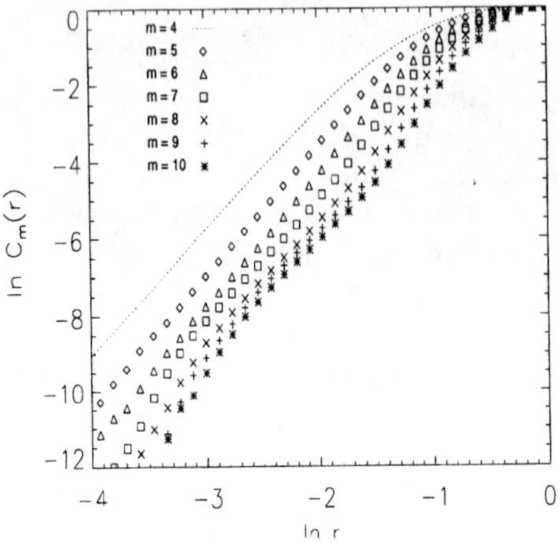

FIGURE 2. The natural logarithm of the correlation sum $C_m(r)$ versus $\ln r$ (normalized) obtained for the cleaned experimental signal for various embedding dimensions: $m = 4$ (dotted curve), $m = 5$ (diamonds), 6 (triangles), 7 (squares), 8 (crosses), 9 (plus), and 10 (stars).

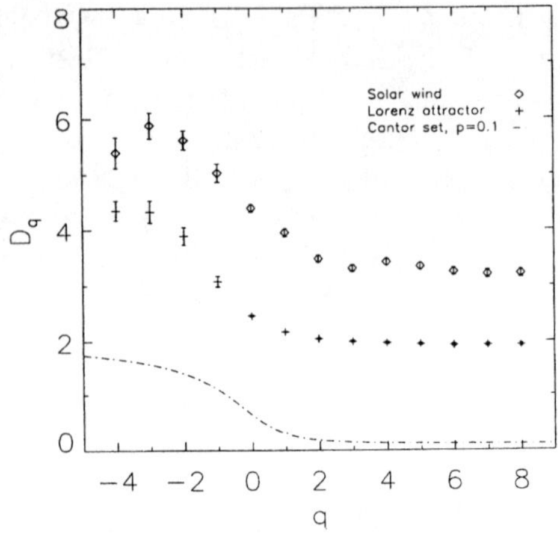

FIGURE 3. The generalized dimensions D_q in Eq. (1) as a function of q. The correlation dimension is $D_2 = 3.5 \pm 0.1$, see Table 1. The values calculated numerically for the Lorenz model (plus signs) and analytically for the weighted Cantor set with $p = 0.1$ (dash-dotted).

In this case the results can be obtained analytically; for any q in Eq. (1) one has

$$(1-q)D_q = log_3[p^q + (1-p)^q]. \tag{4}$$

The difference of the maximum and minimum dimension, associated with the least dense and most dense points on the attractor, correspondingly, is $D_{-\infty} - D_{\infty} = log_3(1/p - 1)$ and in the limit $p \to 0$ this difference rises to infinity. Hence, the parameter p can be regarded as a degree of multifractality. The results calculated for $p = 0.1$ in Eq. (4) are shown in Figure 3 by a dash-dotted line. We see that the multifractal spectrum of the solar wind ($D_q - 3$) is roughly consistent with that for the multifractal measure on the self-similar weighted Cantor set, with a single weighting parameter p. The obtained value of this parameter demonstrates that some cubes that cover the attractor of this dynamical system are visited one order of magnitudes more frequently than some other cubes, as is illustrated in our previous paper, see Figure 5 of Ref. [9].

We also estimate the Kolmogorov correlation entropy, K_2, and the largest positive Lyapunov exponent, λ_{max}. The vertical spacings between the parallel lines in Figure 2 averaged in the saturation region $8 \leq m \leq 10$ are taken as K_2 yielding the value of ≈ 0.1 (per delay time τ). Using the algorithm of Ref. [20], one obtains the magnitude of $\lambda_{\text{max}} \sim 0.1$ in the same units as for K_2 (base e). In general, the entropy K_q is at most the sum of the positive Lyapunov exponents $\sum \lambda_i$, e.g., [2, 3]. The value of the Lyapunov exponent is consistent with the Kolmogorov ($q = 2$) entropy, which should be its lower bound: $K_2 \leq \sum \lambda_i$ (positive). The time over which the meaningful prediction of the behavior of the system is possible is roughly $\sim 1/\lambda_{\text{max}}$, e.g., [2, 3]. Hence the predictability of the system is limited to hours.

The obtained measures of the attractor have been subjected to the surrogate data test [15]. As has been demonstrated in Figure 8 of Ref. [9], if the original data are indeed deterministic, analysis of these surrogate data will provide values that are statistically distinct from those derived for the original data, see also Table 1. In particular, the slope of the correlation sum increases with m (no saturation), and Lempel-Ziv complexity calculated for shuffled-data becomes clearly 1.0, as it should be for a purely stochastic system. Again, we have found that the solar wind data are sensitive to this test.

In conclusion, we have shown that the moving average filter removes some amount of noise, which is sufficient to calculate the generalized dimensions of the solar wind attractor. The obtained multifractal spectrum is consistent with that for the multifractal measure on the self-similar weighted Cantor set with a degree of multifractality of $p \sim 0.1$. The obtained characteristics of the attractor are significantly different from those of the surrogate data. Thus these results show multifractal structure of the solar wind in the inner heliosphere. Hence we suggest that there exists an inertial manifold for the solar wind, in which the system has *multifractal* structure, and where noise is certainly not dominant.

This work was supported by the Scientific Research Committee Grant 2 P03C 001 16.

REFERENCES

1. Hentschel, H. G. E., and Procaccia, I., Physica D, **8**, 435–444 (1983); Grassberger, P., Phys. Lett. A, **97**, 227–230 (1983); Halsey, T. C., Jensen, M. H., Kadanoff, L. P., Procaccia, I., and Shraiman, B. I., Phys. Rev. A, **33**, 1141–1151 (1986).
2. Ott, E., *Chaos in Dynamical Systems*, Cambridge University Press, Cambridge, 1993.
3. Ott, E., Sauer, T., and Yorke, J. A., *Coping with Chaos*, Wiley, New York, 1994.
4. Mandelbrot, B. B., Multifractal measures, especially for the geophysicist, in *Pure and Applied Geophysics*, **131**, 5–42, Birkhäuser Verlag, Basel (1989).
5. Falconer, K., *Fractal Geometry: Mathematical Foundations and Applications*, Wiley, New York, 1990.
6. Kurths, J., and Herzel, H., Physica D, **25**, 165–172 (1987).
7. Burlaga, L. F., Geophys. Res. Lett., **18**, 69–72 (1991).
8. Schwenn, R., Large-scale structure of the interplanetary medium, in *Physics of the Inner Heliosphere*, edited by R. Schwenn and E. Marsch, Springer-Verlag, Berlin, 1990, Vol. 20, pp. 99–182.
9. Macek, W. M., Physica D, **122**, 254–264 (1998).
10. Macek, W. M., Testing for an attractor in the low-speed solar wind flow, in *Solar Wind Nine*, edited by S. R. Habbal, R. Esser, J. V. Hollweg, and P. A. Isenberg, American Institute of Physics, New York, 1999, Vol. 471, pp. 251–254.
11. Macek, W. M., and Obojska, L., Fractal analysis of the low-speed solar wind flow, in ESA Symposium Proceedings on *Environment Modelling for Space-based Applications*, edited by A. Hilgers and T. -D. Guyenne, ESTEC, Noordwijk, 1996, *SP-392*, pp. 247–252.
12. Macek W. M., and Obojska, L., Chaos, Solitons & Fractals, **8**, 1601–1607 (1997); **9**, 221–229 (1998).
13. Macek W. M., and Redaelli, S., Phys. Rev. E, **62**, 6496–6501 (2000).
14. Albano, A. M., Muench, J., Schwartz, C., Mees, A. I., and Rapp, P. E., Phys. Rev. A, **38**, 3017–3026 (1988).
15. Theiler, J., Eubank, S., Longtin, A., Galdrikian, B., and Farmer, J. D., Physica D, **58**, 77–94 (1992).
16. Grassberger P., and Procaccia, I. , Physica D, **9**, 189–208 (1983).
17. Kaspar, F., and Schuster, H. G., Phys. Rev. A, **36**, 842–848 (1987).
18. Grassberger P., and Procaccia, I., Phys. Rev. A, **28**, 2591–2593 (1983).
19. Theiler, J., Phys. Rev. A, **34**, 2427–2432 (1986).
20. Kantz, H., Phys. Lett. A, **185**, 77–87 (1994).

A Wavelet-based method for multifractal analysis of rough surfaces : Applications to high-resolution satellite images of cloud structure

A. Arneodo*, S. G. Roux† and N. Decoster**

*Centre de Recherche Paul Pascal, Avenue Schweitzer, 33600 Pessac, France.
†Laboratoire de Physique, CNRS UMR 5672, École Normale Supérieure de Lyon, 46 allée d'Italie, 69364 Lyon Cedex 07, France.
**Noveltis, 2 Avenue de l'Europe, 31520 Ramonville Saint Agne, France.

Abstract. We apply the 2D wavelet transform modulus maxima (WTMM) method to high-resolution LANDSAT satellite images of cloudy scenes. The computation of the $\tau(q)$ and $D(h)$ multifractal spectra of the radiance fields confirms the relevance of the multifractal description to account for the intermittent nature of marine stratocumulus clouds. This analysis reveals that with the available set of experimental data, there is no way to discriminate between various phenomenological cascade models recently proposed to account for intermittency and their log-normal approximations. We emphasize the log-normal random $\mathcal{W}$-cascade model on separable wavelet orthogonal basis introduced in (N. Decoster, S.G. Roux, A.Arneodo, Eur. Phys. J. B **15**, 739(2000)), as a very attractive model (at least as compared to the models commonly used in the literature) of the cloud architecture. Finally, we comment on the multifractal properties of marine stratocumulus radiance fields comparatively to previous experimental analysis of velocity and temperature fluctuations in high Reynolds number turbulence.

INTRODUCTION

The problematic of nonlinear variability over a wide range of scales has been considered for a long time with respect to the highly intermittent nature of turbulent flows in fluid dynamics [1, 2]. Special attention has been paid to their asymptotic and possibly universal behavior when the dissipation length goes to zero, *i.e.*, when the Reynolds number goes to infinity. Besides wind-tunnel and laboratory (grid, jet, ...) experiments, the atmosphere is a huge natural laboratory where high Reynolds number (fully developed) turbulent dynamics can be studied. Clouds, which are at the source of the hydrological cycle, are the most obvious manifestation of the earth's turbulent atmospheric dynamics [3, 4]. By modulating the input of solar radiation, they play a critical role in the maintenance of the earth's climate [5]. They are also one of the main sources of uncertainty in current climate modeling [6]. During many years, the lack of data hindered our understanding of cloud microphysics and cloud-radiation interactions. Until quite recently, the internal structure of clouds was probed by balloons or aircrafts that penetrated the cloud layer, revealing an extreme variability of 1D cuts of some cloud fields, *e.g.* liquid water content (LWC) [7, 8]. Indeed, during the past fifteen years, vast amounts of data on the distribution of atmospheric liquid water from a variety of sources were collected and analyzed in many different ways. All these data contain information on spatial and/or temporal cor-

CP622, *Experimental Chaos: 6th Experimental Chaos Conference*, edited by S. Boccaletti et al.

relations in cloudiness, enabling the investigation of scale invariance over a range from a few centimeters to hundred of kilometers. An attractive alternative to *in situ* probing is to use high-resolution satellite imagery that now provides direct information about the fluctuations in liquid water concentration in the depths of clouds [9, 10, 11, 12, 13, 14, 15]. These rather sophisticated remote sensing systems called "millimeter radars" are actually sensitive not only to precipating rain drops but also to suspended cloud droplets. Spectral analysis of the recorded 2D radiance field [12, 13, 14, 15] confirms previous 1D findings that make it likely that cloud scenes display scaling over a wide range of scales.

Fractal analysis of atmospheric data has gained considerable momentum since Lovejoy's seminal paper [3] on the area-perimeter relation for clouds and rain. Since then, such morphological studies have become rather popular, following a path open by Mandelbrot in his famous books [16]. Starting from the middle eighties, high-resolution cloud liquid water data became available, confirming the fractal nature of the cloud structure. Most preliminary analysis of these complex 1D signals focused on the characterization of scale invariance in Fourier space [7, 8, 9, 10, 11, 12, 13, 14, 15]. In particular, the fluctuations of liquid water density in clouds were shown to have power-law energy spectra, $S(k) \sim 1/k^{\beta}$, over a large range of scales, from tens of meters to tens of kilometers. Let us point out that the estimated spectral exponent $\beta \sim 1.4-1.7$, is close to the prediction $\beta = 5/3$ of Corssin [17]-Obukhov [18] phenomenology for a passive scalar in fully developed 3D turbulence as well as to Kraichnan [19] prediction for 2D turbulence. But as discussed in Ref. [20], power-spectral analysis has been heavily criticized for its intrinsic inability to capture multifractal scaling. From the measurement of the spectral exponent $\beta = 2H+1$, one gets some estimate of the so-called Hurst exponent, $H \sim 0.20-0.35$, from which one learns that the fluctuating field is singular and non stationary ($0 < H < 1$). Unfortunately, the intermittent nature of the recorded liquid water data required new concepts and more elaborated technical tools. Again, one has to give credit to Lovejoy and co-workers [21, 22, 23, 24, 25] for applying the multifractal description to atmospheric phenomena. Using the trace moment and double trace moment techniques, they have brought experimental evidence for multiple scaling (or in other words, the existence of a continuum of scaling exponent values) in various geophysical fields. More recently, Davis and co-workers [8, 26] have used the structure function method to study LWC data recorded during ASTEX and FIRE programs. Both these analyses converge to the conclusion that the internal marine stratocumulus (Sc) structure is multifractal over at least three decades in scales. Quite similar multifractal behavior has been reported by Wiscombe *et al* [27] when analazing liquid water path (LWP) data (*i.e.*, column integrated LWC), from the Atmosperic Radiation Measurement (ARM) archive. Even though all these studies seem to agree, at least as far as their common diagnostic of multifractal scaling of the cloud structure, they all concern 1D data. To our knowledge, the stucture function method has been also applied to 1D cuts of high-resolution satelitte images [15], but we are not aware of any result coming out from a specific 2D analysis.

In this paper, our first goal is to take advantage of the 2D wavelet transform modulus maxima (WTMM) method introduced in Ref. [20] and further tested on synthetic multifractal surfaces in Ref. [28], to carry out a multifractal analysis of high-resolution satelitte images of Sc cloudy scenes [29, 30, 31]. Our objective is not only to extend previous analysis of 1D profiles to full 2D image processing, but also to remedy to the insufficiencies of the structure function method which, as pointed out in Ref. [32], fails to achieve complete multifractal characterization. The 2D WTMM method is likely to provide reliable quantitative estimates of the entire $\tau(q)$ and $D(h)$ multifractal spectra, within the perspective of confirming the intermittent nature of the internal cloud structure. Although this result would not be a surprise in such a highly turbulent environment, the possibility of comparing quantitatively the statistical scaling properties of the optical depth or the radiance fields with those of the velocity or temperature fluctuations in high Reynolds turbulent flows, might be very instructive as far as the influence of atmospheric dynamics on the liquid water distribution in clouds.

A MULTIFRACTAL FORMALISM BASED ON THE 2D WAVELET TRANSFORM

Analyzing wavelets for multi-scale edge detection

In recent years, there has been increasing interest in the application of the continuous wavelet transform to image processing [33, 34]. The edges of the different structures that appear in an image are often the most important features for pattern recognition. Hence, in computer vision [35], a large class of edge detectors look for points where the gradient of the image intensity has a modulus which is locally maximum in its direction. As originally noticed by Mallat and collaborators [36, 37], with an appropriate choice of the analyzing wavelet, one can reformalize the Canny's multi-scale edge detector [38] in terms of a 2D wavelet transform. The general idea is to start by smoothing the discrete image data by convolving it with a filter and then to compute the gradient on the smoothed signal.

Let us consider two wavelets that are, respectively, the partial derivative with respect to x and y of a 2D smoothing function $\phi(x,y)$:

$$\psi_1(x,y) = \frac{\partial\phi(x,y)}{\partial x} \text{ and } \psi_2(x,y) = \frac{\partial\phi(x,y)}{\partial y} . \tag{1}$$

We will assume that ϕ is a well localized (around $x = y = 0$) isotropic function that depends on $|\mathbf{x}|$ only. In this work, we will mainly use the gaussian function:

$$\phi(x,y) = e^{-(x^2+y^2)/2} = e^{-|\mathbf{x}|^2/2} , \tag{2}$$

as well as the isotropic mexican hat:

$$\phi(\mathbf{x}) = (2-\mathbf{x}^2)e^{-|\mathbf{x}|^2/2} . \tag{3}$$

The corresponding analyzing wavelets ψ_1 and ψ_2 are illustrated in Figure 1. They have one and three vanishing moments when using respectively the gaussian function and the mexican hat as smoothing function.

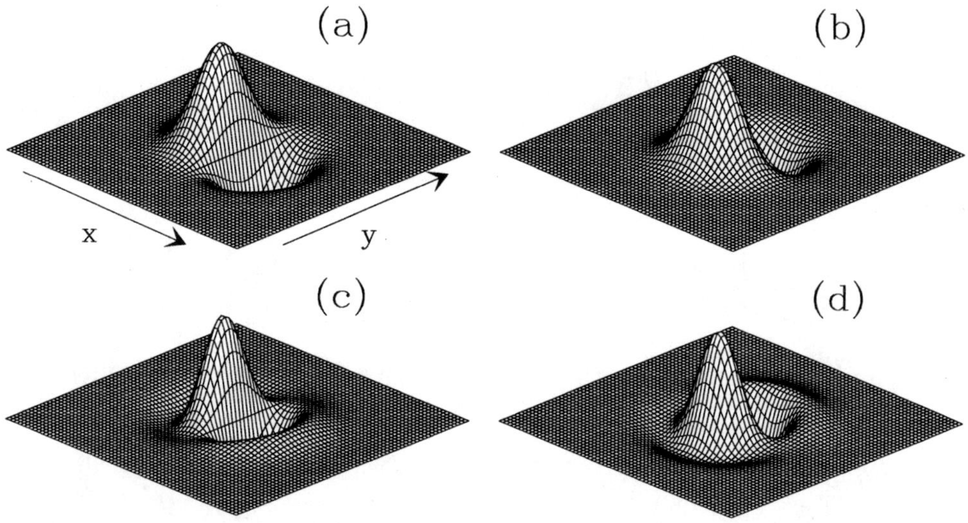

FIGURE 1. The analyzing wavelets ψ_1 and ψ_2 defined in equation (1). First-order analyzing wavelets obtained from a gaussian smoothing function ϕ (Eq. (2)): (a) ψ_1; (b) ψ_2. Third-order analyzing wavelets obtained from the isotropic mexican hat smoothing function ϕ (Eq. (3)): (c) ψ_1; (d) ψ_2.

For any function $f(x,y) \in L^2(\mathbb{R})$, the wavelet transform with respect to ψ_1 and ψ_2 can be expressed in a vectorial form:

$$\mathbf{T}_{\boldsymbol{\psi}}[f](\mathbf{b},a) = \left\{ \begin{array}{l} T_{\psi_1}[f] = a^{-2} \int d^2\mathbf{x}\, \psi_1\big(a^{-1}(\mathbf{x}-\mathbf{b})\big) f(\mathbf{x}) \\ T_{\psi_2}[f] = a^{-2} \int d^2\mathbf{x}\, \psi_2\big(a^{-1}(\mathbf{x}-\mathbf{b})\big) f(\mathbf{x}) \end{array} \right\} \tag{4}$$

Then, after a straightforward integration by parts, one gets:

$$\begin{aligned} \mathbf{T}_{\boldsymbol{\psi}}[f](\mathbf{b},a) &= a^{-2}\boldsymbol{\nabla} \left\{ \int d^2\mathbf{x}\, \phi\big(a^{-1}(\mathbf{x}-\mathbf{b})\big) f(\mathbf{x}) \right\} , \\ &= \boldsymbol{\nabla} \left\{ T_\phi[f](\mathbf{b},a) \right\} , \\ &= \boldsymbol{\nabla} \left\{ \phi_{\mathbf{b},a} * f \right\} . \end{aligned} \tag{5}$$

If $\phi(\mathbf{x})$ is simply a smoothing filter like the Gaussian function (Eq. (2)), then equation (5) amounts to define the 2D wavelet transform as the gradient vector of $f(\mathbf{x})$ smoothed by dilated versions $\phi(a^{-1}\mathbf{x})$ of this filter.

In the following, we will mainly use the representation involving the modulus and the argument of the wavelet transform:

$$\mathbf{T}_{\boldsymbol{\psi}}[f](\mathbf{b},a) = \big(\mathcal{M}_{\boldsymbol{\psi}}[f](\mathbf{b},a), \mathcal{A}_{\boldsymbol{\psi}}[f](\mathbf{b},a)\big) , \tag{6}$$

with

$$\begin{aligned}\mathcal{M}_\psi[f](\mathbf{b},a) &= \left|\mathbf{T}_\psi[f](\mathbf{b},a)\right| , \\ &= \left[\left(T_{\psi_1}[f](\mathbf{b},a)\right)^2 + \left(T_{\psi_2}[f](\mathbf{b},a)\right)^2\right]^{1/2}\end{aligned} \tag{7}$$

and

$$\mathcal{A}_\psi[f](\mathbf{b},a) = \mathrm{Arg}\left(T_{\psi_1}[f](\mathbf{b},a) + iT_{\psi_2}[f](\mathbf{b},a)\right) . \tag{8}$$

Singularity detection and processing with the wavelet transform modulus maxima

The multifractal formalism accounts for possible fluctuations of the local regularity of a rough surface as defined by the Hölder (local roughness) exponent $h(\mathbf{x})$ of the function $f(\mathbf{x})$ whose graph defines the rough surface under study [20] :

$$f(\mathbf{x}+\mathbf{l}) - f(\mathbf{x}) \sim |\mathbf{l}|^{h(\mathbf{x})} , \tag{9}$$

for $|\mathbf{l}| \to 0$ (we refer the reader to Ref. [20] for a more rigorous definition of the Hölder exponent). The so-called $D(h)$ singularity spectrum is defined as the Haussdorf dimension of the set of points $\mathbf{x}$ where the local roughness exponent $h(\mathbf{x})$ is h. As pointed out in Ref. [20], a very efficient way to perform point-wise regularity analysis is to use the wavelet transform modulus maxima.

In the spirit of Canny edge detection [38], at a given scale a, the wavelet transform modulus maxima (WTMM) are defined as the points $\mathbf{b}$ where the wavelet transform modulus $\mathcal{M}_\psi[f](\mathbf{b},a)$ (Eq. (7)) is locally maximum along the gradient direction given by the wavelet transform argument $\mathcal{A}_\psi[f](\mathbf{b},a)$ (Eq. (8)). These modulus maxima are inflection points of $f * \phi_a(\mathbf{x})$. As illustrated in Figure 2, these WTMM lie on connected chains hereafter called *maxima chains* [20, 29, 30]. In theory, one only needs to record the position of the local maxima of $\mathcal{M}_\psi$ along the maxima chains together with the value of $\mathcal{M}_\psi[f]$ and $\mathcal{A}_\psi[f]$ at the corresponding locations. At each scale a, our wavelet analysis thus reduces to store those WTMM maxima (WTMMM) only. They indicate locally the direction where the signal has the sharpest variation. This orientation component is the main difference between 1D and 2D wavelet transform analysis. These WTMMM are disposed along connected curves across scales called *maxima lines* [20, 28, 30]. We will define the WT skeleton as the set of maxima lines that converge to the (x,y)-plane in the limit $a \to 0^+$. This WT skeleton is likely to contain all the information concerning the local Hölder regularity properties of the function f under consideration. Indeed one can prove [20] that, provided the first n_ψ moments of ψ are zero, then :

$$\mathcal{M}_\psi[f]\left(\mathcal{L}_\mathbf{x}(a)\right) \sim a^{h(\mathbf{x})} , \tag{10}$$

along a maxima line pointing to the point $\mathbf{x}$ in the limit $a \to 0^+$, where $h(\mathbf{x})$ ($< n_\psi$) is the local Hölder exponent of f. Moreover, along this maxima line, the wavelet transform argument evolves towards the value :

$$\mathcal{A}_{\psi}[f](\mathcal{L}_{\mathbf{x}}(a)) = \pi + \theta_{\mathbf{x}} , \tag{11}$$

in the limit $a \to 0^+$, where $\theta_{\mathbf{x}}$ is nothing but the direction of the largest variation of f around $\mathbf{x}$, *i.e.* the direction to follow from $\mathbf{x}$ to cross the maxima line at a given (small) scale. From the maxima line $\mathcal{L}_{\mathbf{x}}(a)$, one thus gets the required amplitude as well as directional informations to characterize the local Hölder regularity of f at $\mathbf{x}$.

The 2D wavelet transform modulus maxima method

The WTMM method consists in defining the following partitions functions directly from the WTMMM that belong to the wavelet transform skeleton [20]:

$$\mathcal{Z}(q,a) = \sum_{\mathcal{L} \in \mathcal{L}(a)} \left(\sup_{(\mathbf{x},a') \in \mathcal{L}, a' \leq a} \mathcal{M}_{\psi}[f](\mathbf{x},a') \right)^q , \tag{12}$$

where $q \in \mathbb{R}$. As compared to classical box-counting techniques [39, 40], the analyzing wavelet ψ plays the role of a generalized "oscillating box", the scale a defines its size, while the WTMM skeleton indicates how to position our oscillating boxes to obtain a partition (of $S = \cup S_h$) at the considered scale. Without the "sup" in equation (12), one would have implicitely considered a uniform covering with wavelets of the same size a. As emphasized in Refs [41, 42, 43], the "sup" can be regarded as a way of defining a "Hausdorff like" scale-adaptative partition which will prevent divergencies to show up in the calculation of $\mathcal{Z}(q,a)$ for $q < 0$.

Now, from the deep analogy that links the multifractal formalism to thermodynamics [42, 43], one can define the exponent $\tau(q)$ from the power-law behavior of the partition function:

$$\mathcal{Z}(q,a) \sim a^{\tau(q)} , \; a \to 0^+ , \tag{13}$$

where q and $\tau(q)$ play respectively the role of the inverse temperature and the free energy. The main result of the wavelet-based multifractal formalism is that in place of the energy and the entropy (*i.e.*, the variables conjugated to q and τ), one has the Hölder exponent h (Eq. (9)) and the singularity spectrum $D(h)$. This means that the $D(h)$ singularity spectrum of f can be determined from the Legendre transform of the partition function scaling exponent $\tau(q)$:

$$D(h) = \min_q \left(qh - \tau(q) \right) . \tag{14}$$

From the properties of the Legendre transform, it is easy to convince oneself that *homogeneous* (monofractal) fractal functions that involve singularities of unique Hölder exponent $h = \partial\tau/\partial q$, are characterized by a $\tau(q)$ spectrum which is a *linear* function of q. On the contrary, a *nonlinear* $\tau(q)$ curve is the signature of nonhomogeneous functions that display *multifractal* properties, in the sense that the Hölder exponent $h(\mathbf{x})$ is a fluctuating quantity that depends upon the spatial position $\mathbf{x}$ (in other words the local roughness exponent is fluctuating from point to point). Let us note that the scaling exponent $\tau(q)$ has well-known meaning for some specific values of q : (i) $-\tau(0)$ is the fractal dimension of the set of singularities of f; (ii) $\tau(1)$ is related to the fractal dimension of the rough surface $\mathcal{S}$ ($z = f(x,y)$): $d_f(\mathcal{S}) = \max(2, 1-\tau(1))$; (iii) $\tau(2)$ is related to the scaling exponent β of the spectral density $S(\mathbf{k}) = |\hat{f}(\mathbf{k})|^2 \sim |\mathbf{k}|^\beta$ with $\beta = 4 - \tau(2)$.

TEST APPLICATIONS OF THE 2D WTMM METHOD TO SYNTHETIC ROUGH SURFACES

Isotropic fractional Brownian surfaces

Since its introduction by Mandelbrot and Van Ness [44], the fractional Brownian motion (fBm) has become a very popular model in signal and image processing [16, 45]. In one dimension, fBm has proved useful for modeling various physical phenomena with long-range dependence, *e.g.*, "$1/f$" noises. The generalization of Brownian motion to more than one dimension was first considered by Levy [46]. The generalization of fBm follows along similar lines. A 2D fBm $B_H(\mathbf{x})$ indexed by $H \in]0,1[$, is a process with stationary zero-mean Gaussian increments that can be used to generate random self-affine surfaces with known statistical properties :

$$\tau(q) = qH - 2\,. \tag{15}$$

By Legendre transforming $\tau(q)$ according to equation (14), one gets the following expression for the singularity spectrum :

$$\begin{aligned} D(h) &= 2 && \text{if } h = H\,, \\ &= -\infty && \text{if } h \neq H\,. \end{aligned} \tag{16}$$

Thus fBm surfaces are the representation of homogeneous stochastic monofractal functions characterized by a singularity spectrum which reduces to a single point.

We have tested the 2D WTMM method described in the previous section, on 32 realizations of a 2D fBm process with $H = 1/3$. We have wavelet transformed 32 (1024×1024) images of $B_{H=1/3}$ with an isotropic first-order analyzing wavelet. To master edge effects, we then restrain our analysis to the 512×512 central part of the wavelet transform of each image. In Figure 2 is illustrated the computation of the maxima chains and the WTMMM for an individual image at three different scales. In Figure 2b is shown the convolution of the original image (Fig. 2a) with the isotropic gaussian smoothing filter ϕ (Eq. (2)). According to the definition of the wavelet transform modulus maxima, the maxima chains correspond to well defined edge curves of the smoothed image. The local maxima of $\mathcal{M}_\psi$ along these curves are located at the points where the sharpest intensity variation is observed. The corresponding arrows clearly indicate that locally, the gradient vector points in the direction (as given by $\mathcal{A}_\psi$) of maximum change of the intensity surface. When going from large scale (Fig. 2d) to small scale (Fig. 2c), the characteristic average distance between two nearest neighbour WTMMM decreases

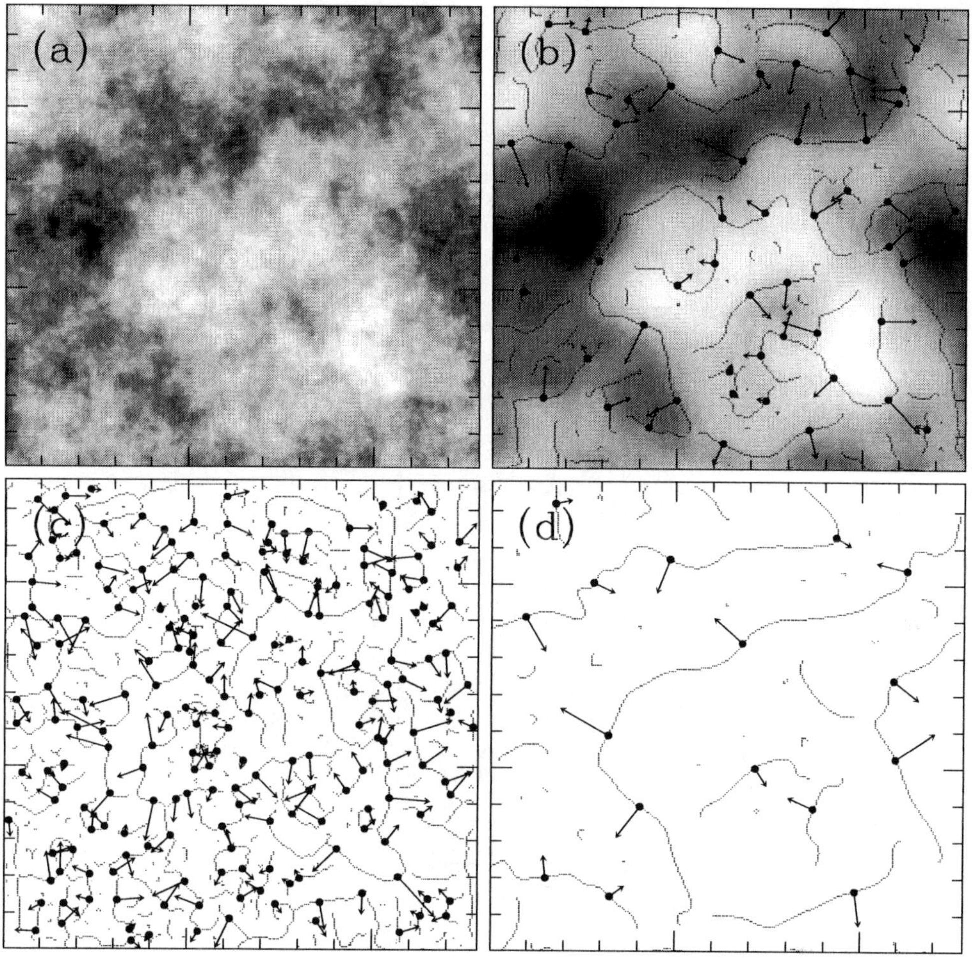

FIGURE 2. 2D wavelet transform analysis of $B_{H=1/3}(\mathbf{x})$. ψ is a first-order radially symmetric analyzing function (see Fig. 1). (a) 32 grey-scale coding of the central 512×512 portion of the original image. In (b) $a = 2\sigma_W$, (c) $a = 2^{0.1}\sigma_W$ and (d) $a = 2^{1.9}\sigma_W$, are shown the maxima chains; the local maxima of $\mathcal{M}_\psi$ along these chains are indicated by (•) from which originate an arrow whose length is proportional to $\mathcal{M}_\psi$ and its direction (with respect to the x-axis) is given by $\mathcal{A}_\psi$. In (b), the smoothed image $\phi_{\mathbf{b},a} * B_{1/3}$ (Eq. (5)) is shown as a grey-scale coded background from white (min) to black (max). $\sigma_W = 13$ (pixels) is the characteristic size of ψ at the smallest resolved scale.

like a. This means that the number of WTMMM and in turns, the number of maxima lines, proliferates across scales like a^{-2}. The corresponding wavelet transform skeleton is shown in Figure 3. As confirmed just below, when extrapolating the arborescent structure of this skeleton to the limit $a \to 0^+$, one recovers the theoretical result that the support of the singularities of a 2D fBm has a dimension $d_F = 2$, *i.e.*, $B_{H=1/3}(\mathbf{x})$ is nowhere differentiable [16, 45, 46].

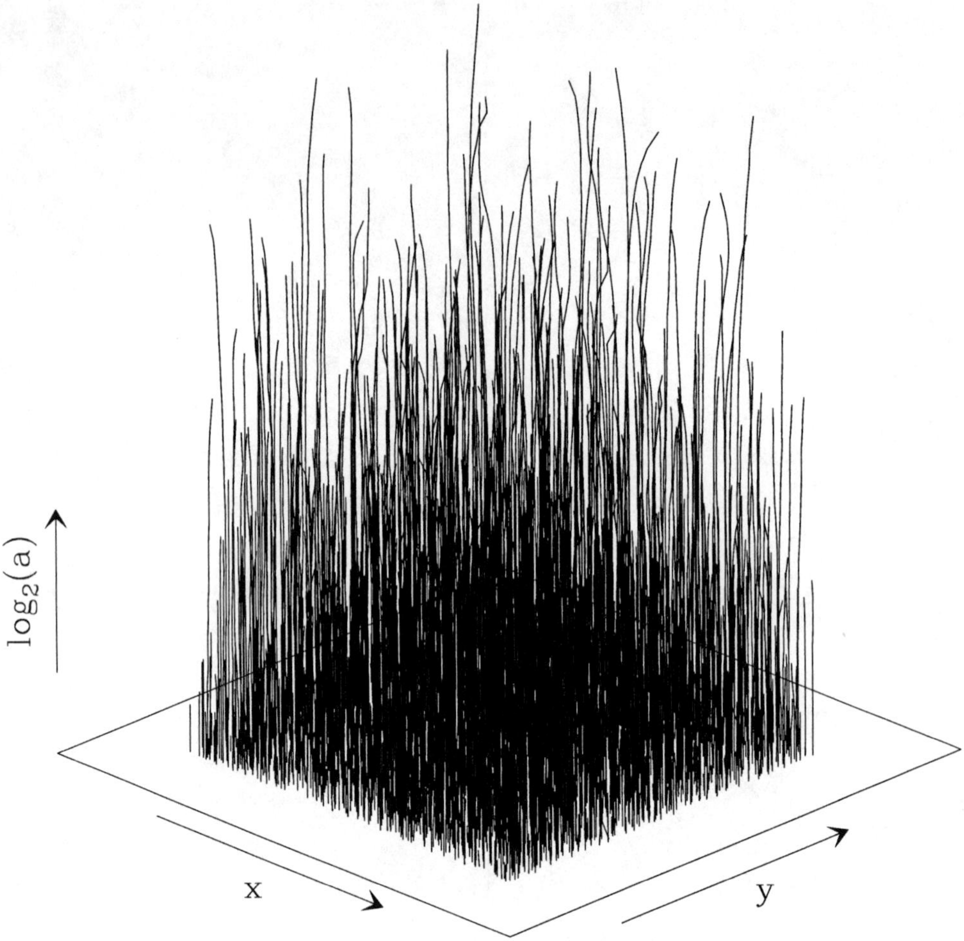

FIGURE 3. Wavelet transform skeleton of the 2D fBm image shown in Figure 2a. This skeleton is defined by the set of maxima lines obtained after linking the WTMMM detected at different scales. Same analyzing wavelet as in Figure 2.

In Figure 4 are reported the results of the computation of the $\tau(q)$ and $D(h)$ spectra using the 2D WTMM method. As shown in Figure 4a, the average partition function $\mathcal{Z}(q,a)$ (over 32 images of $B_{1/3}(\mathbf{x})$) displays a remarkable scaling behavior over more than 3 octaves when plotted *versus* a in a logarithmic representation (Eqs. (12) and (13)). Moreover, for a wide range of values of $q \in [-4,6]$, the data are in good agreement with the theoretical $\tau(q)$ spectrum (Eq. (15)). When proceeding to a linear regression fit of the data over the first two octaves, one gets the $\tau(q)$ spectra shown in Figure 4b for three values of the fBm index $H = 1/3$, $1/2$ and $2/3$. Whatever H, the data systematically fall on a straight line, the signature of homogeneous (monofractal) scaling properties. In Figure 4c are reported the corresponding estimates of $D(h)$, from Legendre transforming

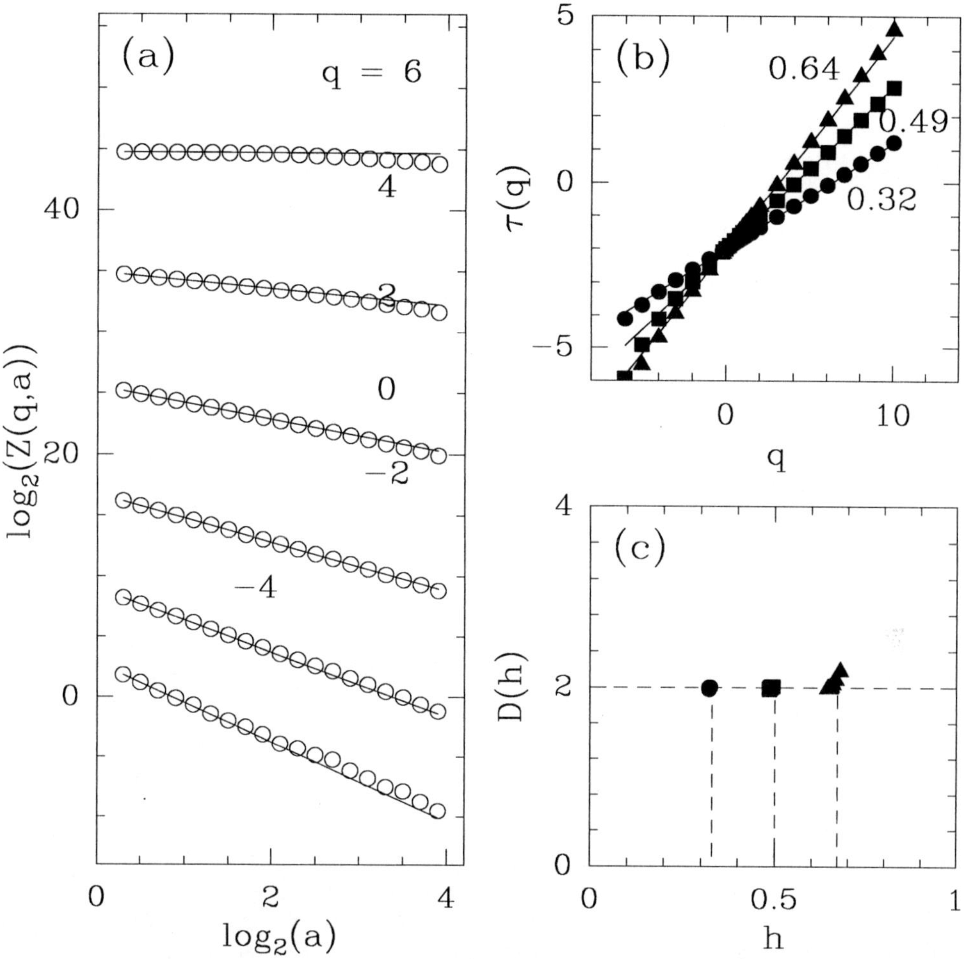

FIGURE 4. Determination of the $\tau(q)$ and $D(h)$ spectra of 2D fBm with the 2D WTMM method [20]. (a) $\log_2 \mathcal{Z}(q,a)$ *vs* $\log_2 a$; the solid lines correspond to the theoretical predictions $\tau(q) = qH - 2$ (Eq. (15)) with $H = 1/3$. (b) $\tau(q)$ *vs* q for $H = 1/3$ (•), $1/2$ (■) and $2/3$ (▲); the solid lines correspond to linear regression fit estimates of H. (c) $D(h)$ *vs* h. Same analyzing wavelet as in Figure 2. These results correspond to averaging over 32 (1024 × 1024) fBm images. a is expressed in σ_W units.

the $\tau(q)$ data (Eq. (14)). The results obtained for both the $\tau(q)$ and $D(h)$ spectra are thus in remarkable agreement with the theoretical predictions given by equations (15) and (16) respectively. The 2D WTMM method can thus be considered as having successfully passed the test of homogeneous fBm rough surfaces.

When further investigating the joint probability distribution function $P_a(\mathcal{M},\mathcal{A})$ as computed from the wavelet transform skeleton of 32 (1024 × 1024) images of $B_{1/3}(\mathbf{x})$, one can check that this pdf actually factorizes $P_a(\mathcal{M},\mathcal{A}) = P_a(\mathcal{M})P_a(\mathcal{A})$, where $P_a(\mathcal{M}) = \int d\mathcal{A}\, P_a(\mathcal{M},\mathcal{A})$ and $P_a(\mathcal{A}) = \int d\mathcal{M}\, P_a(\mathcal{M},\mathcal{A})$ are shown in Figure 5 for

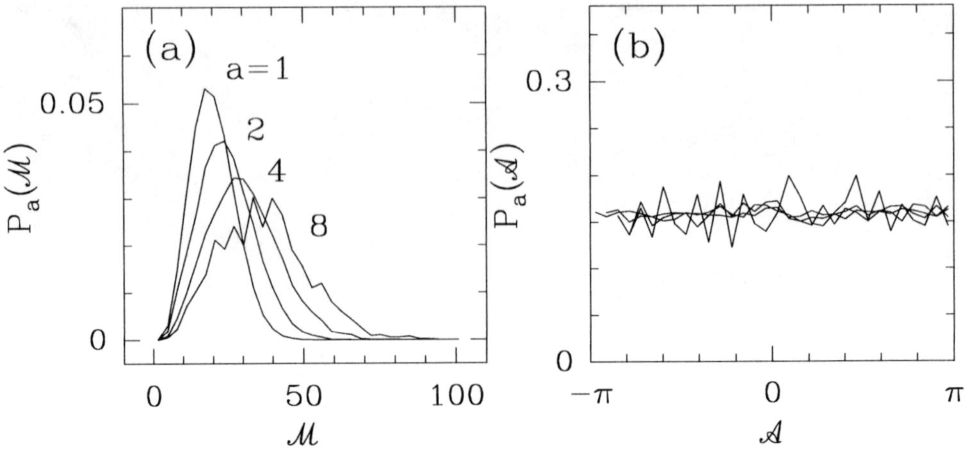

FIGURE 5. Pdf's of the WTMMM coefficients of $B_{1/3}(\mathbf{x})$ as computed at different scales $a = 1$, 2, 4 and 8 (in σ_W units). (a) $P_a(\mathcal{M})$ *vs* $\mathcal{M}$. (b) $P_a(\mathcal{A})$ *vs* $\mathcal{A}$. ψ is the first-order analyzing wavelet shown in Figure 1. These results correspond to averaging over 32 (1024 × 1024) fBm images.

four different values of the scale parameter. This factorization means that the scale-invariance properties are statistically independent of the angular information contained in $P_a(\mathcal{A})$. As seen in Figure 5a, $P_a(\mathcal{M})$ decreases fast to zero at zero. This explains that when concentrating on the wavelet transform skeleton, the integral in the r.h.s. of equation (12) does not diverge when considering negative q values. This remark is at the heart of the 2D WTMM method; by allowing us to compute the $\tau(q)$ spectrum for negative as well as positive q values, the 2D WTMM method is a definite step beyond the 2D structure function method which is intrinsically restricted to positive q values. The corresponding pdf's $P_a(\mathcal{A})$ are represented in Figure 5b. $P_a(\mathcal{A})$ clearly does not evolve across scales. Moreover, except some small amplitude fluctuations observed at the largest scale, $P_a(\mathcal{A}) = 1/2\pi$ is a flat distribution as expected for statistically isotropic scale-invariant rough surfaces.

Isotropic multifractal synthetic rough surfaces

In Figure 6 is illustrated the computation of the maxima chains and the WTMMM for an individual image of a multifractal rough surface generated with the log-normal $\mathcal{W}$-cascade model described in Ref. [28]. This model has two parameters which are fixed at the value $m = -0.38 \ln 2$ and $\sigma^2 = 0.03 \ln 2$. From the WT skeletons of 32 (1024 × 1024) images like the one in Figure 6a, one computes the average of the partition functions $\mathcal{Z}(q,a)$. As shown in Figure 7a, when plotted *versus* the scale parameter a in a logarithmic representation, these average partition functions display a rather impressive scaling behavior over a range of scales of about 4 octaves (*i.e.*, $\sigma_W \lesssim a \lesssim 16\sigma_W$, where $\sigma_W = 13$ pixels). Let us point out that scaling of quite good quality is found for a rather wide range of values of q: $-6 \lesssim q \lesssim 8$. When processing to a linear regression fit of

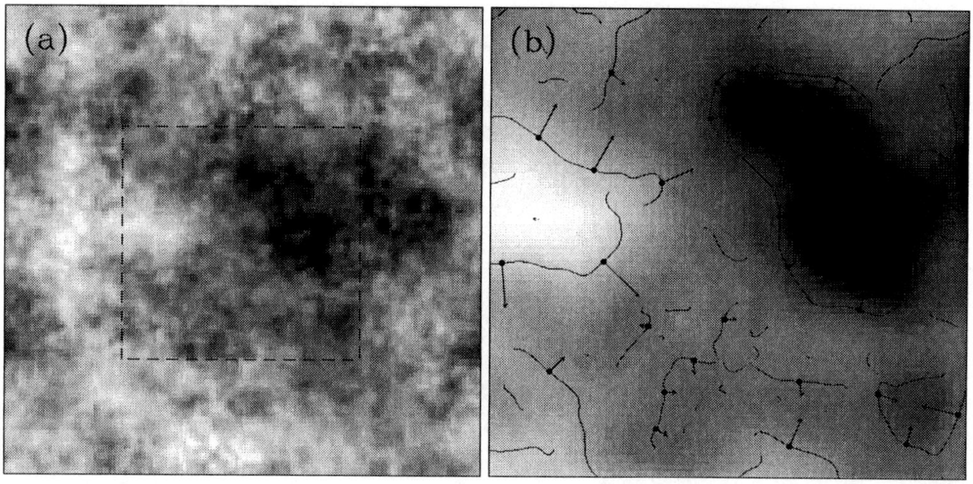

FIGURE 6. 2D wavelet transform analysis of a multifractal rough surface generated with the log-normal $\mathcal{W}$-cascade model with the parameter values $m = -0.38\ln 2$ and $\sigma^2 = 0.03\ln 2$ [28]. ψ is the first-order radially symmetric analyzing wavelet shown in Figure 1. (a) 32 grey-scale coding of the original (1024×1024) image. (b) Maxima chains and WTMMM for the central (512×512) part of the original image (dashed square in (a)) as computed at the scale $a = 2^{2.9}\sigma_W$. The smoothed image $\phi_{\mathbf{b},a} * f$ is shown as a grey-scale coded background from white (min) to black (max).

the data over the first four octaves, one gets the $\tau(q)$ spectrum (○) shown in Fig. 7b. For the range of q values where scaling is operating, the numerical data are in remarkable agreement with the theoretical nonlinear $\tau(q)$ spectrum [28] :

$$\tau(q) = -\frac{\sigma^2}{2\ln 2}q^2 - \frac{m}{\ln 2}q - 2\,. \tag{17}$$

Similar quantitative agreement is observed on the $D(h)$ singularity spectrum in Figure 7c which displays also a parabolic shape as predicted theoretically :

$$D(h) = -\frac{(h+m/\ln 2)^2}{2\sigma^2/\ln 2} + 2\,. \tag{18}$$

The multifractal rough surfaces under study, display intermittent fluctuations corresponding to Hölder exponent values ranging from $h_{\text{min}} = 0.034$ to $h_{\text{max}} = 0.726$. Unfortunately, to capture the strongest and weakest singularities, one needs to compute the $\tau(q)$ spectrum for very large values of $|q|$. This requires the processing of many more images of much larger size, which is out of our current computer capabilities. Note that with the statistical sample studied here, one has $D(h(q=0) = 0.38) = 2.00 \pm 0.02$, which allows us to conclude that the rough surfaces under consideration are singular everywhere. The results reported in Figure 7 demonstrate that the 2D WTMM method is a very efficient tool to resolve multifractality.

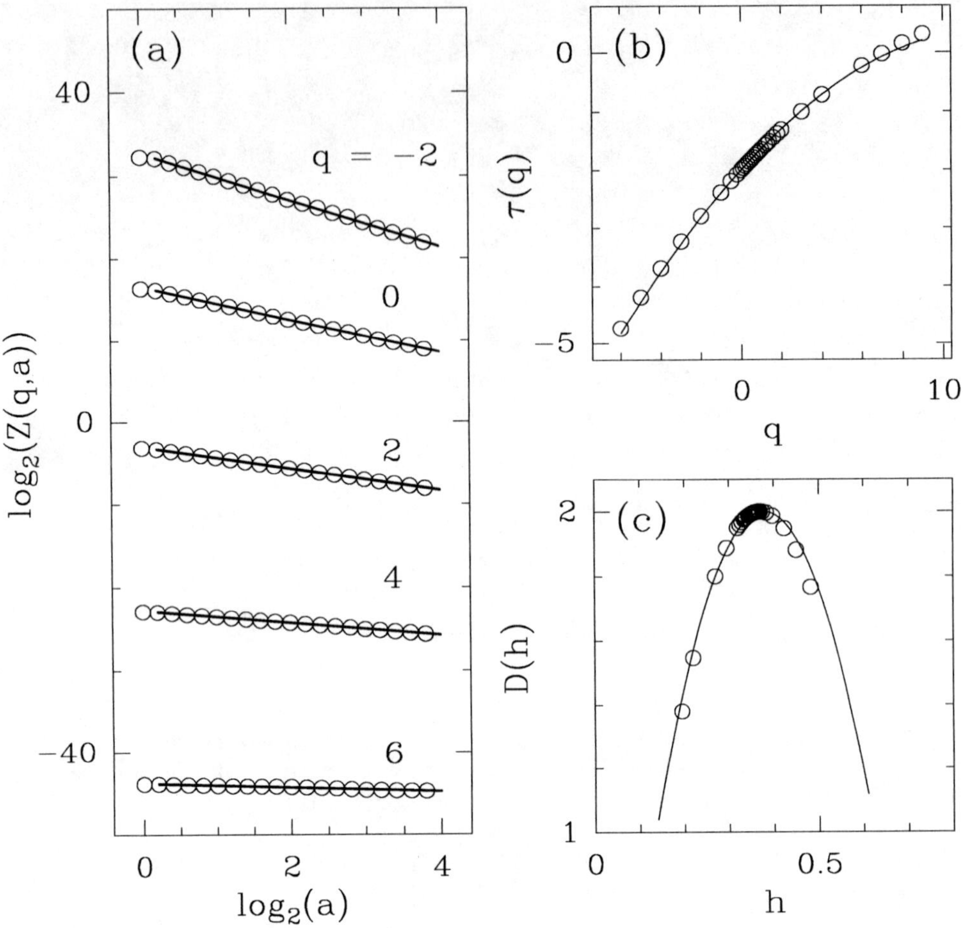

FIGURE 7. Determination of the $\tau(q)$ and $D(h)$ spectra of multifractal rough surfaces generated with the log-normal random $\mathcal{W}$-cascade models (o), using the 2D WTMM method [28]. ψ is the first-order radially symmetric analyzing wavelet used in Figure 6. (a) $\log_2 \mathcal{Z}(q,a)$ *vs* $\log_2 a$; the solid lines correspond to linear regression fit of the data over the first four octaves. (b) $\tau(q)$ *vs* q as obtained from linear regression fit of the data in (a) over the first four octaves. (c) $D(h)$ *vs* h, after Legendre transforming the $\tau(q)$ curve in (b). In (b) and (c), the solid lines represent the theoretical log-normal spectra (Eqs (17) and (18)).

APPLICATION OF THE 2D WTMM METHOD TO LANDSAT IMAGES OF STRATOCUMULUS CLOUDS

Landsat data of marine stratocumulus cloud scenes

Over the past fifteen years, Landsat imagery has provided the remote sensing community at large with a very attractive and reliable tool for studying the Earth's environment [9, 10, 11, 12, 13, 14, 15]. One of the main advantages of high-resolution satellite imagery is its rather low effective cost as compared to outfitting and flying research aircraft. Moreover, this very well calibrated instrument offers the possibility to reach unusual high spatial, spectral and radiometric resolutions [15]. Indeed, the five-channel Multi-Spectral Scanning (MSS, Landsat 1-3) radiometer resolves features down to 80 m with 7 bits of dynamic range. The seven-channel Thematic Mapper (TM, Landsat 4-5) camera does even better with a resolution of 30 m and a dynamic range of 8 bits. Both these instruments are essentially proportional to nadir-viewing radiance at satellite level ($\simeq$ 800 km). Scores of Landsat completely or partially cloudy scenes have thus been acquired, mainly for the purpose of characterizing cloud morphology. Mainly two types of statistical analysis have been applied so far to Landsat imagery: spectral analysis of the 2D radiance field [12, 13, 14, 15, 47] and joint area and perimeter distributions for ensembles of individual clouds [9, 10, 11] defined by some threshold in radiance. One of the most remarkable properties of Landsat cloud scenes is their statistical "scale-invariance" over a rather large range of scales, which justifies why fractal and multifractal concepts have progressively gained more acceptance in the atmospheric scientist community [4].

Off all cloud types, marine stratocumulus (Sc) are without any doubt the ones which have attracted the most attention, mainly because of their first-order effect on the Earth's energy balance [4, 15, 47]. Being at once very persistent and horizontally extented, marine Sc layers carry considerable weight in the overall reflectance (albedo) of the planet and, from there, command a strong effect on its global climate [5]. Furthermore, with respect to climate modeling [6] and the major problem of cloud-radiation interaction [13, 15, 48, 49], they are presumably at their simplest in marine Sc which are relatively thin ($\sim$ 300-500 m), with well-defined (quasi-planar) top and bottom, thus approximating the plane-parallel geometry where radiative transfer theory is well developed [15, 48, 49, 50]. However, because of its internal homogeneity assumption, plane-parallel theory shows systematic biases due to the huge internal variability of marine Sc [8, 26].

Figure 8 is a grey-scale rendering of a small part of the entire ($\simeq 196 \times 168$ km^2) original cloudy Landsat 5 scene captured with the TM camera (1 pixel = 30 m) in the 0.6-0.7 μm channel (*i.e.* reflected solar photons as opposed to their counterparts emitted in the thermal infrared) during the first ISCCP (International Satellite Cloud Climatology Project) Research Experiment (FIRE) field program [51], which took place over the Pacific Ocean off San Diego in summer 1987. In order to master edge effects in the 2D WT computation, we actually select 32 overlapping 1024×1024 pixels^2 subscenes in this cloudy region. The overall extend of the explored area is about 7840 km^2. Figure 8a shows a typical (1024×1024) portion of the original image where the eight-bit grey scale coding of the quasi-nadir viewing radiance clearly reveals the presence of some

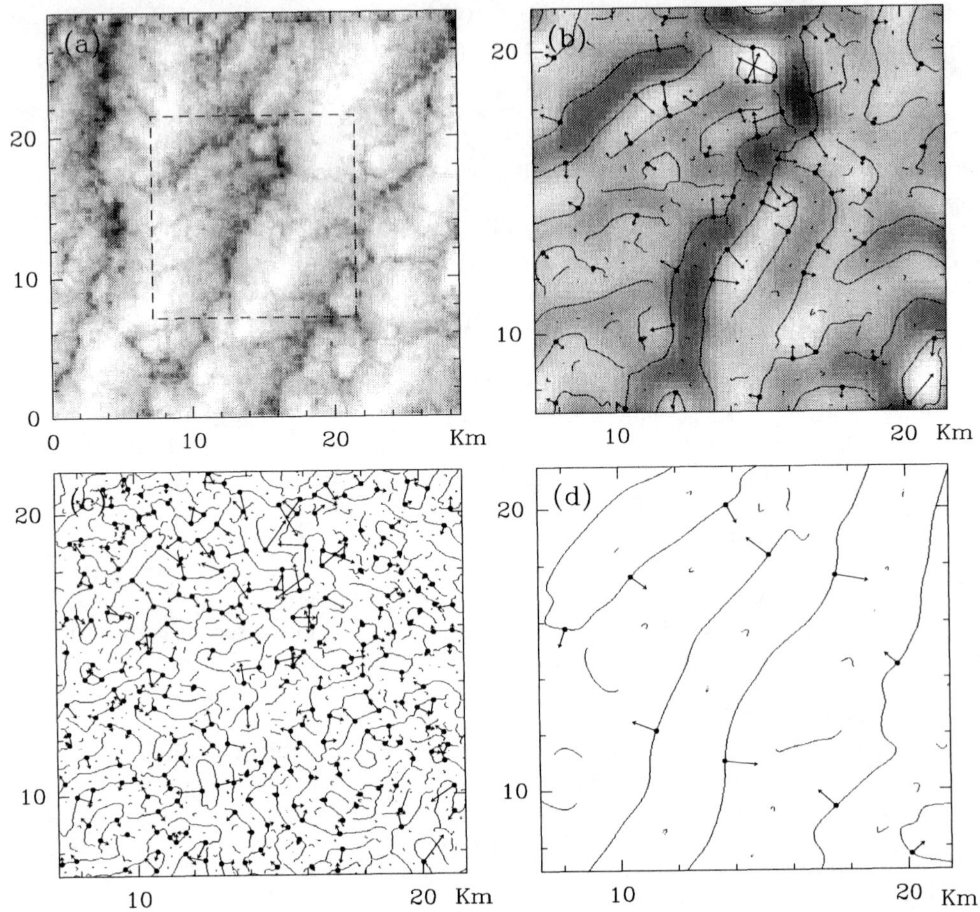

FIGURE 8. 2D wavelet transform analysis of a Landsat image of marine Sc clouds [31]. $\psi(\mathbf{x})$ is the first-order radially symmetric analyzing wavelet shown in Figure 1. (a) 256 grey-scale coding of a (1024 × 1024) portion of the original radiance image. In (b) $a = 2^{2.9}\sigma_W$, (c) $a = 2^{1.9}\sigma_W$ and (d) $a = 2^{3.9}\sigma_W$ (where $\sigma_W = 13$ pixels $\simeq 390$ m), are shown the maxima chains; the local maxima of $\mathcal{M}_\psi$ along these chains are indicated by (•) from which originates an arrow whose length is proportional to $\mathcal{M}_\psi$ and its direction (with respect to the x-axis) is given by $\mathcal{A}_\psi$; only the central (512 × 512) part delimited by a dashed square in (a) is taken into account to define the WT skeleton. In (b), the smoothed image $\phi_{\mathbf{b},a} * I$ is shown as a grey-scale coded background from white (min) to black (max).

anisotropic texture induced by convective structures which are generally aligned to the wind direction.

The computation of the power spectral density reveals a rather nice power-law behavior $S(k) \sim k^{-\beta}$ from $0.2 - 0.4$km up to $\simeq 6$km, with $\beta = 2.72 \pm 0.08$. This value of the spectral exponent is quite compatible with previous estimates obtained for 1D LWC data [7, 12, 13, 14, 15, 47], namely $\beta_{LWC} = 1.4\text{-}1.7 \simeq \beta_{2D} - 1$.

Numerical computation of the multifractal $\tau(q)$ and $D(h)$ spectra using the 2D WTMM method

In Figure 8 is illustrated the computation of the maxima chains and the WTMMM for the studied marine Sc [31] scene. In Figure 8b is shown the convolution of the original radiance field (Fig. 8a) with the isotropic gaussian smoothing filter ϕ (Eq. (2)). According to the definition of the WTMM, the maxima chains correspond to well defined edge curves of the smoothed image. The local maxima of $\mathcal{M}_\psi$ along these curves are indicated by (•) from which originates an arrow whose length is proportional to $\mathcal{M}_\psi$ and its direction (with respect to the x-axis) is given by $\mathcal{A}_\psi$. After linking these WTMMM across scales, one constructs the WT skeleton from which one computes the partition functions $\mathcal{Z}(q,a)$ (Eq. (12)). As reported in Figure 9a, the partition functions averaged over the 32 overlapping (1024 × 1024) images (•), display some well-defined scaling behavior over the first three octaves, *i.e.* over the range of scales 390 m $\lesssim a \lesssim$ 3120 m, when plotted *versus* a in a logarithmic representation. Indeed the scaling deteriorates progressively from the large scale side when one goes to large values of $|q| \gtrsim 3$. As reported in Ref. [31], besides the fact that we are suffering from insufficient sampling, the presence of localized Dirac like structures is likely to explain the fact that the observed cross-over to a steeper power-law decay occurs at a smaller and a smaller scale when one increases $q > 0$. Actually for $q \gtrsim 3$, the cross-over scale $a^* \lesssim$ 1200 m becomes significantly smaller than the so-called integral scale which is approximately given by the characteristic width $\lambda \simeq$ 5-6 km of the convective rolls (Fig. 8a). When processing to a linear regression fit of the data in Figure 9a over the first octave and a half (in order to avoid any bias induced by the presence of the observed cross-over at large scales), one gets the $\tau(q)$ spectrum (•) shown in Figure 9b. In contrast to the previously studied fractional Brownian rough surfaces, this $\tau(q)$ spectrum unambiguously deviates from a straight line. When Legendre transforming this nonlinear $\tau(q)$ curve, one gets the $D(h)$ singularity spectrum reported in Figure 9c. Its characteristic single humped shape over a finite range of Hölder exponents is a clear signature of the multifractal nature of the marine Sc radiance fluctuations.

In Figure 9 are also shown for comparison the results (○) obtained when applying the 2D WTMM method with a third-order ($n_\psi = 3$) radially symmetric analyzing wavelet (the smoothing function ϕ being the isotropic 2D mexican hat). As seen in Figure 9a, the use of a wavelet which has more zero moments seems to slightly improve scaling. But the estimate of the $\tau(q)$ and $D(h)$ spectra yields consistent results in the range $-3 \lesssim q \lesssim 5$. Moreover these spectra are very well fitted by the theoretical quadratic spectra (Eqs (17) and (18)) predicted for log-normal $\mathcal{W}$-cascades [31].

WTMMM probability density functions

In Figures 10a and 10b are respectively shown the pdf's $P_a(\mathcal{M})$ and $P_a(\mathcal{A})$ for three different values of the scale parameter $a = 2^{0.3}\sigma_W$ (480 m), $2^{1.3}\sigma_W$ (960 m) and $2^{2.3}\sigma_W$ (1920 m). First let us concentrate on the results shown in Figure 10b for $P_a(\mathcal{A})$. This distribution is clearly scale dependent with some evidence of anisotropy enhancement

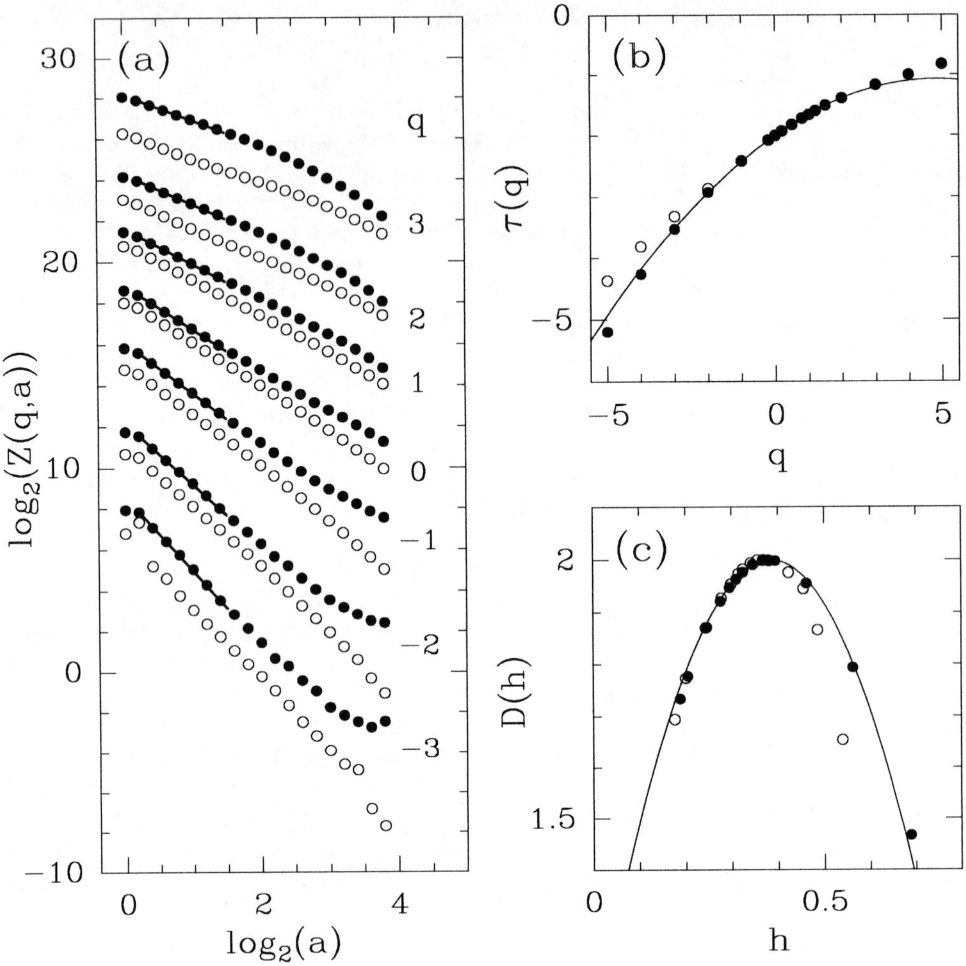

FIGURE 9. Determination of the $\tau(q)$ and $D(h)$ spectra of radiance Landsat images of marine Sc [31]. The 2D WTMM method is used with either a first-order (•) or a third-order (∘) radially symmetric analyzing wavelet (see Fig. 1). (a) $\log_2 \mathcal{Z}(q,a)$ *vs* $\log_2 a$; the solid lines correspond to linear regression fits of the data over the first and a half octave. (b) $\tau(q)$ *vs* q as obtained from a linear regression fit of the data in (a). (c) $D(h)$ *vs* h, after Legendre transforming the $\tau(q)$ curve in (b). In (b) and (c), the solid lines correspond to the theoretical multifractal spectra for log-normal $\mathcal{W}$-cascades with parameter values $m = -0.38\ln 2$ and $\sigma^2 = 0.07\ln 2$ (Eqs (17) and (18)).

when going from small to large scales, in particular when one reaches scales which become comparable to the characteristic width of the convective structures (*i.e.*, a few kilometers wide). Two peaks around the values $\mathcal{A} \simeq -\pi/6$ and $5\pi/6$ become more and more pronounced as the signature of a privilegied direction in the analyzed images. As one can check from a visual inspection of Figure 8a, this direction is nothing but the perpendicular to the mean direction of the convective rolls that are generally aligned to

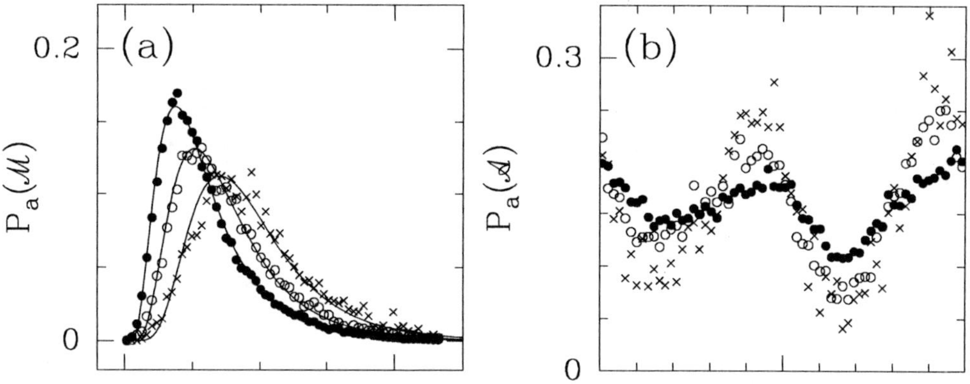

FIGURE 10. Pdf's of the WTMMM coefficients of the 32 (1024 × 1024) radiance Landsat images as computed with a first-order radially symmetric analyzing wavelet: (a) $P_a(\mathcal{M})$ *vs* $\mathcal{M}$; (b) $P_a(\mathcal{A})$ *vs* $\mathcal{A}$; the symbols correspond to the following scales $a = 2^{0.3}\sigma_W = 480$ m (•), $2^{1.3}\sigma_W = 960$ m (○) and $2^{2.3}\sigma_W = 1920$ m (×).

the wind direction.

However, as seen in Figure 11, when conditioning the pdf of $\mathcal{M}$ by the argument $\mathcal{A}$, the shape of this pdf turns out to be independent of the considered value of $\mathcal{A}$, as long as the value of the scale parameter a remains small as compared to the characteristic width of the convective structures. The observation that the joint probability distribution actually factorizes, *i.e.*, $P_a(\mathcal{M},\mathcal{A}) = P_a(\mathcal{M})P_a(\mathcal{A})$, is the signature that $\mathcal{M}$ and $\mathcal{A}$ are likely to be independent [30, 31]. This actual decoupling of $\mathcal{M}$ and $\mathcal{A}$ means that the multifractal properties of the marine Sc radiance fluctuations are contained in the way the shape of the pdf of $\mathcal{M}$ evolves when one decreases the scale parameter a. In other words the anisotropy induced by the wind direction does not affect the multifractal scaling properties of the cloud structure.

DISCUSSION

To summarize, we have reported the results of the experimental application of the 2D WTMM method [20, 28] to radiance Landsat images of marine Sc clouds collected during FIRE program [30, 31]. The computation of the $\tau(q)$ and $D(h)$ spectra have definitely quantified the multifractal scaling properties of the radiance fields. Besides the presence of rather isolated localized downward spike events [31], the internal cloud structure is clearly intermittent and displays rather convincing isotropic scaling over a range of scale that extends from our WT microscope resolution ($\sigma_W = 13$ pixels $= 390$ m) up to $L/2$, where $L \sim 5$-6 km is some integral scale which is apparently dictated by the average characteristic width of the convective rolls that modulate the amplitude of radiance fluctuations.

Let us point out that a similar 1D WTMM analysis of the velocity fluctuations in high Reynolds number turbulence has come to conclusions very close to those of the present study [52, 53]. Besides the presence of rather localized Dirac like structures that witness

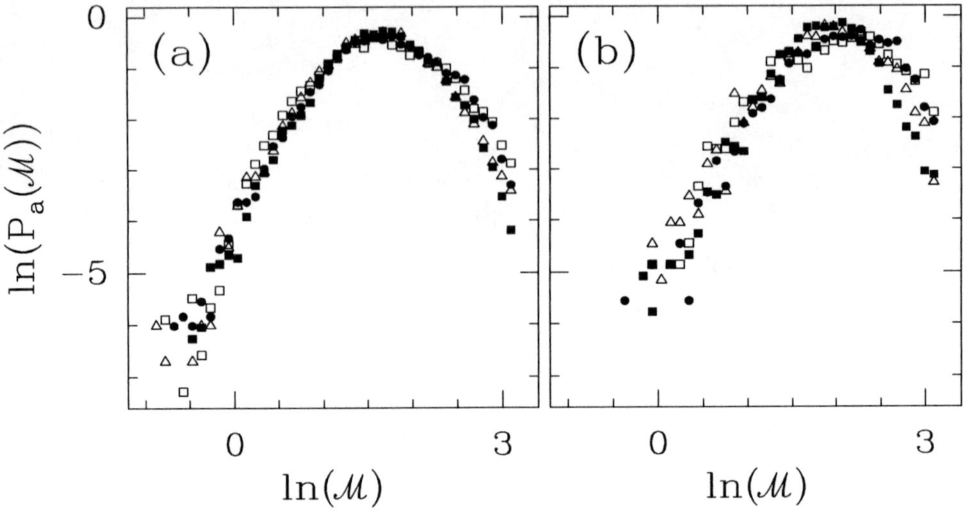

FIGURE 11. Pdf's of the WTMMM coefficients of the 32 (1024 × 1024) radiance Landsat images as computed with a first-order radially symmetric analyzing wavelet. Pdf's of $\mathcal{M}$ when conditioned by $\mathcal{A}$. The different symbols correspond to fixing $\mathcal{A} \pmod{\pi}$ to $0 \pm \pi/8$ (∘), $\pi/4 \pm \pi/8$ (□), $\pi/2 \pm \pi/8$ (△) and $3\pi/4 \pm \pi/8$ (■). (a) $a = 2^{0.3}\sigma_W = 480$ m; (b) $a = 2^{1.3}\sigma_W = 960$ m.

to the probing of vorticity filaments [54], the multifractal nature of turbulent velocity is likely to be understood in terms of a log-normal cascading process which is expected to be scale-invariant in the limit of very high Reynolds numbers [52, 53]. In Figure 12 are shown for comparison the results obtained in Figure 9 for the $D(h)$ singularity spectrum of the radiance Landsat images together with the $D(h)$ data extracted from the 1D analysis of a turbulent velocity signal recorded at the Modane wind tunnel ($R_\lambda \simeq 2000$) [52, 53] (indeed $D(h)+1$ is represented for the latter in order to compare 1D to 2D data). The turbulent velocity $D(h)$ spectrum significantly differs from the results obtained for the marine Sc cloud. They have a common feature, *i.e.*, the Hölder exponent the most frequently encountered in the radiance field $h = m/\ln 2 = h(q = 0) = \partial\tau/\partial q|_{q=0} = 0.38 \pm 0.01$ is undistinguishable from the corresponding exponent $h = h(q=0) = 0.39 \pm 0.01$ found for the turbulent velocity field. Note that these values are significantly larger than the theoretical value $h = 1/3$ predicted by Kolmogorov in 1941 [55] to account for the observed $k^{-5/3}$ power-spectrum behavior. The main difference comes from the intermittency parameter which is much stronger for the cloud, $\sigma^2/\ln 2 = 0.08 \pm 0.01$ ($n_\psi = 1$) or $\sigma^2/\ln 2 = 0.06 \pm 0.01$ ($n_\psi = 3$) than for the turbulent velocity, $\sigma^2/\ln 2 = 0.036 \pm 0.004$. This is the signature that the radiance field is much more intermittent than the velocity field: the $D(h)$ singularity spectrum for the former is unambiguously wider than the corresponding spectrum for the later. (Let us mention that insufficient sampling is probably at the origin of the underestimation of the intermittency parameter $c_2 = \sigma^2/\ln 2$ when using a third-order analyzing wavelet. This means that one must not pay too much attention to the decreasing right-hand part of the $D(h)$ cloud data in Figure 12b.) For the sake of comparison, we have also reported in Figure 12,

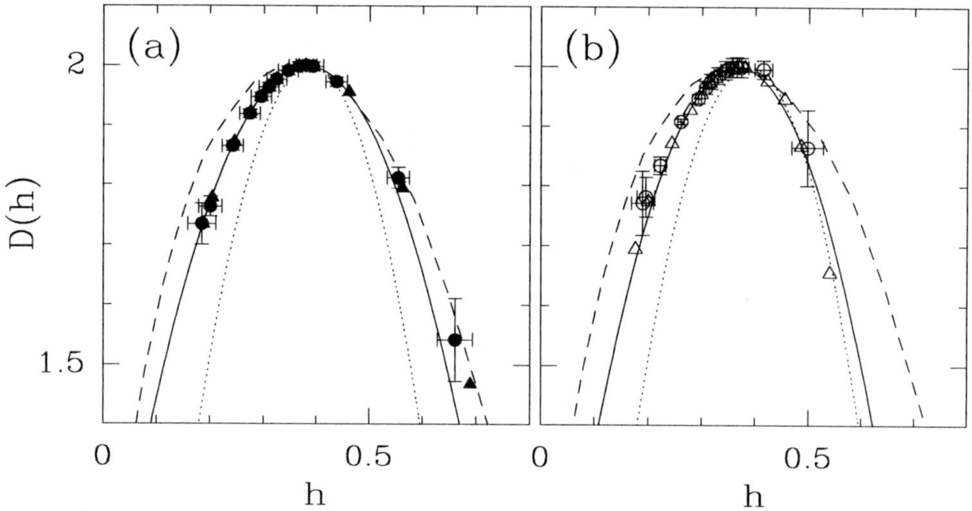

FIGURE 12. $D(h)$ singularity spectrum of the radiance Landsat images as computed with the 2D WTMM method using a first-order (a) or a third-order (b) radially symmetric analyzing wavelet. Same computations as in Figures 9c. The solid lines correspond to the log-normal quadratic spectrum with $m = -0.38 \ln 2$, $\sigma^2 = 0.08 \ln 2$ (a) and $m = -0.366 \ln 2$, $\sigma^2 = 0.06 \ln 2$ (b). The $D(h)$ singularity spectrum of velocity (dotted lines) and temperature (dashed lines) fluctuations in fully developed turbulence are shown for comparison.

the multifractal $D(h)$ spectrum of the temperature fluctuations recorded in a $R_\lambda = 400$ turbulent flow [56]. The corresponding single humped curve is definitely much wider than the velocity $D(h)$ spectrum and it is rather close to the data corresponding to the marine Sc radiance field. It is well recognized however that liquid water is not really passive and that its identification with a passive component in atmospheric dynamics offers limited insight into cloud structure since, by definition, near-saturation conditions prevail and latent heat production affects buoyancy. So cloud microphysical processes are expected to interact with the circulation at some, if not all, scales [57]. Nevertheless, our results in Figure 12 tell us that from a multifractal point of view, the intermittency captured by the Landsat satellite looks statistically equivalent to the intermittency of a passive scalar in fully-developed 3D turbulence. The fact that the internal structure of Sc cloud somehow reflects some statistical properties of atmospheric turbulence is not such a surprise in this highly turbulent environment.

Finally, with respect to the issue of cloud modeling, it comes out quite naturally from the WTMM analysis of marine Sc Landsat data, that the 2D random $\mathcal{W}$-cascade models introduced in Ref. [28], are much more realistic hierarchical models than commonly used multifractal models like the fractionally integrated singular cascade [22, 25, 47] or the bounded cascade models [58]. We are quite optimistic in view of using the log-normal $\mathcal{W}$-cascade models with realistic parameter values for radiation transfer simulations. To our opinion, random $\mathcal{W}$-cascade models are a real breakthrough, not only for the general purpose of image synthesis, but more specifically for cloud modeling. It is likely that better cloud modeling will make further progress in our understanding of cloud-radiation interaction possible.

ACKNOWLEDGMENTS

We are very grateful to R.F. Cahalan, A. Davis, A. Marshak, L. Oreopoulos and J.F. Muzy for helpful discussions. We are very indebted to Y. Gagne, Y. Malecot and S. Ciliberto for the permission to use their experimental turbulent signals. This work was supported by NATO (Grant n^o CRG 960176) and was performed while S.G. Roux held a National Research Council–NASA/GSFC Research Associateship.

REFERENCES

1. Monin, A. S., and Yaglom, A. M., *Statistical Fluid Mechanics*, vol. 2, MIT Press, Cambridge, MA, 1975.
2. Frisch, U., *Turbulence*, Cambridge Univ. Press, Cambridge, 1995.
3. Lovejoy, S., *Science*, **216**, 185 (1982).
4. Wilkinson, G. G., Kanellopoulos, J., and Megier, J., editors, *Fractals in Geoscience and Remote Sensing*, Image Understanding Research Series, vol.1, ECSC-EC-EAEC, Brussels, Luxemburg, 1995.
5. Ramanatahn, V., Cess, R. D., Harrison, E. F., Minnis, P., Barkston, B. R., Ahmad, E., and Hartmann, D., *Science*, **243**, 57 (1989).
6. Cess, R. D., et al., *Science*, **245**, 513 (1989).
7. Davis, A., Marshak, A., Wiscombe, W. J., and Cahalan, R. F., *J. Atmos. Sci.*, **53**, 1538 (1996).
8. Marshak, A., Davis, A., Wiscombe, W. J., and Cahalan, R. F., *J. Atmos. Sci.*, **54**, 1423 (1997).
9. Welch, R. M., Kuo, K. S., Wielicki, B. A., Sengupta, S. K., and Parker, L., *J. Appl. Meteor.*, **27**, 341 (1988).
10. Cahalan, R. F., and Joseph, J. H., *Mon. Wea. Rev.*, **117**, 261 (1989).
11. Sèze, G., and Smith, L., "", in *Proc. Seventh Conf. on Atmospheric Radiation, Amer. Meteor. Soc., San Francisco, CA*, 1990, p. 47.
12. Cahalan, R. F., and Snider, J. B., *Remote Sens. Environ.*, **28**, 95 (1989).
13. Baker, H. W., and Davies, J. A., *Remote Sens. Environ.*, **42**, 51 (1992).
14. Lovejoy, S., Schertzer, D., Silas, P., Tessier, Y., and Lavallée, D., *Ann. Geophysicae*, **11**, 119 (1993).
15. Davis, A., Marshak, A., Cahalan, R. F., and Wiscombe, W. J., *J. Atmos. Sci.*, **54**, 241 (1997).
16. Mandelbrot, B. B., *Fractals : Form, Chance and Dimensions*, Freeman, San Francisco, 1977; *The Fractal Geometry of Nature*, Freeman, San Francisco, 1982.
17. Corssin, S., *J. Appl. Phys.*, **22**, 469 (1951).
18. Obukhov, A., *Izv. Akad. Naut. SSSR, Ser. Geogr. I. Geofiz.*, **13**, 55 (1949).
19. Kraichnan, R. H., *Phys. Fluids*, **10**, 1417 (1967).
20. Arneodo, A., Decoster, N., and Roux, S. G., *Eur. Phys. J. B*, **15**, 567 (2000).
21. Schertzer, D., and Lovejoy, S., *Phys. Chem. Hyd. J.*, **6**, 623 (1985).
22. Schertzer, D., and Lovejoy, S., *J. Geophys. Res.*, **92**, 9693 (1987).
23. Schertzer, D., and Lovejoy, S., *Fractals in Geoscience and Remote Sensing*, p. 11, vol. 1 of [4] (1995).
24. Lovejoy, S., and Schertzer, D., *Fractals in Geoscience and Remote Sensing*, p. 102, vol. 1 of [4] (1995).
25. Schertzer, D., Lovejoy, S., Schmitt, F., Ghigisinskaya, Y., and Marsan, D., *Fractals*, **5**, 427 (1997).
26. Davis, A., Marshak, A., Wiscombe, W. J., and Cahalan, R. F., *J. Geophys. Res.*, **99**, 8055 (1994).
27. Wiscombe, W. J., Davis, A., Marshak, A., and Cahalan, R. F., in *Proc. of the Fourth Atmospheric Radiation Measurement (ARM) Science Team Meeting, Charleston, U.S. Dept. of Energy*, 1995, p. 11.
28. Decoster, N., Roux, S. G., and Arneodo, A., *Eur. Phys. J. B*, **15**, 739 (2000).

29. Arrault, J., Arneodo, A., Davis, A., and Marshak, A., *Phys. Rev. Lett.*, **79**, 75 (1997).
30. Arneodo, A., Decoster, N., and Roux, S. G., *Phys. Rev. Lett.*, **83**, 1255 (1999).
31. Roux, S. G., Arneodo, A., and Decoster, N., *Eur. Phys. J. B*, **15**, 765 (2000).
32. Muzy, J. F., Bacry, E., and Arneodo, A., *Phys. Rev. E*, **47**, 875 (1993).
33. Meyer, Y., and Roques, S., editors, *Progress in Wavelets Analysis and Applications*, Editions frontières, Gif-sur-Yvette, 1993.
34. Mallat, S., *A Wavelet Tour in Signal Processing*, Academic Press, New York, 1998.
35. Marr, D., *Vision*, W. H. Freemann and Co, San-Francisco, 1982.
36. Mallat, S., and Zhong, S., *IEEE Trans. on Pattern Analysis and Machine Intelligence*, **14**, 710 (1992).
37. Mallat, S., and Hwang, W. L., *IEEE Trans. on Information Theory*, **38**, 617 (1992).
38. Canny, J., *IEEE Trans. Patt. Anal. Machine Intell.*, **8**, 679 (1986).
39. Farmer, J. D., Ott, E., and Yorke, J. A., *Physica D*, **7**, 153 (1983).
40. Argoul, F., Arneodo, A., Elezgaray, J., Grasseau, G., and Murenzi, R., *Phys. Rev. A*, **41**, 5537 (1990).
41. Muzy, J. F., Bacry, E., and Arneodo, A., *Phys. Rev. Lett.*, **67**, 3515 (1991).
42. Muzy, J. F., Bacry, E., and Arneodo, A., *Int. J. of Bifurcation and Chaos*, **4**, 245 (1994).
43. Arneodo, A., Bacry, E., and Muzy, J. F., *Physica A*, **213**, 232 (1995).
44. Mandelbrot, B. B., and Van Ness, J. W., *S.I.A.M. Rev.*, **10**, 422 (1968).
45. Peitgen, H. O., and Saupe, D., editors, *The Science of Fractal Images*, Springer Verlag, New York, 1987.
46. Lévy, P., *Processus Stochastiques et Mouvement Brownien*, Gauthier-Villars, Paris, 1965.
47. Davis, A., Marshak, A., Wiscombe, W. J., and Cahalan, R. F., *Topics in Nonstationary Analysis*, World Scientific, Singapore, 1996, p. 97.
48. Davis, A., Lovejoy, S., and Schertzer, D., *Scaling, Fractals and Nonlinear Variability in Geophysics*, Kluwer, Dordrecht, 1991, p. 303.
49. Cahalan, R. F., Ridgway, W., Wiscombe, W. J., Gollmer, S., and Harshvardan, M., *J. Atmos. Sci*, **51**, 3776 (1994).
50. Cahalan, R. F., *Advances in Remote Sensing and Retrieval Methods*, Deepak Pub, Hampton, 1989, p. 371.
51. Cox, S., McDougal, D., Randall, D., and Schiffer, R., *Bull. Amer. Meteor. Soc.*, **68**, 114 (1987).
52. Arneodo, A., Manneville, S., and Muzy, J. F., *Eur. Phys. J. B*, **1**, 129 (1998).
53. Arneodo, A., Manneville, S., Muzy, J. F., and Roux, S. G., *Phil. Trans. R. Soc. London A*, **357**, 2415 (1999).
54. Roux, S. G., Muzy, J. F., and Arneodo, A., *Eur. Phys. J. B*, **8**, 301 (1999).
55. Kolmogorov, A. N., *C. R. Acad. Sci. USSR*, **30**, 301 (1941).
56. Ruiz-Chavarria, G., Baudet, C., and Ciliberto, S., *Physica D*, **99**, 369 (1996).
57. Meong, C. H., Cotton, W. R., Bretherton, C., Chlond, A., Khairoutdinov, M., Krueger, S., Lewellen, W. S., McVean, M. K., Pasquier, J. R. M., Rand, H. A., Siebesma, A. P., Stevens, B., and Sykes, R. I., *Bull. Am. Meteor. Soc.*, **77**, 261 (1996).
58. Marshak, A., Davis, A., Cahalan, R. F., and Wiscombe, W. J., *Phys. Rev. E*, **49**, 55 (1994).

NEUROSCIENCE AND PHYSIOLOGY

Control of Seizing Neuronal Networks

Bruce J. Gluckman

Krasnow Institute and Department of Physics and Astronomy
George Mason University, Fairfax, VA 22030-4444

Abstract. Earlier work [1] applied chaos control algorithms to modulate the timing of neuronal population burst events with the idea that changing such interictal dynamic could be used to prevent seizures. This work utilized classical neurophysiological stimulation techniques, specifically pulse-like current injections, which were limited to exciting new burst events. Building on this work, we have shown that a broadly applied electric field can be used as a more versatile control parameter for interacting with neuronal neurons, able to proportionately excite or suppress average neuronal excitability of large groups of neurons. I'll discuss both the physics underlying the electric field interaction with neurons, and our recent work [2] interacting with and controlling seizure activity in live neuronal networks.

In 1994, S. Schiff, *et al.* [1], demonstrated the first application of chaos control to neuronal activity. This was, to our knowledge, the first demonstration of a nonlinear adaptive control scheme for spontaneously active neural networks. In this work, the timing of population bursts from hippocampal slices, which is typically quite varied from event to event, was stabilized using current pulses to the Schaffer collateral tract of the network. The timing of the control pulses was determined based on an algorithm derived from the original Ott-Grebogi-Yorke (OGY) chaos control algorithm [3] for stabilizing unstable periodic orbits (UPOs). UPOs are deterministic structures in the underlying dynamics of a chaotic system. Although the Schiff *et al.* work demonstrated dynamical control, it had two basic shortcomings. First, the stimuli applied could not be classified as small nor continuously graded perturbations to a parameter of the system. Each current pulse immediately elicited a new burst event, and the degree to which it altered the system-state was uncharacterized. The second shortcoming was the degree of rigor for detecting the existence of UPOs in the unperturbed dynamics.

In an effort to answer the second of these shortcomings, we have developed [4,5] a sophisticated algorithm that allows us to identify UPOs from short noisy time series with confidence limit estimates with respect to null hypotheses. In So, *et al.* [6], we have more conclusively demonstrated UPO existence in activity measured from single cells, small networks, and intact cortex.

In order to find an alternate perturbation, one that would approximate a continuous system parameter with graded response to changes, we have been investigating the effects of an applied electric field on neuronal network activity. It has long been

CP622, *Experimental Chaos: 6th Experimental Chaos Conference*, edited by S. Boccaletti et al.

known that electric fields affect neuronal excitability. In 1927, Rushton showed that an electric field could be used to influence the threshold for excitation of a frog sciatic nerve [7]. He found that the reduction in threshold was directly related to the cosine of the angle between the field orientation and the nerve. In 1940, Katz & Schmitt investigated the interaction of two nearby nerve fibers [8]. They observed that action potential propagation along one fiber would affect the propagation of action potentials along the next, and concluded that this effect could be explained by their shared electric field. These studies were followed by measurements of electric field modulation of neuronal firing rate [9]. Further studies showed externally applied electric fields could suppress or enhance stimulus evoked neuronal activity in laboratory experiments [10-12]. Nicholson and colleagues [13-15] worked out the physics of this interaction in detail. Basically, the field polarizes a neuron, shifting the transmembrane potential at the site of action potential initiation either closer or farther away from threshold. The degree of polarization depends on the alignment of the field with the bulk of the neuron's dendrites and the amplitude of the field.

We have demonstrated [16] that continuous (DC) electric fields of modest amplitude can be used to entirely suppress epileptiform activity in hippocampal slices (Fig. 1). A specially designed chamber and associated electronics were constructed in order to apply a uniform field across the slice. Slices were arranged in the field so that on average the pyramidal neurons were aligned with their primary dendrite to soma axis parallel to the field.

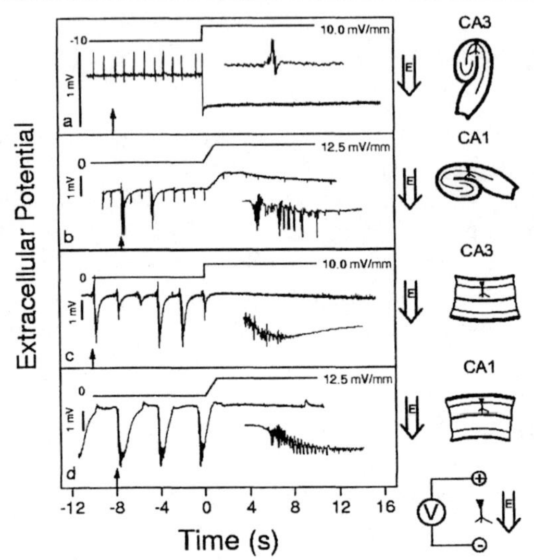

Figure 1: Suppression of epileptiform network activity (bottom trace, each panel) as a function of applied electric field (upper trace, each panel) for different functional regions and slice preparations of rat hippocampus. The inset at right is a single event in with an expanded time base. (a) CA3 region, transverse slice. (b) CA1 region, transverse slice. (c) CA3 region, longitudinal slice. (d) CA1 region, longitudinal slice. (Adapted from Reference [16]).

Differential DC recordings of the activity were made with both the measurement and reference electrodes aligned on nearly the same isopotential of the applied field in order to minimize recording artifacts from the applied field. This was combined with a concerted effort in the design of the field application electronics to isolate the measurement ground from the field application ground to further minimize common mode signals. The result was a system in which small potential variations due to neuronal activity were measurable in the presence of considerable electric fields with minimal stimulus artifacts.

Examples of suppression are shown in Fig. 1 for different neuronal layers and slice preparations. In each panel, the upper trace is the applied field, the lower trace is recorded activity, and the inset is a single event (marked by an arrow) at an expanded time base. In each case, the activity was readily suppressed with the applied field. To our knowledge, this is the first demonstration of electric field suppression of spontaneous epileptiform activity.

The effect of the electric field on neuronal firing rate is not simply a threshold type interaction. Instead, it is a graded response, proportional to the amplitude and sign of the applied field. This graded response is illustrated in Fig. 2. Here, the mean burst rate for the same network as in Fig. 1a is shown as a function of time in response to different field amplitudes and sign. For positive field amplitudes, the larger the field, the slower the event rate. Likewise, for negative field amplitudes, the larger the field the higher the event rate.

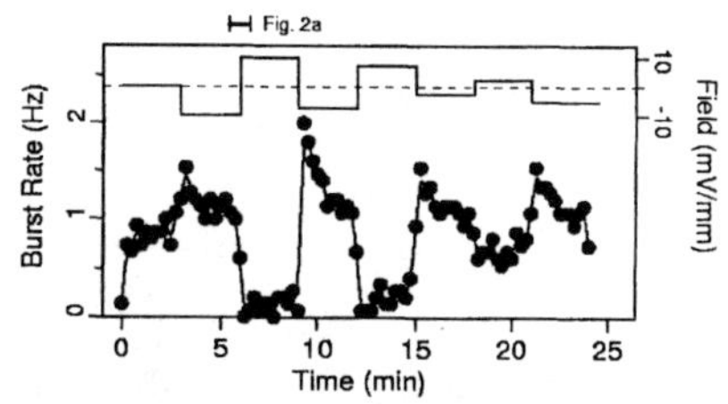

Figure 2: Burst rate (lower trace) as a function of applied field amplitude (upper trace). Data for the same slice and region as Fig. 1(a). A graded response is observed as a function of field amplitude. (Adapted from Ref. [16]).

In a novel experimental application of this technology, we used electric field modulation to produce the first demonstration of stochastic resonance (SR) in mammalian brain [17,18]. SR is a phenomenon in which the response of a nonlinear system to a signal can be enhanced by noise. As part of the investigation and explanation of this work, we also modified a computational model of a network of pyramidal neurons to realistically include the applied electric field. Using this computational model, we were able to reproduce in the model the effects observed in the actual experiment [18].

In the experiment of Fig. 2, the effect of the electric field diminished during steady state (DC) application. Much of this decrement can be attributed to a long-term nonstationarity of the field amplitude, caused by slow electrode polarization. We have redesigned the electronics to eliminate this effect. The design, utilizes four electrodes – two for applying the field and two for measuring its magnitude – in order to eliminate the slow electrode polarization. To our knowledge, this is the first implementation of this simple solution for large-scale field application in neurophysiological endeavors. Despite this modification in chamber design, when we repeat these experiments we continue to observe some decay in the effectiveness of the applied field. This residual effect can be attributed to a combination of tissue polarization and neuronal accommodation. All three of these effects must be taken into account in the engineering design factors when applying this technology to neural prosthetics.

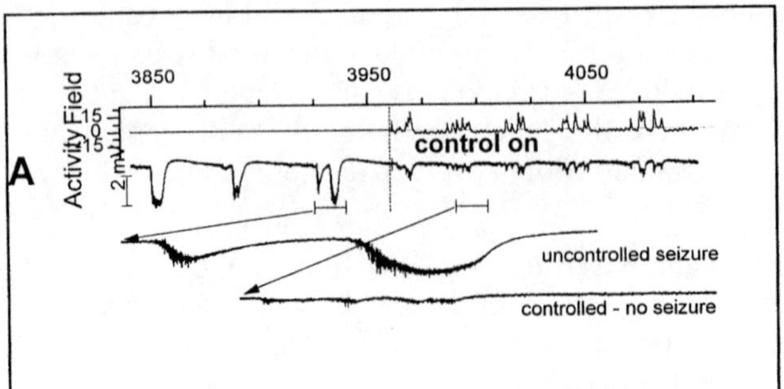

Figure 3: Adaptive electric field feedback control of seizure activity in in-vitro hippocampal slice activity. In A are shown measured field potentials (lower trace) and the applied field (upper trace). The large deflections in field potential prior to control being turned on are seizure-like events. All such events were averted once control was turned on, though some activity continued. Five seconds of activity is shown below for an uncontrolled seizure and controlled dynamics. No seizures were observed for the 16 minute duration of applied control in contrast to the uncontrolled seizure rate of rate of approximately once per minute (Adapted from Gluckman, et al., 2001). Note, the field potentials were recorded with microelectrodes and contain a activity frequencies from DC-1000 Hz, and therefore look different from standard human EEGs of seizures

In a recent paper [2], we demonstrated that a relatively simple proportional feedback algorithm based on a measure of the activity level and incorporating the elements described above is quite successful in controlling seizures in in-vitro experiments. We used the signal power in the frequency range of 100-300 Hz as a measure of activity, and developed an algorithm that both targeted a median activity level and minimized the average field applied. This successfully redirected the group network behavior from seizing approximately once a minute to not seizing at all, while allowing local neuronal activity to continue with minimal interference. An example of such control is shown in Fig. 3. Differences observed in the system dynamics under different control algorithms are detailed in Reference [2], and we hope will lead to better understanding of both the phenomena underlying seizure and how to control them

We have now begun investigating electric field modulation in acute whole-animal models. The physics of the interaction is unchanged in going from slice to whole brains – an electric field aligned along the long-axis of the neurons (perpendicular to the cell body layer) will polarize the neurons, thereby making them all slightly more or slightly less excited. An example of clear electric field modulation of activity is shown in Fig. 4 from intact hippocampus. Rats are deeply anesthetized with a ketamine-xylazine mixture, a craniotomy performed, and the left neocortex removed to expose the hippocampus. A small amount of kainic acid was then micro-injected into the hippocampus to make it epileptigenic. Finally, electric field generating plates were placed, along with bipolar recording electrodes.

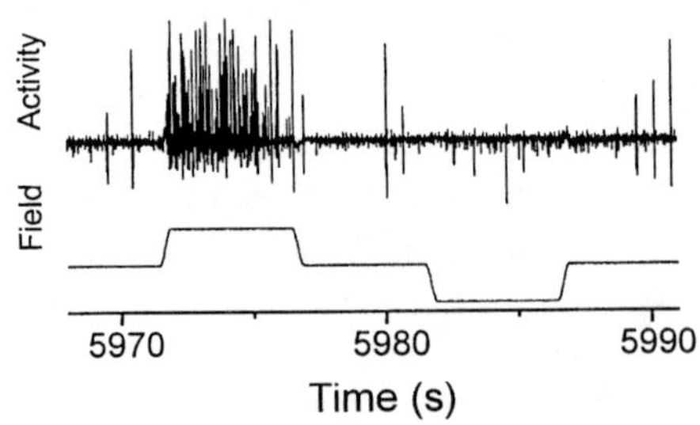

Figure 4: Electric field modulation of interictal activity in intact hippocampus during acute whole animal trial.

Special field-generating electronics were designed to create the electric field, along with differential recording preamplifiers for measurement in the presence of the field. As expected, clear modulation of ongoing activity in intact brain is observed with electric field application.

In conclusion, we have been working over the past few years to develop techniques for interacting with live neuronal network dynamics using electric fields. The field interaction works by polarizing the large outflow neurons, effectively making them slightly more or less excitable in proportion to the polarization. This modulation can be continuously applied over a broad range of time scales, from milliseconds to minutes, without interfering with the capability to monitor ongoing activity. Therefore, this tool provides a very precise and sensitive tool for modulating, probing, and controlling neuronal dynamics.

ACKNOWLEDGMENTS

The above research was funded under support by the National Institutes of Mental Health, the Whitacker Foundation, and the National Science Foundation.

REFERENCES

1. Schiff SJ, Jerger K, Duong DH, Chang T, Spano ML and Ditto WL. *Nature* **370**, 315-620 (1994).
2. Gluckman BJ, Nguyen H, Weinstein SL and Schiff SJ. *J. Neuroscience* **21**, 590-600 (2001).
3. Ott, E, Grebogi, C, and Yorke, JA. *Phys. Rev. Lett.* **64**, 1196-1199 (1990).
4. So, P, Ott, E, Schiff, SJ, Kaplan, DT, Sauer, T, and Grebogi, C. *Phys. Rev. Lett.* **76**, 4705-4708 (1996).
5. So, P, Ott, T, Sauer, T, Gluckman, BJ, Grebogi, C, and Schiff, SJ. *Phys. Rev. E* **55**, 5398-5417 (1997).
6. So, P, Francis, J, Netoff, TI, Gluckman, BJ, and Schiff, SJ. *Biophys. J.* **74**, 2776-2785 (1998).
7. Rushton, WAH. *J. Physiol. (Lond.)* **63**, 357-377 (1929).
8. Katz, B and Schmitt, OH. *J. Physiol. (Lond.)* **63**, 357-377 (1940).
9. Terzoulo, CA, and Bullock, TH. **Proc. Nat. Acad. Sci. 42**, 687-694 (1956).
10. Jefferys, JGR. *J. Physiol. (Lond.)* **319**, 143-152 (1981).
11. Bawin, SM, Sheppard, AR, Mahoney, MD, Abu-Assal, M, and Adey, WR. *Brain Res.* **362**, 350-354 (1986).
12. Bawin, SM, Abu-Assal, ML, Sheppard, AR, Mahoney, MD, and Adey, WR. *Brain Res.* **399**, 194-199 (1986).
13. Tranchina, D, and Nicholson, C. *Biophys. J.* **50**, 1139-1159 (1986).
14. Chan, CY, Nicholson, C. *J Physiol (Lond)* **371**, 89-114 (1986).
15. Chan, CY, Houndsgaard, J, and Nicholson, C. *J Physiol (Lond)* **402**, 751-771 (1988).
16. Gluckman, BJ, Neel, EJ, Netoff, TI, Ditto, WL, Spano, ML, and Schiff, SJ. *J. Neurophysiol.* **76**, 4202-4205 (1996).
17. Gluckman, BJ, Netoff, TI, Neel, EJ, Ditto, WL, Spano, ML, and Schiff, SJ. *Phys. Rev. Lett.* **77**, 4098-4101 (1996).
18. Gluckman, BJ, So, P, Netoff, TI, Spano, ML, Schiff, SJ. *Chaos* **8**, 588-598 (1998).

Statistical Versus Individual Forecasting Of Life-Threatening Cardiac Arrhythmias

Niels Wessel*, Udo Meyerfeldt†, Christine Ziehmann*, Alexander Schirdewan†, Jürgen Kurths*

* *University of Potsdam, Postfach 601553, 14415 Potsdam, Germany,*
† *Franz-Volhard-Hospital, Humboldt-University, Wiltbergstrasse 50, 13125 Berlin, Germany*

Abstract. Ventricular tachycardia or fibrillation (VT) as fatal cardiac arrhythmias are the main factors triggering sudden cardiac death. The objective of this investigation is to find early signs of sustained VT in patients with an implanted cardioverter-defibrillator (ICD). These devices are able to safeguard patients by returning their hearts to a normal rhythm via strong defibrillatory shocks; additionally, they are able to store at least 1000 beat-to-beat intervals immediately before the onset of a life-threatening arrhythmia. We study these 1000 beat-to-beat intervals of 63 chronic heart failure ICD patients before the onset of a life-threatening arrhythmia and at a control time, i.e. without VT event. To characterize these rather short data sets, we calculate heart rate variability (HRV) parameters from time and frequency domain, from symbolic dynamics as well as the finite-time growth rates. We find that no linear parameter shows significant differences in HRV between the VT and the control time series. However, the nonlinear parameters detected a significant increase in short phases with low variability before the onset of VT ($p<0.05$, for time series with less than 10% ectopy). Finally, we are investigating whether these results may lead to individual predictions of VT.

INTRODUCTION

Implantable cardioverter defibrillators (ICD) are a safe and effective treatment for ventricular tachycardia (VT) [1,2]. Third generation ICD offer not only important advances in arrhythmia treatment, but also permit a correct characterization of the rhythm leading to intervention [3,4]. This feature has been extremely helpful in systematically investigating the natural history of spontaneous VT. Previous VT studies depended on randomly recorded episodes during Holter monitoring. These studies captured only relatively small numbers of VT episodes from patients with various heart diseases [5,6]. In contrast, ICD studies have examined the electrocardiographic initiating patterns, especially the occurrence of premature beats and special beat sequences [7-9]. These studies were based on short term intracardiac ICD electrocardiogram storage. Newer ICD systems offer not only a short term intracardiac electrocardiogram, but also an RR interval memory up to 1024 consecutive beat-to-beat intervals before the start of VT. This allows a heart rate variability (HRV) analysis of many spontaneous VT episodes to estimate the influence of the autonomic nervous system on the origin of VT. Autonomic nervous system tone has a significant bearing on VT development [10,11]. The hypothesis of our study was that VT can be predicted by RR interval analysis. By identification of

CP622, *Experimental Chaos: 6th Experimental Chaos Conference*, edited by S. Boccaletti et al.

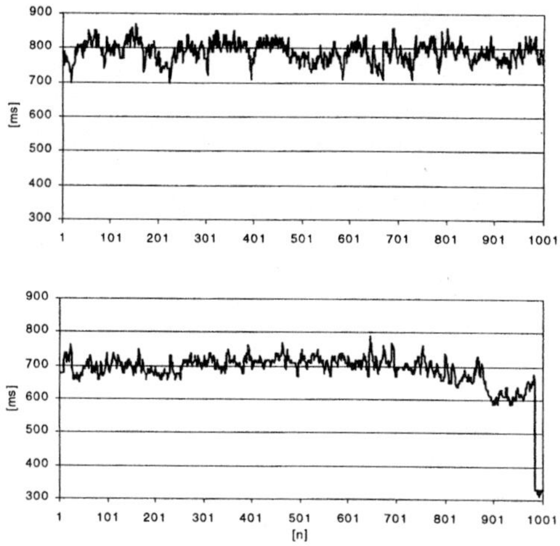

FIGURE 1. 1000 beat-to-beat intervals before a sustained VT (below) and the respective control time series (above) from the same patient. For HRV analysis all beat-to-beat intervals of the VT itself at the end of time series are removed from the tachograms.

specific RR-interval trigger patterns, we assume it may be possible to identify specific VT predictors in patients with an ICD for new preventative strategies. We therefore calculate different linear and nonlinear HRV parameters from ICD stored VT episodes [12-15].

METHODS

Patients

We study the ICD stored 1000 beat-to-beat intervals before the onset of 67 VT episodes and at 47 control intervals without VT in 46 ICD patients of the Franz-Volhard-Hospital with severe congestive heart failure. We analyze the time series just before the onset of VT and at an arbitrarily selected control period without an arrhythmic event stored just before a regular ICD follow up examination. Figure 1 shows an example of 1000 beat-to beat intervals prior to a sustained VT as well as a control tachogram from the same patient.

Standard Heart Rate Variability Analysis

To detect early signs of life-threatening VT, we apply a multiparametric analysis. Before starting the analysis, ventricular premature beats and artifacts usually should be removed from the time series to construct the so-called "normal-to-normal" beat time series (NN). We use the adaptive filtering algorithm [16] for preprocessing of the data. The following standard HRV parameters from time domain [12] are calculated from all corrected time series: meanNN, sdNN, pNN50 and rmssd. In addition, the standard parameters VLF, LF, HF, and P are calculated from the frequency domain [12]. The spectra are estimated using a fast Fourier transformation. To avoid any "leakage" effect, a Blackman Harris window function is applied. The following ratios are included in the analysis: VLF/P, HF/P, and LF/HF, as well as LFn the normalized low frequency band. These standard parameters of HRV analysis are based on linear techniques. To classify dynamic changes in the time series, we present the following nonlinear concepts of symbolic dynamics and the finite-time growth rates outlined below.

Symbolic Dynamics

The first step of this concept is the transformation of the time series into symbol sequences with symbols from a given alphabet. Some details are lost in this process; however, the advantage is that the coarse dynamic behavior can be analyzed [14,15]. The used parameter Polvar10 (**Pro**bability of **l**ow **var**iability, **10** ms difference) characterizes short phases of low variability from successive symbols of a simple alphabet, consisting of only the symbols 0 and 1, where 0 stands for a small difference of less than 10 ms between two successive RR intervals (10 ms is the resolution of the defibrillators used in this study), and where 1 represent cases when the difference between two successive RR intervals exceeds this limit, specifically given the time series $x_1, x_2, \ldots, x_N$ one obtains the symbol series s_n

$$s_n = \begin{cases} 1: & |x_n - x_{n-1}| \geq 10\,\text{ms} \\ 0: & |x_n - x_{n-1}| < 10\,\text{ms}. \end{cases} \tag{1}$$

Words consisting of the unique symbols all 0 or all 1 were counted. To obtain a statistically robust estimate of the word distribution, we restricted ourselves to the 64 words defined by six consecutive symbols. Polvar10 represents the probability of the word 000000 occurring and thus detects even intermittently decreased HRV. We introduced this parameter with the knowledge that a decreased HRV is a risk marker in cardiac patients [12]. The standard parameters of HRV analysis such as sdNN and rmssd are not able to detect short phases with a decreased HRV.

Finite Time Growth Rates

The second approach to characterize dynamic behavior are the finite time growth rates which are derived from the concept of Lyapunov exponents [17,18]. For the calculation of the growth rates, the state of the system is reconstructed using delay coordinates [19]. In this phase space, the actual dynamic behavior can be represented by a point and the time evolution of the system by a trajectory. To analyze the stability or predictability of a current situation over a given duration T, its time evolution is compared with the time evolution of the closest analogous situations found in the past, namely, one searches the nearest spatial neighbor $\overline{X}_k$ of the situation X_k in the reconstructed space and observes how these two trajectories diverge in time [12]. From the original distance of both states and the distance of the evolved states X_k^T and $\overline{X}_k^T$ after T steps we calculate the finite-time growth rate

$$\lambda_k^{(n,\tau,T)} = \frac{1}{T} \ln \frac{\left\| X_k^T - \overline{X}_k^T \right\|}{\left\| X_k - \overline{X}_k \right\|}, \text{ k=0,..,N-(n-1)}\tau, \qquad (2)$$

where n is the embedding dimension and τ the delay. This value quantifies the local short-term predictability for each situation or point X_k. If the values are positive, the distances increase with time and the state is unstable. The larger this $\lambda_k^{(n,\tau,T)}$, the less predictable is the situation. Negative values reflect highly predictable situations. As we have a value of $\lambda_k^{(n,\tau,T)}$ for each X_k, we can consider it as a growth rate time series. Averaging over k leads to the average growth rate $\lambda^{(n,\tau,T)}$ which quantifies a global short-term predictability. We determined the rate of growth from the average behavior of the five nearest neighbors to reduce noise effects. The embedding dimension is 9 and the delay time is 1, i.e. a state is defined by 9 consecutive values of the time series. The average finite time growth rate for an evolution time $T=1$ is denoted as Fitgra9 (**Fi**nite **t**ime **gr**owth **ra**tes, dimension **9**). For heart rate time series this parameter can be interpreted as a measure of regularity. The smaller this value, the larger is the number of epochs with a regular or predictable short term dynamics in the HRV time series indicating a loss of short term variability.

RESULTS

Comparing the HRV parameters of all 67 VT and 47 control time series, only the mean sinus rhythm cycle length shows differences, the time series leading to VT had a significantly shorter cycle length (meanNN 694.4±138.1 ms), than the control time series (meanNN 760.8±140.2 ms). The ectopy time, calculated as the sum of the coupling interval and the following pause of all premature beats, does not differ between the VT and control time series.

Since the filter methods in time series with a large number of premature beats are relatively ineffective we calculate the HRV parameters in a second step for 36 VT and 34 control time series with less than 10% of premature beats, as given in table 1. The time and frequency domain HRV parameters show no significant differences between VT and control time series. However, the parameters of nonlinear HRV analysis

TABLE 1. The results of 34 control and 36 VT filtered time series, with less than 10 % premature beats (p – significance, n.s. not significant p≥0.05, Mann-Whitney U-test).

	Control (n=34)	VT (n=36)	p
meanNN	731.73±122.24	683.32±157.24	n.s.
sdNN	49.00±25.13	49.09±35.83	n.s.
pNN50	0.018±0.022	0.014±0.025	n.s.
rmssd	16.50±8.37	13.74±7.79	n.s.
VLF/P	0.52±0.17	0.54±0.15	n.s.
LH/HF	3.28±1.54	3.93±4.02	n.s.
HF/P	0.04±0.04	0.05±0.04	n.s.
LFn	0.73±0.12	0.72±0.15	n.s.
Polvar10	0.05±0.05	0.13±0.17	<0.05
Fitgra9	0.17±0.06	0.14±0.04	<0.05

Polvar10, characterizing short phases of low variability, and Fitgra9, the finite-time growth rates, exhibit significant differences.

Figure 2 demonstrates one possible onset of VT, where the 1000 RR-intervals before different VT's of the same patient are shown. The increase in heart rate and thus the decrease in HRV in these cases is obvious. The onset of these VT's is triggered by sinus tachycardia and not by ventricular premature contractions (VPC). A second possible onset (not shown here) is a VT initiation based on a special 'short-long-short' beat sequence [8]. In this case a VPC is followed by a normal beat and a second VPC. We find this initiation pattern in different VT episodes of the same patient – the 'short-long-short' behavior, however, also occurs in between the time series (thus not directly leading to VT). However, characteristically for both onsets is the intermittently decreased HRV.

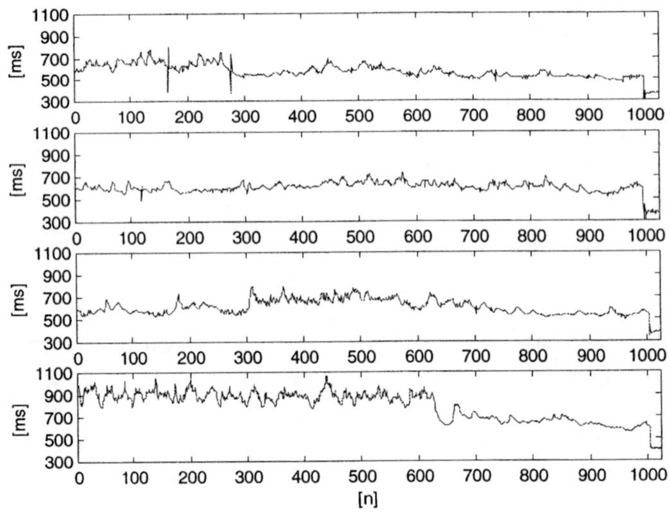

FIGURE 2. Four different ways to VT of the same patient. Depicted are the 1000 RR-intervals before the onset of different VT, which are triggered by sinus tachycardia in all of the cases.

DISCUSSION

We identify significant differences in the dynamic behavior of beat-to-beat intervals between the VT and control time series by means of nonlinear dynamics. Especially, the parameters Polvar10 and Fitgra9 reflect increased short phases with low variability in patients with congestive heart failure. This result demonstrates that a loss of short-term variability precedes the onset of VT [12]. Similar findings were reported by Mäkikallio and Skinner, who estimated other parameters of nonlinear dynamics during Holter monitoring in post myocardial infarction patients [6,20]. In accordance with other authors reporting HRV analyses before VT during Holter monitoring [5-6,21-23], we can not find significant differences in classical time or frequency domain parameters in our HRV analyses.

Considering these results, we notice that all patients received the currently available best medical therapy, especially with high percentage of beta blockers and ACE inhibitors employed in our study. Despite of the favorable effects of beta blockers on heart rate variability [24,25] it is none-the-less possible to identify the periods preceding life-threatening VT in a high risk patient group by the described methods of nonlinear dynamics. This identification, however, only works for time series with low ectopy and for a statistical framework. Individual predictions are hard to be done – we are working on this topic and trying other promising nonlinear methods to enable individual forecasts.

Limitations of our study are the relatively small number of time series and the subsequently limited statistical analysis in terms of subdivisions concerning age, sex, and heart disease. Thus, these results must be validated with a larger data base. The first studies are now in progress to prevent VT by atrioventricular stimulation to shorten the pause after premature beats [26]. Moreover, heart rate dynamics prior to VT onset in terms of drug treatment are currently being investigated [27]. Our data offer the possibility of developing automatic ICD algorithms based on nonlinear dynamic HRV parameters. This approach may permit the automatic short–term prediction of VT and the initiation of earlier modes of prevention.

REFERENCES

1. Mirowski, M., Reid, P.R., Mower, M.M., et al. *N Engl J Med* **303**, 322-324 (1980)
2. The antiarrhythmic versus implantable defibrillators (AVID) investigators, *N Engl J Med* **337**, 1576-1583 (1997)
3. Marchlinski, F.E., Callans, D.J., Gottlieb, C.D., et al. *J Cardiovasc Electrophysiol* **6**, 832-851 (1995)
4. Gronefeld, G.C., Mauss, O., Li, Y.G., et al. *J Cardiovasc Electrophysiol* **11**, 1208-1214 (2000)
5. Berger, M.D., Waxman, H.L., Buxton, A.E., et al. *Circulation* **78**, 885-892 (1988)
6. Makikallio, T.H., Koistinen, J., Jordaens, L., et al. *Am J Cardiol* **83**, 880-884 (1999)
7. Roelke, M., Garan, H., McGovern, B.A., et al. *J Am Coll Cardiol* **23**, 117-122 (1994)
8. Meyerfeldt, U., Schirdewan, A., Wiedemann, M., et al. *Eur Heart J* **18**, 1956-1965 (1997)
9. Saeed, M., Link, M.S., Mahapatra, S., et al. *Am J Cardiol* **85**, 580-587 (2000)
10. Schwartz, P.J., La Rovere, M.T., Vanoli, E. *Circulation* **85**, I77-I91 (1992)
11. Myerburg, R.J., Kessler, K.M., Castellanos, A. *Circulation* **85**, I2-I10 (1992)
12. Wessel, N., Ziehmann, C., Kurths, J., et al. *Phys Rev E* **61**, 733-739 (2000)
13. Task Force: Heart rate variability, *Circulation* **93**, 1043-1065 (1996)
14. Kurths, J., Voss, A., Witt, A., et al. *CHAOS* **5**, 88-94 (1995)
15. Voss, A., Kurths, J., Kleiner, H.J., et al. *Cardiovasc Res* **31**, 419-433 (1996)
16. Wessel, N., Voss, A., Malberg, H., et al. *Herzschr Elektrophys* **11**, 159-173 (2000)
17. Oseledec, V.I. *Transactions of the Moscow Mathematical Society* **19**, 197-231 (1968)
18. Wolf, A., Swift, J.B., Swinney, H.L., et al. *Physica D* **16**, 285-317 (1985)
19. Takens, F. *Lecture Notes in Mathematics* **898**, 366-381 (1981)
20. Skinner, J.E., Pratt, C.M., Vybiral, T. *Am Heart J* **125**, 731-743 (1993)
21. Huikuri, H.V., Valkama, J.O., Airaksinen, K.E., et al. *Circulation* **87**, 1220-1228 (1993)
22. Shusterman, V., Aysin, B., Gottipaty, V., et al. *J Am Coll Cardiol* **32**, 1891-1899 (1998)
23. Vybiral, T., Glaeser, D.H., Goldberger, A.L., et al. *J Am Coll Cardiol* **22**, 557-565 (1993)
24. Lin, J.L., Chan, H.L., Du, C.C., et al. *Am Heart J* **137**, 658-865 (1999)
25. Pousset, F., Copie, X., Lechat, P., et al. *Am J Cardiol* **77**, 612-617 (1996)
26. Fromer, M., Wietholt, D. *Am J Cardiol* **83**, 45D-47D (1999)
27. Pruvot, E., Thonet, G., Vesin, J.M., et al. *Circulation* **101**, 2398-2404 (2000)

Bones And Nonlinear Dynamics – The Quantification Of Architecture

Wolfgang Gowin*, Peter Saparin*, Dieter Felsenberg*, Jürgen Kurths†

*Center of Muscle and Bone Research, Department of Radiology and Nuclear Medicine, University Hospital Benjamin Franklin, Free University Berlin, Hindenburgdamm 30, 12203 Berlin, Germany
†Center for Dynamics of Complex Systems, Institute of Physics, University of Potsdam, Am Neuen Palais 10, 14415 Potsdam, Germany

Abstract. In order to quantify the internal architecture of a bone in a holistic manner based on radiological images, the methodology of nonlinear dynamics was applied. Image processing algorithms, an expansion of symbolic dynamics, and five measures of complexity were introduced to quantify the trabecular part of human lumbar vertebral bodies. Healthy vertebral bones have a complex and ordered architecture with a high degree of spatiodynamics. Pathology changes the architecture significantly and can be quantified by measures of complexity.

INTRODUCTION

Mass, architecture, form, and function of biological objects or tissues depend and are interrelated in manifold complex ways on each other. The bone is a paradigmatic example of a highly complex system with a rich hierarchical organization of its architecture. Metabolic bone diseases can change the mass and the internal structure of bone, leading to osteopenia and in later stages to osteoporosis. Similar changes can be observed when the biomechanical load is decreased, for example after paralysis, in situations requiring immobilization, when the physical activity is reduced, or during space flights. The results are a reduced bone mass and changes of the trabecular arrangement leading to an increased risk of fracture. This relationship works as well in the other direction. Increased physical activity or higher biomechanical loads induce a higher bone mass and more complex architectural structures resulting in increased bone strength, reducing the risk of fracture.

Osteopenia is the most often found bone disease and is prevalent in about 54 % of postmenopausal white women in the United States [1]. When using the WHO-criteria, 16 % of all white women in the US age 50 years or above suffer from fractures caused by osteoporosis [1]. The estimated cost of all osteoporotic fractures was $13.8 billion in the US in 1995 [2]. Vertebral osteoporotic fractures are more common in people younger than 70 years and cannot be treated surgically. It can be assumed that osteoporosis will become a global problem over the next half century [3]. Diagnostic procedures are necessary to choose the appropriate treatment, or to observe bone mass and structure loss before they arise to prevent osteoporotic fractures successfully.

Osteoporosis is diagnosed by clinical examination, including a radiological check-up of the bone status. However, the measurement of a low bone mineral density

CP622, *Experimental Chaos: 6th Experimental Chaos Conference*, edited by S. Boccaletti et al.

(BMD) offers no information about the architectural composition of the remaining structural elements of the bone and is only a non-specific factor for the diagnosis of osteoporosis. New methodologies are needed to quantify structural information of the bone [4,5] because the distribution of mass within the bone is imperative to acquire for the assessment of the integrity and for an estimate of the fracture risk.

Present experimental non-invasive radiation-safe roentgenological procedures to quantify bone are fractal [6-8], texture [9], and pattern analyses [10]. These methods examine profiles of bone sections [10], or assess projected surface patterns [6-9]. None of these procedures have been applied to routine patient examinations, and their sensitivity or specificity in regards to pathological structural changes other than osteoporosis have not been reported yet.

We introduced a method to quantify the structure of vertebral trabecular bone based on CT-images. The procedure is founded on the proposition that the architecture of a bone is not only complicated but follows the rules of a complex system. The method measures the relationships of all geometrical structural parts and variations in such a way that the result is a numerical holistic representation of the architectural composition of the bone. These measures of complexity were applied to quantify architectural properties of segmented trabecular bone of human lumbar vertebrae.

MATERIALS AND METHODS

Material

57 non-fractured human lumbar vertebrae L4 were examined on a CT-scanner (Siemens AG; Somatom Plus S). They were scanned airtight surrounded by water to simulate conditions similar to routine patient examinations. Continuous transaxial slices of 1 mm thickness through the entire vertebra were acquired in high-resolution mode (HRCT; pixel size: 0.182x0.182 mm). A transaxial midvertebral slice in 10 mm thickness was obtained from each specimen in conjunction with a calibration phantom (QCT-mode; pixel size: 0.323x0.323 mm) and provided an assessment of the trabecular BMD. An image matrix of 512 x 512 pixels was applied in both procedures. The L4-series contained 16 normal (BMD > 100 mg/cm^3), 16 osteopenic (BMD = 80-100 mg/cm^3), and 25 osteoporotic (BMD < 80 mg/cm^3) vertebrae. 28 specimens were from females (mean age 72 years), 29 from males (mean age 67 years).

Methods

The image preprocessing technique consists of a noise reduction and of two segmentation procedures. The vertebral bodies are segmented from the surrounding soft tissue and then fragmented into their cortical and trabecular parts. Orientation and size independent algorithms based on topological properties were used.

In order to analyze 2D-images, the technique of symbolic dynamics [11,12] was expanded and developed into a procedure to encode the x-ray attenuation of the image with five different symbols and to quantify the spatial arrangement of the symbols.

The detailed information of each pixel's attenuation value is reduced while the invariant robust properties of the spatial dynamics and the architecture remain intact. Thus, the regularity, symmetry, homogeneity, global organization of structure, and its orderly composition can be measured without jeopardizing the results.

Each symbol represents a statical or dynamical defined amount of the original x-ray attenuation. A landscape terminology is implemented for easier understanding of the pixel intensities represented as symbols. There are 3 static elements: "L" for lake, "V" for valley, "H" for highland, and 2 dynamic elements "I" for incline, and "C" for cliff. The symbols "I" and "C" are found in areas with densely intertwined trabeculae. They represent transitions of different levels of attenuation. The symbol "L" is a representation of the soft marrow tissue located between the trabeculae. The symbols "H" and "V" represent the static elements of the architecture characterized by high and medium values of attenuations. Further details were previously reported [13,14].

After the refinement of the procedure and an additional preliminary study had shown that the proposed technique is stable for the orientation, the range of BMD, and for a reasonable dispersion of noise influence ($\leq$ 20 CT-numbers) [13]; the method was transferred to 10 mm thick QCT-slices. The procedures were then repeated and the results compared. Significant changes between the average of the sum of ten continuous 1 mm thick slices obtained above and below the midvertebral level in HRCT-mode and a single 10 mm thick midvertebral slice were not found. We concluded that the measures of complexity are robust for the expected noise level of CT-images in 1 mm or 10 mm thickness.

The local probability density of symbols and local connectivity produced the basis for the quantification of different aspects of the underlying bone structure. Conventional approaches, such as Shannon entropy or generalized entropy, are unable to assess the shapes of distributions with a small number (five) of bins, and they are unsuitable for the evaluation of non-smooth changes of the distribution shapes [15]. Therefore, five new measures of complexity were introduced to quantify the structural composition of the symbol encoded images [13,14]:

1. The Index of Global Ensemble (IGE) is an indicator of the dynamical distribution of trabeculae and marrow tissue within the investigated bone and expresses the overall spatial dynamics of the trabecular composition.

2. The Structure Complexity Index (SCI) quantifies the complexity of the spatial arrangement of symbols.

3. The Trabecular Net Index (TNI) evaluates the homogeneity of the trabecular architecture. TNI is an expression of the richness of trabecular ramifications. In contrast to IGE and SCI, TNI characterizes the organization of hard tissue solely.

4. The Structure Disorder Index (SDI) quantifies the disorder and the regularity of the trabecular architectural composition.

5. The measure called "Size of maximal L-block" represents the void of hard structural elements within the bone. The regions of soft marrow tissue are encoded by the symbol "L". A L-block is the area composed by pixels which are connected to each other and are all encoded by the same symbol "L". The largest area of all L-blocks is taken for this negative (void space) architectural measurement and normalized by the total trabecular bone area.

RESULTS

IGE, TNI, and SCI increase nearly exponentially with BMD. The increase of IGE is very rapid, so that a comparable depiction was only achieved on a logarithmic scale, see figure 1. Specimens with a BMD below 40 mg/cm^3 demonstrated low complexity measurements in IGE, TNI, and SCI. This is confirmed in other measures of complexity (SDI and Size of maximal L-block). IGE is more sensitive to changes in the vertebrae than BMD, due to its much greater range. It still provides structural information for normal trabecular bone when the SCI, a local level measurement, shows already a saturation. TNI behaves similar to the BMD.

SDI, as a measurement of disorder, provides additional value: the higher the SDI the more disorder is measured in the structural composition.

The size of the maximal L-block is a direct measure of the substitution of marrow cells for lost trabeculae. Similar to IGE, this measure of the space between the structural compositional elements is more sensitive to changes within the vertebra than BMD, to which it is related in reverse order.

Although the BMD data followed the known age-dependency, the measures of complexity were unable to show a significant dependency on age or gender.

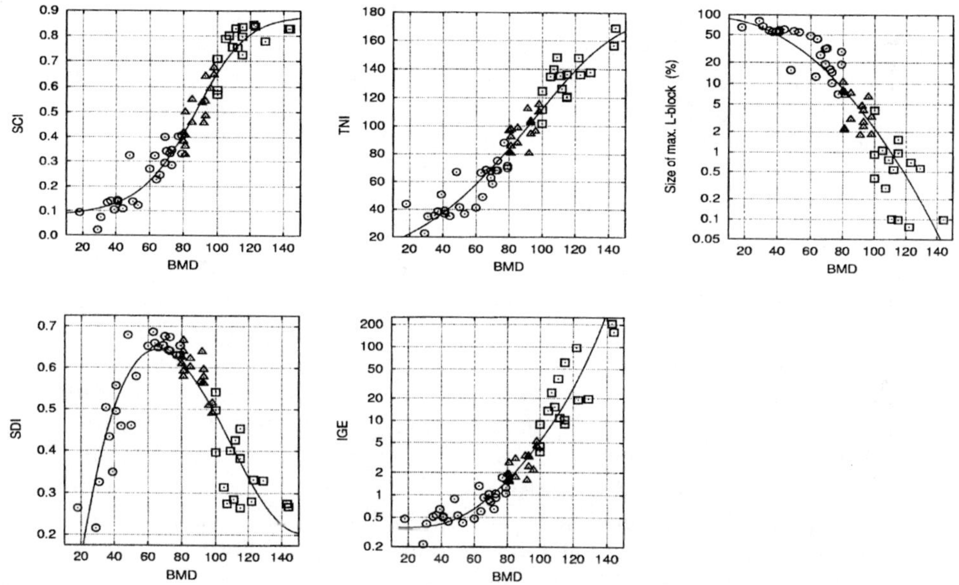

FIGURE 1. Measures of complexity (SCI, TNI, Size of max. L-block, SCI, IGE) versus BMD of 57 human lumbar L4 vertebrae. Circle: osteoporotic, triangle: osteopenic, squares: normal specimens.

CONCLUSION

The structural architecture of a natural system is a result of its collective internal properties, purpose, and its application within the external environment. Real structures are consequential nonequilibrium, history- and path-dependent, and constantly dynamic. The structure of the vertebral trabecular bone is a typical example of such a structure meeting the definition of complexity. We suggest that such a structure shall be quantified with measures of complexity. We developed appropriate measures of complexity and applied these parameters to a non-invasive QCT-procedure of lumbar vertebrae. The advantage is the acquisition of two different bone measures: BMD for bone mass and complexity measurements for bone structure. Both complement and depend on each other.

The inability to find a significant age or gender dependency of the measures of complexity can be explained with the argument that the complexity of the trabecular structure depends on its architectural construction which in turn depends on age and gender independent biomechanical adaptation [16]. A structural composition relies probably more on the interstratification of its structural elements than on the density of the material used. Since vertebrae are utilized in all ages in the same manner by females as well as males, it is not surprising to see that structural parameters based on the complexity of the trabecular architecture are less depended on age and gender than the material sensitive measure of BMD.

This assumption is supported by the saturation of SCI in normal specimens (fig. 1). The higher the value of SCI, the more complex and regionally different is the structural composition of the trabecular network. The more material is build into the given space of the vertebra the less different will be the regional variance. The complexity will reach its limit. This is shown by the plateau of the SCI through increased BMD.

TNI was introduced to evaluate the orderliness of the hard tissue elements and the homogeneity of the trabecular net. A high TNI is an indicator of a richly interconnected trabecular bone. It is expected that this parameter will reach a plateau as well at higher levels.

IGE quantifies the ensemble of symbols used in the architectural composition. The higher the IGE-value the more interstratifications between all symbols are found and / or the distribution of the attenuation over the given volume is more varied.

The differences in the structural composition in the different regional parts of the bone are quantified by SDI. It is depreciated in vertebrae with an ordered architectural composition, whereas a less regular structure leads to an increased SDI-value. A high SDI is a sign of a disordered architecture. In the viewpoint of nonlinear dynamics, it is fascinating to see that the vertebral cancellous bone of all osteopenic (BMD 80 - 100 mg/cm^3) and moderate osteoporotic specimens (BMD 40 - 80 mg/cm^3) demonstrate a high SDI-value. This transition could be a criterion for the activity of the disease.

Since the size of the maximal L-block is a normalized measurement of the void space between the hard bone structural elements, it has been a scrutiny parameter during the development of the measures of complexity for trabecular bone. It decreases exponentially with increasing BMD, IGE, TNI, and SCI.

It can be concluded that normal vertebral trabecular bone has a complex ordered architecture with highly developed spatial dynamics. Highly osteoporotic bone is simple structured but to an orderly degree. The spatiodynamics are strongly suppressed, the trabecular ramifications are low. Osteopenic trabecular bone behaves as being in a transition phase. The structure has changed significantly from the healthy state. The architecture is much less complex, is disordered, and the trabecular network is disrupted and partly filled with soft tissue.

Measurements of complexity provide an estimation of structural quantity in the cancellous part of vertebral bones. The pathological loss of structure may be more important to the integrity of the bone than the pathological loss of mass. Other studies have shown that the decrease of structural quality results in a much greater diminution of strength than the apparent reduction of BMD can predict [17,18]. The proposed measures of complexity point into the same direction. Further experimental studies are necessary to verify the biomechanical correlation.

ACKNOWLEDGMENTS

This work was supported by a grant from Roche Pharmaceuticals, Germany. The specimens were provided by the EU BIOMED-Project (# BMH1-CT92-0296).

REFERENCES

1. Melton, L.J., "Epidemiology of Fractures," in *Osteoporosis: Etiology, Diagnosis, and Management*, edited by B.L. Riggs, L.J. Melton III, Philadelphia: Lippincott-Raven Pub., 1995, pp. 225-247.
2. Ray , N.F., Chan, J.K., Thamer, M., Melton, L..J., *J Bone Miner Res* **12**, 24 -35 (1997).
3. Cooper, C., and Melton, L.J., "Magnitude and Impact of Osteoporosis and Fractures," in *Osteoporosis*, edited by R. Marcus, D. Feldman, J. Kelsey, San Diego: Academic Press, 1996, pp. 419-434.
4. Feldkamp, L.A., Goldstein, S.A., Parfitt, A.M., Jesion, G., Kleerekoper, M., *J Bone Miner Res* **4**, 3-11 (1989).
5. Beck, T.J., *Radiology* **199**, 612-614 (1996).
6. Majumdar, S., Weinstein, R.S., Prased, R.R., Genant, H.K., *Calcif Tissue Int* **52**, 168 (1993).
7. Buckland-Wright, J.C., Lynch, J.A., Rymer, J., Fogelman, I., *Calcif Tissue Int* **54**, 106-112 (1994).
8. Benhamou, C.L., Lespessailles, E., Jacquet, G., Harba, R., Jennane, R., Loussot, T., Tourliere, D., Ohley, W., *J Bone Miner Res* **9**, 1909-1918, (1994).
9. Ito, M., Ohki, M., Hayashi, K., Yamada, M., Uetani, M., Nakamura, T., *Radiology* **194**, 55-59 (1995).
10. Klepper, R., Johner, C., Hesch, R.D., *Fortschr Röntgenstr* **159**, 337-342 (1993).
11. Hao, B.L., *Physica D* **51**, 161-176 (1991).
12. Rapp, P.E., Goldberg, G., Albano, A.M., Janicki, M.B., Murphy, D., Niemeyer, E., Jimenez-Montano, M.A., *Int J Bifurcation and Chaos* **3**, 525-542 (1993).
13. Saparin, P.I., Gowin, W., Kurths, J., Felsenberg, D., *Phys Rev E* **58**, 6449-6459 (1998).
14. Gowin, W., Saparin, P.I., Kurths, J., Felsenberg, D., *Tech Health Care* **6**, 373-390 (1998).
15. Saparin, P.I., Gowin, W., Kurths, J., Felsenberg, D., "Processing of vertebral CT-images and quantification of their structure by measures of complexity," in *Applied Nonlinear Dynamics And Stochastic Systems Near The Millennium,* edited by J. B. Kadtke, A. Bulsara, Woodbury, NY: American Institute of Physics, 1997, pp. 347-352.
16. Thompson, D.W., *On Growth and Form*, The complete revised edition, Mineola, NY: Dover Publications, 1992.
17. Hayes, W.C., Bouxsein, M.L., "Biomechanics of cortical and trabecular bone: implications for assessment of fracture risk," in *Basic Orthopaedic Biomechanics*, edited by V. C. Mow, W. C. Hayes, 2nd ed., Philadelphia, PA: Lippincott-Ravens, 1997, pp. 69-111.
18. Jensen, K.S., Mosekilde, L., Mosekilde, L., *Bone* **11**, 417-423 (1990).

Nonlinear Dynamics of a Small Biological Neural Network

Allen I. Selverston, Mikail I. Rabinovich, and Henry D.I. Abarbanel

Institute for Nonlinear Science, University of California, San Diego, Gilman Drive La Jolla, CA 92093-0402 USA, aselverston@ucsd.edu

Abstract. We examine the role of chaos in a small biological central pattern generating network in which all of the neurons and their connections are known. The lobster stomatogastric ganglion has 30 neurons, 14 of which form the pyloric central pattern generator (CPG). This CPG produces a three phase rhythm at a frequency of about 1 Hz. Although CPGs of this type can produce such rhythms entirely independent of sensory feedback or commands from higher centers, they generally produce oscillatory synchronized activity when modulated by chemical substances acting as hormones or released directly from neurons that have inputs to the ganglion. Sensory input acts to control the output of the CPG on a cycle-by-cycle basis. We demonstrate first of all that individual identified neurons that have been isolated from their synaptic inputs are low dimensional and behave chaotically over a large range of their operating regimes. The neurons can be regularized by connecting them to other neurons in the circuit with chemical or electrical synapses. We have modeled individual and small ensembles of pyloric neurons with both Hodgkin-Huxley and Hindmarsh-Rose type models. The latter has also been implemented in analogue hardware to construct electronic neurons. These electronic neurons are very similar to the biological neurons in their dynamical properties and when interfaced to the biological neurons, form hybrid circuits that can function normally with the electronic neurons taking the place of missing or damaged biological neurons. Supported by NIH, ONR, DOE and CIA.

INTRODUCTION

Common rhythmic behaviors such as locomotion are shaped by motor patterns that are organized into bursts of impulses to antagonistic muscles. Such bursts are generated by neurons whose membrane potentials oscillate. It is thought by many that similar oscillatory neural activity also underlies sensory processing in the olfactory [1] and visual systems [2]. Oscillatory neural activity at 30-60 Hz has been suggested as a way to synchronize spikes occurring in spatially separate areas of the central nervous system [3]. There are two possible mechanisms that the nervous system could use for the generation of bursts. Single neurons could have the intrinsic conductances necessary to generate what are called bursting pacemaker potentials. , a mechanism that has been shown to occur in a variety of vertebrate and invertebrate neurons. These so called "endogenous bursters "can be integrated into circuits that control pairs of muscles to generate rhythmic behaviors. In fact there are only a few instances where truly endogenous bursters have been

CP622, *Experimental Chaos: 6th Experimental Chaos Conference*, edited by S. Boccaletti et al.

demonstrated. Most neurons are silent or tonic when completely isolated and to oscillate require the presence of neuromodulatory chemical substances.

An alternate theory suggests that synaptic connections, particularly inhibitory connections between neurons, can generate oscillatory behavior. For many of the invertebrate central pattern generators (CPGs) which have been described, the most common mechanism appears to be a combination of these two possibilities although some investigators still insist that one mechanism or the other alone is sufficient.

The neurophysiological approach for trying to understand how rhythmic behaviors are generated and controlled has focused on the physiological and biochemical analysis of neural circuits. The general idea is that by knowing the properties of the neurons comprising a central pattern generating network and then working out the synaptic and electrical connections between them, a qualitative description of the mechanism can be made. Unforseen however was the fact the even in very small networks, the nonlinear interactions between elements and indeed even within single neurons makes a rigorous explication that has predictive value impossible without specialized modeling methods that take into account the nonlinear dynamical properties of the system.

In this paper we demonstrate that the application of such nonlinear modeling techniques to the study of a small CPG network suggests a vital role for dynamical chaos in motor function. We demonstrate that individual neuronal elements in this CPG have intrinsic low dimensional chaotic behavior but when shaped into functional neural circuits by chemical neuromodulators, are capable of producing periodic synchronized bursts of motor impulses to the appropriate muscles. We demonstrate how stable rhythmic motor patterns can emerge from circuits of neurons that are inherently irregular by utilizing the chemical and electrical synapses that bind them together.

THE STOMATOGASTRIC SYSTEM

To examine the nonlinear properties of the brain would be a heroic undertaking because of its size and complexity. We have instead chosen a small invertebrate CPG network , an ensemble of neurons that produces rhythmic motor patterns without any feedback from sensory receptors. The stomatogastric ganglion (STG) contains only thirty neurons yet produces two complex patterns, the gastric and the pyloric. All of the neurons for both patterns have been identified and their connections established [4]. In this report we consider only the pyloric CPG, responsible for producing rhythmic contractions and dilations of the lower portion of stomach at a frequency of about one Hz.. The musculature of the lobster stomach is striated and unlike our stomach muscles, requires patterned output from the central nervous system to operate. The three phase pyloric pattern is generated by only 14 neurons in the STG. These neurons fire in properly phased bursts of action potentials provided neuromodulatory inputs from several higher ganglia are intact.

The neural circuits in the STG represent a complex system that is small enough to be experimentally tractable but large enough to be interesting for the study of pattern formation and it's dynamical components. The circuit shown can generate many variations of the pyloric and gastric rhythms depending on it's neuromodulatory environment. The modulators can be delivered by specific neurons or be carried in the blood as hormones. They act via G-protein second messenger systems to alter ionic conductances in the neuron membrane in a way that can alter the synaptic and cellular properties --thus functionally "rewiring" the anatomical circuits. In addition the CPGs receive sensory modulation from receptors in the stomach wall and associated muscles.These receptors have been found to act like typical vertebrate stretch receptors, basically adjusting muscle tension to load. Like many invertebrates the oscillatory properties of the pattern are a result of both intrinsic cellular properties and synaptic interactions. The cellular properties give rise to sine-wave like membrane oscillations with bursts of spikes being generated when the membrane potential crosses threshold. The frequency of the burst pattern is largely determined by this cellular property. Since all of the neurons in both pyloric and gastric circuits burst when modulated, the CPGs can, for some purposes, be considered systems of coupled oscillators. The phase relationships between the bursts is governed by the synaptic interactions between individual neurons, so slight changes in synaptic time course or in post-inhibitory rebound, can have major effects on phase relationships.

The connectivity within the stomatogastric system can be seen to be made up almost exclusively of chemical inhibitory synapses (Fig 1). These use acetylcholine and glutamate as neurotransmitters. There is also extensive electrical coupling between neurons, some of it rectifying, but virtually no excitatory synapses. The neurons are all individually identifiable and each neuron is itself a complex system that bursts chaotically when isolated and connected to higher ganglia which supply modulatory drive.

About twenty naturally occurring compounds have been found in lobsters that can modulate the pyloric and gastric CPGs. Mixtures of different neuromodulatory substances would likely produce mixed effects that would hard to interpret although this is no doubt how they actually are presented *in vivo*. But to study their action, individual neuromoodulatory compounds can be applied at known concentrations to ganglia that have been isolated from all other neuromodulatory compounds, especially those coming from descending pathways, so that the effect of individual substances can be determined free of complications that would result if a mixture were present. The neuromodulators which have been tested can restructure both rhythms by modifying such properties as bursting, firing threshold, IPSP amplitude, sign and time course. This important point is often not considered when modeling neural networks which are usually regarded hard wired circuits whose physiological properties remain constant. In the experiments reported here we keep the neuromodulatrory environment constant by leaving three "higher" ganglia, known to contain neuromodulators, attached to the stomatogastric ganglion.

One important advantage of small networks like the pyloric is that single neurons or pairs of neurons can be synaptically isolated so that their dynamical properties can be studied free from the influence of extraneous inputs. We isolate cells using a combination of synaptic blockers such as picrotoxin to block glutamatergic synapses and a cell-killing technique to remove cholinergic synapses from the neurons to be studied [5].

Control of synaptic strength and electrical coupling is achieved *via* microelectrodes. These electrodes record the membrane potential intracellularly and by interfacing a computer-operated current -passing circuit between the electrode and the amplifier, can pass current into a cell in a controlled way. . This device, called a dynamic clamp, takes the current output from one cell, modifies it according to the program, and injects a new current into the electrically or chemically coupled postsynaptic cell [6]. Using this device, one can increase or decrease the natural electrical coupling between cells or even reverse the current flow. Similarly, one can add current to or subtract current from chemical synapses thus making them artificially stronger or weaker.

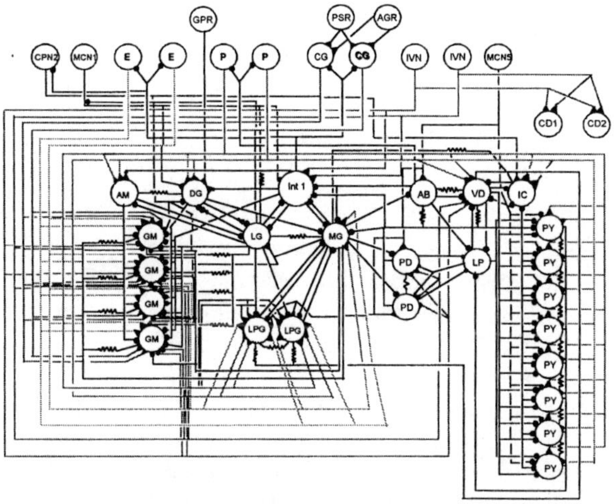

Gastric Mill and Pyloric CPGs of Lobster

Figure 1. Complete circuit diagram of the stomatogastric system. Large circles represent identified neurons. Black dots are chemical inhibitory synapses and black triangles are chemical excitatory synapses. Resistors represent electrical coupling between neurons. The two CPGs, gastric and pyloric are the lower left and lower right group respectively while the neurons along the top half are various identified sensory, neuromodulatory and command elements.

COMPUTATIONAL AND ELECTRONIC MODELING TOOLS

Because our initial analysis of biological neurons had shown they appeared to operate with only three or four degrees of freedom, we chose initially to use a familiar simplified model put forward by Hindmarsh and Rose [7]. This type of phenomenological model is useful because it is more nimble than a conductance –based model, particularly when the number of neurons in the model starts to grow, however it suffers from the inability to directly utilize experimental data which is usually in the form of measured conductances. The general form of this model contains three terms:

$$dx/dt = ay(t) + bx^2(t) - cx^3(t) - dz(t) + I \qquad (1)$$

$$dy/dt = e - fx^2(t) - f(t) \quad (2)$$
$$dz/dt = \mu[-z(t) + S(x(t) + h)] \quad (3)$$

The first three equations are from the original model where x(t) corresponds to membrane voltage, y(t) represents a "fast" current and by making $\mu<<1$, z(t) a "slow" current. The first three equations (the 3-D model) can produce several modes of spiking-bursting activity including a regime of chaos that appears similar to that seen in biological neurons. However the parameter space for the chaotic behavior is much more restricted than we observe in real neurons.The chaotic regime is greatly expanded by incorporation of the fourth term into the model:

$$dw/dt = v[-kw(t) + r(y(t) + l)] \quad (4)$$

Adding this term w(t) to introduce an even slower process ($v<\mu<<1$) is intended to represent intracellular Ca dynamics. Note this also adds a –gw(t) to the the second equation. When this last term is taken into account, the model produces simulations of intracellular activity that are more similar to the biological observations.We, as yet, do not know if the w(t) term actually represents Ca kinetics in pyloric neurons and experiments are currently under way to measure Ca transients in neurons using Ca imaging techniques.

Because of its relative simplicity, the HR model has been extremely useful in constructing an analogue implementation – an electronic neuron – that could perform the computations necessary to emulate stomatogastric neurons in real time. However the down side of such a simplified model is that it is difficult to compare with biological neurons which are made up of a multitude of individual conductances and compartments. Since the experimental manipulation of these parameters is crucial in establishing physiological mechanisms, we also developed a Hodgkin-Huxley (H-H) type model that is more biologically realistic than that of Hindmarsh and Rose. In this second model [8], the neuron is represented by two compartments, one for the lumped neuropilar processes and soma and the other for the axon. The currents we have used are based on previous descriptions but are restricted to those that we consider to be the most important in generating slow wave and spiking currents. In all, five currents are present in the soma-neuropil compartment (I_{Ca1} low voltage activated calcium current, I_{Ca2} high voltage activated calcium current, I_h hyperpolarization-activated inward current, $I_{K(Ca)}$ calcium dependent potassium current, and a leak current I_L). The axonal compartment contains the well-known Hodgkin-Huxley currents for spiking - I_{Na} , I_K , and a leak current I_L . The soma –neuropil compartment also incorporates intracellular dynamics based on the model of Li et al. [9]. In this model, cytosolic Ca is determined by influx across the plasma membrane and by uptake and release from the ER and extrusion by a plasma membrane pump and a plasma membrane Na-Ca exchanger.

ELECTRONIC NEURON MODELS

In a third form of modeling we have used three and four dimensional Hindmarsh –Rose models, to construct low dimensional analog electronic neurons (ENs) whose properties emulate the membrane voltage characteristics of individual stomatogastric neurons [10]. The ENs are simple analog circuits which integrate the differential equations representing fast and slow subcellular mechanisms that produce the characteristic regular/chaotic spiking-bursting behavior of these cells. A schematic diagram for the EN is shown in Fig. 2. Our strategy for building an analog device instead of using numerical integration of the mathematical model on a PC or DSP board was due to the fact that digital integration of the H-H equations is a slow procedure associated with the two or three different time scales in the model We also plan to eventually construct real-time networks of the entire pyloric CPG and for a large number of neurons and for this, analog implementation is a necessity.

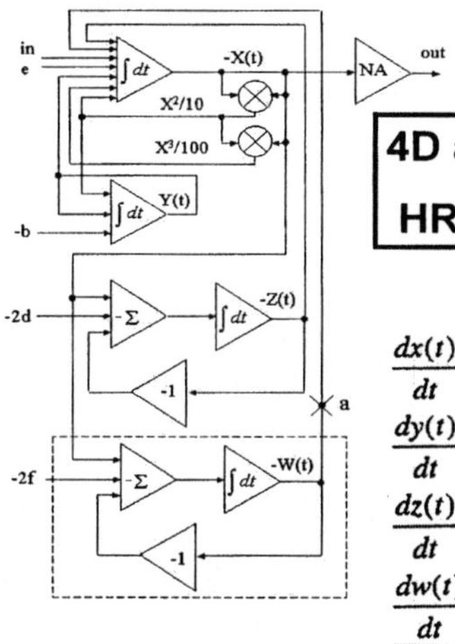

4D analog circuit for HR neuron

$$\frac{dx(t)}{dt} = ay(t) + bx^2(t) - cx^3(t) - dz(t) + I$$

$$\frac{dy(t)}{dt} = e - fx^2(t) - y(t) - gw(t)$$

$$\frac{dz(t)}{dt} = \mu(-z(t) + s(x(t) + h))$$

$$\frac{dw(t)}{dt} = \nu(-kw(t) + r(y(t) + l))$$

Figure 2. Block diagram or the four dimensional HR + Ca^{++} electronic neuron used in our experiments. These neurons were based on the equations shown in the text and were designed to replicate the behavior of individual isolated neurons from the lobster STG. In our experiments the electronic neurons were coupled to each other electrically as well as via an electronic implementation of inhibitory and excitatory chemical synapses.

DYNAMICAL PROPERTIES OF ISOLATED NEURONS

We selected the LP neuron of the pyloric CPG for a detailed analysis because it appeared to be typical of many STG neurons and a large amount of physiological data about this particular neuron already existed [10]. Intracellular time series lasting several hours were sampled at a data rate of 2kHz. The Fourier power spectrum for this time series was broadband without separate peaks as is typical in cases of a chaotic time series, and did not reveal anything about the dynamics underlying the intracellular voltage signal. A nonlinear analysis however suggested that by using the method of global false nearest neighbors [11], the active degrees of freedom for the LP neuron are only four or five. This is despite the fact that each neuron uses many more than four or five processes to meet it's normal metabolic and information handling needs.

The low dimensionality of the neurons suggestsed we could make models of neuronal function with only five degrees of freedom. This finding opened up several options in terms of modeling. We could use rather simple phenomenological models that capture the dynamics of the neuron but would be inappropriate for modeling experimental data. Alternatively, we could utilize the more complex conductance-based models. We started by using a modified Hindmarsh-Rose formalism [7] as a good phenomenological model that captures the dynamical properties of biological neurons. We have also performed computer simulations with a conductance-based model [8] that also has dynamical properties similar to stomatogastric neurons.

When the analysis of the LP time series was complete, it showed that the isolated neuron had a positive global Lyapunov exponent, a hallmark of chaotic behavior in a dynamical system. In another study in which we made slight perturbations to the orbit, we found that there were two positive global Lyapunov exponents, one zero exponent, suggesting the behavior is governed by ordinary differential equations, and two negative exponents. A fractal dimension of approximately 4.4 was establishedbased on these global exponents. Chaotic neurons in invertebrates have been reported previously [12] [13], but their analysis was much less detailed. We were particularly careful in these studies to insure that the neuron being analyzed was synaptically isolated by means of pharmacological blockade and photoinactivation of presynaptic neurons when necessary. In contrast to the chaotic activity observed in the isolated state (Fig. 3), the LP, along with other pyloric neurons display very regular spiking-bursting activity in well –modulated normal preparations.. So although the pyloric circuit is made up of unstable neurons, they operate in a reliable and predictable fashion, robust enough to resist strong perturbations. The machinery for producing rhythmic motor patterns must contain a dependable system for providing impulses to to the appropriate muscles in the correct temporal sequence. It is therefore extremely important to determine the mechanisms involved in achieving a degree of regularity and stability in a system composed of inherently unstable elements without losing the ability of the CPG to respond to sensory and command elements, or their remarkable ability to bifurcate into functionally different circuits that produce different stable patterns under the influence of neuromodulators.

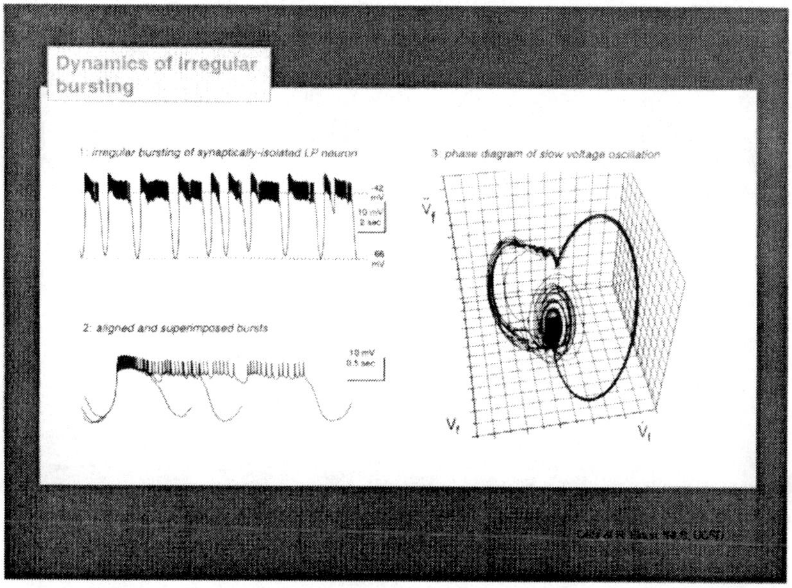

Figure 3. Irregular spontaneous bursting in a synaptically isolated LP neuron still receiving modulatory input from the commissural ganglia. (1) Excerpt from an extended time series showing a typical irregular bursting pattern. (2) Three bursts aligned by their first spike and superimposed. (3) Phase space diagram of slow voltage oscillations from a time- series of a similarly isolated LP. The membrane potential [V(t)] was low-pass filtered to remove the spikes. The plot of V(t) against its first and second derivatives reconstructs an attractor showing, in addition to noise, apparent deterministic structure. The motion in time is counter-clockwise. The large orbit corresponds to the overall slow wave (from burst to burst), the smaller spiral to faster oscillations on the plateau (during each burst).

The circuit in Fig. 1 contains three fundamental mechanisms that could be involved in regularizing the output patterns generated by the circuitry:

- electrical connections between neurons that allow a direct spread of current from one cell to another,
- chemical inhibitory synapses
- chemical excitatory synapses.

To study the role each plays in synchronization and regularization, we used computer controlled artificial electrical and chemical synapses to produce "adjustable" connections for all three types of connection between isolated pairs of identified neurons.

Regularization by Electrotonic Coupling

We studied the dynamical effects of adding artificial electrical coupling with different strengths in parallel to the natural coupling between the two PD neurons. The effective coupling could be altered by injecting equal and opposite current into each cell via microelectrodes. The natural coupling of these neurons was sufficient to synchronize bursts but not individual spikes. However if the two PDs were depolarized slightly so that they fired tonically, the natural coupling was sufficient to synchronize the spikes implying a shunting action of slow membrane conductances during burst generation. Opposing the flow of current that occurs during natural coupling *via* the dynamic clamp causes the bursting to become more asynchronous finally leading to out-of-phase bursting and incresased regularization. In this first case the net negative coupling approximates reciprocal inhibition. For most coupling strengths the increased coupling caused chaotic synchronization.

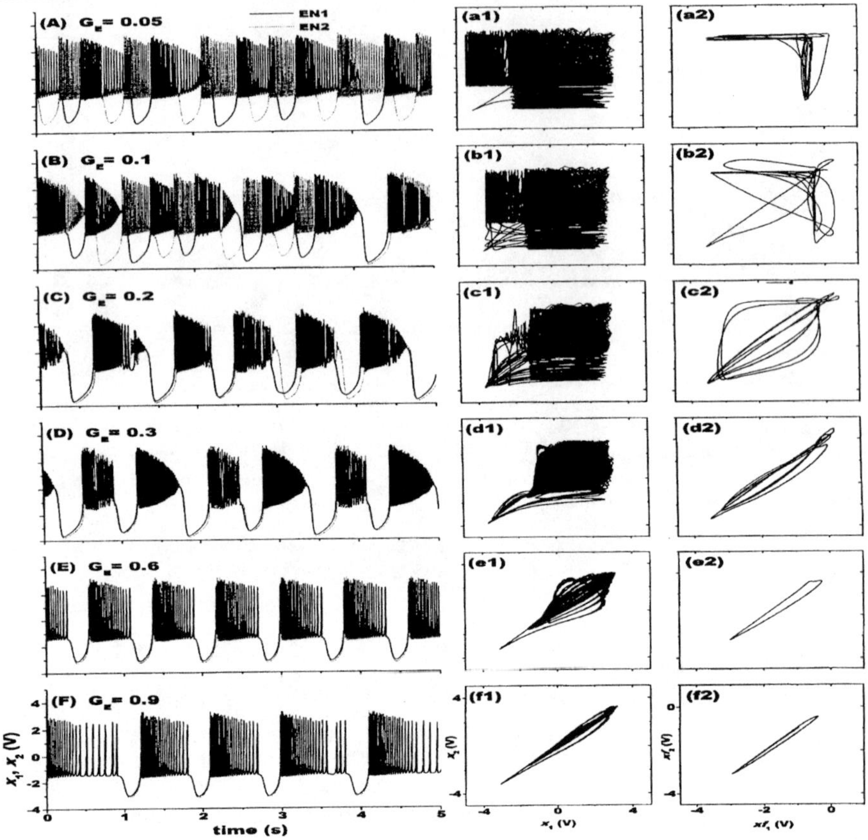

Figure 4. Positive electrical coupling of two chaotic ENs. Characteristic time series of the membrane potentials $x_1(t)$, $x_2(t)$ [A-F] as synaptic conductance G_E is varied. Phase portraits (a1-f1) are shown on the right unfiltered and after low pass filtering (a2-f2). (A) intermittent out-of-phase activity (B) nearly independent chaotic spiking-bursting pattern (C) chaotic oscillations with most bursts synchronized (D) periodic oscillations with partial synchronization of the ENs, bursts not synchronized (E) periodic oscillations with complete synchronization of the ENs (F) chaotic but completely synchronized oscillations.

Similar experiments were performed using electronic neurons whose dynamical properties were very similar to those of the biological neurons (Fig. 4). When the coupling strength $G_E \approx 0.5$, the two neurons were effectively uncoupled and displayed independent chaotic oscillations. However as the strength of the coupling increased, some regions of synchronized bursting began to appear including regions where intermittent anti-phase bursting could be observed. As the coupling was made stronger the bursting activity became regular and went from a region of partial synchronization to a region of total synchronization. Between values of .8 and 1, there was total synchronization in the spiking bursting activity and the oscillations were chaotic. For negative electrotonic coupling, something that does not occur in nature, the oscillations were mostly chaotic and the bursts occured in antiphase. As with experiments on biological neurons, there was a sharp phase transition between nearly synchronous and fully synchronous behavior . One observed a wider range of stable states between coupled neurons than expected and because they appeared to bifurcate into these regions, they were likely due to the intrinsic dynamics of the neurons.

Regularization and Synchronization by Chemical Synapses

We examined the role of inibitory and excitatory chemical synapses by first mimicing their actions on postsynaptic neurons with short positive and negative pulses of current (Fig. 5). We found that depolarizing current pulses were poor regulators of the neurons irregular activity but that hyperpolarizing pulses were able to exert a strong, frequency-dependent regularizing action [14]. The LP was synaptically isolated from other neurons in the pyloric circuit by photoinactivating the PD and VD neurons (because they are cholinergic and cannot be specifically blocked pharmacologically) and blocking the remaining glutamatergic synapses with picrotoxin . The CGs were left connected to the stomatogastric ganglion via the stomatogastric nerve so the isolated LP would receive the same neuromodulators it receives under normal experimental conditions. Long time series traces (5-6 minutes) of membrane voltage were digitized and the degree of irregularity in the observed oscillations was determined from the power spectra and a measure of the entropy for the harmonics of the power spectra for frequencies up to 40 Hz (Fig. 5d-f).

Fast Fourier spectra were calculated from the digitized voltage signals without filtering and the power in harmonics up to 40 Hz., P(n) for n =20,,976, were converted to probability values as $p(n) = P(n)/\Sigma P(n)$. An entropy function (in bits),

$S = - \Sigma_n p(n) \log_2 [p(n)]$, was then evaluated. S gives a quantitative measure of the complexity of the power spectrum associated with the observed time series. By this measure, a power spectrum with a single peak (a linear periodic system) yields S = 0 bits.

A spectrum with two isolated peaks (a linear quasiperiodic system with two possible states) S=1 bit A flat or white noise spectrum extending to 40 Hz. (or n = 20,976) yields S = $\log_2$ 20,976 = 14.35 bits. Low values of S, relative to 14.35 bits, are then interpreted as associated with lower complexity in the oscillations of the system.

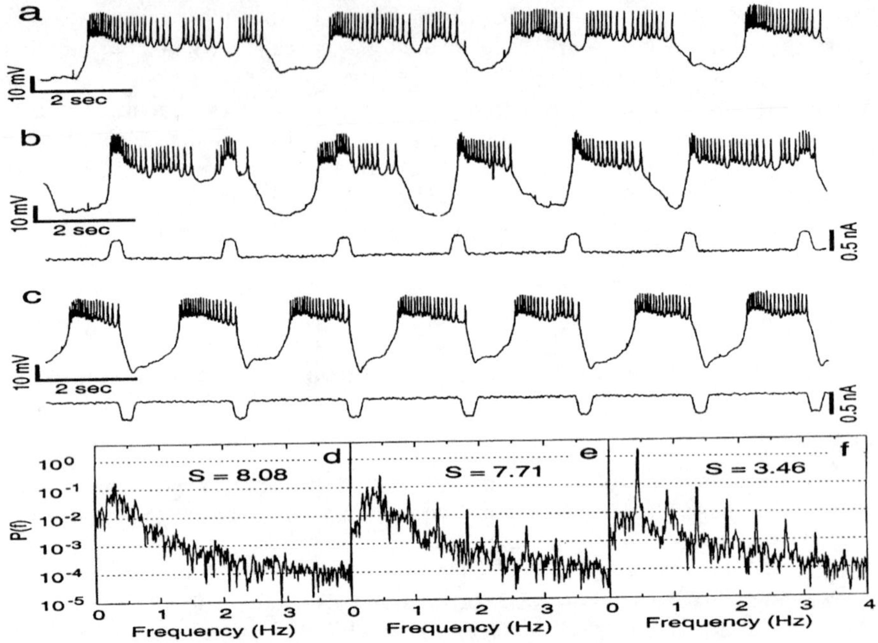

Figure 5. Regularization of irregular bursting activity by injection of periodic curent pulses. Recordings (A-C) and analysis (D-F) of voltage activity in an LP neuron after removal of chemical synaptic input from other pyloric neurons. (A) Free running control activity. (B) Forcing by periodic depolarizing current pulses (frequency, 0.45 Hz). (C) Forcing by periodic hyperpolarizing current pulses at the same frequency. (D) Fourier power spectrum for control activity. (E) Spectrum for depolarizing pulses. (F) Spectrum for hyperpolarizing pulses. Entropy values (S) are given in bits. Conditions: VD and both PDs killed; PTX; 1mM 4-AP.

An isolated but still modulated LP neuron behaves irregularly with an entropy S value of 8.08 bits and a power spectrum which shows a wide distribution and a broad peak. When current pulses were applied at a slightly higher frequency than the free-running neuron, 0.45 Hz., to approximate excitatory input, there were changes in the timing and duration of the bursts as might be expected, but little change in the irregularity The entropy dropped only to 7.71 and there was little change to the power spectrum aside from the addition of local peaks at the stimulus frequency and it's multiples. Hyperpolarizing current pulses meant to mimic inhibitory input had a much more powerful effect on regularizing the bursts. Entropy was reduced to 3.46 bits and the power spectrum was concentrated into sharp peaks at the applied frequency and its harmonics. The mechanism underlying the stabilization appears to be due to the hyperpolarizing current terminating the plateau potentials before instabilities in the burst phase occur. The initial stable phase of the burst can also be extended until the arrival of the next inhibitory input.

Regularization of LP by biological pacemaker inhibition

We also examined the way in which the LP neuron could be regularized by biological inhibitory input. To do this, we experimentally produced a subcircuit consisting of the pacemaker group consisting of the electrotonically - coupled AB and PD neurons, and the LP neuron. The VD neuron was photoinactivated and removed from the network and all other synapses once again blocked with picrotoxin.. This left the PDs with their inhibitory inputs to the LP intact and by direct current injection we could alter the frequency of bursting in the AB- PD pacemaker group. If we examine the effects of the PD forcing at different frequencies it is clear that the spontaneous pacemaker input regularized the free-running chaotic activity of LP in a frequency-dependent manner.

During spontaneous activity of the subcircuit, the LP displayed a power spectrum with strong peaks at the PD spontaneous frequency and its harmonics. The LP entropy was reduced to S = 5.1 bits. The LP was also able to follow at lower frequencies but as the forcing frequency was reduced, the LP power spectra broadened and lost structure. When the PDs were shut off entirely, the LP activity displayed a broadband structure and the entropy increased to S = 7.9 bits, essentially the same as above.

Burst Synchronization by reciprocal inhibition

Can bursting neurons synchronize each other ? When PD and LP inhibit each other via their normal connectivity which is reciprocal inhibition, they adopted antiphase coordination and regularized their bursting (Fig.6). With all natural synapses blocked, and with simulated synapses inserted via the dynamic clamp, we could show that only reciprocal inhibition (not any other pattern of connectivity) could produce regularization (unpublished data).

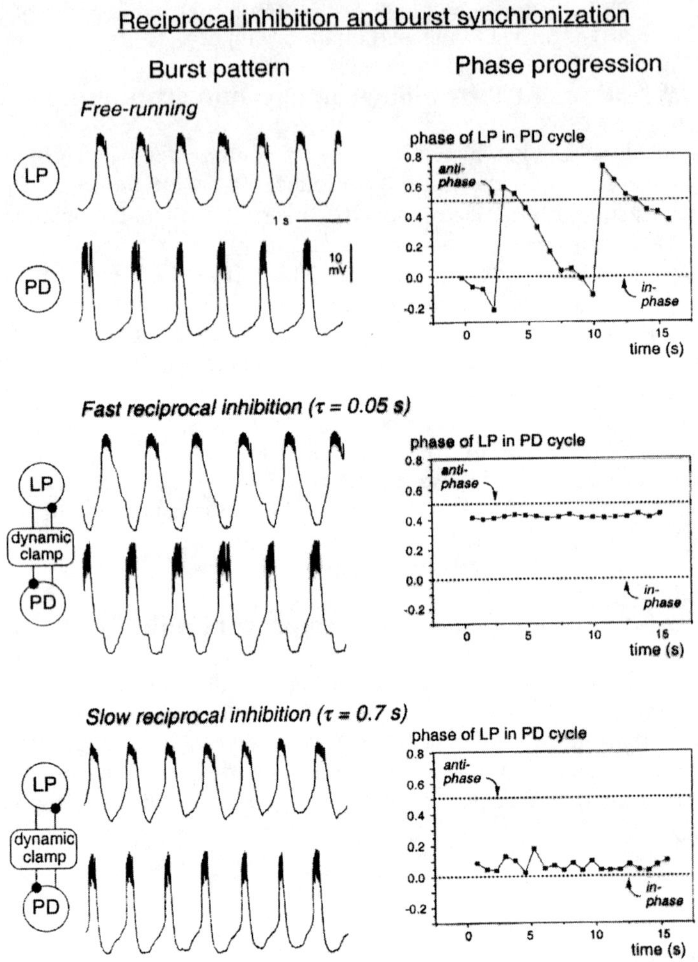

Figure 6. Reciprocal synaptic inhibition synchronizes the chaotic bursting of LP and a single PD neuron. Intrinsic bursting of LP and PD neurons recorded (at different times in the same experiment) while isolated from circuit interactions. The same two neurons coupled by reciprocal inhibition regularize out of phase. The LP-to-PD connection is inserted via a dynamic clamp. Conditions: AB,VD and one PD photoinactivated; PTX. With very slow synapses, the reciprocal inhibition can synchronize the two neurons in phase with one another.

Hybrid Silicon-Biological Circuits

One way to test the verisimilitude of our ENs was to interface them with single biological neurons or with the entire pyloric circuit. It would then be possible in principle to see if the EN could substitute for the real neuron under a variety of perturbations as well as in steady state conditions. The pyloric system is exceptionally well-suited for these experiments. The 14 neuron pyloric network has one interneuron, the AB, that is connected to the two PD neurons via electrotonic junctions Together these three neurons appear to form the pacemaker for the pyloric system, setting the frequency and periodicity of the pyloric pattern. The AB is the strongest neuron of the three in this respect and because it is robust and highly periodic it has been difficult to manipulate the rhythm when it is intact. The AB can be removed from the pyloric circuit by photoinactivation and glutamatergic transmission from other pyloric neurons onto the PDs was blocked with picrotoxin. A three dimensional EN was connected to the two PD neurons via an artificial electrical synapse which allowed us to dynamically interact with the biological neurons rather than simply providing them with current commands through a microelectrode.

When the AB was removed from a reduced circuit leaving only the PDs still connected to the VD neuron, the normal oscillatory pattern was immediately disrupted and the PD and VD fired tonically and irregularly. If the EN was adjusted to fire in bursts similar to those observed in the AB and then connected to the PDs by an artificial electrical synapse, the PD now fired synchronously with the EN (Fig. 7). When the EN is also made to fire irregularly and then connected to the chaotically firing PDs, the ENs and the PDs immediately went into synchronous in- phase bursting/firing activity. Similarly, if the electrical coupling was made negative, bursting/firing activity also ensued but now was out of phase. If the glutamatergic activity was blocked, the whole rhythm became severely disrupted immediately after the AB pacemaker neuron was photoinactivated. After about one hour the pyloric rhythm again became regularized although not as strongly as before. Replacing the destroyed AB with an artificial AB could rescue the circuit and the entire rhythm returned to its normal state.. Experiments were also carried out using a 4D EN with similar results.

Previous efforts using hybrid circuits have been made using a digital version of a Hodgkin-Huxley model which ran on a DSP board of a PC and was used to replace PD, LP and PY neurons. This hybrid circuit demonstrated that many aspects of the pyloric rhythm could be accurately reproduced [15]. This laboratory subsequently developed VLSI devices for implementing HH type models and has used them in hybrid circuits with results similar to ours but we have found that thus far such complex neurons are not needed in order to retain the output properties and dynamical features of the network. Furthermore the VLSI neurons simply drove the networks they were connected to whereas ours received current from the neurons in the circuit in a way that influenced their output.

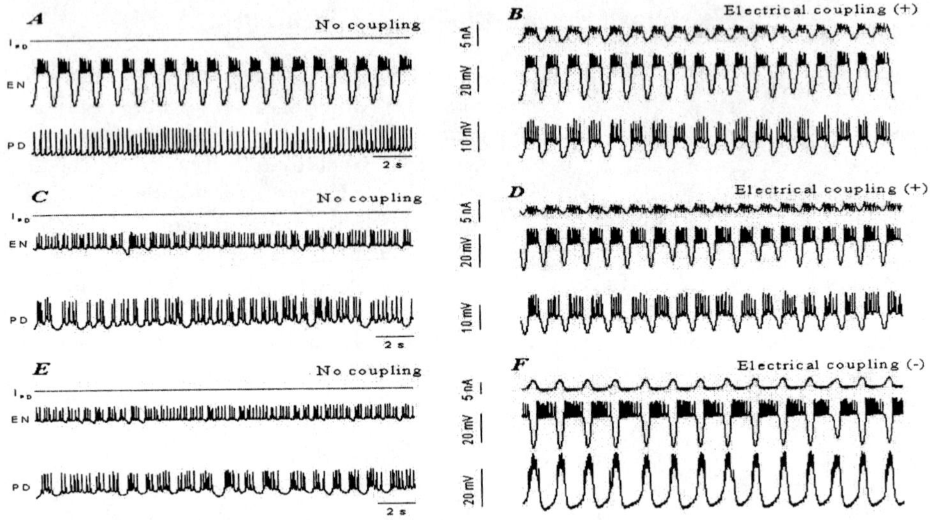

Figure 7. Hybrid circuit made up of an electronic(EN) and biological neuron (PD) coupled electrically produces unanticipated results. In A, a bursting EN coupled positively to a tonically firing PD produces a chaotic firing pattern (B). However when both the EN and the PD fire tionically and irregularly (C &E) bot h neurons burst periodically in phase when coupled positively (D) and out of phase when coupled negatively (F).

Modeling with hybrid circuits is an interesting way of testing how well an electronic neuron can interact with a small number of neurons in a well-characterized biological CPG. So far our silicon neuron having only four degrees of freedom appears to be capable of behaving indistinguishably from its biological counterparts. We therefore appear to have captured enough of the salient features of real neurons to allow their substitution into our small biological network <u>or</u> we have not yet tested the electronic neuron hard enough to expose its frailties. We find it remarkable however that our ENs are not only able of driving the biological neurons, but are influenced by them as well i.e., they actually become part of the CPG circuitry.

CONCLUSIONS

Our analysis of several components of a small neural network have produced a rather clear indication of the biological regulatory mechanisms that are used to regularize and synchronize the output of the system. We have found:

- Synaptically isolated neurons generally display intrinsic chaotic activity that has a dynamic origin.

- Coupling neurons together with electrical synapses or with chemical excitatory synapses will produce chaotic synchronization.
- Chemical inhibitory synapses will regularize chaotic bursting and spiking activity.
- Regularization of chaotic activity occurs with negative electrical coupling because it acts like a slow inhibitory chemical synapse.
- Reciprocal excitatory synapses, or reciprocal excitatory/inhibitory synapses will never regularize.

While the technology for determining if these same principles apply to the more complex systems found in vertebrates is still not available, every indication is that the same mechanisms for producing regularization and synchronization are available.

It is likely that isolated neurons are chaotic because of the many fast and slow processes occuring in the cell and not for purposes of of information processing.

Systems of such chaotic neurons explore a wider region of phase space thus confering greater flexibility and noise tolerance. Finally, because systems of chaotic neurons can bifurcate, they can easily switch the systems output from one stable state to another.

One additional point. An often neglected aspect of studies on the dynamics of motor pattern formation is the role of muscle properties. While there are many types of muscle and neuromuscular arrangements in the animal kingdom, one common feature of muscle is that it can only shorten and not lengthen. Tension to move an appendage therefore can only be produced during the shortening phase. To move the appendage in the opposite direction force must be applied in opposite direction by the antagonist muscle. This suggests that as motor systems evolved to control antagonistic pairs they likely would have developed mechanisms for producing alternating bursts of activity. If the alternating bursts were extremely periodic, there would be little tolerance for noise but more importantly if the system was perturbed it would have little flexibility in its response. Because the same muscles that produced oscillatory behavior also had to produce graceful episodic movements, control mechanisms would have to be developed to damp out underlying oscillatory activity. A large portion of the brain is devoted to doing this and pathologies of these areas often unmask this underlying oscillatory activity. If it were evolutionarily advantageous for motor systems to be built with an underlying base of oscillatory activity, it is not hard to imagine how similar activity became incorporated into other CNS functions as well.

ACKNOWLEDGEMENTS

The authors thank Rob Elson, Ramon Huerta, Pablo Varona, Reynaldo Pinto and Attila Szücs for many stimulating discussions and experimental data. This work was supported by ONR grant # N00014-00-1-0181 & NIH grant # NS 40110-01A2.

REFERENCES

[1] Laurent, G., Dynamical representation of odors by oscillating and evolving neuronal assemblies. Trends Neurosci., 1997. 19:489-496.

[2] Gray, C.M., Konig, P., Engle, A.K. Singer, W., Oscillatory responses in cat visual cortex exhibit inter-columnar correlation which reflects global stimulus properties. Nature, 1989. 338:334-337.

[3] Singer, W., Search for Coherence. Concepts in Neuroscience, 1990. 1:1-26.

[4] Selverston, A.I. and Moulins, M., (1987) The Stomatogastric Nervous System, Springer-Verlag, N.Y.

[5] Miller, J.P. and A.I. Selverston, Rapid killing of single neurons by irradiation of intracellularly injected dye. 1979. 206: p. 702-704.

[6] Sharp, A., L.F. Abbott, and E. Marder, Artificial electrical synapses in oscillatory neurons. 1992. In press.

[7] Hindmarsh, J.L., Rose, R.M., A model of the nerve impulse using two first-order differential equations. Nature, 1982. 296:162-164

[8] Falke, M., Huerta, R., Rabinovich, M.I., Abarbanel, H.D.I., Elson, R.C., Selverston, A.I., Modeling observed chaotic oscillations in bursting neurons: the role of calcium dynamics and IP3. Biol. Cyber., 2000. 82: p. 517-527.

[9] 8, Y.-X., Stojilkovic, S.S., Keizer, J., Rinzel, J., Sensing and refilling calcium stores in an excitable cell. Biophys J., 1997. 72: p. 1080-1091.

[10] Golowasch, J. and E. Marder, Ionic currents of the lateral pyloric neuron of the stomatogastric ganglion of the crab. J. Neurophysiol, 1992. 67: p. 318-331.

[11] Abarbanel, H.D.I. Analysis of observed chaotic data (1996) Springer Verlag, N.Y.

[12] Hayashi, H. and S. Ishizuka, Chaotic nature of bursting discharges in the Onchidium pacemaker neuron. J. Theor. Biol., 1992. 156: p. 269-291.

[13] Mpitsos, G.J., Evidence for chaos in spike trains of neurons that generate rhythmic motor patterns. Brain Res. Bull., 1988. 21: p. 529-538.

[14] Elson, R.C., Huerta, R., Abarbanel, H.D.I., Rabinovich, M., Selvertston, A.I. Dynamic control of irregular bursting in an identified neuron of an oscillatory circuit. J. Neurophysiol., 1999. 82:115-122.

[15] Renaud-LeMasson, S., G. LeMasson, S., Marder, L., Abbott, L.F., Hybrid circuits of interacting computer model and biological neurons, in Advances in Neural Information Processing Systems 5, S.J. Hanson, J.D. Cowan, andC.L. Giles, Editor. 1993, Morgan Kaufman: San Mateo, CA. p. 813-819.

Homoclinic Bifurcation in a Thermally Sensitive Neuron

Ulrike Feudel*, Alexander Neiman†, Xing Pei†, Winfried Wojtenek† and Frank Moss†

*Institute for Chemistry and Biology of the Marine Environment, Carl von Ossietzky University of Oldenburg, PF 2503, 26111 Oldenburg, Germany
†Center for Neurodynamics, University of Missouri at St. Louis, St. Louis, MO 63121, USA

Abstract. A wide class of sensory neurons demonstrates spontaneous oscillatory activity. Moreover, some thermosensitive neurons, for example, electroreceptors of the paddlefish, dogfish, the warm and cold receptors of rat and cat and the caudal photoreceptor of the crayfish, display complicated bifurcation sequences of the spike train patterns as the control parameter, e. g. the temperature, changes. Recent experiments also revealed the existence of low-dimensional chaotic behavior of some thermoreceptors. We study a rather unusual behavior of a bursting neuron exhibiting an explosion of interspike intervals at a certain temperature value. This phenomenon can be qualitatively demonstrated with electro-physiological experiments with the caudal photoreceptor of the crayfish. Furthermore we investigate a modified Hodgkin-Huxley model to understand the experimentally observed abrupt increase of the interspike intervals. We identify this transition with a homoclinic bifurcation of a saddle-focus equilibrium state which is embedded in the chaotic attractor of the system.

INTRODUCTION

Oscillating activity is characteristic for a large class of sensory neurons. The experimentally observed neuronal impulse patterns are of different types ranging from regular or irregular single-spike activity (tonic firing) to more or less rhythmic grouping of impulses (burst discharges). Such temporal patterns are partly ascribed to synaptic interactions within neuronal networks but can also result from the nonlinear internal dynamics of individual neurons. Here we present observations from electrophysiological measurements taken from crayfish caudal photoreceptor. Additionally we analyse the impulse pattern modulation of a Hodgkin-Huxley type model which originally was developed to simulate peripheral cold receptors discharges [1]. These receptors encode environmental temperatures into neuronal impulse trains and thereby exhibit a fascinating array of different types of firing patterns which, without doubt, originate from internal dynamics of single receptors.

Because of the small size and sparse distribution of cold receptors intracellular recordings of the membrane potential or ionic currents are not possible so far. The temporal patterns of the discharges, however, clearly indicate that the impulses are generated by an oscillating membrane process of systematic temperature dependencies. This can most

CP622, *Experimental Chaos: 6th Experimental Chaos Conference*, edited by S. Boccaletti et al.

easily be seen at mid-temperatures where rhythmic burst discharges are triggered. However, there is also an essential contribution of noise which becomes of particular functional significance especially towards higher temperatures where oscillations with skippings occur [2]. This phenomenon can spread all over the complete encoding range in other thermosensitive neurons, e.g. shark electroreceptors [3]. Moreover, there are other types of irregular discharges, preferably at lower temperatures, which at least partly seem to result from low-dimensional dynamics [1].

More detailed analysis of the impulse pattern indicate that the oscillations arise from subthreshold-mechanisms which are independent of spike-generation. For example, when one or more oscillation cycles are without spike-generation, the oscillations obviously continue with an almost unchanged rhythm and, at a given temperature, the period of the burst discharges is almost unaffected by random variations of the number of spikes per burst [2, 4]. The assumption of subthreshold, spike-independent oscillations is additionally supported by electrophysiological recordings under the variation of extracellular calcium concentrations and applications of calcium channel blockers. These experiments indicate that low-threshold calcium channels contribute to the oscillating process, but mainly seem to exert repolarizing effects, probably via Ca-activated K-channels, with neglectable contribution to depolarization [5].

There is a variety of different models for the generation of spike-triggering oscillations, most of them focus on specific types of patterns of central neurons (e.g. [6, 7]), others also consider the transitions between regular and chaotic patterns (e.g. [8]) or even directly refer to cold receptor data (e.g. [9]). The model which will be examined here has been developed on the background of the above described experimental findings, i.e. it considers subthreshold oscillations in addition to spike-generation. It uses a Hodgkin-Huxley-type approach which, however, is drastically simplified and reduced. Nevertheless, all relevant physiological parameters mentioned above do still have their explicit counterpart in the equations. This model, with addition of noise, exhibits all major characteristics of stationary cold receptor discharges [1] and, with slight modifications, can simulate principal neuromodulatory properties also observed in other types of temperature sensitive receptors and neurons [10, 11].

Usually, the only observable which is accessible in electrophysiological experiments is the interspike intervals recorded from a spike train of action potentials. These interspike intervals can be regarded as direct measurements of the Poincaré return time [12, 13]. From this point of view measurement techniques usually used in the study of neuro-biological systems are very convenient to investigate homoclinic bifurcations. Such bifurcations can be qualitatively identified by a sudden increase in the Poincaré return time of orbits close to the homoclinic one. Thus phjysiological evidence of this bifurcation can be expected in experiments where a sudden explosion of interspike intervals can be observed.

The dynamics of biological neurons is always contaminated with high-dimensional random processes, or "noise". For subthreshold regimes noise might play a constructive role leading to more coherent behavior of neuron systems via mechanisms of stochastic [14] or coherent [15] resonances. However, noise usually destroys the fine bifurcation structure of dynamical system, and, therefore, turns the bifurcation analysis into a hard statistical problem. Here we address these difficulties by studying the behavior of a modified Hodgkin-Huxley (MHH) model [1], with and without added noise, comparing the results with those observed in electrophysiological experiments.

EVIDENCE OF EXPLOSIONS OF INTERSPIKE INTERVALS IN PHYSIOLOGICAL EXPERIMENTS

In this section, we report experimental findings showing qualitative evidence of a homoclinic bifurcation. The measurements are obtained from extracellular recordings of the discharges from the caudal photoreceptor cell embedded in the crayfish sixth, or terminal, ganglion. In particular, we seek evidence of an explosion of the interspike time intervals using temperature as the control parameter. Moreover, one must search in sensory neurons which are known to exhibit subthreshold oscillators similar to those described by the MHH model.

The preparation has been described previously [16] but will be very briefly reviewed here. Crayfish (*Procambarus clarkii*) of approximately one year of age were obtained. The abdomen and tailfan were isolated quickly and fixed ventral side up in a dissection chamber filled with crayfish saline solution. The exoskeleton and muscles were removed in order to expose the sixth or terminal ganglion. The sixth ganglion containing both caudal photoreceptors was removed along with its associated bundle of axons (connective between the 5th and 6th ganglion). Nerve roots providing input to the 6th ganglion were cut away. The ganglion preparation was transferred to a recording chamber where the temperature of the saline bath was controlled. An extracellular electrode was attached to the CPR axon, and long sequences of action potentials (spike trains) were recorded at various bath temperatures.

Our data are presented as the record of interspike time intervals versus temperature as shown by the three panels in Fig. 1. These records were obtained from three separate animals. All three show at least qualitative evidence of the homoclinic bifurcation in the form of an explosion of the interspike time intervals that occurs at a more-or-less identifiable critical temperature. The best indication is shown in Fig. 1 (a), where the onset of the explosion occurs with decreasing temperature at approximately 8.3^oC. This recording is particularly interesting, because it clearly shows the initial period doubling bifurcation at about 13.5^oC followed by what appears to be a chaotic region. Moreover, this region includes at least two periodic windows at $T \approx 10.8^oC$ and 12.1^oC. Similar evidence of a period doubling bifurcation followed by an explosion of interspike intervals at low temperatures has been obtained for the rat cold receptor neuron, as shown in Fig. 6 of Ref.[17]. The recordings shown in Fig. 1 (b) and (c) are less clear, but both show the explosion upon increasing temperature. Figure 1 (b) also shows some evidence of the period doubling bifurcation at $T \approx 9.2^oC$.

Not all CPR cells show this behavior. We tested 16 CPR cells from 16 different animals. Of these only 4 showed evidence of the bifurcation. Moreover, we emphasize that the physiological evidence presented here is only qualitative. Nevertheless it is indicative and suggests further experiments as well as an explanation for time interval explosions observed in pacemaker cells grounded in the nonlinear dynamics of the underlying chaotic attractor.

Upon the basis of this physiological evidence for a homoclinic bifurcation we analyse

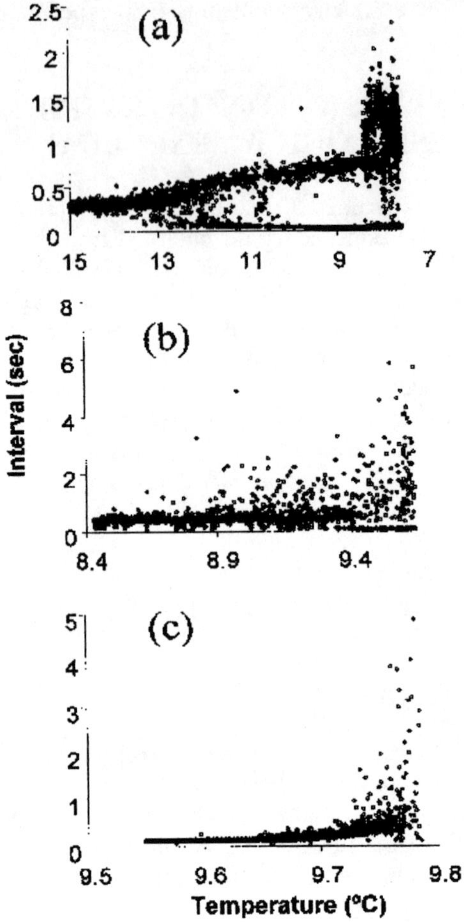

FIGURE 1. Interspike time interval recordings versus temperature of the saline bath for three different crayfish caudal photoreceptor cells.

next a Hodgkin-Huxley type model, which possesses the essential features of the spike train patterns observed. The aim of this study is to investigate the conditions under which such bifurcation can occur in thermosensitive neurons.

BIFURCATION ANALYSIS OF THE MODIFIED HODGKIN-HUXLEY NEURON MODEL

The MHH type model of thermally sensitive neurons used for our study has been proposed by Braun et al. [1] and mimics all spike train patterns observed in electroreceptors from dogfish [3], catfish [10], from facial cold receptors [10] and hypothalamic neurons of the rat [18]. All these neurons can be characterized by spontaneous, noisy oscillations which are reflected in the spike trains. The classical HH model[19] has been modified in the following sense: on the one hand, it has been simplified (e.g. no inactivation terms), but on the other hand, it has been extended by two additional slow currents according to the experimental findings of spike-independent oscillations (see above). By adding a stochastic term it produces more relastically looking simulations, e.g. makes the transitions between different impulse patterns more smooth, and is necessary to account for subthreshold oscillations with skippings when high temperature ranges are considered (which, however, are not in the scope of this paper). The MHH model (for a full description cf. Ref.[1]) is given by the usual equation for the membrane potential,

$$C_M \frac{dV}{dt} = -I_l - I_d - I_r - I_{sd} - I_{sr} + \xi(t), \tag{1}$$

where C_M is the membrane capacitance, V is the membrane potential, I_l is the leakage current, which can be written as $I_l = g_l(V - V_l)$. I_d and I_r are simplified fast Hodgkin-Huxley currents representing Na and K channels, respectively, which account for spike generation, I_{sd} and I_{sr} are the slow currents which account for the oscillations, and $\xi(t)$ is the noise. The voltage-dependent currents I_d, I_r, I_{sd} and I_{sr} are dependent on the voltage V and the temperature T.

$$I_k = \rho(T) g_k a_k (V - V_k), \tag{2}$$

where the index k stands for d, r, sd or sr currents, $\rho(T)$ is the temperature dependent scaling factor for the ionic currents, g_k is the maximum conductance and V_k is the reversal potential. The activation variable is a_k are also described by the differential equations. Thus our final model consists of 4 differential equations for the membrane potential V and the 3 activation variables a_r, a_sd and a_sr, while a_d is assumed to behave quasistationary. This assumption relies on the fact that the sodium current changes much faster than the other ionic currents.

The topological structure of the chaotic attractor of this system has been studied in Ref.[20], while several specific dynamical properties of this model, for example, the period doubling cascade to chaos and specific resonance phenomena have been studied in [21]. We focus here on global bifurcation phenomena in the MHH model. In particular we provide numerical evidence for the occurrence of a homoclinic bifurcation in the modified Hodgkin-Huxley model, show its robustness to noise, and also demonstrate similar dynamic behavior in electrophysiological experiments with the crayfish caudal photoreceptor. For this purpose the modified MHH model has been analysed numerically firstly without noise and secondly under the influence of noise. For an easy comparison with electrophysiological experiments, we use the interspike intervals, τ_n, as a state variable. The interspike intervals are registered by the membrane potential crossing a threshold (at $-20mV$) with positive derivative (Poincaré surface of section). Note that

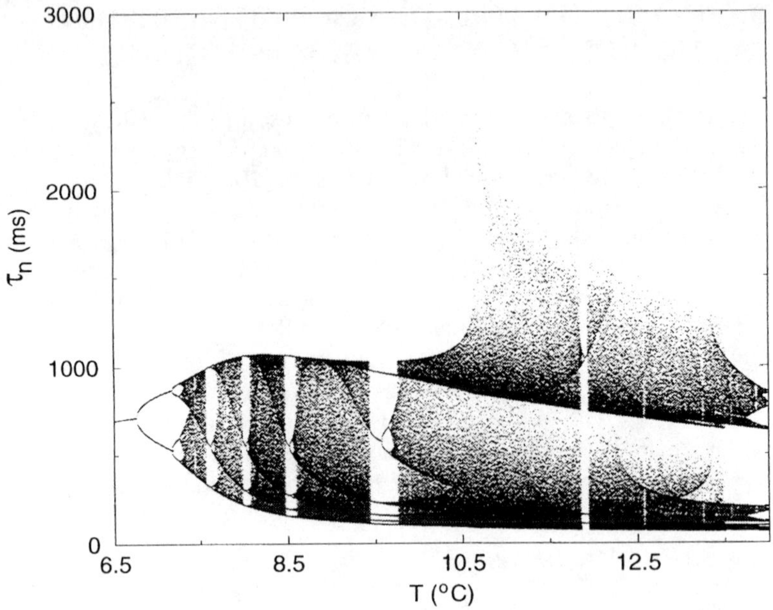

FIGURE 2. Bifurcation diagram of the MHH system. The deterministic case.

for oscillatory neuron models they can be associated with the instantaneous period of the action potential.

One possible way to study the generic properties of low-dimensional chaotic systems is the analysis of bifurcation sequences of limit sets under the variation of a control parameter. The bifurcation diagram of the MHH is shown in Fig. 2 and represents a classical route to chaos through a period-doubling cascade [22] up to a temperature $T \approx 10^oC$. The first period doubling bifurcation occurs at $T_1 \approx 6.765^oC$, the second at $T_2 \approx 7.195^oC$. Finally for $T > T_{ch} \approx 7.31^oC$ the system becomes chaotic. Inside the chaotic region we observe several periodic windows opened by a saddle-node bifurcation and closed by a global bifurcation, namely an interior crisis. The focus of our interest is the abrupt increase of the duration of the interspike intervals in the region $10.658^oC < T < 10.9^oC$ (see Fig. 2). For $T < T_{cr} = 10.6589^oC$ the duration of the interspike intervals is not larger than 1600 ms. At $T = T_{cr}$ an abrupt explosion of the interspike intervals occurs. The signature of this transition in phase space is the appearance of a new type of orbits within the attractor.

This new type of orbits suggests that this transition is connected with a bifurcation of the chaotic attractor. Such bifurcations are often related to global bifurcations, which are, in general, mediated by the stable and unstable manifolds of certain periodic orbits of saddle type. Examples for such global bifurcations are homoclinic bifurcations

[23, 24, 25] and different types of crises [26]. Homoclinic bifurcations are connected to the formation of a homoclinic orbit (or separatrix loop) associated with a saddle equilibrium state. Depending on the eigenvalues of the equilibrium state one distinguishes saddle-node, saddle-focus or bifocus type scenarios [23]. One of the characteristic features of a homoclinic bifurcation is the exponential growth of the times needed to return to a Poincaré surface of section for orbits in the neighborhood of the homoclinic orbit. Thus, the divergence of the return time can be used as an indicator for the appearance of a homoclinic bifurcation. Let us discuss this explosion in the interspike intervals in terms of a homoclinic bifurcation. As mentioned above, long time intervals between consecutive returns to a Poincaré section can be considered as an indicator for the approach of a homoclinic orbit. Such a homoclinic orbit is associated with an equilibrium state of saddle type and its stable and unstable manifolds. In our case this saddle equilibrium is embedded in the already existing chaotic attractor. It has two stable and two unstable directions. The eigenvalues, corresponding to the stable directions, are real ($\lambda_1 < 0, \lambda_2 < 0$) while the eigenvalues, corresponding to the unstable directions, are complex ($\gamma_{1,2} = \rho \pm i\omega, \rho > 0$). Thus we have an equilibrium state of saddle-focus type. Moreover the eigenvalues satisfy the Shilnikov condition [23] $-\lambda > \rho$ at a temperature value $T = 10.7456^oC$.

In the bifurcation a separatrix loop is built up, where the outgoing path is tangent to the unstable manifold while the returning path is tangent to the stable manifold of the equilibrium state. Nearby trajectories on the attractor approach the equilibrium state along its stable manifold. The motion slows down so that the trajectory spends a long time in the neighborhood of the equilibrium state before it is ejected along the unstable manifold. The same happens to saddle limit cycles embedded in the attractor. If they are close to the homoclinic orbit then at least one point of the saddle limit cycle is close to the saddle equilibrium state. As a consequence the Poincaré return time of these saddle limit cycles, that is their period, becomes also very large. The motion on the saddle limit cycle slows down approaching the vicinity of the equilibrium state along its stable manifold. For the homoclinic orbit itself the return time is infinitely long.

The computation of this homoclinic orbit using direct numerical methods or perturbation methods is rather difficult since the saddle-focus has a two-dimensional stable and two-dimensional unstable manifold with two stable real eigenvalues very close to each other, a situation which cannot be easily reduced to the usual Shilnikov scenario in R^3. Therefore, we can show the evidence of the homoclinic bifurcation only by indirect indicators. Two different indicators have been used. Firstly, we computed different saddle cycles of low period and looked at their distance to the saddle-focus. It turned out that at the temperature $T = 10.7456^oC$ the saddle cycles are extremely close to the saddle-focus, while they have been rather far from it before the expected bifurcation point.

Secondly, we use the fact that the phase trajectory approaches the vicinity of the equilibrium state the motion of the system slows down, so that the phase trajectory spends a long time in the neighborhood of the equilibrium state. This gives rise to very long interspike intervals. To support this argument we have computed the dependence of the minimal distance between the saddle-focus and the trajectory on the chaotic attractor depending on the control parameter and the maximal residence time of the phase trajectory spent in a small neighborhood of the equilibrium state. This way we

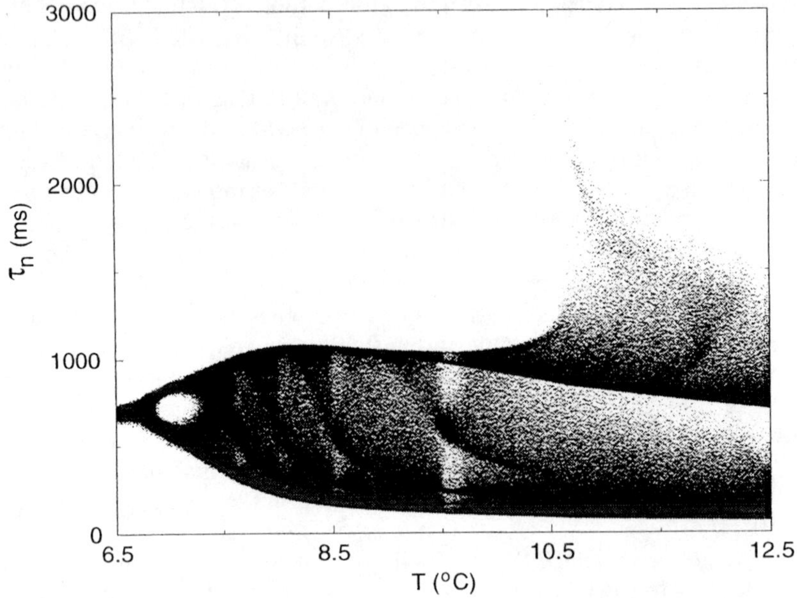

FIGURE 3. Bifurcation diagram of the stochastic MHH system with $D = 0.001$.

were able to identify a clear maximum of the residence times close to the saddle-focus. The location of this maximum suggests that the homoclinic bifurcation occurs at $T \approx 10.7456^oC$.

Since all biological systems are inherently noisy, it is necessary to test how robust the above described transition is against noise. For this purpose we include an additive white Gaussian noise $\xi(t)$, $<\xi(t)>= 0$, $<\xi(t)\xi(t+s)>= 2D\delta(s)$, where D is the intensity, into the first equation of the MHH for the membrane potential and study numerically the corresponding set of stochastic differential equations. It turns out, that the system is extremely sensitive to this noise. The bifurcation diagram of the noisy system is shown in Fig. 3 for $D = 0.001$. Although almost all fine structure of the bifurcation diagram is blurred by the noise, the transition studied above in a noiseless system is still well pronounced: again at a certain value of the temperature there is an explosion of interspike intervals. However the transition appears to be smoothed and the critical parameter values are shifted towards smaller values of temperature. Finally we would like to mention that close to the homoclinic bifurcation one can observe an intermittent process.

SUMMARY

We have presented electrophysiological experiments exhibiting evidence of an explosion of interspike intervals as a response on the variation of a control parameter, namely the temperature. The observed behavior resembles qualitatively the dynamics obtained from studies of homoclinic bifurcations. Therefore, we have analyzed the bifurcation structure of a modified Hodgkin-Huxley type model to explain the explosion of interspike intervals in terms of a homoclinic bifurcation. There are a number of arguments which support this explanation, namely the exponential growth of the Poincaré return times of saddle periodic orbits with different periods and their closeness to the saddle equilibrium state embedded in the attractor. We obtained an accumulation of these saddle periodic orbits as we approach the homoclinic bifurcation. Close to the expected homoclinic orbit we find an intermittency between stretches of trajectories with only short interspike intervals and excursions to long interspike intervals. The described bifurcation picture suggests the following mechanism for the observed explosion of the interspike intervals. As we approach the homoclinic bifurcation of the saddle equilibrium state, the phase trajectory on the chaotic attractor spends a considerable amount of time in the vicinity of the saddle-focus equilibrium point. This leads to the appearance of very long interspike intervals. At the onset of this transition the phase trajectories reach the neighborhood of the saddle-focus very rarely, so that the long interspike intervals appear very seldom too. With the increase in the control parameter, the system approaches the homoclinic bifurcation, the phase trajectories come closer and more frequently to the saddle-focus. As a result a larger number of the long interspike intervals appears and the mean length between bursts decreases. The divergent behavior of interspike intervals can be observed for both cases of increasing and decreasing temperature, when the control parameter approaches the homoclinic bifurcation. The existence of such a global bifurcation and its drastical consequence, that is the explosion of the interspike intervals, suggests that the temporal code might be an appropriate encoding scheme for thermoreceptors. Specifically, the occurrence of long interspike intervals probably indicates the approach to painful temperatures, while the explosion of interspike intervals, which, as we have shown, occurs due to the homoclinic bifurcation, physiologically means a blockage of neuron activity. Furthermore, this transition is robust against noise and is observable in the experiment. However, further experiments are necessary to understand the physiological significance of the phenomenon studied here.

ACKNOWLEDGMENTS

This work was supported by the U. S. Office of Naval Research, The Fetzer Institute, the Department of Energy and the German Research Foundation (DFG).

REFERENCES

1. Braun, H., Huber, M., Dewald, M., Schäfer, K., and Voigt, K., *Intern. J. Bifurc. and Chaos*, **8**, 881 (1998).

2. Braun, H., Bade, H., and Hensel, H., *Pflügers Arch*, **386**, 1 (1980).
3. Braun, H., Wissing, H., Schäfer, K., and Hirsch, M., *Nature*, **367**, 270–273 (1994).
4. Schäfer, K., Braun, H., and Rempe, L., *Experientia*, **47**, 47 (1991).
5. Schäfer, K., Braun, H., and Isenberg, C., *J. Gen. Physiol.*, **88**, 557 (1986).
6. Toth, T., and Crunelli, V., *NeuroReport*, **3**, 65 (1992).
7. Wang, X., *NeuroReport*, **5**, 221 (1993).
8. Chay, T., Fan, Y., and Lee, Y., *Int J Bifurcation & Chaos*, **5**, 595 (1995).
9. Longtin, A., and Hinzer, K., *Neural Comput.*, **8**, 215 (1995).
10. Braun, H., Schäfer, K., Voigt, K., Peters, R., Bretschneider, F., Pei, X., Wilkens, L., and Moss, F., *J. Comp. Neurosci.*, **4**, 335 (1997).
11. Huber, M., Krieg, J., Dewald, M., Voigt, K., and Braun, H., *Biosystems*, **48**, 95 (1998).
12. Sauer, T., *Phys. Rev. Lett.*, **72**, 3811 (1994).
13. Janson, N., Pavlov, A., Neiman, A., and Anishchenko, V., *Phys. Rev. E*, **58**, R4 (1998).
14. Wiesenfeld, K., and Moss, F., *Nature (London)*, **373**, 33 (1995).
15. Pikovsky, A., and Kurths, J., *Phys. Rev. Lett.*, **78**, 775 (1997).
16. Pei, X., Wilkens, L., and Moss, F., *J. Neurophysiol.*, **76**, 3002 (1996).
17. Braun, H., Dewald, M., Voigt, K., Huber, M., Pei, X., and Moss, F., *J. Comp. Neurosci.*, **26**, 79 (1999).
18. Dewald, M., Braun, H. A., Voigt, K., Pei, X., and Moss, F., *J. Comp. Neurosci.* (2001), submitted.
19. Hille, B., *Ionic Channels of Excitable Membranes, 2nd ed.*, Sinauer, Sunderland, MA, 1992.
20. Gilmore, R., Pei, X., and Moss, F., *Chaos*, **9**, 812 (1999).
21. Braun, W., Eckhardt, B., Braun, H., and Huber, M., *Phys. Rev. E*, **62**, 6352–6360 (2000).
22. Ott, E., *Chaos in dynamical systems*, Cambridge University Press, 1993.
23. Shil'nikov, L., *Math. USSR Sb.*, **10**, 91–102 (1970).
24. Tresser, C., *Ann. Inst. H. Poincaré*, **40**, 441–461 (1984).
25. Glendinning, P., "", in *New Directions in Dynamical Systems*, edited by T. Bedford and J. Swift, Cambridge University Press, 1987.
26. Grebogi, C., Ott, E., and Yorke, J., *Phys. Rev. Lett.*, **48**, 1507 (1982).

Periodic Forcing of Congenital Nystagmus

Richard A. Clement*, Richard V. Abadi†, David S. Broomhead¶ and Jonathon P. Whittle‡

**Visual Sciences Unit, Institute of Child Health, University College London, 30 Guilford Street, London WC1N 1EH*
†Department of Optometry and Neuroscience, UMIST, PO Box 88, Manchester M60 1QD.
¶Department of Mathematics, UMIST, PO Box 88, Manchester M60 1QD.
‡The University of Sheffield, Department of Ophthalmology and Orthoptics, Royal Hallamshire Hospital, Glossop Road, Sheffield S10 2JF.

Abstract. Congenital nystagmus is an involuntary oscillation of the eyes, which is associated with reduced visual acuity. We recover periodic orbits from the eye movement recordings and then explore the possibility of using periodic forcing to stabilise the waveforms, so that the retinal image movement is minimised and visual acuity development can be enhanced in children.

INTRODUCTION

Congenital nystagmus is an oculomotor disorder which manifests itself within a few months of birth and persists through life. Fixation is disrupted by predominantly horizontal eye movements, which can be either approximately sinusoidal, or can consist of alternating fast and slow phases. Although visual acuity in subjects with congenital nystagmus is variable, most adults with nystagmus do not have driving vision.

The aim of our project is to investigate new approaches to improving vision in congenital nystagmats, based on recently developed ideas about the behaviour of neuronal systems [1]. We assume that the behaviour of the oculomotor system underlying nystagmus is based around a limited number of periodic orbits, so that identification of such orbits provides a method of characterising the system. This approach has great implications for the control of the system behaviour, since both periodic and chaotic behaviours can be controlled by making small perturbations to move the state of the system from close to one periodic orbit to another.

IDENTIFICATION OF PERIOD ONE ORBITS

A first pass at identifying the presence of periodic orbits is to search for close returns in the time series of measurements x[1], x[2],...x[N] [2]. In fig. 1 an example of a nystagmus waveform is shown, together with a plot of $|x[i] - x[i+p]|$. The difference between offset data points is plotted as a grey level function such that zero difference is shown as black. Approximately horizontal, parallel black lines are

CP622, *Experimental Chaos: 6th Experimental Chaos Conference*, edited by S. Boccaletti et al.

clearly visible in the close returns plot, separated by intervals of 0.25 seconds, implying an underlying period 1 orbit.

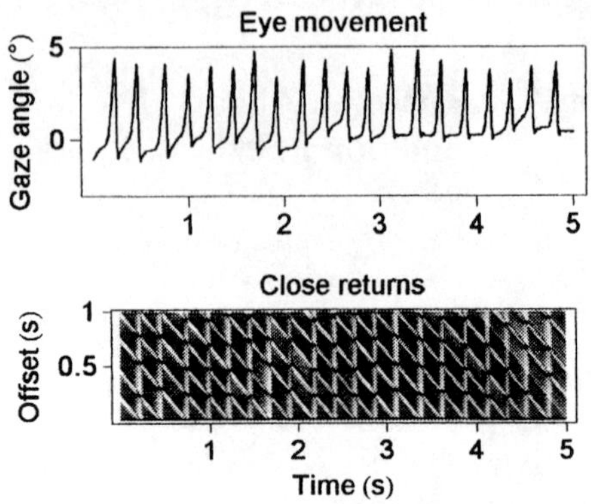

FIGURE 1. The upper plot is an example of a congenital nystagmus eye movements. The lower plot shows the close returns of the eye movements, with zero difference plotted in black.

A more accurate estimate of the period of the orbit can be obtained by analysing the recurrence properties of the system [3]. A previous investigation of the stability of gaze holding in congenital nystagmus revealed that the system is effectively 3 dimensional close to the direction of fixation [4]. We reconstructed the phase space trajectories of normalised data using the method of delays with $\mathbf{x}[n] = (x[n], x[n+3], x[n+6])$, and searched for points which were within a 0.25 radius of each other. Peaks were found in the histogram of frequency of recurrent points against the time offset between the samples at multiples of 0.25 seconds. Fig. 2 shows the results for the data shown in figure 1, together with the waveform cycle which most closely matches the period of 0.25 seconds.

Another approach to identification of periodic orbits is based on reconstructing the dynamics of the systems from intervals between events. A recently developed technique for finding periodic orbits in noisy experimental data involves transforming the data so that it is concentrated on the periodic orbits, which can be identified by sharp peaks in a histogram of the transformed data [5]. Fig. 3 shows the results of applying this techniques to our data. The dimension of the system was reduced by 1 by thresholding the recorded waveforms and calculating the inter threshold crossing intervals, so we used the version of the data transformation algorithm which is appropriate for 2 dimensional systems.

Although we found statistically significant evidence for periodic one orbits within congenital nystagmus waveforms, this behaviour could arise from a deterministic system behaving chaotically, or from a periodic system with noise. Attempts to fit a local linear model to the fixed point did not show a minimum in the one step prediction error as the neighbourhood size was increased, and so eigenvalues could not be extracted reliably. Our data sets consist of several hundred intervals, whereas those for which linear models have been established typically consist of several thousand intervals [6,7].

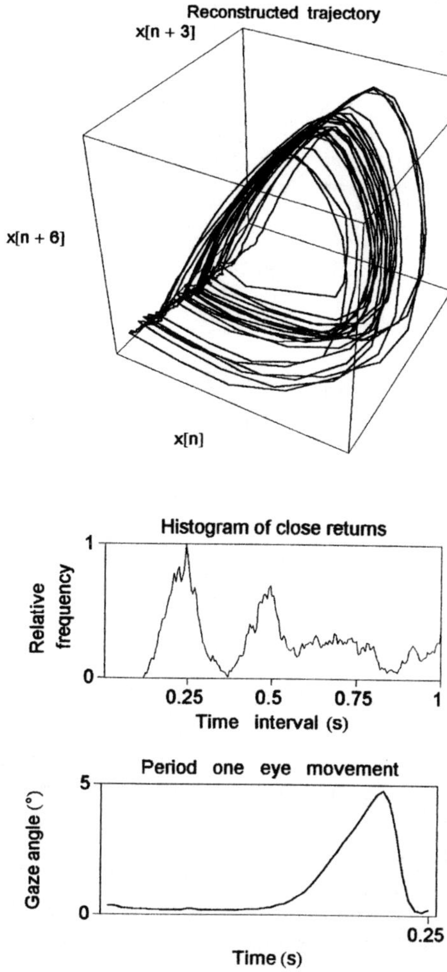

FIGURE 2. Identification of a period one orbit by the method of close returns. The upper figure is a reconstruction of the phase space trajectory of the data show in figure 1. The middle figure reveals that points on the trajectories are close to each other most often when they are multiples of 0.25 seconds apart in time. The lower figure shows the cycle of the waveform which most closely returns to itself after the period.

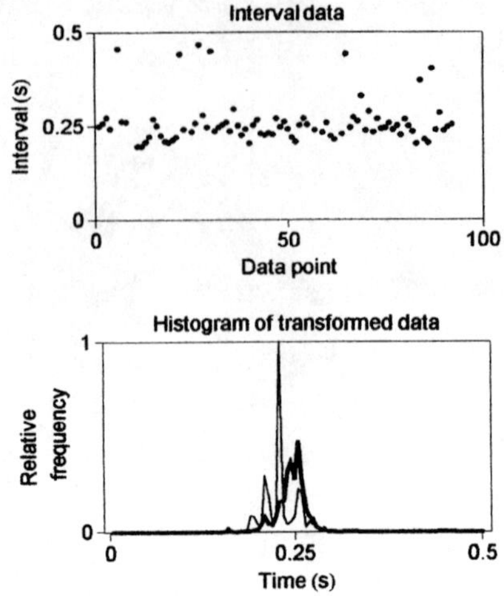

FIGURE 3. Identification of period one orbits by the fixed point technique. The interval data obtained from the eye movement recording is shown in the upper plot. The lower plot is a histogram of the transformed data, which shows a peak at the interval of 0.23 s. The thick line is the average histogram of the transformed surrogate data.

PERIODIC FORCING

Although the lack of evidence for a chaotic system ruled out the application of chaos control techniques, periodic pacing can be applied effectively to a system behaving chaotically or a periodic system with noise [8]. Direct access to the mechanism of the oculomotor system was not possible, but it was possible to alter the visual stimulus. We presented the subject with a target which moved periodically with the waveform associated with her periodic orbit and found eye movements which were periodic and synchronised with the forcing function. Fig. 4 shows the eye movement recording becoming synchronised with the stimulus in the middle of the trace. Interestingly, in the latter part of the trace the subject shows rapid return eye movements in both directions, so the stimulus and eye movement system are not simply operating independently at different frequencies. However, the bidirectional waveform precludes the use of any analysis based on simply thresholding the data.

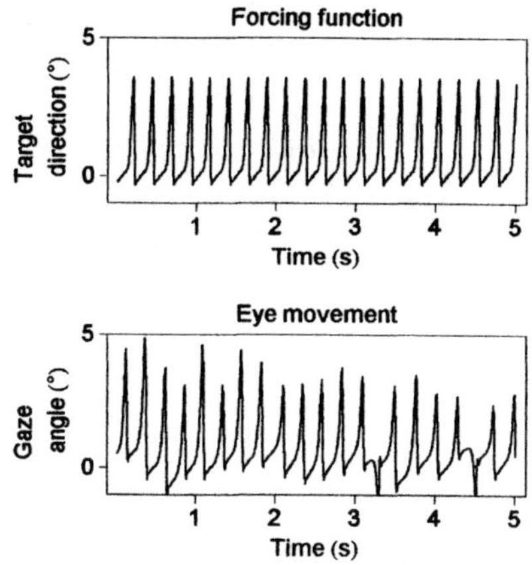

FIGURE 4. Periodic forcing of congenital nystagmus. The stimulus is shown in the upper figure and the eye movements of the subject are shown in the lower figure.

IMPLICATIONS

The underlying periodicity of congenital nystagmus is well established from spectral analysis of eye movement time series [9]. The advantage of the nonlinear dynamics approach is that it enables identification of the portions of the waveforms where the system is closely following a periodic orbit. Clinically this is useful, because it enables one to isolate the characteristic waveforms from the range of oscillatory behaviours shown by any one subject. A longstanding criticism of clinical classifications of congenital nystagmus waveforms is that they do not directly relate to the mechanism of the condition [10].

The finding that the period one behaviour continues with periodic forcing is promising, because it suggests that congenital nystagmus eye movements will match the movements of targets which move with the waveform associated with the period one orbit of the subject. If the techniques for isolating period one orbits can be used in small children (who usually produce very short and very noisy data sets), the periodic forcing approach will enable the clearest possible images to be presented to the developing visual system.

REFERENCES

1. So, P., Francis, J.T., Netoff, T.I., Gluckman, B.J. and Schiff, S.J., Biophys. J. 74: 2776-2785 (1998).
2. Mindlin, G.B., Solari, H.G., Natiello, M.A., Gilmore, R. and Hou X.-J., J. Nonlinear Sci. 1: 147-173 (1991).
3. Tufillaro, N.B., Abbott, T. and Reilly, J., An Experimental Approach to Nonlinear Dynamics and Chaos. Addison-Wesley, Redwood City, California, 1992, pp.242-249.
4. Abadi, R.V., Broomhead, D.S., Clement, R.A., Whittle, J.P. and Worfolk, R. Exp. Brain Res. 117: 335-361(1997).
5. So, P., Ott, E., Sauer, T., Gluckman, B.J., Grebogi, C. and Schiff, S.J. Phys. Rev. E 55: 5398-5417 (1997).
6. Pei, X. and Moss, F. Nature 379: 619-621. (1996).

7. Braun, H.A., Dewald, M., Schäfer, K., Voigt, K., Pei, X., Dolan, K. and Moss, F. *J. Comp. Neurosci.* **7**:17-32 (1999).
8. Christini, D.J., and Collins, *J.J. Phys. Rev. E.* **52**: 5806-5809 (1995).
9. Reccia, R., Roberti, G. and Russo, P., *J. Biomed. Eng.* **12**: 39-45. (1990).
10. Abadi, R.V. and Dickinson, C.M., *Documenta Ophthalmol.* **64**:153-167. (1986).

Complexity of EEG-signal in Time Domain - Possible Biomedical Application

Wlodzimierz Klonowski, Elzbieta Olejarczyk, Robert Stepien

Lab. of Biosignal Analysis Fundamentals, Institute of Biocybernetics and Biomedical Engineering, Polish Academy of Sciences, PL-02109 Warsaw, Poland, e-mail: wklon@hrabia.ibib.waw.pl

Abstract. Human brain is a highly complex nonlinear system. So it is not surprising that in analysis of EEG-signal, which represents overall activity of the brain, the methods of Nonlinear Dynamics (or Chaos Theory as it is commonly called) can be used. Even if the signal is not chaotic these methods are a motivating tool to explore changes in brain activity due to different functional activation states, e.g. different sleep stages, or to applied therapy, e.g. exposure to chemical agents (drugs) and physical factors (light, magnetic field). The methods supplied by Nonlinear Dynamics reveal signal characteristics that are not revealed by linear methods like FFT. Better understanding of principles that govern dynamics and complexity of EEG-signal can help to find 'the signatures' of different physiological and pathological states of human brain, quantitative characteristics that may find applications in medical diagnostics.

INTRODUCTION

System Theory and Nonlinear Dynamics are very interdisciplinary. Nonlinear Dynamics applies to systems that are not in equilibrium - from a single protein macromolecule with intrinsic physical instability (in scientific literature called *Klonowski-Klonowska conformon* [1]) and self-oscillating cross-linked (bio)polymeric materials [2], to nonlinear dose-effect relationship in Medicine (called *hormesis* [3]), and human brain [4].

Processes taking place in complex systems often produce dynamical signals that can be registered in a form of time series. It is not only by accident that colloquially one speaks about *'social organisms'* - the same methods that are used for analysis of complex systems in natural sciences may also be adapted for socio-economic systems' analysis. These systems also produce some specific 'signals' and complexity analysis of these signals can help to reveal 'the signatures' of different *'physiological'* and *'pathological' states of society and government.*

EEG-signal generated by human brain may have a fractal or multifractal temporal structure, while being inhomogeneous and non-stationary. That is why methods of Nonlinear Dynamics may be (and should be) used to analyze EEG-data. We search for characteristics that might be calculated relatively easily, facilitating assessment of patient's state and of the impact of a prescribed therapy, and eventually helping doctors in diagnostics.

CP622, *Experimental Chaos: 6th Experimental Chaos Conference*, edited by S. Boccaletti et al.

IS EEG-SIGNAL CHAOTIC?

EEG-signal appears to be irregular and unpredictable. It used to be thought of as a random one. But since Nonlinear Dynamics has introduced deterministic chaos as a separate class of dynamics this theory and the new methods of nonlinear analysis found various applications in analysis of EEG and other biosignals.

Among characteristic properties of chaotic dynamics are finite correlation dimension, positive largest Lyapunov exponent (LE), and good short-time predictability. In our previous works we found the largest LE of EEG-signals to be positive (cf. [5]). However LE is very sensitive to noise and long stationary signal epochs are needed to calculate this quantifier, leading to false rejections of the signal's stochastic origin hypothesis [6]. On the other hand, some stochastic processes that produce time series with power law spectra may show finite correlation dimension [7].

Recently we analyzed 16-channels EEG of 21 healthy human subjects under resting condition [8], performing the test for nonlinearity by surrogate data method with the *null hypothesis*: "These EEG-data were generated by a linear stochastic process (LSP)" and with nonlinear prediction error (NPE) of either simple nonlinear prediction or of S-map forecasting (short for Sequential Locally Weighted Global Linear Maps, SLWGLM [9]) as a *discriminant statistic*. The null hypothesis is rejected if the value of the discriminant statistic differs between the original data and the surrogate data, which are series of artificially generated data sharing some linear properties with the original time series (cf. [6]).

The results show that on each of 16 channels the null hypothesis that EEG signal is generated by LSP should be rejected for some subjects, but in most cases (319 out of 336 'person-channels') the null hypothesis can be accepted (Fig. 1).

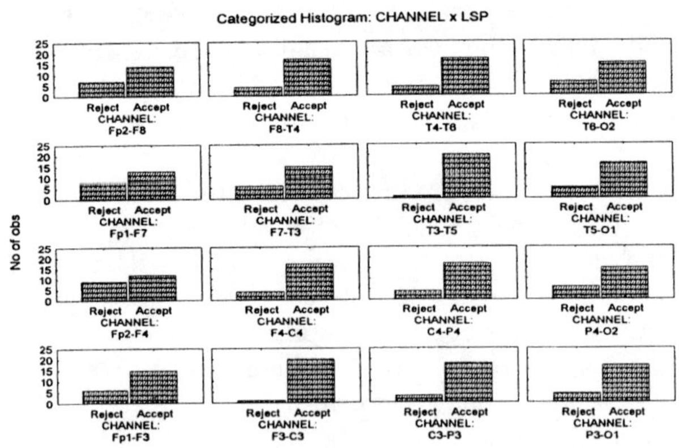

Figure 1. Results of the nonlinearity test by S-map prediction for 21 healthy subjects on 16 channels

However, it should be emphasized, that EEG dimensional complexity could be used as a *relative* index of signal complexity regardless of the nature of the underlying dynamics ([10]-[12]) i.e. even if the analyzed signal is not chaotic.

SIGNAL COMPLEXITY

Signal complexity [13] can be analyzed either in frequency domain, or in time domain, or in the system's phase space. Analysis in the frequency domain requires Fourier or wavelet transform of the signal, while analysis in the phase space requires embedding of the data in a multidimensional space.

Complexity in time domain may be characterized by *fractal dimension*, D_f. It characterizes complexity of the curve representing the signal on a plane, e.g. that of the line traced out on a moving paper tape by a pen of a classical signal recorder (nowadays replaced by a computerized registration of signal's amplitude as a function of time). Similarly like the fractal dimension of a coastline, D_f has its values always between 1 and 2. D_f calculated in time domain **should not be confused with fractal dimension of the attractor in the system's phase space** [14]. Calculation of fractal dimension in time domain is quick and easy, it can be done in real.

Fractal Dimension as a Measure of Signal Complexity

The problem with *fractal dimension* is that the same notion is used to denote many, often very different quantities [10]. E.g. a physicist while hearing the words 'fractal dimension' almost automatically imagines an attractor in a system's phase space. It is in the phase space where Nonlinear Dynamics and Deterministic Chaos Theory are intimately related to Fractal Geometry, since systems that give rise to deterministic chaos have *chaotic* or *strange attractors* which are characterized by fractal (here meaning 'not-integer') dimension. Reconstruction of system's attractor from a signal generated by this system is based on the Taken's method of delayed coordinates and poses several methodological problems – the data characterizing the system (like amplitudes of the considered signal in discrete moments of time) have to be embedded in a multidimensional space of properly chosen dimension; also time delay parameter has to be properly chosen to find the balance between the irrelevance and the redundancy [12,15].

But an object in Euclidean space, like a coastline of an island, may also be characterized by a fractal dimension. In a rough sense, it measures 'how many points' belong to a given set. A plane is 'larger' than a line, while Sierpinski triangle sits somewhere in between these two sets. On the other hand, all three of these sets have the same number of points in the sense that each set is uncountable. Somehow, though, fractal dimension captures the notion oh 'how large a set is' while its fractional part captures the notion of *'how complex a set is'*. Similarly, fractal dimension in time domain, or rather its fractional part, characterizes *signal's complexity* – the greater it is the more complex the signal under consideration is. Fractal dimension of a signal provides insight into how elaborate the process that generated the signal might have been, since the larger the dimension the larger the number of degrees of freedom likely have been involved in that process.

Fractal Dimension Calculated in Time Domain

Correlation dimension of a reconstructed system's attractor, D_2, has been used as a relative measure of complexity (cf. [12]). Instead of D_2 we prefer to use EEG-signal fractal dimension calculated in time domain, D_f, using e.g. the algorithm proposed by Higuchi (cf. [4],[10]). Calculation of signal fractal dimension in time domain needs no embedding of the data in a phase space, so there is no need of elaborated procedures to choose proper embedding dimension and proper time delay.

Calculation of D_f is simple and quick – raw EEG-data are transformed into $D_f(t)$ curve, one local D_f value for as few (or as many) as 100 data points, leading to 100-fold compression of the original data. $D_f(t)$ correlates with power spectrum of the EEG-signal - for the intervals for which fractal dimension is below the average D calculated for the whole signal, the power spectrum is moved towards lower frequencies; for those intervals for which fractal dimension is above the average D, the power spectrum is moved towards higher frequencies [14].

We have demonstrated that D_f may help to assess the effectiveness of phototherapy in patients with Seasonal Affective Disorder (SAD, better known as Seasonal Depression) [16], to assess effects of chemotherapy [4], to assess influence of magneto-stimulation on healthy subjects [17].

For EEG registered during sleep (together with other measured biosignals, like ECG, EOG, etc. called *polysomnogram* [18]) $D_f(t)$ on one selected channel enables categorization of 4 sec. long time segments into (at least three out of five) different sleep stages, i.e. it may be helpful for specialists in constructing so called *hypnograms*, which are important tools in diagnostics of sleep pathologies and in assessment of chemotherapy with sleep-helping drugs. The method is prone to signal artefacts – the mean values of D_f in various sleep stages are practically the same, no matter if calculated from the raw signal or from the signal with artefacts removed by a EEG-specialist [11].

For computerized construction of the hypnogram we propose to use EEG-signal on channel C3 (using standard 10/20 electrode placing system nomenclature) relative to the average local value (CSD, current source density) – from the unipolar EEG-signal on given electrode one subtracts the average value of the signal on four nearest-neighbor electrodes

$$C3_{CSD} = [(C3 - F3) + (C3 - P3) + (C3 - T3) + (C3 - Cz)] / 4 \qquad (1)$$

We calculated Higuchi's fractal dimension of $C3_{CSD}$ signal of 15 healthy subjects for each 4 second-long epoch categorized by a specialist to one of five sleep stages or as waked; then we calculated the average value of fractal dimension in each sleep stage for each person. The results show that this may be a promising method for computerized construction of hypnograms (Fig. 2). It is important to stress that even a specialist cannot categorize stage 3 from stage 4 and stage 1 from REM based on EEG only; for that purpose other biosignals registered in the polysomnogram, such as EOG, are necessary.

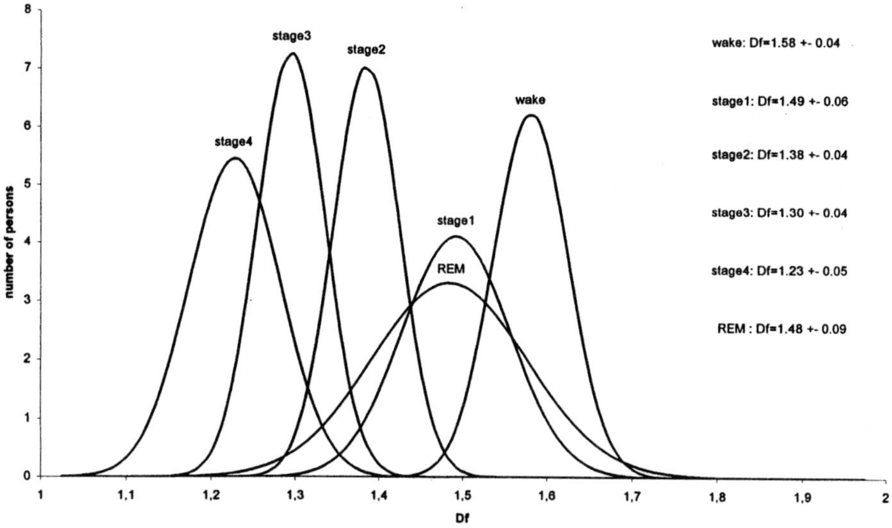

Figure 2. Fractal dimension of $C3_{CSD}$ EEG-signal in various sleep stages in healthy subjects.

DISCUSSION

It is feasible to use fractal dimension as a tool to characterize the complexity of electroencephalographic time series (EEG-signals). The EEG appears to have complexity properties which are sensitive to the clinical state of the subject.

Tools derived from Nonlinear Dynamics enable to detect orderly dynamical patterns hidden in seemingly random EEG-signal time series. Complexity of EEG-signal should be regarded, at least partially, as a measure of lawful and deterministic regulatory processes of brain activity. We can use such measures as D_f to quantify the patterns in EEG and characterize different brain states, physiological and pathological ones, correspondingly.

To be able to analyze properties of brief duration (transients) it is necessary to make calculations directly in the time domain, which is analogous to calculate fractal dimension of a curve representing the signal on a paper tape. Complexity analysis in time domain allows investigating relevant EEG 'events' shorter than those detectable by means of other linear and non-linear techniques. Local (in time) changes of fractal dimension of EEG-signal due to different 'events' (for example FD-ratio i.e. eyes opening/eyes closing ratio we have proposed [16]) are even more suitable for assessment than average value D and correlate well with some psychological tests used by doctors.

Calculation of fractal dimension of EEG signals does make sense if D_f is not treated as an absolute measure but as a *comparative measure.* We need to stress once again that fractal dimension D_f should not be mistaken with fractal dimension of an attractor, which is calculated in the system's phase space. As a comparative measure of signal complexity, D_f may be used to assess changes in EEG-signals even if the signals are stochastic and not chaotic. It enables doctors to assess changes in EEG due e.g. to applied therapy, and may be helpful also in diagnostics.

ACKNOWLEDGEMENTS

The authors wish to thank Prof. Waldemar Szelenberger, Head of the 1st Dept. of Psychiatry, Medical Univ. of Warsaw, and to his Coworkers for supplying EEG-data.

REFERENCES

1. Klonowski, W., *Computers and Chemistry*, in press (2001).
2. Klonowski, W. in *Self-Organization Of Complex Structures - From Individual to Collective Dynamics*, edited by F.Schweitzer, Ed., Gordon and Breach Publ. 235-239 (1997).
3. Klonowski, W., *Biocyb.Biomed.Eng.* **19**, 41-55 (1999).
4. Klonowski, W., Stepien, R., Olejarczyk, E., Jernajczyk, W., Niedzielska, K., and Karlinski, A., *Medical & Biological Eng. & Computing*, **37**, Suppl. 2: 436-437 (1999).
5. Stepien, R., and Klonowski, W., "Application of standard chaotic quantifiers for chemotherapy assessing by EEG-signal analysis," in *Modeling and Simulation: a Tool for the Next Millenium*, edited by H.Szczerbicka, 13th European Simulation Multiconference ESM'99, Warsaw, 1999, pp. II:476-479.
6. Theiler, J., Eubank, S., Longtin, A., Galdrikian, B., and Farmer, J.D., *Physica D* **58**, 77-94 (1992).
7. Osborne, A.R., and Provenzale, A., *Physica D* **35,** 357-382 (1989).
8. Stepien, R., *Acta Neurobiol. Exp.* submitted.
9. Sugihara, G., *Trans.R.Soc.Lond. A* **348**, 477-495 (1994).
10. Klonowski, W., "Signal and Image Analysis Using Chaos Theory and Fractal Geometry", Proceedings of the 6th Conference GKPO'2000, Podlesice, 2000, *Machine Graphics & Vision* **9** (1/2) 403-431 (2000).
11. Klonowski, W., Olejarczyk, E., and Stepien, R., *Technology and Health Care* **9**, 87-90 (2001).
12. Pritchard, W,S., "A Tutorial Review of the Use and Benefits of Non-linear Dynamical Analysis of the EEG", in *Analysis of the Electrical Activity of the Brain,* edited by F.Agelcri, S.Butler, S.Giaquinto and J.Majkowski, J.Wiley & Sons, Chichester, New York, 1997, pp. 3-44.
13. Wackermann, J., *Acta Neurobiol.Exp.* **56**, 197-208 (1996).
14. Klonowski, W., Olejarczyk, E., and Stepien, R., „Nonlinear Quantifiers of EEE-signal Complexity," in: Proceedings of 2000 International Symposium on Nonlinear Theory and its Application (NOLTA'2000), Dresden, 2000, pp.261-264.
15. Radojicic, I., Mandic, D., and Vulic, D., "Searching for an optimal time delay in a state-space reconstruction of a heart rate variability signal," in *IFMBE Proceedings MEDICON 2001,* Pula, 2001, pp. II 893-896.
16. Ciszewski, J., Klonowski, W., Stepien, R., Jernajczyk, W., Karlinski, A., and Niedzielska, K., *Medical & Biological Eng. & Computing,* **37**, Suppl. 1: 359-360 (1999).
17. Klonowski,W., Olejarczyk, E., Stepien,R.: "Nonlinear Dynamics of EEG-signal Reveals Influence of Magnetic Field on the Brain" in: Proceedings of Chicago2000 World Congress on Medical Physics and Biomed. Engineering, 2000, FR-A325-2, on CD only, (abstracts\4048-52180, short_papers\4038-52296); also on www: http://www.ibib.waw.pl/~lbaf .
18. Kaminski, M., Blinowska, K.J., and Szelenberger, W., *Electroenceph.Clin.Neurophys.* **10**, 216-227 (1997).

DATA ANALYSIS

Continuous Nonlinear Delayed-Feedback Dynamics from Noisy Observations

H. U. Voss*, W. Horbelt* and J. Timmer*

**Freiburger Zentrum für Datenanalyse und Modellbildung, Eckerstr. 1, D-79104 Freiburg, Germany*

Abstract. Many nonlinear dynamical systems incorporate feedback mechanisms with a time delay. The identification of continuous time-delay models from observed time series is still an ongoing research topic. We propose a combination of nonparametric modeling based on optimal transformations and maximal correlation, and parametric modeling based on a maximum likelihood method as a strategy to determine nonlinear differential equations from measured time series. This strategy is exemplified on experimental data from an electronic circuit from which we obtain an accurate reconstruction of the observed attractor. Whereas the strength of the nonparametric approach is in finding initial guesses for the model structure and accurate models for low observation noise, the parametric approach may yield accurate estimates also in the presence of strong observation noise.

INTRODUCTION

Systems with a nonlinear time-delayed feedback are quite abundant in nature and technology. Already in linear systems a delayed feedback can cause instabilities leading to complex oscillations. In nature, time delays are mostly caused by finite signal transmission speeds and memory effects. In technology, delayed feedbacks caused e.g. by finite switching speeds are most often unwanted since they can severely disturb a regular motion leading to malfunction. Here we consider scalar delay-differential equations (DDEs) of the form

$$\dot{x} = f(x, x_\tau) \qquad [x \in I\!R,\ x_\tau := x(t-\tau),\ f\,\text{continuous}]\,. \tag{1}$$

It is theoretically known and experimentally confirmed for a long time now that even the solutions of scalar DDEs can be high-dimensional chaotic [1, 2, 3, 4]. This can be understood by considering that delay-differential equations are *functional* differential equations that possess an infinite-dimensional phase space; for a system with time delay τ, a continuum of initial conditions defined on an interval of length τ is required for the unique description of the state. Nevertheless, the theory of delay-induced instabilities is quite well developed for scalar equations [5, 6, 7, 8].

The wide abundance and the peculiar properties of delayed-feedback systems ask for new modeling and analysis methods. Especially for the task of model inference from measured outputs of these systems methods have to be developed that go beyond the now widely used methods for the analysis of low-dimensional chaotic dynamics [9]. In spite of the fact that in recent times there appeared many tools to tackle questions of DDE model inference from time series observations [10, 11, 12, 13, 14, 15, 16, 17],

CP622, *Experimental Chaos: 6th Experimental Chaos Conference*, edited by S. Boccaletti et al.

some questions are still not solved satisfactorily: If no insights into first principles that explain the observed dynamics are available, both the structure and the parameters of the model have to be determined. Even if insights into first principles are available, they may be inaccurate or wrong and may lead, therefore, to a misspecified model structure. Once the model structure is determined, parameter estimation algorithms can be used to estimate the unknown values of the system parameters from measurements of the system's signals. Then the question arises, how to find a method that not only finds the optimal delay and parameters but also a model that optimally reproduces the observed *dynamics* in a robust way, also for chaotic dynamics.

In this contribution we describe a hybrid approach for the estimation of nonlinear DDEs from time series observations, which takes these points into account. First, we perform a nonparametric model search in model structure space rather than an optimization in parameter space. This step is especially necessary if there is no proper parametric model at hand, like in many situations where a complex system with unknown components is considered. For example, the particular kind of nonlinearity leading to chaotic oscillations may be of interest. A search in model structure space can be performed in two ways: Estimation of coefficients of certain basis functions (e.g. coefficients of polynomials) and nonparametrically. Here we chose nonparametric modeling. This procedure does not only give a hint for an appropriate model structure but also estimates of coefficients of a parameterized model. These results are used in a second step as initial guesses for a more refined parameter estimation taking the dynamics of the observed system explicitly into account. We explain and exemplify this strategy on a measured time series from an electronic circuit.

The paper is organized as follows: In the next section we introduce the experiment to which the data analysis methods will be applied, an electronic circuit with time-delayed feedback and chaotic dynamics. Next, we briefly describe the two modeling procedures: To find a model structure, a nonparametric search in structure space is performed using optimal transformations and maximal correlation. To accurately estimate coefficients in this model, parametric modeling based on a least squares objective function is used. Then the hybrid modeling procedure is applied to the circuit data. Finally, the parametric approach is investigated further on the application to noisy observations.

THE MODEL SYSTEM

The time series that we will analyze by the methods described below was taken from an electronic circuit in a chaotic regime [18]. A wiring diagram is shown in Fig. 1. It consists of: (i) a nonlinearity f, built in essence from a transistor and an adjustable amplifier to control the loop gain, (ii) a low-pass filter, consisting of a capacitor and a resistor, and (iii) a commercially available "bucket brigade" delay line. Parts (ii) and (iii) are inspired from the experiment of Kittel et al. [19].

The differential equation describing the dynamics of the system follows from applying Kirchhoff's mesh rules. With x the voltage measured at the node between R_1 and C, the DDE reads

$$\dot{x} = -\alpha x + f(x_\tau) \,. \tag{2}$$

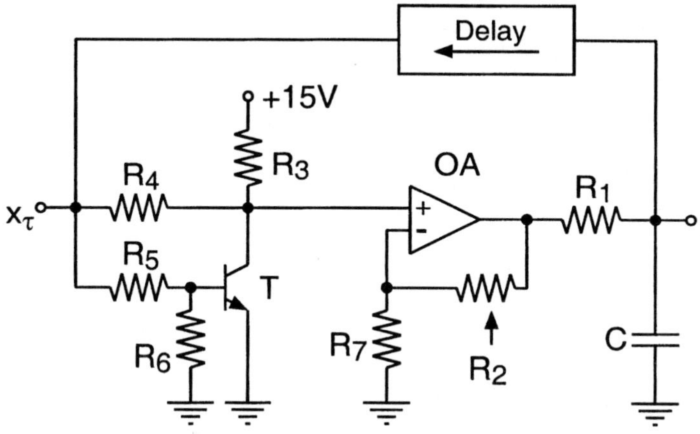

FIGURE 1. The electronic circuit. The values of the electronic components are: Delay line: bucket brigade line MN 3011 with 3328 stages, triggered by MN 3101 (both National Panasonic); OA: LM 324N; C=660 pF; R_1=470 kΩ, R_2=100 kΩ lin., R_3=22 kΩ, R_4=4.7 kΩ, R_5=10 kΩ, R_6=1 kΩ, R_7=47 kΩ, T: BC 238C. The components maintaining operation of the delay line are not shown.

The damping factor α is determined by R_1 and C. It leads to a time scale $\tau_D = \alpha^{-1} = R_1 C \approx 0.31$ ms. The delay of the loop can be adjusted within certain limits and is fixed to $\tau \approx 13.28$ ms. Therefore, one has $\tau \gg \tau_D$, a necessary condition for delay-induced instabilities. The exact form of the nonlinearity f is determined by the characteristic curve of the transistor and the way the resistors R_3 to R_6 are applied.

The adjustable loop gain is used as the bifurcation parameter. In the model (2) this corresponds to a factor contained in the nonlinearity f. In increasing the loop gain of the system, first a fixed-point is observed, then a periodic oscillation, and finally a period-doubling route to chaos. Increasing the loop gain further, the structure of the attractor becomes more complex (corresponding to a higher-dimensional chaotic motion). All this is in accordance with recent theoretical and numerical results [2, 3, 4, 7]. We proceed with the signal as measured shortly after the onset of chaotic dynamics, because this point is easy to reproduce. Care was taken that the signal is indeed chaotic and not merely an orbit of high periodicity; both cases can clearly be distinguished using an oscilloscope. The signal, comprising 10^4 data points, is sampled with a time interval of $\Delta T = 0.01$ ms, and an offset voltage of 6.62 V is subtracted. To obtain a smooth signal from the discrete values of the A/D converter and to remove some spiky artifacts caused by the bucket brigade delay line, the signal is slightly low-pass filtered by a Butterworth filter. A sample of the measurements and an attractor reconstruction are given in Figs. 2a and 2b, respectively.

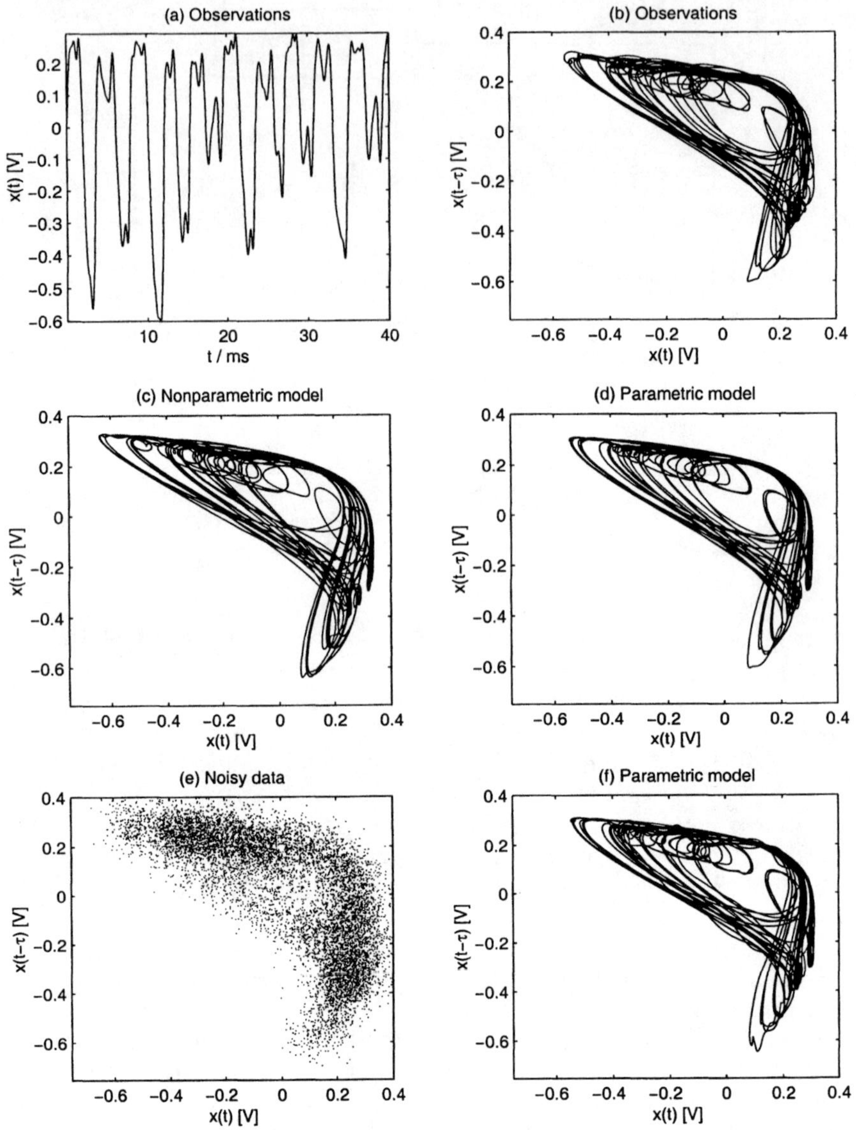

FIGURE 2. The observed time series and fitted model trajectories:
(a) Part of the experimentally observed data sample.
(b) The whole data set used for model fitting, shown as a return map with a delay of $\tau = 13.28$ ms. The same delay is used in all subsequent plots.
(c) The "free-run" reconstructed attractor of the nonparametric model, i.e., a numerical simulation of system (2) with the nonlinearity of Fig. 3c.
(d) Estimated attractor of the parametric model.
(e) Noise-disturbed data.
(f) Estimated attractor of the parametric model for the noisy data.

METHODS

Nonparametric regression and optimal transformations

We start with an outline of the method to search for an appropriate model structure. This task can best be performed in a nonparametric way due to the high flexibility of nonparametric function estimators. This has been published [20, 21] and applied [22, 23, 24] earlier and is, therefore, reviewed only briefly.

We use the concepts of optimal transformations and maximal correlation. They provide a nonparametric framework to detect and determine nonlinear relationships in data sets. Let x_0 and x_1 denote two zero-mean random variables and

$$R(x_0, x_1) = \frac{E[x_0 x_1]}{\sqrt{E[x_0^2]\,E[x_1^2]}} \tag{3}$$

their linear (normalized) correlation coefficient, where $E[.]$ is the expectation value. The basic idea of this approach is to find transformations $\Phi_0(x_0)$ and $\Phi_1(x_1)$ such that the absolute value of the correlation coefficient between the transformed variables is maximized. More specifically, the so-called *maximal correlation* [25, 26, 27]

$$\Psi(x_0, x_1) = \sup_{\Phi_0^*, \Phi_1^*} |R(\Phi_0^*(x_0), \Phi_1^*(x_1))| \tag{4}$$

is calculated. The supremum is taken over all finite-variance Borel measurable functions Φ_0^* and Φ_1^*. The functions $\Phi_0(x_0)$ and $\Phi_1(x_1)$ for which the supremum is attained are called *optimal transformations*. Generalizing the concept of linear correlation, where the linear correlation coefficient $R(x_0, x_1)$ quantifies linear dependences, $\Psi(x_0, x_1)$ quantifies nonlinear dependences of the form

$$\Phi_0(x_0) = \Phi_1(x_1) + \eta\,, \tag{5}$$

where η denotes independent normally distributed noise with zero mean and variance σ^2. Especially, if there is complete statistical dependence [27], i.e., $\sigma^2 = 0$ and x_0 is a function of x_1 or vice versa, the maximal correlation attains unity. The same holds for the more general dependence of Eq. (5) with $\eta = 0$. Here we are mainly interested in the estimation of optimal transformations in multivariate regression problems. These can be written, using the maximal correlation, [1] as

$$\Psi(x_0, \ldots, x_k) = \sup_{\Phi_0^*, \ldots, \Phi_k^*} |R(\Phi_0^*(x_0), \sum_{i=1}^{k} \Phi_i^*(x_i))|\,. \tag{6}$$

---

[1] This is not the most general possibility for a definition of a generalized measure of dependence for vectorial random variables. For such concepts see also [28, 29, 30]. However, since the most important properties of the maximal correlation [31] are still valid in this expression, in the following we will call $\Psi(x_0, \ldots, x_k)$ also "maximal correlation."

By definition, the value of $\Psi(x_0, \dots, x_k)$ is restricted to the interval $[0,1]$ and constitutes a generalized measure of statistical dependence between random variables.

The regression functions involved can be estimated from data in a nonparametric way by the *Alternating Conditional Expectation* (ACE) algorithm of Breiman and Friedman [32]. In this algorithm the optimal transformations are estimated by local smoothing of the data using kernel estimators. We use the ACE algorithm since it "is one of the most powerful and universal tools for the analysis of multivariate data due to its ability to recover frequent types of nonlinear structure" [33]. For a description, we refer to the original paper of Breiman and Friedman or the reviews in Härdle [34] and Schimek [35]. Brief descriptions are also given in [20, 21, 36]. A MATLAB- and a C-program can be obtained from the authors or from the web site `http://www.fdm.uni-freiburg.de/~hv`.

For fitting a DDE, we take the rather general description

$$\Phi_0(\dot{x}) = \Phi_1(x) + \Phi_2(x_\tau) , \tag{7}$$

where again $x_\tau = x(t-\tau)$ is used. The arguments $\dot{x}$, x, and x_τ are taken or estimated from the data, and the functions Φ_0 to Φ_2 are then estimated by the ACE algorithm. This model can easily be extended to incorporate higher derivatives and multiple delay terms [20]. To find the optimum time delay τ, for each possible value of τ the maximal correlation is estimated. The argument of the maximum value of all these values then yields an estimate for the delay τ. The optimal transformations corresponding to this estimate are used as model functions.

Parametric modeling

Consider a time-continuous, dynamical process described by a set of delay differential equations (DDEs) with a single time lag τ and a set of unknown parameters $\boldsymbol{p} \in I\!R^p$ (including τ):

$$\dot{x} = f(t, x, x(t-\tau), \boldsymbol{p}), \qquad (t \in (0,T]) \tag{8}$$

$$x(t) = x_0(t, \boldsymbol{p}) \qquad (t \in [-\tau, 0]). \tag{9}$$

Contrary to initial *values* in the case of ordinary differential equations, an initial *curve* $x_0([-\tau,0])$ must be specified in order to define a unique solution of the DDE. The initial curve is generally unknown. It must be parameterized appropriately, since only a finite number of degrees of freedom are identifiable from the measured data. These parameters are also contained in $\boldsymbol{p}$. To sum up, the model trajectory $x(t,\boldsymbol{p})$ is defined as a solution of Eqs. (8) and (9), depending on the parameters $\boldsymbol{p}$.

An experiment gives access to a time series $\{y_i\}$, which is a measurement of the trajectory at discrete times t_i via the observation function g:

$$y_i = g(x(t_i, \boldsymbol{p}), \boldsymbol{p}) + \eta_i \qquad (i = 1..N) . \tag{10}$$

Here, η_i again denote independent normally distributed random numbers with zero mean and variance σ_i^2, accounting for measurement noise. In the example described later on, $g(x,\boldsymbol{p})$ is identical to x.

We aim at identifying the vector $\hat{\boldsymbol{p}}$, for which the model trajectory is closest to the observed dynamics y_i. To this end, the objective function $\chi^2(\boldsymbol{p})$ is defined as the sum of squared residues between the data and the model, weighted with the inverse variances of the data:

$$\chi^2(\boldsymbol{p}) = \sum_{i=1}^{N} \left(\frac{y_i - g(x(t_i,\boldsymbol{p}),\boldsymbol{p})}{\sigma_i} \right)^2 . \tag{11}$$

The parameter estimate $\hat{\boldsymbol{p}}$ is identified as the vector minimizing $\chi^2(\boldsymbol{p})$. Under the given assumption of normally distributed noise, $\hat{\boldsymbol{p}}$ is the maximum likelihood estimator of the true parameter vector. The minimum of $\chi^2(\boldsymbol{p})$ is determined in an iterative process using a generalized Gauss-Newton method. The quality of the starting guess for $\boldsymbol{p}$ is essential for the success of the optimization. The initial curve is represented by a cubic spline function, which is uniquely defined by its values at a number of spline knots and its first derivatives at the two end points $t = -\tau$ and $t = 0$ [37]. By choosing an appropriate density of the spline knots, the number of degrees of freedom of the initial curve can be adapted to the problem. The integration of the DDE is done by the code RETARD [38]. Details of the method and aspects of implementation are described in [39]. Applications to ordinary differential equations can be found in [36, 40].

APPLICATION TO THE ELECTRONIC CIRCUIT

Nonparametric model fit

To recover the driving system's nonlinearity f and the damping factor α from the measured data, the method of optimal transformations and maximal correlation is applied. The time delay is known from the experiment to be $\tau \approx 13.28$ ms. We test for all time delays $\tau \in [0.20, \ldots, 20.00]$ ms, proceeding in steps of 0.01 ms. This yields a maximum value for the maximal correlation of $\Psi = 0.98$ at $\hat{\tau} = 13.22$ ms, close to the experimentally estimated value of $\tau = 13.28$ ms. (Due to oversampling, the trajectory is rather smooth, and for $\tau < 0.20$ ms the dynamics can be described *locally* even more accurate, yielding $\Psi \approx 1$ for very small τ.) For this value of τ, the resulting optimal transformations are depicted in Fig. 3: $\Phi_0(\dot{x})$ (Fig. 3a) is estimated to be approximately the identity transformation, in accordance with the model. $\Phi_1(x)$ (Fig. 3b) turns out to be a linear function $-\alpha x$; the damping factor α is estimated from a subsequent fit to $\Phi_1(x)$ to $\hat{\alpha} = 3.24$ ms^{-1}, in accordance with the experimental value of $\alpha \approx 3.22$ ms^{-1}. The function $\Phi_2(x_\tau)$ represents the estimated nonlinearity $f(x_\tau)$ (Fig. 3c). A subsequent polynomial fit yields $\hat{f}(x_\tau) = 0.85 - 3.63x_\tau \ -13.95x_\tau^2 - 10.44x_\tau^3$ [2]. Finally, we use the so-obtained polynomial model for a numerical "free-run" simulation. The resulting model attractor (Fig. 2c) resembles closely the observed attractor (Fig. 2b).

[2] For the sake of readability, in the following we suppress the physical dimensions of the coefficients. These are determined by the dimension of f which is V/ms.

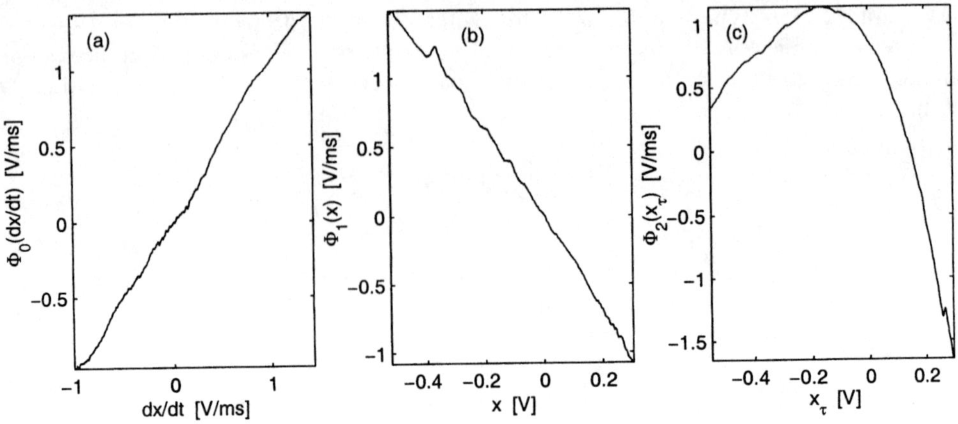

FIGURE 3. The estimated optimal transformations.

Parametric model fit

Motivated by the nonparametric results, we parameterize the nonlinear function f by the third order polynomial

$$f(x_\tau) = a_0 + a_1 x_\tau + a_2 x_\tau^2 + a_3 x_\tau^3. \tag{12}$$

The parametric methods described above were applied to Eqs. (2) and (12) to estimate the time lag τ, the damping factor α and the coefficients a_i from the measured time series. The starting guesses for all parameters were taken from the results of the nonparametric method. The resulting estimates are $\tau = 13.30 \pm 0.03$ ms, $\alpha = 3.69 \pm 0.45$ ms^{-1}, $a_0 = 0.88 \pm 0.14$, $a_1 = -3.97 \pm 0.52$, $a_2 = -15.21 \pm 2.27$, and $a_3 = -11.48 \pm 3.92$. An attractor of the simulated model is shown in Fig. 2d, resembling closely the attractor reconstructed from the data (Fig. 2b).

Finally, we disturb the observations with white Gaussian noise. The standard deviation of the noise is chosen to be 20% of the standard deviation of the observations. The fine structures of the attractor are destroyed completely by this procedure (Fig. 2e). They can be revealed to a good approximation by a parametric model fit to the disturbed data (Fig. 2f). We remark that to obtain results like these, a good starting guess for the coefficients has to be given.

Unlike in the nonparametric model study above, where the model has been estimated without utilizing the quadratic distance between the model trajectory and the observations, in the maximum likelihood approach the model is estimated by searching for the most likely trajectory that fits both model and observations. This enables one to directly compare the *estimated trajectories* with the observations, which is done in Fig. 4. Hardly any difference can be seen in the entire interval. Figure 4 reveals that the estimated trajectory approximates the true trajectory very well; the additive observational noise has been almost completely eliminated by this nonlinear filtering process.

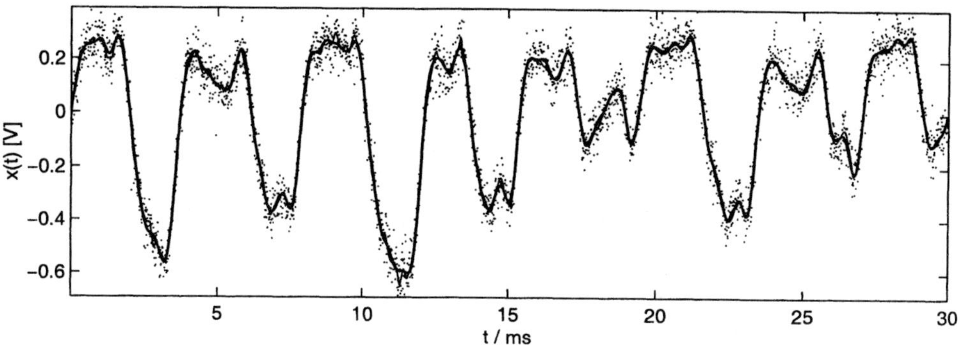

FIGURE 4. Comparison of the noise-disturbed time series (dots), the best fit trajectory to these data (thin line, partly covered), and part of the true time series (bold line).

DISCUSSION

We proposed a hybrid approach of nonparametric and parametric modeling of delayed-feedback systems by the analysis of observed time series produced by these systems, and exemplified this strategy on an electronic circuit application. We combined both approaches since each has certain advantages and drawbacks:

Nonparametric modeling by optimal transformations requires the estimation of time derivatives from the data and does not explicitly take into account the fact that the underlying model trajectory is continuous. Therefore this approach alone is not optimized with respect to accurate dynamical modeling. Its power is the ability to search for a suitable model structure and to suggest a specific parameterized model. On the other hand, the likelihood-based approach to parameter estimation in differential equations is feasible if a parametric model is on-hand. Model misspecifications can be detected but not compensated. Exploiting the continuity of the underlying trajectory, it yields extraordinarily accurate estimates of the parameters.

Note that both approaches do not need an attractor reconstruction by a delay embedding. Thus, all problems associated with the delay reconstruction of a chaotic and noisy phase space, like finding an optimal embedding window and the "curse of dimensionality," are of no relevance here. Most importantly, considerably fewer amounts of data as generally needed for a useful delay reconstruction have to be provided, and the hybrid method is robust against high levels of observation noise.

In the discussed application the results are extremely accurate considering that an experimental time series is used and that the global dynamics of a highly nonlinear system could be reproduced. The accurate agreement between the dynamics of a measured chaotic time series and an estimated model is a highly nontrivial result [9] in nonlinear modeling.

ACKNOWLEDGMENTS

We acknowledge the help of H. Ernst (IMTEK, University of Freiburg) in the data acquisition process.

REFERENCES

1. Ikeda, K., Daido, H., and Akimoto, O., *Phys. Rev. Lett.*, **45**, 709–712 (1980).
2. Farmer, J., *Physica D*, **4**, 366–393 (1982).
3. Ikeda, K., and Matsumoto, K., *Physica D*, **29**, 223–235 (1987).
4. Dorizzi, B., et al., *Phys. Rev. A*, **35**, 328–339 (1987).
5. Hale, J., and Verduyn Lunel, S., *Introduction to Functional Differential Equations*, Springer, New York, 1993.
6. Gopalsamy, K., *Stability and Oscillations in Delay Differential Equations of Population Dynamics*, Kluwer, Dordrecht, 1992.
7. Wischert, W., et al., *Phys. Rev. E*, **49**, 203–219 (1994).
8. Diekmann, O., et al., *Delay Equations*, Springer, New York, 1995.
9. Kantz, H., and Schreiber, T., *Nonlinear Time Series Analysis*, Cambridge University Press, Cambridge, 1997.
10. Baker, C., and Paul, C., *SIAM J. Sci. Comp.*, **18**, 305–314 (1997).
11. Bocharov, G., and Romanyukha, A., *Appl. Num. Math.*, **15**, 307–326 (1994).
12. Bünner, M., Popp, M., Meyer, T., Kittel, A., Rau, U., and Parisi, J., *Phys. Lett. A*, **211**, 345–349 (1996).
13. Hegger, R., Bünner, M., Kantz, H., and Giaquinta, A., *Phys. Rev. Lett.*, **81**, 558 (1998).
14. Ellner, S., Kendall, B., Wood, S., McCauley, E., and Briggs, C., *Physica D*, **110**, 182–194 (1997).
15. Marchuk, G., *Mathematical Modelling of Immune Response in Infectious Diseases*, vol. MIA Vol. 395, Kluwer Academic Publishers, Dordrecht, 1997.
16. Marshall, J., "Identification strategies for time-delay systems", in *Delay Differential Equations and Dynamical Systems*, edited by S. Busenberg and M. Martelli, Springer, 1991, pp. 141–150.
17. Mittler, J., Sulzer, B., Neumann, A., and Perelson, A., *Math. Biosc.*, **152**, 143–163 (1998).
18. Voss, H., *To appear in Int. J. Bifurcation and Chaos* (2001).
19. Kittel, A., Parisi, J., and Pyragas, K., *Physica D*, **112**, 459–471 (1998).
20. Voss, H., and Kurths, J., *Phys. Lett. A*, **234**, 336–344 (1997).
21. Voss, H., and Kurths, J., *Chaos, Solitons & Fractals*, **10**, 805–809 (1999).
22. Voss, H., Schwache, A., Kurths, J., and Mitschke, F., *Phys. Lett. A*, **256**, 47–54 (1999).
23. Timmermann, A., Voss, H., and Pasmanter, R., *J. Phys. Oceanogr.*, **31**, 1579–1598 (2001).
24. Voss, H., "Analyzing Nonlinear Dynamical Systems with Nonparametric Regression", in *Nonlinear Dynamics and Statistics*, edited by A. Mees, Birkhäuser, Boston, 2001, pp. 413–434.
25. Gebelein, H., *Zeitschrift für angew. Math. und Mech.*, **21**, 364–379 (1941).
26. Hirschfeld, H., *Proc. Camb. Phil. Soc.*, **31**, 520–524 (1935).
27. Rényi, A., *Probability Theory*, Akadémiai Kiadó, Budapest, 1970.
28. Wysocki, W., *Zastosow. Mat.*, **21**, 225–233 (1991).
29. Wysocki, W., *Zastosow. Mat.*, **21**, 211–224 (1991).
30. Pompe, B., *Journal of Statistical Physics*, **73**, 587–610 (1993).
31. Rényi, A., *Acta. Math. Acad. Sci. Hungar.*, **10**, 441–451 (1959).
32. Breiman, L., and Friedman, J., *J. Am. Stat. Assoc.*, **80**, 580–598 (1985).
33. Buja, A., *Ann. Stat.*, **18**, 1032–1069 (1990).
34. Härdle, W., *Applied Nonparametric Regression*, Cambridge University Press, Cambridge, 1990.
35. Schimek, M., editor, *Smoothing and Regression. Approaches, Computation and Application*, Wiley & Sons, New York, 2000.
36. Timmer, J., Rust, H., Horbelt, W., and Voss, H., *Physics Letters A*, **274**, 123–134 (2000).
37. Press, W., et al., *Numerical Recipes*, Cambridge University Press, Cambridge, 1992.

38. Hairer, E., Nørsett, S. P., and Wanner, G., *Solving Ordinary Differential Equations I*, Springer, Berlin, 1987.
39. Horbelt, W., *Maximum likelihood estimation in dynamical systems*, Ph.D. thesis, University of Freiburg (2001), http://webber.physik.uni-freiburg.de/~horbelt/diss/.
40. Horbelt, W., Timmer, J., Bünner, M., Meucci, R., and Ciofini, M., *Phys. Rev. E*, **64**, 016222 (2001).

A New Surrogate for Experimental Data Analysis

Kevin T. Dolan[1] and Mark L. Spano[2]

[1]Center for Neurodynamics, University of Missouri, St. Louis, MO 63121
[2]NSWC, Carderock Laboratory, Code 645, W. Bethesda, MD 20817

Abstract. We show here that the commonly used surrogate generating techniques based on phase randomization produce a population of surrogates that are not consistent with the null hypothesis that they are designed to test. We present a new surrogate generating method, based on digital filtering techniques. This surrogate algorithm has significant advantages over the most commonly used techniques, in that it provides a more robust statistical test by producing an entire population of surrogates that are consistent with the null hypothesis. The new surrogate is tested against an autoregression process and the Rössler system.

INTRODUCTION

Performing statistical analysis on experimental data from nonlinear systems often involves testing a null hypothesis based on some nonlinear measure. Whether the measure in question is a traditional nonlinear measure such as the correlation dimension [1,2] or the Lyapunov exponent [3,4], or is an experimentally motivated measure like the frequency of encounters with unstable periodic orbits [5,6], surrogate data is usually required to perform the test. Using surrogate data allows us to determine the probability distributions of these measures empirically. This is of particular importance when using nonlinear measures, as the probability distributions can rarely be determined analytically.

The most commonly used techniques for generating surrogate data for the statistical analysis of nonlinear processes include random shuffling of the original time series, Fourier-transformed surrogates [7], amplitude adjusted Fourier-transformed (AAFT) surrogates [7,8], and iterated AAFT surrogates [9]. Each of these methods has strengths and weaknesses with respect to the power of the null hypothesis that can be rejected, the computational complexity of the algorithm and the accuracy with which the method models the statistical properties of the data being analyzed. In this paper, we will discuss some of the difficulties that occur when using these types of surrogates, and we present a new surrogate algorithm that addresses these problems.

COMMONLY USED SURROGATE TECHNIQUES

The simplest algorithm for generating surrogate data is random shuffling. A random permutation of the original data file is made. All of the original data are

CP622, *Experimental Chaos: 6th Experimental Chaos Conference*, edited by S. Boccaletti et al.

included in a completely random order. This method guarantees that the surrogate data will be consistent with the null hypothesis of a δ-correlated random process, while exactly preserving the distribution of the original data. This algorithm is of limited usefulness because the only statistical property of the original data that is preserved is the distribution. For example, Theiler *et al.* showed that shuffled surrogates are unsuitable for nonlinearity tests based on calculations of the correlation dimension [7].

Theiler *et al.* went on to describe an algorithm for testing the null hypothesis of a linear stochastic process that can be applied to data that is normally distributed. This method takes the discrete Fourier-transform of the original data and then assigns a random phase to each positive frequency component. The negative frequency components are assigned corresponding phases to assure that the inverse Fourier-transform will be real. The result is a time series with the same Fourier amplitudes as the original data, but which is in every other respect randomized.

This phase randomization technique is not suitable when the original data is not normally distributed, since the Fourier-transformed surrogate will then have a different distribution from the original data. The amplitude adjusted Fourier transformed (AAFT) surrogate was developed to resolve this limitation [7,8]. These surrogates attempt to test the null hypothesis that the original data is from a linear stochastic process that has undergone a static nonlinear transformation. The algorithm for generating this type of surrogate data is described as follows:

(1) The original data is *rescaled* to a normal distribution. This is done by generating a time series of Gaussian white noise and sorting it according to the ranking of the original data.

(2) A Fourier-transformed surrogate of the rescaled data is constructed.

(3) The final surrogate is scaled to the distribution of the original data by sorting the original data according to the ranking of the Fourier-transformed surrogate.

This algorithm guarantees that the distribution will be preserved, and approximately preserves the power spectrum as well. The ability of this algorithm to properly preserve the power spectrum depends on the existence of a continuous static transformation from the original distribution to a normal distribution. Since the underlying distribution of the original data is, in general, unknown, the transformation must be determined empirically. The rescaling in step (1) constitutes an empirical transformation from the distribution of the original data to that of a specific finite realization of a Gaussian distributed process. The rescaling in step (3) then transforms from an independent realization of a different Gaussian distributed process back to the original data distribution. The forward and backward transformations are not exact inverses of each other, and this results in changes to the power spectrum of the final surrogate. The result of this alteration is to whiten the power spectrum of the original data [9]. Of course, the amount of whitening that occurs depends on both the length of the time series (for long time series the distribution will be very nearly continuous, reducing the effect) and on the degree to which the original distribution is non-Gaussian.

A method was developed by Schreiber [9] that addresses the issue of power spectrum whitening by performing a series of iterations in which the power spectrum of an AAFT surrogate is adjusted back to that of the original data before the distribution is rescaled back to that of the original data. The process can be described as follows:

(1) An AAFT surrogate is generated using the procedure above. The Fourier amplitudes calculated in step (2) are recorded for later use.

(2) The surrogate is Fourier-transformed, and its Fourier amplitudes are adjusted back to the amplitudes recorded in step (1).

(3) The result of step (2) is inverse Fourier-transformed and rescaled back to the original data distribution as in step (3) of the AAFT algorithm.

(4) Steps (2) and (3) are repeated until the whitening of the power spectrum is sufficiently small.

Schreiber showed that, for a nonlinearly-transformed autoregression process, the iteration procedure will converge towards the Fourier amplitudes of the original data until a saturation point is reached, where the Fourier amplitude adjustment is so small that the rescaling puts the data into the exact order it had before the amplitude adjustment [9].

DIFFICULTIES WITH CURRENT ALGORITHMS AND A NEW SURROGATE

Clearly, for surrogate data to be used to properly test a null hypothesis, it is necessary that the surrogate data be consistent with the null hypothesis. It is clear that by preserving the distribution and power spectrum of the original data while randomizing the data in all other respects, a surrogate file generated by the iterated AAFT algorithm will be consistent with the null hypothesis that the data were drawn from a nonlinear static transformation of a linear stochastic process. But in order to reject a null hypothesis, it is not sufficient for each surrogate file to individually satisfy the null hypothesis. The entire population of surrogates must satisfy the null hypothesis. For our example here, this means that all the surrogates must be statistically independent realizations of a nonlinear static transformation of a linear stochastic process.

The iterated AAFT method does not satisfy this requirement. A set of statistically independent realizations of a nonlinear static transformation of a linear stochastic process will all have slightly different power spectra estimates. Given a power spectrum estimate P_k where k is the frequency bin, the variance of P_k over the set of surrogates is given by:

$$\sigma_k^2 = P_k^2, \tag{1}$$

where σ_k^2 is the variation in the power from one realization of the process to the next in any given frequency bin [10]. It is because of this high variability that power spectra are typically averaged over several frequency bins, sacrificing frequency resolution for a more accurate power spectrum estimate. Eq. (1) then becomes

$$\sigma_i^2 = \frac{1}{N} P_i^2, \tag{2}$$

where N is the number of frequency bins averaged and i denotes the new larger bins.

The iterated AAFT method will produce a population of surrogates that, by construction, all have exactly the same power spectrum estimates. These populations of surrogates are therefore not consistent with the null hypothesis. It is clear that this issue can result in erroneous results for statistical tests. If the measured statistic being analyzed is sensitive to variations in the power spectrum of the data, then it is possible that the variance of the statistic will be significantly smaller for the population of surrogates than it would be if the null hypothesis were actually true, resulting in a false-positive statistic. These iterated AAFT surrogates therefore do not properly test the null hypothesis.

Digitally Filtered Shuffled Surrogates

We will now present a new surrogate algorithm that resolves the issues described above. This new algorithm, which we will refer to as the digitally filtered, shuffled (DFS) surrogate algorithm, convolves a random permutation of the data with a response function based on the estimated power spectrum of the data. Since each surrogate is made from a different random permutation, each surrogate is, by construction, a statistically independent realization of the process. The procedure for generating DFS surrogates is as follows:

(1) Estimate the power spectrum of the data using overlapping windowed Fourier transforms. The data should be de-meaned for this calculation.

(2) Calculate a response function by taking the square root of the power spectrum estimate and taking its inverse Fourier transform.

(3) Make a random permutation of the data (a shuffled surrogate as described above).

(4) Digitally filter the shuffled surrogate by convolving it with the response function. The shuffled surrogate should be de-meaned before filtering.

(5) Rescale the filtered surrogate back to the original data distribution. The rescaling can be done as for the AAFT surrogates.

By using an overlapping windowed Fourier transform to estimate the power spectrum; the accuracy of the estimate can be greatly improved. This is done by calculating the Fourier transform of N samples of length L, each overlapping by 50% [11]. These samples are multiplied by a windowing function such as the Welch window [12]. Any other favorite windowing function may be substituted for the Welch function.

$$w_i = 1 - \left(\frac{2i-(L-1)}{(L+1)} \right)^2, \tag{3}$$

where i ranges from 0 to $L-1$. This windowing function goes to zero at both ends, reducing any high-frequency artifacts due to lack of periodicity. The power spectrum estimate made with this technique will thus have a variance given by

$$\sigma_i^2 = \frac{11}{9N} P_i^2 . \tag{4}$$

Because a de-meaned, shuffled surrogate has a white power spectrum, the convolution with the response function will result in a time series with the appropriate power spectrum. This also allows variations in the power spectrum from one surrogate realization to the next, since each surrogate is made from a statistically independent shuffled surrogate.

Rescaling and Power Spectrum Whitening

As with the AAFT surrogates, the process of rescaling to the original data distribution alters the power spectrum. An iterated method, similar to what is used for the iterated AAFT surrogates, can be used to reduce the whitening. The procedure for this is to adjust repeatedly the Fourier amplitudes and subsequently to rescale the result. The amplitudes to which we should adjust are not the Fourier amplitudes of the original data, but instead of the surrogate just before the first rescaling in step (4).

COMPARISON OF METHODS

To demonstrate that the DFS algorithm is properly preserving the linear properties of the data we will initially look at a simple nonlinearly transformed auto-regression process,

$$\begin{aligned} x_{n+1} &= 0.7x_n + \xi_n \\ y_n &= x_n^3 \end{aligned}, \tag{5}$$

where ξ_n is Gaussian white noise.

First, we will establish that the DFS algorithm is producing the proper degree of variability from one surrogate file to the next. To do this we generate a time series from the above AR process of length 2048 points. We then generate a population of 100 DFS surrogates. Next, we introduce the following measure of spectral variability,

$$\alpha = \frac{1}{N}\sum_{i=1}^{N}\frac{\left(P_i - \overline{P}_i\right)^2}{\overline{P}_i^2}, \tag{6}$$

where N is the number of frequency bins (in this case 1025), P_i is the power in the ith frequency bin of the surrogate, and $\overline{P}_i$ is the average power in the ith frequency bin, averaged over the other 99 surrogate files. This yields a value of α for each surrogate.

If our hypothesis that each surrogate is an independent realization of the same process is true, then we can predict the following,

$$\left\langle \left(P_i - \overline{P}_i\right)^2 \right\rangle = \overline{P}_i^2 . \tag{7}$$

This implies that the expectation value of α should be 1. Fig. (1a) shows the distribution of α for the 100 DFS surrogates. Fig. (1b) shows the distribution for 100 AAFT surrogates, and Fig. (1c) shows the measure α for 100 independent realizations of the actual AR process as a control example.

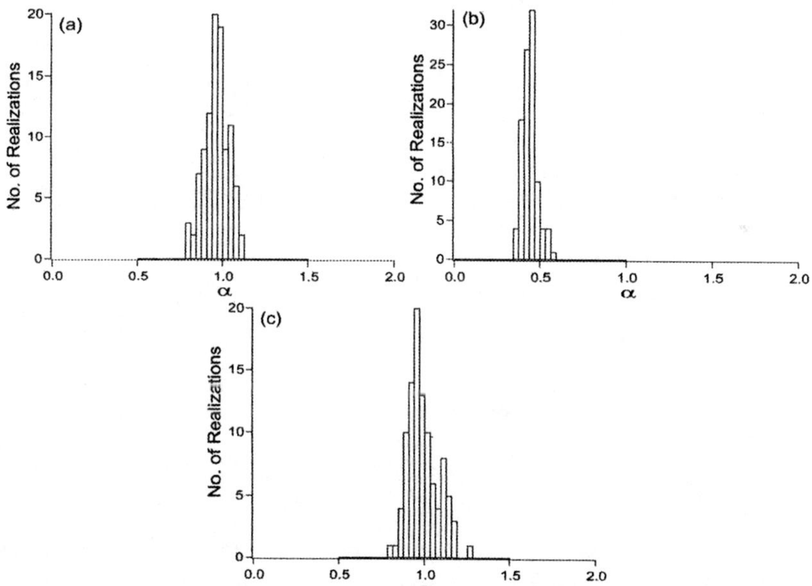

FIGURE 1. Histograms of the measure α for the DFS surrogates (a), AAFT surrogates (b), and 100 realizations of the actual AR process (c).

We can see from Fig. (1) that the DFS surrogates preserve the appropriate degree of variability between surrogates whereas the AAFT surrogates show a much lower degree of variability. In fact, it is clear from the description of the AAFT algorithm that the only reason α is not zero for all of the AAFT files is that the whitening effect described above introduces a small amount of variability.

This suggests the question, how well have the AAFT and DFS surrogate algorithms preserved the power spectrum of the data? We have calculated the power spectrum of the original AR data, as well as the spectra of the AAFT and DFS surrogates. The results are shown in Fig. (2).

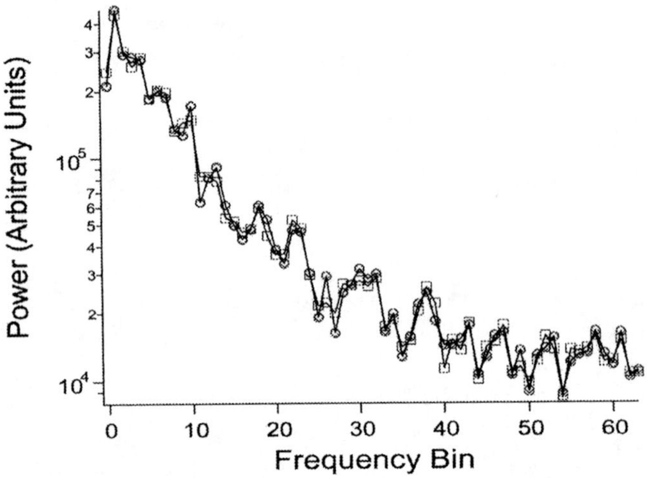

FIGURE 2. Power spectrum estimate of the AR process (circles), DFS surrogates (squares) and AAFT surrogates (diamonds). A log scale was used to show variations more precisely. 64 frequency bins were used to average out statistical fluctuations.

We see that both surrogates properly preserve the power spectrum of this data. Although the power spectrum whitening effect due to the rescaling is non-zero for this data, it is insignificant. This is not always the case though, particularly when the data is strongly non-Gaussian.

To demonstrate the iterated version of the DFS algorithm, we look at a classic nonlinear system, the Rössler system [13].

$$\begin{aligned} \dot{x} &= -(y+z) \\ \dot{y} &= x + ay \\ \dot{z} &= b + z(x-c) + D\xi(t) \end{aligned} , \qquad (8)$$

where $a = b = 0.2$, $c = 5.7$. Dynamical noise was added, where $\xi(t)$ is Gaussian white noise with zero mean and unit variance, and the intensity $D = 0.01$. The Rössler system is chaotic for these parameter values. A Poincaré section of this data was made by measuring y at every positive zero crossing of x. 2048 such crossings were recorded. Fig. (3) shows the distribution of this data.

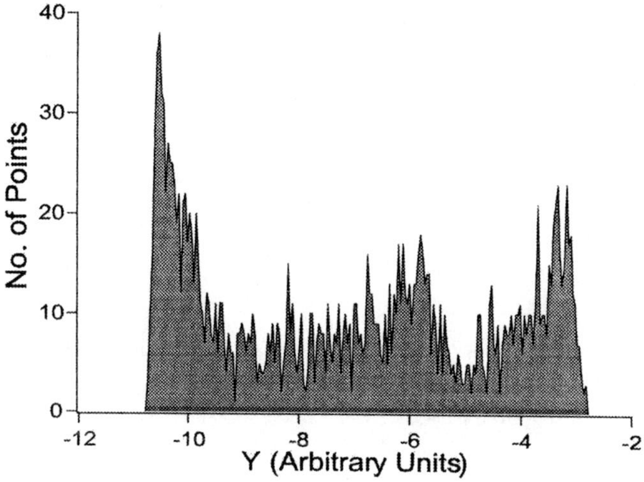

FIGURE 3. Distribution of Rössler section data: Note the highly non-Gaussian, multi-modal distribution.

Once again 100 surrogates of both the DFS and AAFT type were made. Fig. (4) shows the power spectrum of the surrogates compared to that of the original data.

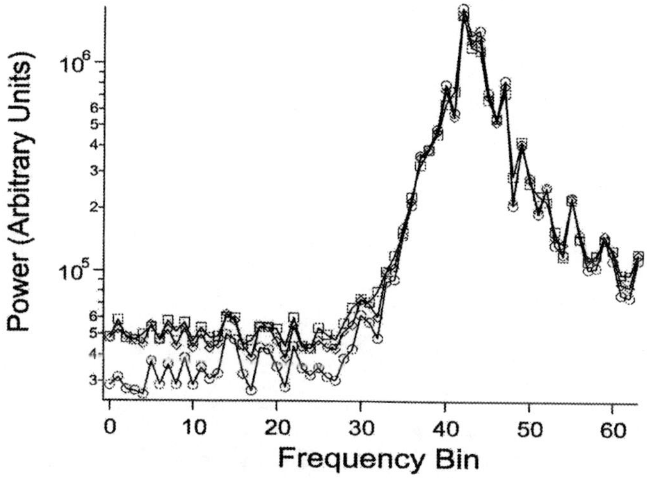

FIGURE 4. Power spectra of the original Rössler section data (circles), DFS surrogates (squares) and AAFT surrogates (diamonds). Note the significant systematic increase in power at the lower frequencies due to the power spectrum whitening effect.

In this case, the whitening due to the rescaling is not negligible. We therefore apply the iterated technique described above. Fig. (5) shows the spectra of 10 iterations of the DFS and AAFT algorithms. Note that both surrogates converge towards the proper spectrum at about the same rate.

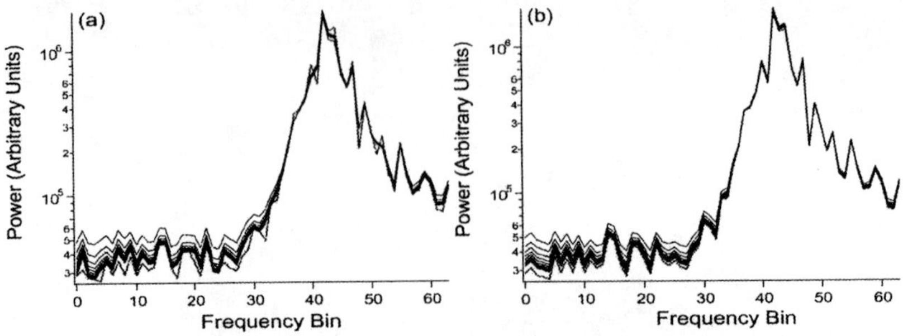

FIGURE 5. Power spectra for Iterated DFS surrogates (a) and iterated AAFT surrogates (b). The dotted line represents the power of the actual data, and the solid lines show consecutive iterations of the surrogate algorithms.

This demonstrates that the iterated version of the DFS algorithm can compensate for the power spectrum whitening caused by the rescaling. But what effect has this iterated process had on the variability of the surrogates? We now apply the measure defined in Eq. (6) to each iteration of the surrogate process. Fig. (6) shows the results of this analysis.

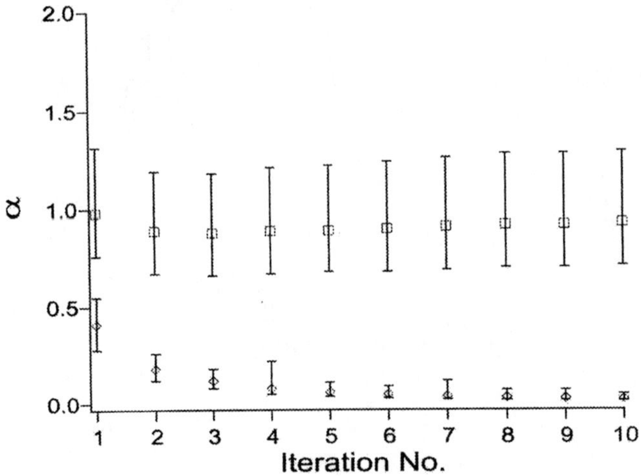

FIGURE 6. The variability measure α for each iteration of the DFS (squares) and AAFT (diamonds) algorithms. The error bars represent the 99% confidence interval given by the smallest and largest values of α for each iteration.

The iterated DFS algorithm preserves the proper variability throughout the iteration process. As we would expect, the variability of the AAFT surrogates goes to zero as we iterate, since the variability is due entirely to the whitening in this case.

DISCUSSION

We have clearly shown that the AAFT and iterated AAFT algorithms do not properly test the null hypothesis of a nonlinear static transformation of a linear stochastic process. The new DFS surrogate algorithm described herein has been demonstrated not only to preserve the power spectrum of the original data as well as the AAFT algorithm, but also to produce the appropriate inter-surrogate variability. Furthermore, the iterated DFS algorithm converges to the proper spectrum just as quickly as the iterated AAFT algorithms, but without sacrificing its variability properties.

It should further be noted that the computational complexity of the DFS and the iterated DFS algorithms are about the same as that of the AAFT and iterated AAFT algorithms. This algorithm can also be modified to operate in the time domain, which may be more efficient for data streams that are much longer than the required frequency resolution.

ACKNOWLEDGMENTS

The authors would like to thank Frank Moss and Alexander Neiman for stimulating discussions. The authors gratefully acknowledge support from the U.S. Office of Naval Research, Physical Sciences Division.

REFERENCES

1. P. Grassberger and I. Procaccia, *Physica D* **9**, 189 (1983).
2. T. Schreiber and H. Kantz, *Chaos* **5**, 133 (1995).
3. V.I. Oseledec, *Trans. Mosc. Math. Soc.* **19**, 179 (1968).
4. H. Kantz and T. Schreiber, *Nonlinear Time Series Analysis*, Publisher ,Cambridge University Press, Cambridge (1997).
5. D. Pierson and F. Moss, *Phys. Rev. Lett.* **75**, 2124 (1995).
6. K. Dolan, A. Witt, M.L. Spano, A. Neiman and F. Moss, *Phys. Rev. E* **59,** 5235 (1999).
7. J. Theiler, S. Eubank, A. Longtin, B. Galdrikain and J.D. Farmer, *Physica D* **58**, 77 (1992).
8. J. Theiler and D. Prichard, *Physica D* **94**, 221 (1996).
9. T. Schreiber and A. Schmitz, *Phys. Rev. Lett.* **77**, 635 (1996).
10. A.V. Oppenheim and R.W. Schafer, *Digital Signal Processing*, Publisher, Prentice Hall, Englewood Cliffs, N.J., (1975).
11. W. H. Press, B. P. Flannery, S. A. Teukolsky, and W. T. Vetterling, *Numerical Recipes in C, The Art of Scientific Computing*, Publisher, Cambridge University Press, Cambridge, (1992).
12. P.D. Welch, in *Modern Spectrum Analysis*, edited by D.G. Childers, Publisher, IEEE Pres, New York, (1978).
13. O.E. Rössler, *Z. Naturforsch. A* **31**, 1168 (1976).

Predicting phase synchronization from non-synchronized chaotic data

Isao Tokuda* and Jürgen Kurths†

*Department of Computer Science and Systems Engineering, Muroran Institute of Technology, Muroran, Hokkaido 050-8585, Japan
†Nonlinear Dynamics, Institute of Physics, University of Potsdam, D-14415, Potsdam, Germany

Abstract. A novel modeling approach is presented for reconstruction of the global behavior of coupled chaotic systems from bivariate time series. We analyze two coupled chaotic oscillators, which are able to phase synchronize due to coupling. It is shown that our technique enables the recovery of the synchronization diagram from only three data sets. This allows the estimate of the relative strength of the coupling and the parameter mismatch of both subsystems.

INTRODUCTION

In the studies of coupled chaotic systems, four basic types of synchronization have been found recently: complete [1], generalized [2], phase [3], and lag synchronization [4]. Among them, we focus on phase synchronization as a novel nonlinear phenomenon which has many important applications (*cf.*: ref. [5] and references therein), including laboratory experiments, such as circuits [6], lasers [7], and plasmas [8], as well as natural systems, such as cardiorespiratory interaction [9, 10], brain activity of Parkinsonian patients [11], paddlefish electrosensitive cells [12], Canadian lynx-hare populations [13], and solar activity [14]. To analyze such experimental data, some special techniques of synchronization analysis have been proposed and it has been shown that they are very efficient even for noisy and instationary data [5, 9, 11, 15]. However, the important problem, how to reconstruct models from such synchronized data, remains open. Therefore, it is of special interest to estimate the system parameters such as the coupling strength of interaction and the frequency difference between the two systems. Such parameters often cannot be measured, in particular in natural systems. To retrieve the synchronization regime of the underlying dynamical system, the reconstruction of a parametrized model family from sets of bivariate recording data is required.

There are several approaches to recover such a family from time series [16, 17]. By using radial basis functions, the bifurcation diagram of a discrete-time dynamical system has been successfully reconstructed, however, under the restriction that the value of the bifurcation parameters that underlie each of the given time series are known a priori [18]. Also, a nonlinear regression technique based on maximal correlation enables, in a nonparametric way, to retrieve the time delay in experimental laser data [19], or to reconstruct a model from binary experimental fluid experiments [20].

In this study, we focus on a special technique that enables to estimate the underlying bifurcation parameters of dynamical systems [21]. By using nonlinear predictors, the

CP622, *Experimental Chaos: 6th Experimental Chaos Conference*, edited by S. Boccaletti et al.

algorithm reconstructs a bifurcation parameter family from only a set of time series associated with different parameter values. This approach has the important practical advantage that no prior knowledge of the parametrized family of the dynamics, in particular the underlying bifurcation parameter values, is necessary. The efficiency of this technique has been shown for prototypical model systems of chaos, such as the Hénon, Rössler and coupled delayed-logistic system [21] and for instationary data [22].

The aim of the present study is to demonstrate a modeling approach applied to coupled chaotic oscillators using bivariate time series. The global behavior of the underlying system is learned from measurements made with different parameter values of the system and the model is then used to reconstruct the synchronization diagram.

MODELING TECHNIQUE

Suppose we have two coupled oscillators whose dynamical states are simultaneously recorded into bivariate time series $\{\xi_1(t), \xi_2(t)\}$. Among several dynamical components, only a single variable $\xi_{1,2}(t)$ is observed from each oscillator. With a change in the system parameters p such as the coupling strength parameter, the coupled system exhibits a variety of dynamical patterns, *e.g.*, regimes of phase synchronization, non-phase synchronization patterns, borderlines between both, etc. From N sets of bivariate time series $\{\xi_1(t, p_i), \xi_2(t, p_i)\}_{i=1,\dots,N}$ recorded with different system parameter values $\{p_i\}_{i=1,\dots,N}$, we want to estimate the corresponding system parameter values. This is equivalent to a practical experimental situation, since in real experimental measurements, it is not unusual to have a variety of data acquired under different recording conditions. Assuming that we have no prior knowledge about the coupled systems, it is very difficult to estimate the exact parameter values from only the measured time series.

The main idea of our approach is, therefore, to estimate system parameters Γ which are qualitatively similar to the original parameters p. For a one-to-one correspondence between Γ and p, Γ gives rise to a parametrized family of nonlinear dynamics $F(\Gamma, \cdot)$ which exhibits qualitatively similar bifurcation phenomena inherent to the original system.

Our algorithm is mainly composed of two steps:
(i) We embed the bivariate time series $\{\xi_1(t), \xi_2(t)\}$ into delay coordinates [23]:

$$\begin{aligned} X_1 &= \{\xi_1(t), \xi_1(t-\tau), \dots, \xi_1(t-(d-1)\tau)\}, \\ X_2 &= \{\xi_2(t), \xi_2(t-\tau), \dots, \xi_2(t-(d-1)\tau)\}. \end{aligned} \tag{1}$$

Within the same parametrized family, two coupled nonlinear predictors $F_{1,2}$ that model the N sets of time series as

$$\begin{aligned} \frac{dX_1}{dt} &= F_1(\Omega, X_1, X_2), \\ \frac{dX_2}{dt} &= F_2(\Omega, X_2, X_1) \end{aligned} \tag{2}$$

are then constructed. This means that we seek the N sets of nonlinear prediction parameters $\{\Omega(p_i)\}_{i=1,\dots,N}$ that correspond to each time series $\{\xi_1(t, p_i), \xi_2(t, p_i)\}_{i=1,\dots,N}$. For the nonlinear predictors $F_{1,2}$, any global functional model, such as polynomial functions [17] or radial basis functions [18] may be used. In this study we exploit 3-layer feed forward neural network [24] having $2d$ neurons in the input layer and d neurons in the output layer for $F_{1,2}$. The generalization capability of the neural network [25] works efficiently for the learning of a dynamical parametrized family.

(ii) By using the singular value decomposition, principal components Γ are extracted from the nonlinear prediction parameters $\{\Omega(p_i)\}_{i=1,\ldots,N}$. The number of the principal components Γ are determined in such a way that the sum of the normalized eigenvalues that represent significances of Γ is greater than 0.95. On this principal component parameter space Γ, we define a parametrized family of coupled nonlinear predictors $F_{1,2}$. We show that the principal component parameter family of nonlinear predictors gives rise to synchronization phenomena qualitatively similar to the original one. The underlying system parameters of the phase synchronization are therefore estimated by using the principal component parameters of the nonlinear predictors.

APPLICATION

Let us apply our technique to the two bidirectionally coupled Rössler oscillators:

$$\begin{aligned} \frac{dx_{1,2}}{dt} &= -\omega_{1,2}y_{1,2} - z_{1,2} + C(x_{2,1} - x_{1,2}), \\ \frac{dy_{1,2}}{dt} &= \omega_{1,2}x_{1,2} + 0.15y_{1,2}, \\ \frac{dz_{1,2}}{dt} &= 0.2 + z_{1,2}(x_{1,2} - 10). \end{aligned} \tag{3}$$

Note that both subsystems are usually not identical, but have a mismatch $\Delta\omega$ in the parameters $\omega_{1,2} = 1 \pm \Delta\omega$. In the absence of coupling (C=0) both oscillators are chaotic. It has been recently shown that such two non-identical Rössler systems are able to phase synchronize due to a weak coupling strength C [3]. A systematic dependence of the synchronization region on the coupling strength C and the parameter mismatch $\Delta\omega$ is shown in Fig. 1. To learn the global behavior of this coupled system [Eqs. (3)] in dependence on both parameters, we simulate 3 sets of bivariate time series $\{\xi_1(t), \xi_2(t)\}_{i=1,2,3}$ associated with 3 different parameter values of $(C, \Delta\omega) = (0.06, 0.045)$, $(0.05, 0.04)$, $(0.06, 0.035)$. For the time series variables $\xi_{1,2}$, we pick the $y_{1,2}$-coordinate from each oscillator.

At the first step of the algorithm, the time lag and the embedding dimension are set as $(\tau, d) = (0.2, 3)$. Next, we model each of the 3 records using the coupled nonlinear predictors Eqs. (2). A 2-dimensional principal component parameters (Γ_1, Γ_2) are extracted from the 3 sets of nonlinear prediction parameters that correspond to the 3 records. Then we perform a linear rescaling of this space, in order to bring these parameters into close correspondence with the system's parameters C and $\Delta\omega$. First, we rewrite both components in polar coordinates $(\Gamma_1, \Gamma_2) = (r\cos\theta, r\sin\theta)$ and determine the optimal angle θ that corresponds to C. The main point of this optimization is to find the θ value with the strongest influence on the coupling parameters $\{\Omega_k : k \in \text{Coupling}\}$ of the nonlinear predictors $F_{1,2}$. This subset of the prediction parameters includes all those which couple $X_{1,2}$ in Eqs. (2). Their influence is measured by the mean variance $\delta\Omega(\theta)$ of these coupling parameters $\{\Omega_k\}$ obtained for all radii $\{r = i/40 : i = -40, -39, \ldots, 40\}$. The mean variance $\delta\Omega$ is then normalized by using 500 sets of the mean variances computed for randomly selected parameters of the nonlinear predictors. From this bootstrapping technique, we calculate Theiler's [26] significance value $S(\theta)$ and get a maximum for $\theta = 0.251$, which is related to $S > 6$, a good level of significance. The so optimized reconstructed synchronization diagram shows a surprising similarity to the original diagram (see Fig. 1). Even the critical lines of the both diagrams, *i.e.*, the borderlines between the synchronized and non-synchronized regimes, are in good agreement as indicated in Fig. 2.

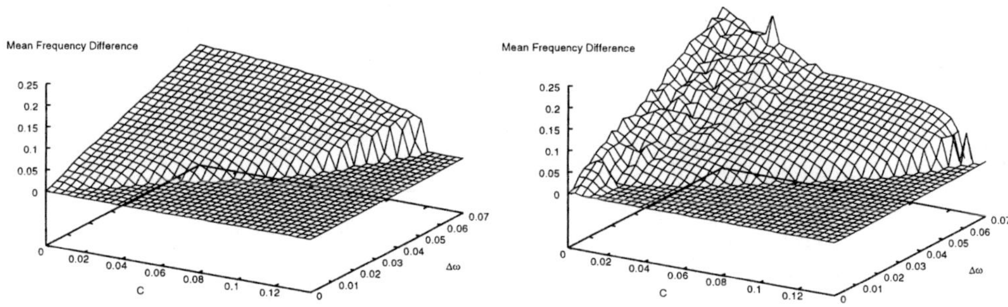

FIGURE 1. Synchronization diagram of the two coupled Rössler oscillators (right) and the coupled nonlinear predictors (left). The mean frequency difference of the two oscillators is computed by changing the coupling strength C and the frequency mismatch $\Delta\omega$.

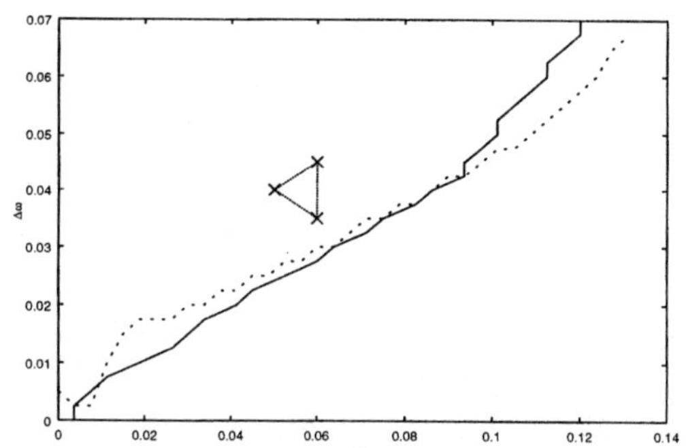

FIGURE 2. Borderlines between regimes of phase synchronization and non-synchronization for the coupled Rössler systems (solid line) and the and the nonlinear model (dotted line). The triangle indicates the locations of the 3 sets of parameter values used for the learning.

It is important to note that the capability of reconstructing the synchronization diagram strongly depends upon the type of the data, *i.e.*, whether or not they are from a phase synchronized regime. In the above experiment, all records used for the model learning are from the non-phase synchronization regime (Fig. 2). We find that the similarity of the reconstructed synchronization diagram to the original one is not as good as when we replace more and more of these records by those from the synchronized region. This can be interpreted as follows: In the phase space of the coupled oscillators, there is a low-dimensional nonlinear manifold on which the phase synchronization is realized. The data from the phase synchronization regime are, therefore, constrained in this nonlinear manifold and they do not provide any information about the attractor in the extended phase space, which is generated by the non-synchronized dynamics. From such phase synchronized data, a prediction of the non-synchronized dynamics seems to be very hard. We, therefore, emphasize that the most efficient way to reconstruct the global behavior of such systems is to use only data from the non-phase synchronized regime. Even when we have to deal with some synchronized data, the results might be improved if we use transient or noisy data because they may contain some dynamical information from outside of the phase synchronization manifold.

CONCLUSION

To conclude, the modeling approach presented in this study enables to reconstruct the synchronization diagram of two coupled chaotic oscillators from only 3 records of bivariate time series. This has been demonstrated for a prototypical model for coupled chaotic systems. Our technique allows to estimate the relative strength of coupling and to evaluate how close a system is to the borderline of phase synchronization as well as non-phase synchronization, even when no further information about the underlying dynamics is available. This should be of importance especially for natural coupled system, *e.g.*, in Neuroscience.

REFERENCES

1. L.M. Pecora and T.L. Carroll, Phys. Rev. Lett. **80**, 2109 (1990).
2. N.F. Rulkov, M.M. Sushchik, L.S. Tsimring, and H.D.I. Abarbanel, Phys. Rev. E **51**, 980 (1995); L. Kocarev and U. Parlitz, Phys. Rev. Lett. **76**, 1816 (1996); K. Josic, Phys. Rev. Lett. **80**, 3053 (1998).
3. M.G. Rosenblum, A.S. Pikovsky, and J. Kurths, Phys. Rev. Lett. **76**, 1804 (1996).
4. M.G. Rosenblum, A.S. Pikovsky, and J. Kurths, Phys. Rev. Lett. **78**, 4193 (1997)
5. J. Kurths (Ed.), *Phase Synchronization and its Applications*, Special Issue in Int. J. Bifurcation and Chaos, **10-11** (2000); A. Pikovsky, M. Rosenblum, and J. Kurths *Synchronization - A Universal Concept in Nonlinear Sciences*, (Cambridge University Press, Cambridge, 2001).
6. U. Parlitz, L. Junge, W. Lauterborn, and L. Kocarev, Phys. Rev. E **54**, 2115 (1996); S. Taherion and Y.-C. Lai, Int. J. Bifurcation and Chaos **11**, 2587 (2000).
7. E. Lariontsev, Int. J. Bifurcation and Chaos **10**, 2441 (2000); E. Allaria, F.T.Arecchi, A. DiGarbo, and R. Meucci, Phys. Rev. Lett. **86**, 787 (2001).
8. C.M. Ticos, E. Rosa, Jr., W.B. Pardo, J.A. Walkenstein, and M. Monti, Phys. Rev. Lett. **85**, 2929 (2000); E. Rosa *et al.*, Int. J. Bifurcation and Chaos **10**, 2551 (2000).
9. C. Schafer, M.G. Rosenblum, J. Kurths, and H.-H. Abel, Nature **392**, 239 (1998).
10. V.S. Anishchenko *et al.*, Int. J. Bifurcation and Chaos **10**, 2339 (2000); A. Stefanovska *et al.*, Phys. Rev. Lett. **85**, 4831 (2000).
11. P. Tass *et al.*, Phys. Rev. Lett. **81**, 3291 (1998).
12. A. Neiman *et al.*, Phys. Rev. Lett. **82**, 660 (1999); A. Neiman *et al.*, Int. J. Bifurcation and Chaos **10**, 2499 (2000).
13. B. Blasius, A. Huppert, and L. Stone, Nature **399**, 354 (1999); B. Blasius and L. Stone, Int. J. Bifurcation and Chaos **10**, 2361 (2000)

14. M. Palus, J. Kurths, U. Schwarz, D. Novotna, and I. Charvatova, Int. J. Bifurcation and Chaos **10**, 2519 (2000).
15. N.B. Janson, A.G. Balanov, V.S. Anishchenko, and P.V.E. McClintock, Phys. Rev. Lett. **86**, 1749 (2001).
16. H.D.I. Abarbanel, *Analysis of Observed Chaotic Data*, (Springer-Verlag, New York, 1996); H. Kantz and T. Schreiber, *Nonlinear Time Series Analysis*, (Cambridge University Press, Cambridge, 1997).
17. J. P. Crutchfield and B. S. McNamara, Complex Systems **1**, 417 (1987).
18. M. Casdagli, Physica **35D**, 335 (1989).
19. H. Voss and J. Kurths, Phys. Lett. A **234**, 336 (1997).
20. H. Voss, P. Kolodner, M. Abel, and J. Kurths, Phys. Rev. Lett., **83**, 3422 (1999).
21. R. Tokunaga, S. Kajiwara, and T. Matsumoto, Physica **79D**, 348 (1994); I. Tokuda, S. Kajiwara, R. Tokunaga, and T. Matsumoto, Physica **95D**, 380 (1996).
22. I. Tokuda, R. Tokunaga, and T. Matsumoto, Physica **135D**, 63 (2000).
23. F. Takens, in *Dynamical Systems and Turbulence*, Lecture Notes in Math., edited by D. A. Range and L.S. Young (Springer, Berlin, 1981), Vol. 898, p. 366; T. Sauer, J.A. York, and M. Casdagli, J. Stat. Phys. **65**, 579 (1991).
24. D.E. Rumelhart, J.L. McClelland, and the PDP Research Group, *Parallel Distributed Processing*, (MIT Press, Cambridge, 1986).
25. G. Cybenko, Math. Control Signals Systems **2**, 303 (1989); K. Funahashi, Neural Networks **2**, 183 (1989); K. Hornik, Neural Networks **2**, 359 (1989).
26. J. Theiler, S. Eubank, A. Longtin, B. Galdrikian, and J.D. Farmer, Physica **58D**, 77 (1992); J. Theiler and D. Prichard, Physica **94D**, 221 (1996).

A surrogate test for pseudo-periodic time series data

Michael Small[†*], Robert G. Harrison[†] and C.K. Tse[*]

[*]*Electronic and Information Engineering, Hong Kong Polytechnic University, Hong Kong*
[†]*Department of Physics, Heriot-Watt University, Edinburgh, United Kingdom*

Abstract. Standard (linear) surrogate methods are only useful for time series exhibiting no pseudo-periodic structure. We describe a new algorithm that can distinguish between a noisy periodic orbit and deterministic non-periodic inter-cycle dynamics. Possible origins of deterministic non-periodic inter-cycle dynamics include: non-periodic linear or nonlinear dynamics, or chaos. This new algorithm is based on mimicking the large-scale dynamics with a local model, but obliterating the fine scale features with dynamic noise. We demonstrate the application of this method to artificial data and experimental time series, including human electrocardiogram (ECG) recordings during sinus rhythm and ventricular tachycardia (VT). The method is able to successfully differentiate between the chaotic Rössler system and a pseudo periodic realization of the Rössler equations with dynamic noise. Application to ECG data demonstrates that both sinus rhythm and VT exhibit nontrivial inter-cycle dynamics.

INTRODUCTION

Experimentally observed time series often exhibit a strong periodic component. In such cases it is common to ask if the dynamical system underlying the process is more complicated than a periodic orbit. For example, an observed oscillatory system may be classified as simply a noisy periodic orbit or a form of pseudo-periodic chaos (such as the chaotic Rössler system). In this paper we introduce an algorithm to differentiate between a time series generated by a noise driven periodic orbit or non-periodic deterministic dynamics. The method we describe is a new surrogate data test [1] that employs local linear modelling methods [2] to mimic the large scale periodic structure of a time series. Fine scale (i.e. long term aperiodic deterministic) features are destroyed by phase space randomisation.

Surrogate data methods [1] allow one to test a given time series $\{y_t\}_t$ against a specified hypothesis $\mathcal{H}_0$. One applies a *surrogate generation algorithm* to generate an ensemble of data sets that are both like the time series data $\{y_t\}_t$ and consistent with the hypothesis under consideration. Surrogate data are what we would expect the data $\{y_t\}_t$ to look like if it was consistent with the hypothesis $\mathcal{H}_0$ [3]. Using this collection of surrogate data sets, one then estimates the distribution of test statistic values (for some test statistic) expected for the hypothesis $\mathcal{H}_0$. This distribution of statistic values is then compared to the value obtained from the original time series data. If the time series data is atypical, then $\mathcal{H}_0$ may be rejected (with an appropriate statistical confidence); otherwise, it may not.

CP622, *Experimental Chaos: 6th Experimental Chaos Conference*, edited by S. Boccaletti et al.

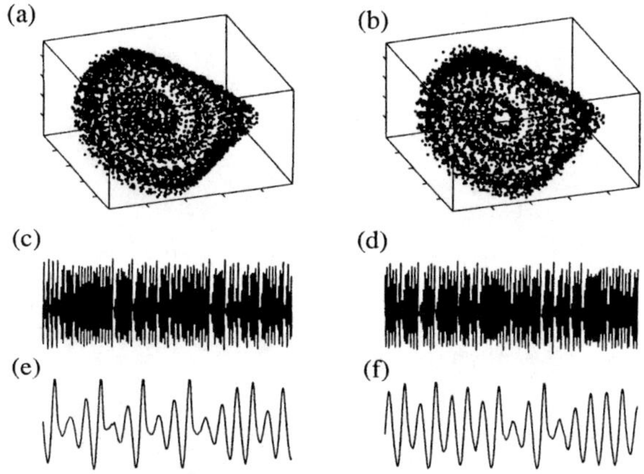

FIGURE 1. Generation of PPS data from a chaotic Rössler time series. The top panels show the reconstructed attractors ($d_e = 3$, $\tau = 8$) for (a) the original data, and (b) the PPS data. Also shown is: (c) the full original and (d) PPS time series, and a short section of (e) original and (f) PPS data. Note that the time series appear virtually indistinguishable and the reconstructed attractors are qualitatively similar (although not identical).

This methodology requires a suitable algorithm for generating surrogate data sets consistent with a hypothesis of interest. The most commonly applied surrogate generation methods allow one to test only simple linear stochastic dynamics [1]. For data that exhibits strong periodic components such hypotheses are trivially false. The surrogate data test we describe here is capable of testing pseudo-periodic time series against the hypothesis of no additional (non-periodic) determinism.

PSEUDO-PERIODIC SURROGATE DATA

Pseudo-periodic surrogate (PPS) data must satisfy two criteria: (i) the PPS data must exhibit the same periodic structure as the original data $\{y_t\}_t$, and (ii) the PPS data must have no other deterministic structure. To achieve these goals we must characterise the dynamics responsible for $\{y_t\}_t$ and discriminate between periodic and non-periodic behaviour. To characterise the underlying dynamics we employ the standard time delay embedding and local modelling techniques (effectively, we employ a local constant model). Randomisation and discrimination between periodic and non-periodic structure is achieved by iterating trajectories randomly among spatial neighbours.

Let $\{y_t\}_{t=1}^{N}$ be a scalar time series of N observations. For *embedding dimension* d_e and *embedding lag* τ we reconstruct the underlying dynamics according to the Takens' embedding theorem [4]:

$$z_t = (y_t, y_{t-\tau}, y_{t-2\tau}, \ldots, y_{t-(d_e-1)\tau})$$

for $t = (d_e - 1)\tau + 1, \ldots, N$. For convenience let us re-index the embedded time series to

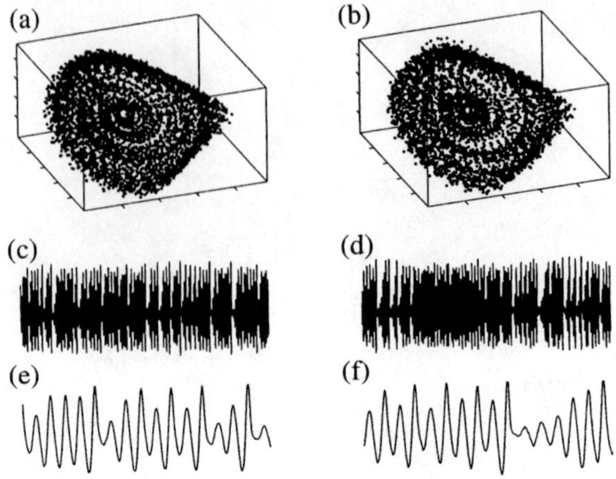

FIGURE 2. Generation of PPS data from a noisy period 6 Rössler time series. The top panels show the reconstructed attractors ($d_e = 3$, $\tau = 8$) for (a) the original data, and (b) the PPS data. Also shown is: (c) the full original and (d) PPS time series, and a short section of (e) original and (f) PPS data. Note that the time series and reconstructed attractors are largely similar to those depicted in figure 1.

be $\{z_t\}_{t=1}^{\tilde{N}}$ where $\tilde{N} = N - (d_e - 1)\tau$. To construct a surrogate we employ the following scheme:

1. Randomly choose an initial condition s_1 where $s_1 \in \{z_t : t = 1, \ldots, \tilde{N}\}$. Set $i = 1$.
2. For s_i choose a neighbour $s_j \in \{z_t : t = 1, \ldots, \tilde{N}\}$ with probability

$$\mathrm{Prob}(s_j = z_t) \quad \propto \quad \exp \frac{-\|z_t - s_i\|}{\rho}.$$

   The parameter ρ is the *noise radius* and will be discussed later.
3. Set $s_{i+1} = s_{j+1}$ and then increment i.
4. Repeat this procedure from step 2 until $i = N$.

The vector time series $\{s_t\}_{t=1}^{N}$ represents a stochastic trajectory on the attractor approximated by $\{z_t\}_{t=1}^{\tilde{N}}$ with the same underlying dynamics (contaminated by noise) as the original system. Let $x_t \in \mathbf{R}$ be the first component of s_t (i.e. $s_t = (x_t, \cdot, \cdot, \ldots)$), then $\{x_t\}_{t=1}^{N}$ is a PPS data set for $\{y_t\}_{t=1}^{N}$. Examples of this process are illustrated in figures 1 and 2.

The above algorithm requires selection of three parameters: embedding dimension d_e, embedding lag τ, and the noise radius ρ. We will not address the issue of selection of the embedding parameters d_e and τ here, for a review see [5]. However, the noise radius ρ must be carefully selected. If ρ is too large then the PPS algorithm will introduce too much randomisation and the surrogates will no longer resemble the data. Conversely if ρ is too small, then the PPS algorithm will introduce insufficient randomisation, and the surrogates will contain non-periodic determinism. We have found that an intermediate value of ρ may be selected by computing the expected number of sequences of points in the surrogate that are identical to the data.

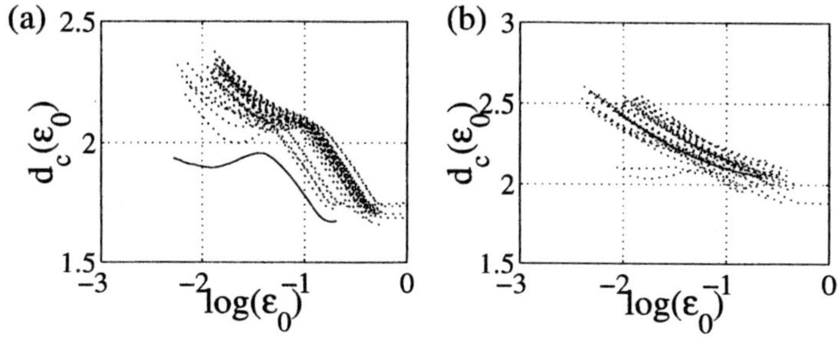

FIGURE 3. Comparison of correlation dimension for original and PPS data for the Rössler system. The correlation dimension estimate $d_c(\varepsilon_0)$ is plotted against $\log(\varepsilon_0)$ (small scales correspond to small $\log(\varepsilon_0)$). The value of $d_c(\varepsilon_0)$ for the data is shown as a solid line, for each surrogate it is plotted as a dotted line. Panel (a) shows correlation dimension estimates for the data shown in figure 1 and 30 PPS data sets. Panel (b) is the same calculation for the data in figure 2. For the chaotic time series (a) the data and surrogates are clearly distinguishable, however for the noise driven periodic system (b) there is no distinction.

EXAMPLES AND APPLICATIONS

All surrogate data methods require a *test statistic* with which to compare data to surrogates. We choose to apply the correlation dimension estimation algorithm described by Judd [6]. This algorithm estimates correlation dimension d_c as a function of viewing scale ε_0, hence $d_c(\varepsilon_0)$. The algorithm has been shown to be robust (*pivotal* in the sense suggested by Theiler [3], see [7]) when employed as a surrogate data test statistic. In the next section we demonstrate the application of this algorithm with an example from the Rössler system. In the following section we present several experimental examples.

The Rössler system

The first of our applications of this algorithm is the Rössler system. For different parameter values this system exhibits broad band chaos, and period 6 behaviour [8]. In both these regimes we integrated this system for 5000 time steps of 0.2 time units. We added Gaussian noise with standard deviation of 0.05 to the each scalar component at each step. These time series, along with PPS data and reconstructed attractors are illustrated in figures 1 and 2. Results of PPS algorithm calculations for these data are depicted in figure 3.

Figure 1 shows that the distinction between the original chaotic attractor, and an attractor reconstructed from a representative surrogate are not obvious to casual inspection. Furthermore, the representative time series appear qualitatively similar. However, the quantitative correlation dimension analysis of figure 3(a) shows clear distinction: the PPS algorithm successfully identifies the deterministic non-periodic dynamics.

The results in figure 3(b) of the analysis of the periodic Rössler data show the opposite situation. In this case the PPS algorithm is unable to distinguish between data and surrogates, and therefore correctly concludes that the data under consideration is indistinguishable from that generated by a noisy periodic orbit. Significantly, the data

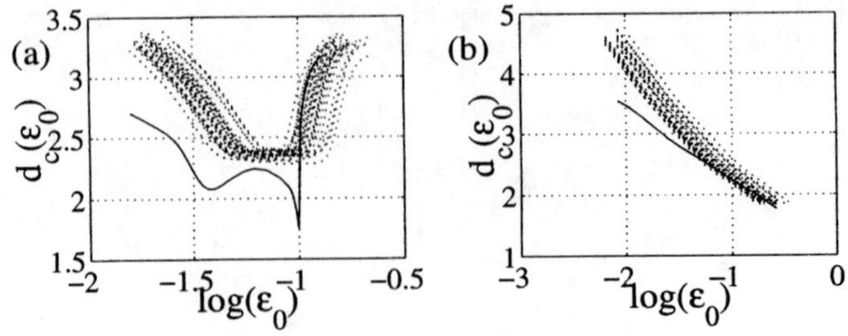

FIGURE 4. PPS data calculation for human ECG data. Panel (a) is the correlation dimension estimates for human ECG during sinus rhythm and 30 PPS data sets. Panel (b) is the correlation dimension estimates for human ECG during VT and 30 PPS data sets. Although this quantitative analysis was able to distinguish data and PPS data they appear qualitatively similar.

and surrogates time series shown in figure 2 are qualitatively similar to one another and also similar to the data in figure 1.

Experimental data

In this section we summarise the application of the PPS algorithm to several experimental time series, including ECG data during sinus rhythm and ventricular tachycardia (VT). Applied to these data, our algorithm demonstrates that sinus rhythm and VT both exhibit deterministic non-periodic inter-cycle dynamics. We also present the results of the application of this method to annual sunspot numbers, NH_3 laser intensity data and monthly reported cases of measles, mumps and chicken pox in New York City (USA).

We repeated the calculation applied to the Rössler data with two recordings of human ECG during sinus rhythm and VT (figure 4). For both sinus rhythm and VT, the 30 PPS data sets were clearly distinguished from the ECG recordings. The null hypothesis of a noise driven periodic orbit should therefore be rejected for this data.

Tablc 1 provides a summary of these results and results for a variety of other experimental systems. The laser data is known to be chaotic [9] and the analysis presented here confirms this. Similarly the result for the sunspot time series is consistent with the literature [10] — the limited amount of data available appears to be consistent with a noisy periodic orbit. The results of our analysis conclude that the measles epidemic time series is a noisy periodic orbit and that mumps and chicken pox contain substantial nonlinearity. This is also consistent with the differing phenomenological origin for these diseases [11].

CONCLUSION

We have described a computationally inexpensive algorithm that may be employed to test for determinism and appears to be robust to a wide variety of parameter values.

TABLE 1. Application of PPS algorithm for various experimental data.

Data source	N	d_e	τ	ρ	Conclusion
ECG Sinus	10000	6	6	0.005	Reject $\mathcal{H}_0$
ECG VT	10000	6	6	0.002	Reject $\mathcal{H}_0$
Sunspots	300	6	3	100	Accept $\mathcal{H}_0$
Laser	10000	6	2	5	Reject $\mathcal{H}_0$
Measles	534	6	4	50000	Accept $\mathcal{H}_0$
Mumps	534	6	4	20000	Reject $\mathcal{H}_0$
Chicken pox	498	6	3	35000	Reject $\mathcal{H}_0$

Using the Rössler system as a test case we were able to successfully differentiate between chaotic and periodic systems contaminated with dynamic noise.

The experimental data used in this paper was deliberately selected to be highly noisy and many of the data sets are particularly short. The PPS time series generated from these data is qualitatively very similar to the data. However, quantitative analysis demonstrated that in some cases the data and surrogates are clearly distinct. In each case, the results of this analysis are consistent with the existing literature. Significantly, the result that human ECG during both sinus rhythm and VT exhibits deterministic non-periodic intercycle dynamics provides further support for previous observations that ECG during sinus rhythm and arrhythmia are nonlinear and deterministic [12, 13].

ACKNOWLEDGMENTS

The research was partially funded through a Research Development Grant by the Scottish Higher Education Funding Council (SHEFC), No. RDG/078. MS is currently funded by a Hong Kong Polytechnic University Postdoctoral Fellowship award.

REFERENCES

1. Theiler, J., Eubank, S., Longtin, A., Galdrikian, B., and Farmer, J. D., *Physica D*, **58**, 77–94 (1992).
2. Sugihara, G., and May, R., *Nature*, **344**, 737–740 (1990).
3. Theiler, J., and Prichard, D., *Physica D*, **94**, 221–235 (1996).
4. Takens, F., *Lecture Notes in Mathematics*, **898**, 366–381 (1981).
5. Abarbanel, H. D., *Analysis of observed chaotic data*, Institute for nonlinear science, Springer-Verlag, New York, 1996.
6. Judd, K., *Physica D*, **71**, 421–429 (1994).
7. Small, M., and Judd, K., *Physica D*, **120**, 386–400 (1998).
8. Thompson, J., and Stewart, H., *Nonlinear dynamics and chaos*, John Wiley & Sons Ltd., Chichester, 1986.
9. Arbarbanel, H. D., Gills, Z., Liu, C., and Roy, R., *Phys Rev A*, **53**, 440–453 (1996).
10. Judd, K., and Mees, A., *Physica D*, **82**, 426–444 (1995).
11. Keeling, M. J., Rohani, P., and Grenfell, B. T., *Physica D*, **148**, 317–335 (2000).
12. Small, M., Yu, D., Simonotto, J., Harrison, R. G., Grubb, N., and Fox, K., *Chaos, Solitons and Fractals* (2001), to appear.
13. Small, M., Yu, D., Harrison, R. G., Robertson, C., Clegg, G., Holzer, M., and Sterz, F., *Chaos*, **10**, 268–277 (2000).

APPLICATIONS

A Dynamical Approach for Modelling and Control of Production Systems

Bernd Scholz-Reiter, Michael Freitag, Alf Schmieder

Department of Planning and Control of Production Systems, University of Bremen, P.O.Box 33 05 60, D-28335 Bremen, Germany

Abstract. Production systems are considered to be dynamical systems. A dynamical approach for modelling and control of production systems evolves out of the context of new developments in production research. A model of an elementary production system is set up in order to demonstrate the idea of control in the introduced approach. Some simulation results are given.

INTRODUCTION

The development of Nonlinear Dynamics and its applications enables us to understand and describe complex systems where linear approaches fail or are too far from reality. Complex dynamic behaviour can occur in relatively simple production systems. Beaumariage and Kempf have shown the sensitive dependence of throughput times on the initial conditions and scheduling rules in a production system model [1]. The origin of this unstable behaviour is not obvious. Other authors find even contrary results [2]. Bartholdi, Bunimovich and Eisenstein have shown, that deterministic models can describe the dynamics of production lines appropriately [3][4]. Understanding the production dynamics is therefore a fundamental part in a dynamical approach for modelling and control of production systems.

Classical production planning and control systems (PPC systems) are based on concepts that do not consider the production system as a dynamical system. Usually, heuristic approaches are preferred in order to simulate the production process and its scheduling and control. But optimisation methods do not provide the controller with good results if there are some changes during the optimisation period [5][6]. In the context of today's highly dynamic market with its rapid changes in demand, flexibility and variety are recognized as increasing measures of modern PPC systems.

In the third section of this paper, a dynamical point of view is introduced, which incorporates dynamical aspects of a concept level of PPC, taking into account functional aspects of PPC and modelling of production systems as well.

CP622, *Experimental Chaos: 6th Experimental Chaos Conference*, edited by S. Boccaletti et al.

PRODUCTION SYSTEMS AND PPC CONCEPTS

Production systems can be classified as capacity oriented production vs. customer oriented production. Customer oriented production systems are considered to have a high degree of flexibility. They need information feedback from the shop floor in order to influence the product flow in short time intervals in a non-periodic, order dependent way to meet the needs of a highly dynamic market. This type of production system is considered in this paper.

The production planning and control in an enterprise consists of the following problems:

- Production program planning
- Material requirements planning
- Throughput time scheduling
- Capacity requirement planning
- Order release
- Order monitoring and control

All of these problems have to be solved by a PPC system, which is able to run the entire manufacturing process. How they are solved, in what order, if local or as part of a global strategy, depends on the PPC concept underlying the PPC system. In general, PPC systems are designed to find optimal solutions for specific PPC problems. The most important optimisation goals are throughput times, total costs, capacity utilization, inventory costs and delivery reliability. Thereby, strategies for different objectives can lead to contrary results, e.g. maximal capacity utilization and minimal throughput times.

Models of production systems are used to put these strategies into practice. They are derived from classical optimisation tasks studied in Operations Research [7]. The model for the production system itself therefore consists of different optimisation problems. There is no interaction of these particular theories. This could be developed in a meta-theory, but in general, the results will be hard to interpret [8].

Most recent PPC systems work in line with successive planning concepts consisting of successive planning steps mentioned above. It is widely used but also criticised because it lacks the interaction between the planning steps [8]. Several approaches try to overcome the difficulties resulting from this planning and control scheme [8][9]. They extend the possibilities of traditional PPC systems towards faster adaptation to changes in demand.

THE DYNAMICAL APPROACH

Why a Dynamical Approach?

Known concepts in PPC are normally founded on models for parts of PPC that solve local problems like the optimisation function of lot sizes [10]. Optimisation models (solved by exact algorithms or heuristic approaches), as discussed in Operations Research, are often not satisfying even at a low level of complexity, e.g. 3 machines, 2 job steps, more than 2 products [10]. They do not describe the qualitative behaviour.

Dynamical models are different from optimisation models as they describe relations between system variables. The main principle is understanding the dynamic behaviour instead of finding solutions for particular problems. To get predictable results, an intrinsic deterministic model is necessary, into which stochastic influences can be incorporated later on. Bartholdi, Bunimovich and Eisenstein have shown that sewing production lines can be described by way of a deterministic model [3][4]. The dynamic behaviour was driven exclusively by deterministic rules.

In a more complex production system, a number of deterministic processes occur. A classification of the functional aspects of PPC is therefore necessary. It will be discussed in the following sections.

The Idea of the Dynamical Approach

Rapid adaptation to frequent changes in demand and product mix is required in a customer oriented production system. The point of view put forward in this paper is the combination of ideas founded on the Nonlinear Dynamics Theory and PPC research into a new approach of modelling and control.

The dynamical approach is a generalized approach which includes modelling of manufacturing processes and a dynamical control mechanism. This includes analysing dynamic behaviour according to Nonlinear Dynamics and solving PPC problems as well. The starting point for the control of chaos is the phase space of the system, which is spread out by the system variables. Several methods for the control of chaos have been developed [11].

The dynamics of particular manufacturing processes is caused by rules and conditions, which are parts of the functional structure of the PPC system. The dynamical approach consists of 3 levels of description:

1. Modelling of the production system
2. Dynamical control
3. Functional structure

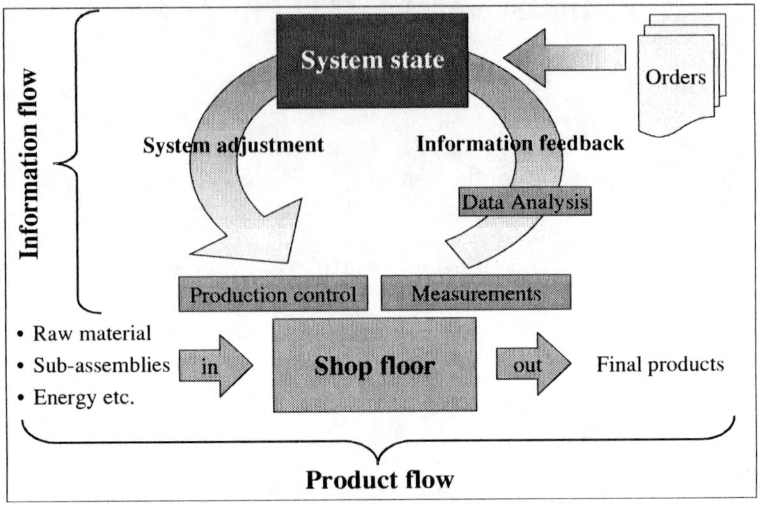

FIGURE 1. The dynamical approach.

Figure 1 shows the interaction of level 1 (product flow) and level 2 (information flow). A system state is defined which represents the actual situation in the production system. It is derived from the state space of the system, which contains the time series of measured data in analogy to the phase space of physical systems. Applications to production systems use variables related to the work content in the system like the so-called "work-in-process" (WIP), a widely used key parameter of manufacturing.

Information feedback from the shop floor and the arrival of new orders update the system state. This can be put into practice by frequent measurements of system variables like buffer levels, WIP of machines or system specific variables. To determine the system state, further analysis of the data is necessary in complex realistic models. This can be done by means of Nonlinear Dynamics Theory. In simple cases, the immediate construction of state spaces from the time series can reveal the dynamics of the system. An example will be given in the last section of this paper.

The control mechanism consists of 3 steps:

1. The System is in equilibrium.
2. Disturbances occur.
3. A new equilibrium is reached after dynamic adjustment of the system.

Orders are regarded as disturbances, leading to a new system state which includes reorganisation of order sequences.

In Level 3, basic functional aspects are collected in independent functional groups that enable the system to work. They are at first a framework for modelling the production system and provide finally possibilities to control the production process by a controller or by the system itself. The latter case is a step towards self-control, which is a fundamental idea in the dynamical approach.

The following functional groups are defined:

- The structure
- The capacity
- The operational rules
- The order release
- The queuing policies

These functional groups generate the dynamics of a production system and enable and influence the product flow through the system. They are discussed in detail in the next section.

The Functional Groups in the Dynamical Approach

The Structure

The functional group "structure" contains information about the number and arrangement of work stations, machines, buffers or storages necessary to describe the manufacturing process sufficiently. The hierarchy of structural information allows a more or less detailed description depending on the modelling aims.

The structure also contains information about the product flow. It shows how different types of products can be put through a system. This becomes relevant, for instance, in parallel manufacturing facilities where it is possible to do the same job at different machines. The structure of the product flow is determined by the layout of the shop floors and by the process plans of every product in the production program. In a typical job shop production system, one finds a complex situation with a number of machines arranged according to their functions, e.g. turning, milling, drilling etc. Therefore, the product flow contains convergences, division and feedback loops.

The structure of the product flow is considered as a function of the structure and state of the production system. Changes in the structure, e.g. caused by machine replacements, immediately influence the structure of the product flow and system state due to the dynamic interactions instead of new planning steps.

The Capacity

The capacity of a production system includes three important aspects: the capacity of space, the capacity of time and the capacity of manufacturing. The capacity of space describes the physical space to store and manufacture raw material, sub-assemblies and final products, e.g. storages and buffers. Capacity of time means the working time in a day, e.g. 8 or 16 hours per working day. The capacity of manufacturing contains the volume of production, product variety, quality measures and other parameters depending on machine parameters and structure.

Capacity is usually considered as a parameter (or a set of parameters) in production systems characterized by upper and lower limits. In a number of models there are no capacity limitations. In the dynamical approach, a detailed description of capacity is necessary to enable the system to adjust itself in a flexible way. Therefore, all three capacities are considered as variables that can be continuously modified.

So the role of storages and buffers could move from a passive box with fixed upper and lower limits to an active element. The flexible handling of the capacity of time (overtime work, part-time work) is a common procedure to balance costs and demand. The capacity of manufacturing is a variable value if at least one of the processes affecting this capacity changes over time. An example for a large scale change is a growing product variety that causes additional setup processes at the machines. This increases the throughput times and so decreases the capacity to manufacture these products. Examples of small scale changes to the capacity of manufacturing are machine breakdowns or missing tools, late deliveries of parts or rush orders.

The Operational Rules

If demand exists and manufacturing orders are waiting to be released and processed, the operational rules generate the dynamics of a production system. Such rules are generally simple handling processes but need to be defined in detail. An example of an operational rule is: what happens in a machine if the input buffer is empty or if the output buffer is full and how does this effect the adjacent machines and buffers.

There are some investigations that identify and eliminate such constraints in a production system [12]. In the majority of cases, the order and material flow and the position of the constraint are considered to be constant and static. This assumption is appropriate for a flow shop production system with a constant demand and a constant product mix. But for a job shop, changing demand and customer-specific products are typical. In this case, the order and material flow and therefore the position of constraints vary over time.

The dynamical approach suggests a dynamic adjustment of the operational rules on the current situation in the production system. An example is given in the last section of this paper.

The Order Release

In conventional PPC systems, the production orders are released after the throughput time scheduling, the capacity planning and the order sequencing. Order sequence and release dates are more or less fixed. Such a production schedule assumes constant throughput times and capacity availabilities during the entire planning period.

If rush orders must be released or resources fail, the fixed production schedule becomes void.

This problem is well-known and has been investigated as the successive concept for PPC fails against the background of a highly dynamic market with a growing product variety, customer specific products and short delivery times [13][14].

The dynamical approach suggests a dynamic order release depending on both the incoming customer orders and the actual situation on the shop floor.

The Queuing Policies

Queuing policies are rules that determine the withdrawal of material from buffer to manufacture on a machine. A well-known and widely used queuing rule is "first in - first out". Thereby, the processing sequence of the orders remains constant. Dynamic queuing rules like "least average static slack" or "least average dynamic slack" cause withdrawal depending on the stage of processing and due-date. But they do not take into account the situation at the work stations.

The dynamical approach allows the choice between different queuing rules. For the control of the order flow under quickly changing conditions, the dependence on the system state will play an important role.

A DYNAMICAL MODEL OF A PRODUCTION SYSTEM

The Boundary Conditions

The following sections will demonstrate the function of operational rules which influence the production process depending on the system state. The operational rules are considered dynamic objects. All other functional groups are kept constant. The goal is to control the production system via its system state by means of parameter changes in the operational rules.

Structure: An elementary production system with two work stations as shown in figure 2 will be considered. The structure of the product flow is linear with one input and one output channel assigned to each work station.

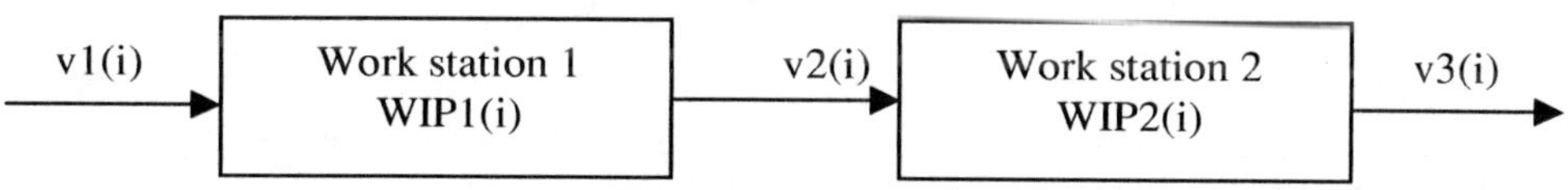

FIGURE 2. Production system model consisting of two work stations with work-in-process WIP(i) and product flow velocities v(i).

Operational rules: The current work-in-process of every work station is represented by the variable WIP(i) where i is the number of discrete time steps. The product flow into and out of the work station is represented by the product flow velocities v(i), which can be interpreted as work per time unit. Therefore, the variation of the work-in-process of the work stations is indicated by

$$\Delta WIP(i) = WIP(i+1) - WIP(i) = \tau \cdot (v_{in}(i) - v_{out}(i)) \tag{1}$$

with τ as time step size determining the grade of discretisation. One approach to dynamic feedback control is the regulation of the product flow depending on the work-in-process:

$$v1(i) = COM1 \cdot (WIP1_{max} - WIP1(i)) \tag{2}$$

$$v2(i) = COM2 \cdot (WIP2_{max} - WIP2(i)) \tag{3}$$

$$v3(i) = COM3 \cdot WIP2(i) \tag{4}$$

The parameters COM are constant, normalized production rates.

Capacity: The capacity of manufacturing is described by the production rates of the work stations (COM), which are kept constant. The work-in-process is limited by the maximum buffer level WIP_{max}

$$0 \leq WIP \leq WIP_{max} \tag{5}$$

Queuing policies: The global queuing policy is „first in – first out“.

Order release: There are no manufacturing orders with due-dates. Only the product flow between two work stations as a part of a more complex production system is considered.

The Dynamics of the Linear Coupled System

Under the assumptions named in the previous section, the work-in-process can be described as iterative mechanisms:

$$WIP1(i+1) = WIP1(i) + \tau \cdot (v1(i) - v2(i)) \tag{6}$$

$$WIP2(i+1) = WIP2(i) + \tau \cdot (v2(i) - v3(i)) \tag{7}$$

The WIP levels change their values according to the sum of preceding WIP levels and inputs and outputs in each iteration step. So the work stations are linearly coupled via the product flow velocity 2. The dynamic behaviour of this system is generated by equations 6 and 7, which can be understood as the operational rules for the manufacturing process.

This system ran in a simulation with the following system parameters and initial conditions:

$\tau = 1$, $WIP1_{max} = WIP2_{max} = 1$, $WIP1(1) = 0.4$, $WIP2(1) = 0.9$, $COM1 = 0.9$, $COM2 = 0.4$, $COM3 = 0.4$.

The system stabilises itself after a few iteration steps (see figure 3 and 4). The state space diagram shows the system state {WIP1(i), WIP2(i)} proceeding from the initial value (0.4, 0.9) to the stabilization point (0.78, 0.5). The dynamic behaviour of the system does not depend on the initial values nor on the relation of the production rates COM.

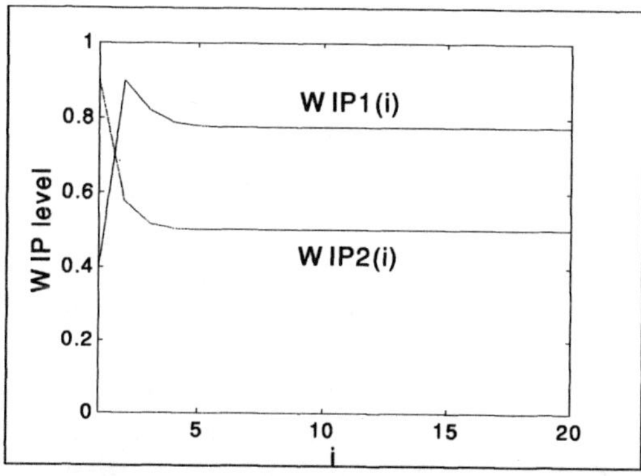

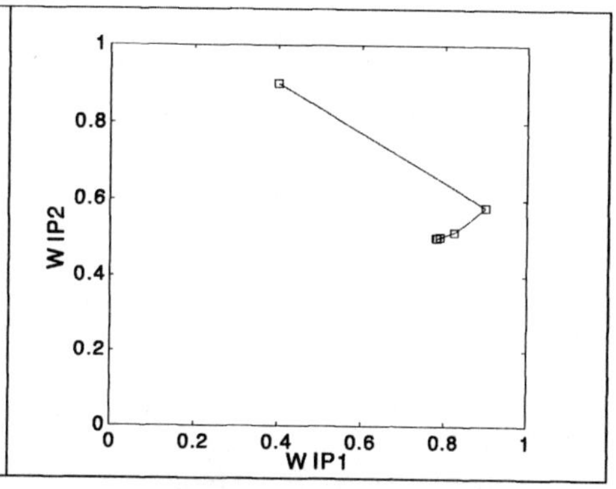

FIGURE 3. WIP levels of the linear coupled system, $WIP1_{eq} = 0.78$, $WIP2_{eq} = 0.5$.

FIGURE 4. State space of the linear coupled system.

The Dynamics of the Non-linear Coupled System

Non-linear operational rules are introduced to get a number of system states needed for faster adaptation of the system. An example is represented by equations 8 and 9.

$$WIP1(i+1) = WIP1(i) + \tau \cdot (v1(i) - X \cdot WIP1(i) \cdot v2(i)) \tag{8}$$

$$WIP2(i+1) = WIP2(i) + \tau \cdot (X \cdot WIP1(i) \cdot v2(i) - v3(i)) \tag{9}$$

Here the control of product flow is non-linear by coupling both buffer levels. Additionally, a coupling parameter X was introduced, which has the dimension of the production rate COM. This coupling parameter allows the controller to choose a tight or loose coupling of the work stations. The dynamic behaviour of this system depends sensitively on the coupling parameter X.

Figure 5 shows the oscillating WIP levels at X = 4. These oscillations are indicated by two fixed points in the state space and a diagonal link between them (see figure 6).

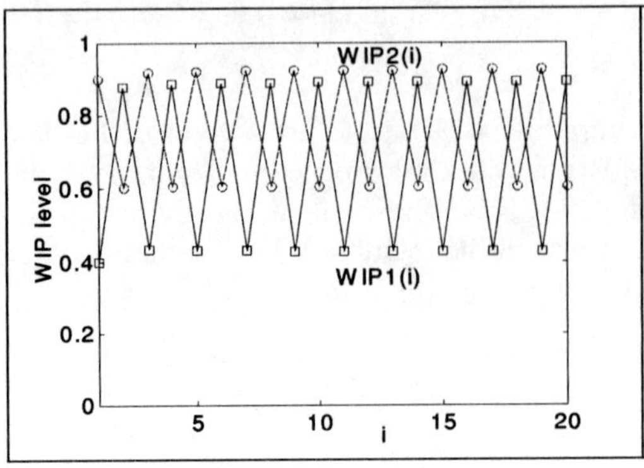

FIGURE 5. WIP levels of the non-linear coupled system with coupling parameter X = 4, $WIP1_{eq} = (0.43, 0.89)$, $WIP2_{eq} = (0.61, 0.93)$.

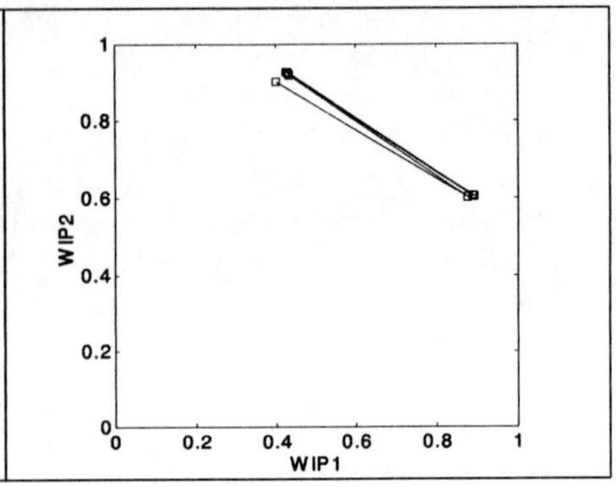

FIGURE 6. State space of the non-linear coupled system with coupling parameter X = 4.

Figure 7 shows two typical bifurcation scenarios, which in this case originate from the coupling of WIP1 and WIP2. Variation of the coupling parameter X shows different qualities of dynamic behaviour. Bifurcation and deterministic chaos occur at large values of X, whereas lower values of X lead to the stabilization of the buffer levels after a small number of iteration steps similar to the linear coupled system (see figure 3 and 4). Depending on the value of the production rate COM2, the coupling parameter X can be used as a control parameter. It permits the controller to operate the system in a stabilized mode or in a multimode. The value of X is limited by the capacity restrictions. Large values deliver results beyond the capacity limits. This results from neglecting the range of the production rates. Interpretation is difficult in this case, but it shows that operation of the system is possible in a wide range of the coupling parameter X.

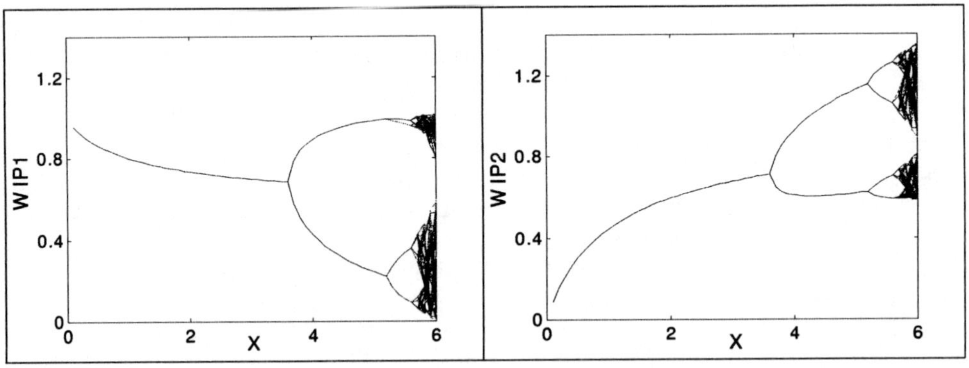

FIGURE 7. WIP levels of the non-linear coupled system depending on the coupling parameter X.

To satisfy the capacity limits in real production systems, additional restrictions were formulated like "stop on overflow" or "stop if buffer empty". These restrictions force the system into a stabilized state (see figure 8).

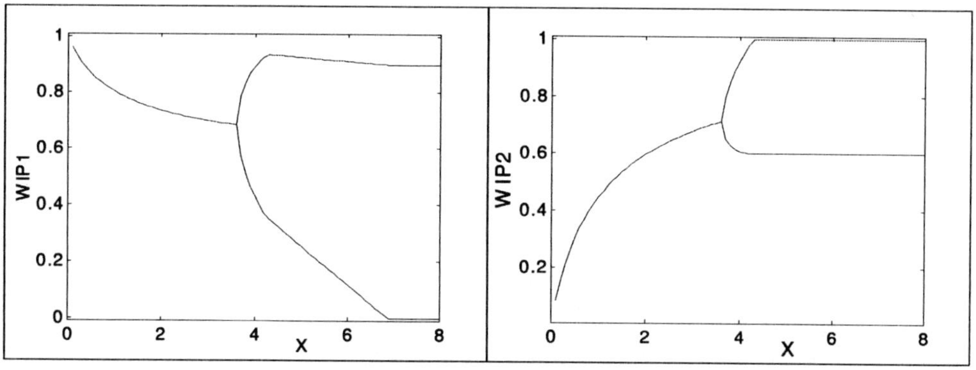

FIGURE 8. WIP levels of the non-linear coupled system with overflow control depending on the coupling parameter X.

Rapid adaptation can be achieved more easily with a non-linear control mechanism. The parameter X can be used to adjust the equilibrium WIP levels in a wide range, allowing the exchange of available capacity between different work stations.

A point of further discussion will be the interpretation of the parameter X regarding capacity control. Changing of X only leads to flow regulation. As in the case of capacity restrictions, limits to the ability of flow changes should be included. This can be done only in a more system specific approach, taking into consideration additional system parameters.

CONCLUSION

A general approach for modelling and control of production systems was introduced that was developed using concepts of Nonlinear Dynamics Theory and PPC research. Manufacturing processes and PPC mechanisms are considered to be a unity that build a dynamical system. The functions of a PPC system have been introduced as functional groups. The possibilities of the dynamical approach have been shown for every group.

A model was discussed in the last section, demonstrating the function of non-linear operational rules, which enable multimode operation (bifurcation, chaos). The application of Nonlinear Dynamics control methods is possible in this case of explicit deterministic chaos.

In most cases, the dynamic behaviour of a production system will not be described only by a set of operational rules. System state definition will be the primary step.

ACKNOWLEDGEMENT

This research was founded by the Volkswagen Stiftung, Germany, under reference number I/77366.

REFERENCES

1. Beaumariage, T., and Kempf, K., "The Nature and Origin of Chaos in Manufacturing Systems", in *IEEE/SEMI Advanced Semiconductor Manufacturing Conference*, 1994, pp. 169-174.
2. Hanson, D., Armbruster, D., and Taylor, T., "On the stability of re-entrant manufacturing systems", in *Proceedings of the 31st MTNS (Mathematical Theory of Networks and Systems)*, edited by Beghi et al, Il Poligrafo, Padova, Italy, 1999, pp. 937-940.
3. Bartholdi, J.J., Bunimovich, L.A., and Eisenstein, D.D., "Dynamics of 2- and 3-worker 'bucket brigade' production lines", in *Operations Research*, **Vol. 47**, No. 3, 1999, pp. 488-491.
4. Bartholdi, J.J., and Eisenstein, D.D., "A production line that balances itself", in *Operations Research*, **Vol. 44**, No.1, 1996, pp. 21-32.
5. Renner, A., *Kostenorientierte Produktionssteuerung*, Vahlen, München, 1991.
6. Bertrand, J.W.M., Wortmann, J.C., and Wijngaard, J., *Production Control*, Elsevier, 1990.
7. Nieswandt, A., *Operations Research*, Oldenbourg, München, 1994.
8. Tempelmeier, H., *Material-Logistik - Grundlagen der Bedarfs- und Losgrößenplanung in PPS-Systemen*, Springer, Berlin, 1995.
9. Turowski, K., *Flexible Verteilung von PPS-Systemen*, Deutscher Universitätsverlag, Wiesbaden, 1997.
10. Domschke, W., Scholl, A., and Voß, S., *Produktionsplanung*, Springer, München, 1997.
11. Ditto, W.L., Spano, M.L., and Lindner, J.F., "Techniques for the control of chaos", *Physica D*, **Vol. 86**, 1995, pp. 198-211.
12. Goldratt, E., and Cox, J., *Das Ziel*, Campus, 2001.
13. Wiendahl, H.-P., *Fertigungsregelung*, Hanser, München, 1997.
14. Scherer, E., *Shop Floor Control - A Systems Perspective. From deterministic Models towards Agile Operations Management*, Springer, Berlin, 1998.

Analysis of the dynamics of transition from deep to shallow penetration CO_2 laser welding

Edvard Govekar*, Janez Gradišek*, Igor Grabec*, Mark Geisel†, Andreas Otto† and Manfred Geiger†

*Faculty of Mechanical Engineering, University of Ljubljana, Slovenia
†Chair of Manufacturing Technology, University of Erlangen–Nuremberg, Germany

Abstract. By varying the output power density of a CO_2 laser beam either a deep penetration or a shallow penetration welding regime can be achieved. The transition between the two regimes is accompanied by abrupt changes in the statistical and dynamical characteristics of emitted light intensity. For a characterisation of the dynamics the coarse-grained entropy rate is applied. A high entropy rate is typical for deep penetration, while a low value is characteristic for a shallow penetration welding regime. Additionally, dynamics of the welding regimes are modelled by the Langevin equation. Analysis reveals that in both welding regimes the dynamics corresponds to stochastic fluctuations around a stable fixed point. It is demonstrated that a stable node corresponds to the deep penetration welding, while a stable spiral corresponds to the shallow penetration welding. The transition between welding regimes can be thus described by a change of the fixed point character.

INTRODUCTION

CO_2 laser beam welding is a joining process in which parts are locally melted by a laser beam. The dynamics of the process are determined by a variety of physical phenomena such as light absorption, heat conduction, melting, evaporation, plasma generation and shielding. Depending on the power density of the laser beam at the work-piece surface two different welding regimes can be achieved: deep penetration (also called a key hole) and shallow penetration (heat conduction). Deep penetration welding occurs at high power densities where sufficient amounts of energy are supplied into the work-piece. This causes intense vaporisation of material in the molten pool, where a capillary known as a keyhole is created. The keyhole enables deep penetration of the laser beam into the material, and related deepening of the molten pool. Shallow penetration welding occurs at low power densities where the supply of energy into the material is less intense and no keyhole is formed. Consequently the laser beam penetration depth is considerably smaller.

Due to the high importance of laser beam welding in manufacturing, analysis and characterisation of the dynamics of this process are very often the subject of theoretical as well as experimental investigations. Experimental analysis of the process dynamics is motivated by the need to support and to verify analytical modelling of the dynamics, especially low-dimensional non-linear modelling [1]. An additional motivation for experimental analysis of laser beam welding dynamics is also the need for reliable in-process monitoring of the process state. An example of practical relevance is the detection of the

CP622, *Experimental Chaos: 6th Experimental Chaos Conference*, edited by S. Boccaletti et al.

transition from deep to shallow welding regime [2].

The subject of the article is experimental analysis of the dynamics of CO_2 laser welding dynamics and characterisation of transition between the welding regimes. The most common physical quantity being measured for the experimental characterisation of CO_2 laser welding dynamics is the intensity of the light emitted at certain wavelength by the welding process. Although deterministic chaos in time series of the emitted light intensity induced by CO_2 laser welding has been reported in the literature [1], our results do not indicate a determinism to such an extent that the process could be described as a chaotic one [3]. Instead of chaotic characteristics, which are sensitive to noise, we consequently use the coarse-grained characteristic for the analysis and characterisation of the experimental time series [4]. It is shown in the article that the coarse-grained entropy rate can be successfully applied for quantitative characterisation of laser welding regimes as well as for the detection of the transition from deep to shallow penetration welding.

To obtain additional qualitative information about the process dynamics we use a method of stochastic data analysis, which assumes that the stochastic dynamics of the process can be described by the Langevin equation [5]. Using this approach it is found that the dynamics of CO_2 welding can be treated as stochastic fluctuations around a stable fixed point. The deep penetration welding regime corresponds to a stable node, while the shallow penetration welding regime corresponds to stable spiral.

EXPERIMENTS

The welding experiments were performed on an industrial 5 kW, CO_2 laser system. Helium was used as the shielding gas. Work-pieces were made of steel sheet 3 mm thick. A photo diode with an optical notch filter was used to detect intensity $I(t)$ of emitted light at a wavelength of 540.4 nm$\pm$10 nm. The signals were recorded by a digital oscilloscope at a sampling frequency of 45 kHz and stored into a PC for off-line analysis.

Different welding regimes can be achieved, depending on laser-beam power density on the work-piece surface. There are three laser welding control parameters that influence the power density of the laser- beam on the work-piece surface: laser-beam power, focus position and feed rate. In our experiments the feed rate v_f and the focus position a were used as control parameters. A transition from a deep to a shallow penetration welding regime was experimentally caused by a slow increase of a selected control parameter, while the other parameters were held constant.

In Fig. 1, the dependence of intensity of emitted light I on increasing feed rate v_f and increasing focus position a are shown. Qualitative comparison of the time series shown in Fig. 1 suggests that the intensity of $I(t)$ depends on the control parameter. In addition to this, a significant change in the course of the intensity envelope at a certain value $\mathcal{T}_h$ of the control parameter is evident, that indicates the transition between the two welding regimes.

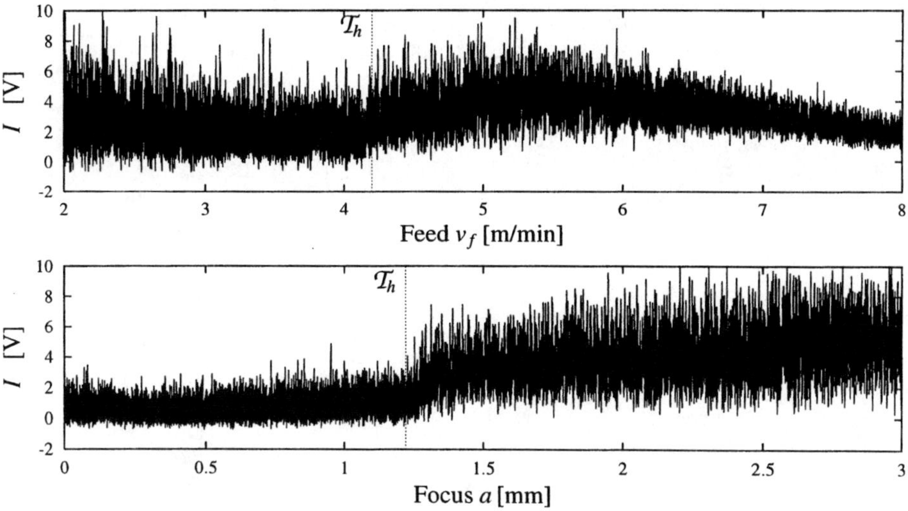

FIGURE 1. Dependence of the emitted light intensity I on the feed rate v_f (above) and the focus position a (below).

Analysis of the light intensity time series

To confirm the above observations the dependence of the statistics of the amplitude distribution of I on the control parameters v_f and a were determined. In Fig 2a and b, the dependence of the mean value μ_I on the feedrat v_f and on the focus position a are shown. If we focus our observation on the vicinity of the transition $\mathcal{T}_h$ we observe that

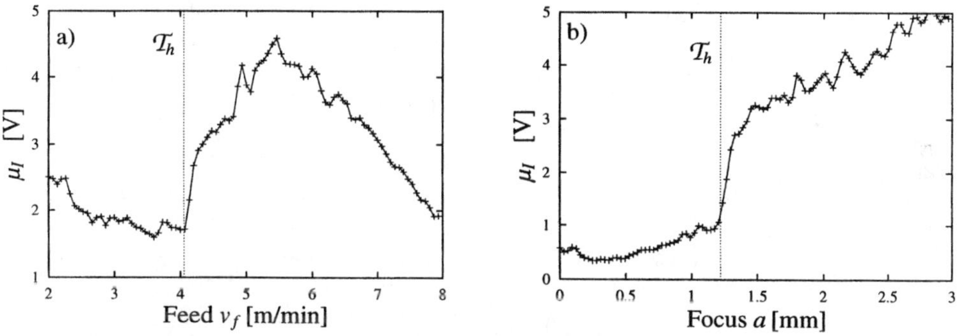

FIGURE 2. Dependence of the mean value μ_I of amplitude distribution of the emitted light intensity I: a) on the increasing feed rate v_f and b) on the increasing focus position a.

the transition from the deep to the shallow penetration welding is related to a dramatic increase of the mean value. Similar changes at the transition are also typical for higher order statistics of $I(t)$ such as skewness and kurtosis [6].

The dynamic properties of the light intensity $I(t)$ were characterised by the coarse-grained entropy rate (CER). This characteristic measures the regularity and predictability of the process [4]. The normalised value of CER, NCER has a span [0,1] [7]. The higher the entropy rate NCER, the lower the predictability of the process. The dependence of NCER on the feed rate and the focus position is shown in Fig. 3. The transition from deep to shallow welding, indicated by $\mathcal{T}_h$, corresponds to an abrupt decrease of NCER. A high value of NCER is typical for deep penetration, a low one for the shallow penetration welding regime. This indicates that in the shallow penetration welding regime the dynamics are more regular than in the deep penetration welding regime. To

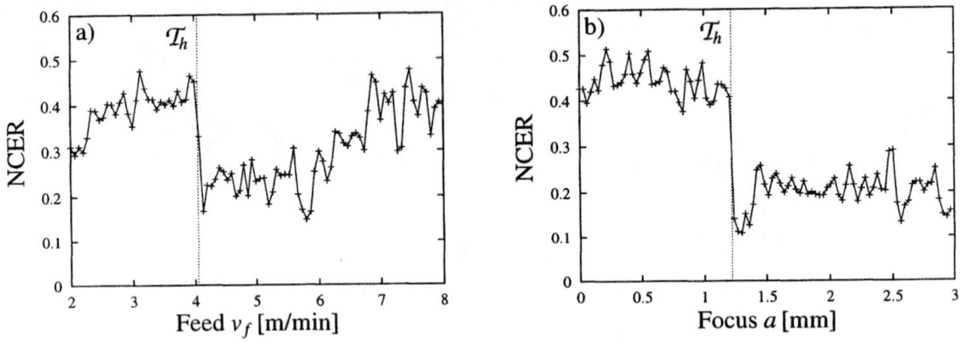

FIGURE 3. Dependence of the normalised coarse-grained entropy rate (NCER) of the emitted light intensity I: a) on the increasing feed rate v_f, and b) on the increasing focus position a.

obtain more insight into the dynamics of both regimes a method for the analysis of stochastic data sets was applied to time series of variable $I(t)$ as described in the following section.

Stochastic analysis of data: deep vs. shallow penetration welding

The above results show that the transition from the deep to the shallow penetration welding corresponds to a significant change in the dynamics. The dynamics of observed welding regimes can be treated as highly stochastic in the case of the deep penetration welding, as more regular in the case of the shallow penetration welding.

Recently, a method for the analysis of such stochastic data sets was proposed elsewhere [5]. The method assumes that the process dynamics contain an additional dynamic noise. The dynamics of such a process can be described by a Langevin equation: $\dot{X}(t) = F[X(t)] + \zeta(t)$, where $X(t)$ is the process state, F is a non-linear function, and $\zeta(t)$ is the noise component. The dynamics of such a stochastic process can also be described by the Fokker-Planck equation, which is determined by the drift $D^{(1)}$ and the diffusion $D^{(2)}$ coefficients. The drift coefficient corresponds to the deterministic component, while the diffusion coefficient is related to the random component (noise) of the process dynamics. On the basis of the relations between the terms of the Langevin and Fokker-Planck equations these coefficients can be estimated directly from the measured time series [5]. By drawing drift coefficients as arrows, the deterministic component of the process dynamics are represented as a vector field or a flow in the phase space. From the vector field qualitative information on the local stability properties of the process in the phase space can be extracted. For a given initial set of conditions a deterministic evolution of the process can also be reconstructed using the estimated drift coefficients.

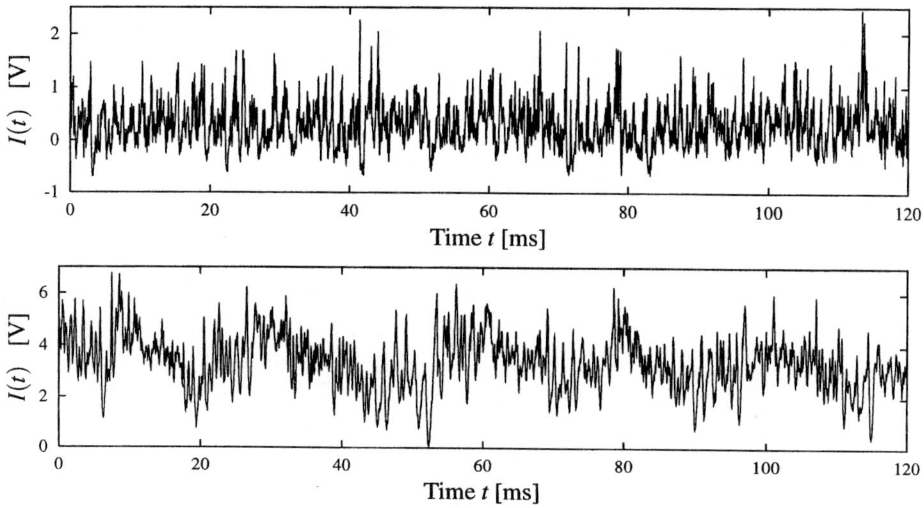

FIGURE 4. Time series of the emitted light intensity I for two different welding regimes achieved by different focus positions a: (above) $a = 0.5$mm, deep welding regime; (below) $a = 2$mm, shallow welding regime. The time series measured during the deep welding regime appear very random, while in the time series of shallow welding regime less irregular, even slightly periodic fluctuations can be observed.

For the time series $I(t)$, portions of which are presented in Fig. 4, the drift coefficients were estimated in a two dimensional phase space reconstructed from the time series using the delay coordinates. In Fig. 5 typical vector fields, with two superimposed

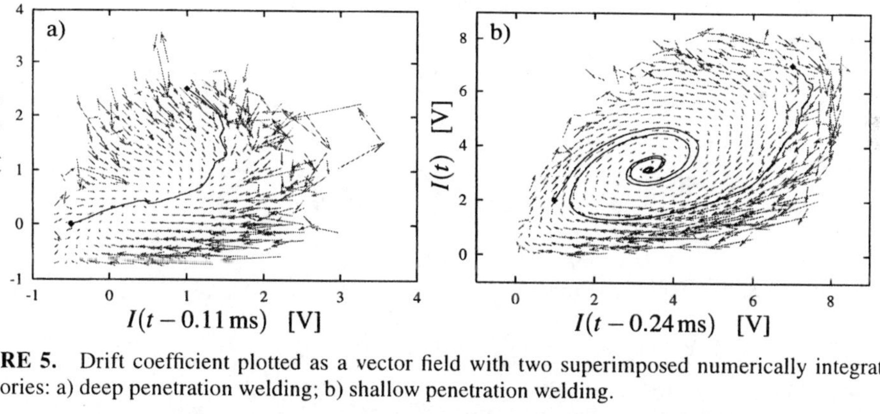

FIGURE 5. Drift coefficient plotted as a vector field with two superimposed numerically integrated trajectories: a) deep penetration welding; b) shallow penetration welding.

trajectories each, are shown. From the vector field, we can see that in both regimes the field arrows point towards the centre of the phase space. This indicates the existence of a stable fixed point for both welding regimes. Closer examination of the drift coefficients reveals that integrated trajectories, which start at the edge of the phase space, approach the attracting centre differently, depending on the welding regime. According to this a fixed point is a stable node in the deep welding regime, (Fig. 5a), while it is a stable spiral (Fig. 5b) in the shallow penetration regime.

CONCLUSIONS

The dynamics of the deep penetration and the shallow penetration CO_2 laser- beam welding regimes were investigated. The dynamics were indirectly described by the time series of light intensity I emitted during the process at a wavelength of 540.4 nm$\pm$10 nm.

The dynamics of both regimes were quantitatively characterised using a normalised coarse-grained entropy rate (NCER). Independent of the applied control parameter an abrupt drop in the NCER value indicates the transition from the deep to the shallow penetration welding regime. A high NCER value is typical for the less regular deep welding regime, while a low value is typical for the more regular shallow welding regime. Besides a quantitative indication of process dynamics, it is also possible to distinguish between both welding regimes and detect the transition based on measured NCER. From our experiments we have estimated that NCER < 0.3 indicates the shallow penetration welding regime.

In addition, the dynamics were characterised by a method of stochastic data sets analysis based on the Langevin and Fokker-Planck equations. Analysis of the drift coefficient of the Fokker-Planck equation reveals that both welding regimes can be described as stochastic fluctuations around a stable fixed point. The fixed point of the deep welding regime resembles a stable node, while the fixed point of the shallow penetration welding regime resembles a stable spiral. Based on our experiments we conclude that the transition from the deep penetration to the shallow penetration welding regime dynamically corresponds to a change in the type of the fixed point from a stable node to a stable spiral.

Acknowledgements

The authors gratefully acknowledge the support of the Volkswagen Foundation. EG also gratefully acknowledges the support of the Alexander von Humboldt Foundation.

REFERENCES

1. Kurzyna, J., *J. Phys. D: Appl. Phys.*, **31**, 680–629 (1998).
2. Nakamura, S., Sakurai, M., Kamimuki, K., Inoue, T., and Ito, Y., *J.Phys. D: Appl. Phys.*, **33**, 2941–2948 (2000).
3. Govekar, E., and Geisel, M., "On Nonlinear Dynamics of Laser Welding Process", in *Technische Anwendungen von Erkenntnissen der nichtlinearen Dynamik*, Frankfurt, Germany, 1999, pp. 209–212.
4. Paluš, M., *Physica*, **D 93**, 64–77 (1996).
5. Friedrich, R., and Peinke, J., *Physica*, **D 102**, 147–155 (1997).
6. Govekar, E., Gradišek, J., Grabec, I., Geisel, M., Otto, A., and Geiger, M., "On characterisation of CO_2 laser welding process by means of light emited by plasma and images of weld pool", in *Investigation of nonlinear dynamic effects in production systems*, edited by F. Klocke, 3-rd International VW Symposium, Cottbus, Germany, 2000.
7. Gradišek, J., Govekar, E., and Grabec, I., *Journal of Sound and Vibration*, **214**, 941–952 (1998).

Experimental Verification of the Performance of Chaos-Based Asynchronous DS-CDMA Systems

F. Agnelli*, G. Mazzini†, R. Rovatti** and G. Setti†

*DI – University of Ferrara, via Saragat 1, 44100 Ferrara, Italy
†DI University of Ferrara, via Saragat 1, 44100 Ferrara, Italy
**DEIS – University of Bologna, viale Risorgimento 2, 40136 Bologna, Italy

Abstract. Recent theoretical investigations concern the application of chaos to DS-CDMA system. The optimum spreading sequences have been analytically characterized to find that they can be generated with negligible approximation by a discrete-time chaotic system. The adoption of these chaos-based spreading sequences can be theoretically anticipated to produce a zero-cost average 15% increase in the number of allocable users or a 15% increase in the signal-to-interference ratio controlling the communication quality. We here confirm theoretical predictions by building a DSP-based prototype system comprising 7 transmitters and a receiver matched with one of the transmitters. It is worthwhile to be noticed, that, although still preliminary, these are the first presented experimental results confirming that the adoption of chaos-based techniques allows to improve the performance of a communication system with respect to currently adopted techniques.

INTRODUCTION

Many attempts have recently been made to exploit the peculiar nature of chaotic systems to improve performances in several engineering problems. As a noteworthy result of this efforts, it has been clarified that, thanks to the mixed deterministic/stochastic nature of chaotic systems, the tasks which are most likely to benefit from chaos-based techniques are those where the statistical properties of the generated signals are the key factor.

Within this framework, spread-spectrum communication has been thoroughly investigated by highlighting the trade-offs that chaos-based techniques may help to address, as well as several foreseeable advantages in their applications. In [1][2][3][6][7] the impact of the adoption of chaos-based sequences in a standard asynchronous DS-CDMA systems is studied when Additive White Gaussian Noise (AWGN) channels are considered and co-channel interference is the dominant cause of nonideality. For such environments, theoretical performance bounds have been obtained which show that performance improvements can be achieved by employing a properly designed family of chaotic systems generating real trajectories that are then quantized and periodically repeated to yield the users signatures. This result has been further strengthened in [4] where it is shown that chaos-based spreading allows to practically reach the absolute minimum of co-channel interference. As such a lower bound is approximately 15% lower than the average interference achievable by classical *m*- or Gold sequences, a definite improvement is demonstrated. Though the users are not synchronized one with the other, synchronization between the transmitter and the receiver for the same user must be guaranteed before a reliable communication link can be established. Few steps in the direction of

CP622, *Experimental Chaos: 6th Experimental Chaos Conference*, edited by S. Boccaletti et al.

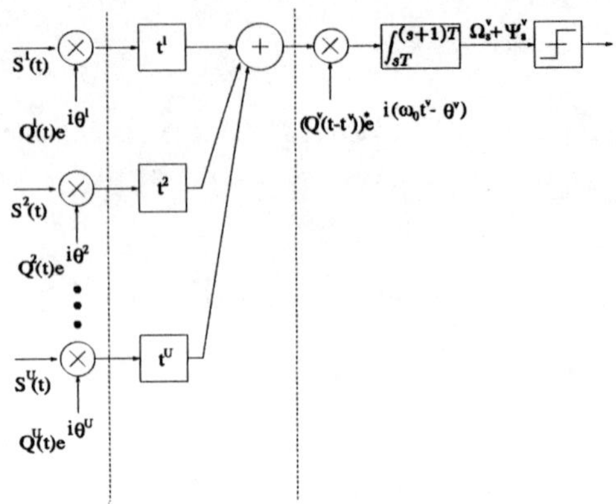

FIGURE 1. Baseband equivalent of an asynchronous DS-CDMA communication systems

analyzing this synchronization mechanism have been taken in [5], where it is found that, even if synchronization can always be ensured, sequence correlation properties affect the speed and the reliability of this mechanism so that chaos-based spreading may, again, offer advantages over classical sequences. A good and fairly complete survey of the results in this field and in others related to chaos-based communication can be found in [8].

Beyond theoretical results, one of the first unavoidable steps towards the real-world exploitation of these approaches is surely the realization of a possible experimental set-up allowing field-validation. We here report the results of such validation along with a sketch of the theoretical background needed to interpret them. To this aim, we first concentrate on the construction of spreading sequences allowing the minimization of co-channel interference We show how this can be reduced to the development and analysis of a suitable family of chaotic maps to be used as sequence generators. Then, details on a low-cost DSP-based DS-CDMA system implementation are given, along with the measured bit error probability when chaos-based sequences are employed for spectrum-spreading. Measured performance results to be in good agreement with the theoretical analysis.

ASYNCHRONOUS DS-CDMA SYSTEMS: DEFINITIONS AND OPTIMAL PERFORMANCE

We consider a standard DS-CDMA system in the worst possible environment with respect to signals synchronization, i.e. when the channel and user mobility joint effects introduce random carrier phases and timing delays for each interfering user.

A simplified baseband equivalent of the system is reported in Figure 1, where a common carrier with pulsation ω_0 is assumed, U users are supposed and t^u and θ^u indicate respectively the absolute delay and carrier phase of the u-th user signal. The last quantities are assumed as uniformly distributed random variables to model transmission from mobile terminals to a fixed base-station. The generic u-th users information signal is $S^u(t) = \sum_{s=-\infty}^{\infty} S_s^u g_T(t - sT)$, where $S_s^u = \pm 1$ are the information symbols and $g_T(t)$ is the rectangular pulse which is 1 within $[0,T]$ and vanishes otherwise. The spreading signal depends on sequences of symbols x_s^u from the alphabet X which are mapped into the set of the L complex roots of the unity by the function Q, and are combined to form the complex signal $Q^u(t) = \sum_{s=-\infty}^{\infty} Q(x_s^u)\; g_{T/N}(t - sT/N)$. The signal $Q^u(t)$ is then multiplied by $S^u(t)$ and transmitted along the (equivalent) channel together with the spread-spectrum signals from the other users. Each user adopts a different spreading code, assigned at the connection start-up.

The receiver is made of a multiplier which combines the incoming signal with a synchronized replica of the spreading sequence of the v-th user and of an integrate-and-dump stage in charge of extracting the information symbol by correlation. Even when thermal noise can be neglected, symbol recovery is hindered by the presence of the current and delayed signals of the other users and by the delayed version of the useful signal. Following the approach in [9][1] we will assume that spreading sequences are periodic of a period equal to the spreading factor N.

The system performance depends on how much these spurious signals affect the integrate-and-dump output, i.e. on how they are correlated with the spreading sequence of the useful user. In the following we will assume that the system performance is mainly limited by co-channel interference, namely we will suppose that spreading-despreading sequence synchronization has been achieved, and that multipath effects are negligible. With this, the receiver output before the hard decision block of the useful v-th user can be expressed as $\Upsilon_s^v = \Omega_s^v + \Psi_s^v$ [1], where it is assumed that the s-th information symbol S_s^v is being decoded and where $\Omega_s^v = S_s^v/2$ is the useful component, and Ψ_s^v is the interference component. We may now compute the useful and interfering component and evaluated the relative power by using the standard Gaussian approximation (SGA), which assumes that co-channel interference can be effectively approximated by a a zero-mean Gaussian random variable [9], so that the only characteristic quantity is its variance [1]

$$\mathbf{E}_{\underline{x}^v}[(\sigma_\Psi^v)^2] = \frac{(U-1)}{24N^3} \sum_{k=1-N}^{N-1} \left\{ 2\mathbf{E}_{\underline{x}^u,\underline{x}^v}^{u \neq v}[|\Gamma_{N,k}(\underline{x}^u,\underline{x}^v)|^2] + \mathrm{Re}\left(\mathbf{E}_{\underline{x}^u,\underline{x}^v}^{u \neq v}[\Gamma_{N,k}(\underline{x}^u,\underline{x}^v)\Gamma_{N,k+1}^*(\underline{x}^u,\underline{x}^v)] \right) \right\}$$

In the above expression, $\Gamma_{N,k}(\underline{x}^u,\underline{x}^v)$ indicates the partial cross-correlation function between the two spreading sequences $\underline{x}^u = \{x_l^u\}$ and $\underline{x}^v = \{x_l^v\}$ defined as $\Gamma_{N,k}(\underline{x}^u,\underline{x}^v) = \sum_{l=0}^{N-k-1} Q(x_k^u)Q^*(x_{l+k}^v)$ if $k = 0,\ldots,N-1$, $\Gamma_{N,-k}^*(\underline{x}^v,\underline{x}^u)$ if $k = -1,\ldots,N$ and vanishing otherwise. With this [1], we obtain $P_{\mathrm{err}} \approx 1/2\,\mathrm{erfc}\sqrt{1/8(\sigma_\Psi^v)^2} = 1/2\,\mathrm{erfc}\sqrt{1/(U-1)R}$ so that the quantity R assumes, in term of signal power, the significance of an *expected interference-to-signal ratio per interfering user*, i.e. the expected degradation in system performance when a new user is added. If all the sequences can be thought of as independent and if they satisfy a mild form of second-order stationarity, namely $\mathbf{E}_{\underline{x}^u}[Q(x_m^u)Q^*(x_n^u)] = \mathbf{E}_{\underline{x}^u}[Q(x_0^u)Q^*(x_{|n-m|}^u)]$, then R can be expressed as a function of the sequences auto-correlation function $A_k = \mathbf{E}_{\underline{x}^u}[Q(x_0^u)Q^*(x_k^u)]$. With this a suitable minimization procedure can be devised [4] obtaining that the unique minimum of R can be obtained when

$$A_k = A_k^{\mathrm{opt}} = (-1)^k \frac{N}{N-k} \frac{r^{k-N} - r^{N-k}}{r^{-N} - r^N} \tag{1}$$

where $r = 2 - \sqrt{3}$. Note how A_k are not only real but also *with alternating signs* so that the positive contributions to R due to $|A_k|^2$ are counterbalanced by the negative contributions of $A_{k-1}A_k$.

By substituting (1) into the expression of R, the minimum interference is $R^{\mathrm{opt}} = \sqrt{3}/3N(r^{-2N}-r^{2N}/r^{-2N}+r^{2N}-2)$ which, for large N, tends to $\sqrt{3}/(3N)$ (actually $|R^{\mathrm{opt}} - \sqrt{3}/(3N)| < 10^{-12}$ for $N \geq 10$) which is less than the classical $R = 2/(3N)$ obtained for ideal random sequences where $A_k = 0$ for $k > 0$.

CHAOS-BASED OPTIMAL SPREADING SEQUENCES

The problem of generating spreading sequences with such a profile is here tackled by means of a chaos-based approach. In particular we will refer to a piecewise-continuous map $M : [0,1] \mapsto [0,1]$ whose iteration $x_{k+1} = M(x_k)$ starts from an initial condition x_0 uniformly distributed in the state space. Since optimal performance can be obtained with sequences whose auto-correlation is real we may restrict to sequences that are real themselves and thus binary. Hence we will assume that $Q : [0,1] \mapsto \{1,-1\}$ is defined so that $Q(x) = -1$ for all $x \in [0,d]$ and $Q(x) = 1$ for all $x \in]d,1]$ for some threshold value d. Spreading sequences are taken to be $y_k = Q(x_k)$ for $k = 0,\ldots,N-1$ and then repeated periodically. The map M will be designed so that $A_k = (-r)^k$ which is a good approximation of (1) for the most significant values of A_k, i.e. those higher in modulus, for which $k \ll N$. Extending such approximation to all the A_k we have

$$R = \frac{1}{2(r^2-1)^3}\left[N^2(r^2-1)^2(r-2r^2) + N(r^5 - 4r^4 + 4r^2 - r) + (r^{2n} - 1)(2r^4 - 2r^3 + 2r^2)\right]$$

which is such that $|R^{\mathrm{opt}} - R| < 10^{-4}$ for $N \geq 10$ guaranteeing a relative error smaller than 0.15%.

At least two kinds of maps have been proposed to achieve this approximation. The first kind of map (see e.g. [8, chap. 3]) is the family of (n,t)-tailed shifts which is defined with n even and $t < n/2$, as

$$M(x) = \begin{cases} (n-t)x \pmod{\frac{n-t}{n}} + \frac{t}{n} & \text{if } 0 \leq x < \frac{n-t}{n} \\ t\left(x - \frac{n-t}{n}\right) \pmod{\frac{t}{n}} & \text{otherwise} \end{cases}$$

so that they are affine in each of the intervals $X_j = [(j-1)/n, j/n]$ and map the intervals from X_1 to X_{n-t} onto $X_{t+1} \cup \cdots \cup X_n$ while the last t intervals are mapped onto $X_1 \cup \cdots \cup X_t$. In this case the threshold d is set to $1/2$.

These maps produce sequences whose autocorrelation function can be proved to be $A_k = (-t/(n-t))^k$. Hence, for any given n, we may chose t as the integer closest to $nr/(1+r)$, the accuracy being obviously improved as $n \to \infty$. For example, for $n = 10$ we may set $t = 2$ and obtain $t/(n-t) = 0.25$ which is an approximation of $r = 0.2679\ldots$.

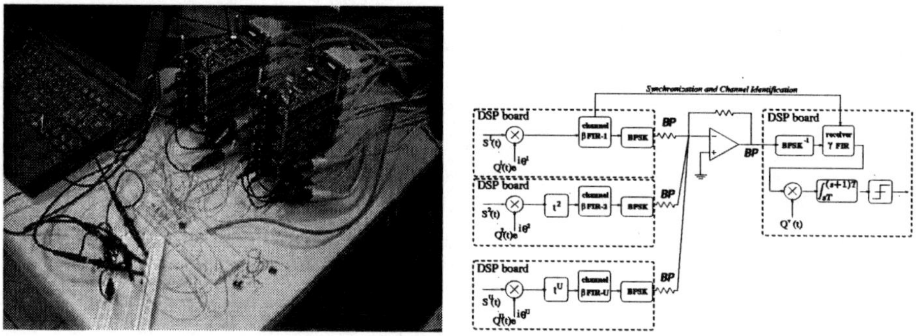

FIGURE 2. Experimental set-up of the DS-CDMA communication system prototype and its schematic representation

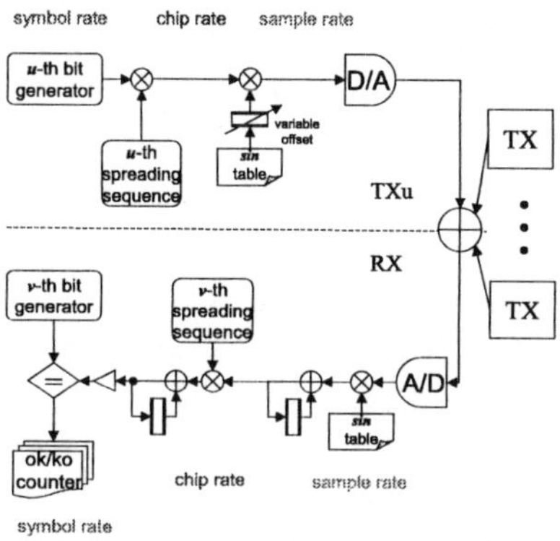

		$P_{\text{err}} \times 10^3$	
	N	Theory	Measure
NC	15	13.5	16.2
OC	15	6.33	6.52
NC	20	2.66	3.81
OC	20	0.98	1.46

(b)

FIGURE 3. (a) Table showing theoretical and measured P_{err} with non-dispersive channel and (b) experimental set-up of the DS-CDMA communication system prototype and its schematic representation

SYSTEM PROTOTYPE AND MEASUREMENT RESULT

With the aim of obtaining a first and effective confirmation of the theoretical results reported in the previous Sections, we realized a low-cost prototype of the communication system using TMS320C542 DSP boards. A picture of the whole system is reported in Figure 2 (a). Figure 2 (b) reports the equivalent structure. Transmitter boards are in charge of generating baseband signals, of their BPSK modulation and of the simulation of the channel. To reduce computational burden the FIR simulating the channel is swapped with the BPSK modulation and operated at baseband level. No external RF circuit has been included in this low-cost implementation. One board is devoted to the implementation of the useful transmitter, one board is devoted to the implementation of the corresponding receiver and of the estimator of the bit error rate while all the other boards can be used to account for multiple access interference.

In Figure 3 (a) we report the detailed signal-level scheme of the computations carried out in each of the blocks in Figure 2 (b). Spreading sequences are generated off-line and stored on the transmitters. A copy of the useful sequences is also stored in the receiver. At symbol rate, information bits are locally generated. They are multiplied by the spreading sequence thus generating data at chip rate that are then multiplied by stored samples of the sinusoidal carrier to generate the passband signals that are added and converted into the analog output of the transmitter. The offsets in the sin table can be adjusted to account for different time-varying delays and carrier phases in the interfering transmitters, they are kept constant in the useful transmitter. If B is the symbol rate, $N \times B$ is the chip rate and, if M is the number of sinusoid samples per chip, $M \times N \times B$ is the sample rate. Since the A/D and D/A conversion is in the audio band we get that the prototype transmits at $B = 25 \times 10^3/(MN) \simeq 6.5$ bits per second. The same structure is replicated for the interfering transmitters in which a randomly variable delay and a random carrier phase are also introduced to account for the asynchronous environment. The delay and carrier phase is obtained in the digital chain by means of the adjustable offsets in the sin table. In the transmitters, digitally generated pseudo-random variables are used to generate information bits, as well as to possibly generate and update interferer delays and carrier phases once every u_i bits. The signal arriving at the receiver is converted into digital and fed into a quadrature demodulator relying again on cos and sin tables and on an accumulator whose output is then fed into the correlation-based despreader and in the decision block.

A prerequisite for correct operation of the receiver is its synchronization with the transmitter along with the availability of a perfect identification of the channel taps. In addition, to ascertain if the received bit is identical to the transmitted one the information stream generated at the transmitter must be also available. In fact, the two bit streams must be compared to compute the bit error rate. As far as all the quantities depending on digital pseudo-random variables are concerned, we use a suitable digital generator addressing the trade-off between randomness and fixed-point computation whose seed is common to both the transmitter and receiver. As far as the time-synchronization is concerned, the D/A converter of the useful transmitter and the A/D converter of the receiver share the same timing signals (a clock `CLK` and a frame sync `FS` signal). At startup, a simple handshake is initiated relying on two more wires carrying interrupt signals from one board to the other (see Figure 4 (a)) to implement a classical *request/acknowledge* (`RQ`/`ACK`) sequence. Figure 4 (b) reports the tracks of the `RQ`, `ACK` signals, along with the output of the transmitter D/A and its synchronization signal `FS`. Since the DSP clock frequency is 40MHz and the time scale of the analog signals is within the μs range, digital events appear as impulses. Nevertheless, it is evident that from the very beginning the `RQ` signal is toggled by the transmitter to test receiver readiness. After internal initialization, the receiver assesses its availability by pulsing the `ACK` signal. This starts the conversion in both the transmitter and receivers. Note that a delay is present between the receiver acknowledge and the actual appearance of a meaningful signal on the transmission line. This is due to the smoothing filter at the output of the D/A and cannot be avoided. To cope with this the receiver is instructed to discard a certain number (determined experimentally) of samples actually postponing synchronization. Conventional sequences and optimized chaos-based will be tested and they are indicated as optimal chaos (OC) and no chaos (NC). Table 3 (b) compares the measured bit error rate when

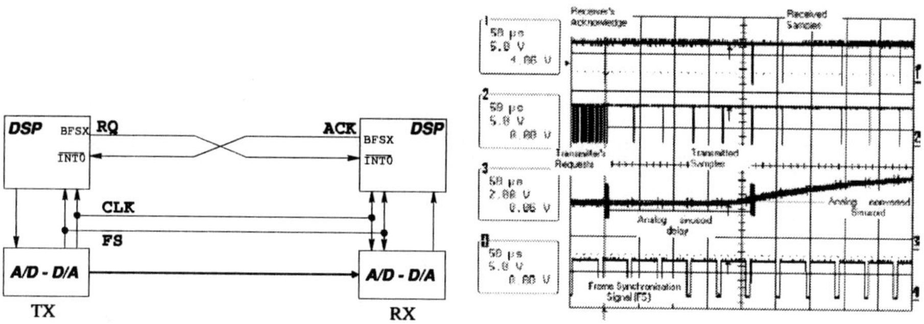

(a) (b)

FIGURE 4. (a) Scheme indicating the signal involved in the handshake operation between transmitter and receiver for initial synchronization (b) corresponding oscilloscope diagrams

the prototype system is operated for 2×10^6 bits for $U = 6, N = 15$ and $N = 20, M = 256$, and $u_i = 10$. For every configuration performance is averaged over 100 sequence sets and considering the average P_{err} experienced by every user.

Results confirm the fact that chaos-based DS-CDMA outperform classical DS-CDMA. Also quantitative theoretical predictions are in good agreement with experimental results.

REFERENCES

1. G. Mazzini, G. Setti, R. Rovatti, "Chaotic Complex spreading Sequences for Asynchronous DS-CDMA - Part I: System Modeling and Results", IEEE TRANS. CIRC. SYS.-I, vol. 44, pp. 937-947, 1997
2. R. Rovatti, G. Setti, G. Mazzini, "Chaotic Complex spreading Sequences for Asynchronous DS-CDMA - Part II: Some Theoretical Performance Bound", IEEE TRANS. CIRC. SYS.-I, vol. 45, pp. 496-506, 1998
3. R. Rovatti, G. Mazzini, G. Setti, "Interference Bounds for DS-CDMA Systems Based on Chaotic Piecewise-Affine Markov Maps", IEEE TRANS. CIRC. SYS.-I, vol. 47, pp. 885-896, 2000
4. G. Mazzini, R. Rovatti, G. Setti, "Interference Minimization by Autocorrelation Shaping in Asynchronous DS-CDMA Systems: Chaos-Based Spreading is Nearly Optimal," *Electronics Letters*, vol. 35, pp. 1054-1055, 1999.
5. G. Setti, R. Rovatti, G. Mazzini "Synchronization Mechanism and Optimization of Spreading Sequences in Chaos-Based DS-CDMA Systems", IEICE TRANS. ON FUNDAMENTALS, vol. E82-A, pp. 1737-1746, 1999
6. R. Rovatti, G. Mazzini, G. Setti, "A Tensor Approach to Higher-Order Expectations of Quantized Chaotic Trajectories - Part I: General Theory and Specialization to Piecewise Affine Markov Systems," IEEE TRANSACTION ON CIRCUITS AND SYSTEMS - I, vol. 47, pp. 1571-1583, 2000.
7. G. Mazzini, R. Rovatti, G. Setti, "A Tensor Approach to Higher-Order Expectations of Quantized Chaotic Trajectories - Part II: Application to Chaos-Based DS-CDMA in Multipath Environments," IEEE TRANSACTION ON CIRCUITS AND SYSTEMS - I, vol. 47, pp. 1584-1596, 2000.
8. M. P. Kennedy, R. Rovatti, G. Setti (Eds.), *Chaotic Electronics in Telecommunications*, CRC Press, Boca Raton, 2000
9. M. B. Pursley, "Performance Evaluation for Phase-Coded Spread-Spectrum Multiple-Access Communication - Part I: System Analysis," IEEE TRANS. COMM., vol. 25, pp. 795-799, 1977

Communication using Synchronization of Chaos in Semiconductor Lasers with optoelectronic feedback

S. Tang*, L. Illing†, J. M. Liu*, H. D. I. Abarbanel** and M. B. Kennel‡

*Department of Electrical Engineering, University of California, Los Angeles, Los Angeles, CA 90095, USA

†Institute for Nonlinear Science and Department of Physics, University of California, San Diego, La Jolla, CA 93093-0402, USA

**Institute for Nonlinear Science, Department of Physics and Marine Physical Laboratory at Scripps Institution of Oceanography, University of California, San Diego, La Jolla, CA 93093-0402, USA

‡Institute for Nonlinear Science, University of California, San Diego, La Jolla, CA 93093-0402, USA

Abstract. A communication scheme based on the synchronization of two chaotic semiconductor lasers is experimentally tested. The Chaos in the single-mode semiconductor lasers is generated by means of an optoelectronic feedback. Synchronization of the chaos is achieved by coupling a fraction of the transmitter's output power into the driving current of the receiver. We present experimental results on the route to chaos and the synchronization of GHz chaotic signals. We then test a proposed communication scheme by successfully transmitting messages.

INTRODUCTION

Communications using chaotic waveforms as "carriers" of information promise possible advantages over traditional communications strategies in the achievement of power efficiency gains and better use of broadband capability. Optical communication utilizing synchronized chaotic lasers with optoelectronic feedback has been demonstrated for quite slow data rates by using chaotic wavelength fluctuations [1]. The first high speed optical chaotic communication has been achieved using fiber ring lasers with a rare earth doped fiber segment as active element [2]. We found in our earlier work on such erbium doped fiber ring lasers [3] that, although observed bit rates up to 250 Mbits/s are interesting in principle, the change of the chaotic waveform in these types of lasers is slow compared to the time it takes the light to make one round trip in a standard laser composed of tens of meters of passive optical fiber. For practical applications, where bit rates of gigabits per second are desirable, one has to consider different active elements in the laser cavity, which give rise to fast chaos. In this paper we investigate semiconductor lasers with optoelectronic feedback. This is one way to generate high-frequency chaos and has in comparison to optical feedback the advantage of being phase insensitive.

CP622, *Experimental Chaos: 6th Experimental Chaos Conference*, edited by S. Boccaletti et al.

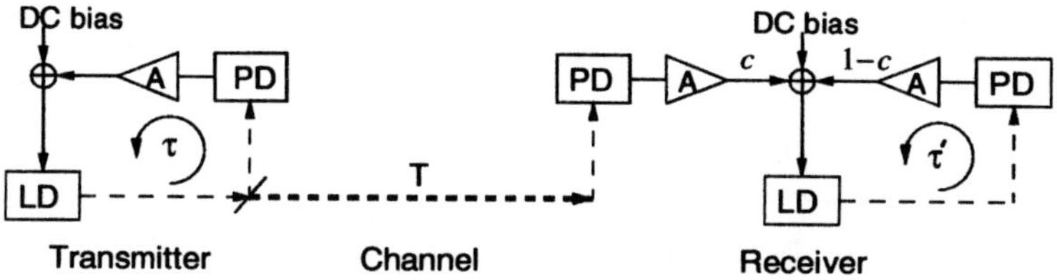

FIGURE 1. Schematic experimental setup for the synchronization of two chaotic semiconductor lasers with delayed optoelectronic feedback. LD: laser diode; PD: photodetector; A: amplifier; τ: feedback delay time; T: transmission time.

CHAOS IN OPTOELECTRONIC FEEDBACK LASERS

A schematic of our experimental setup for the investigation of semiconductor lasers with optoelectronic feedback can be seen in Fig. 1. In this section we are only concerned with the dynamics of the transmitter in Fig. 1. The laser diode is driven by a DC bias current and lases in a single mode. The output light is received and converted to a current by a photodetector. The current is subsequently amplified and added to the bias current, resulting in a delayed positive feedback. The delay time can be varied by changing the distance between the laser diode and the photodetector.

In our numerical investigations of the delay differential equations, which describe the dynamics of the optoelectronic feedback laser, we find a rich bifurcation diagram as the delay time and the feedback strength are varied [4, 5]. In general chaotic regions are interspersed with periodic and quasi-periodic ones and multistability of different types of attractors, e.g. fixed points and limit cycles, are a common feature. We find that the optoelectronic feedback laser enters chaos through the quasi-periodic route. This agrees well with results from analytic bifurcation analyses of the delay differential equations modeling the system [6, 7].

In the experiments we use a single mode DFB laser diode with a wavelength of 1.3 μm from Rockwell. The threshold of the laser is 34 mA and the laser is biased at 45.5 mA. The light is converted to a current by a high-speed InGaAs photodetector (6 GHz bandwidth). The amplifier in the experiment is composed of two stages of Avantek SSF86 amplifiers (0.4 - 3 GHz bandpass). The small signal gain of this two-stage amplifier is about 60 dB. The optical output detected by the photodetector is observed with an HP Agilent 54845A Infiniium oscilloscope with a 1.5 GHz bandwidth and an up to 8 GSa/s sampling rate. The power spectrum is measured with an HP 8592A RF spectrum analyzer. The experiment is performed by changing the delay time through varying the distance between the laser diode and the photodetector.

Fig. 2 shows the experimentally observed time series of the laser output, the corresponding power spectra, and the corresponding Poincaré section at different time delays. The Poincaré section is obtained in a delay-embedding space using the experimental time series. The data shown in Figs. 2(a)–(c) were taken at measured delay times of $\tau \simeq 7.47$ ns, 7.09 ns and 6.92 ns, respectively. The time series in Figs. 2(a)–(c) show a clear transition from regular limit cycle oscillations to quasi-periodic oscillations with

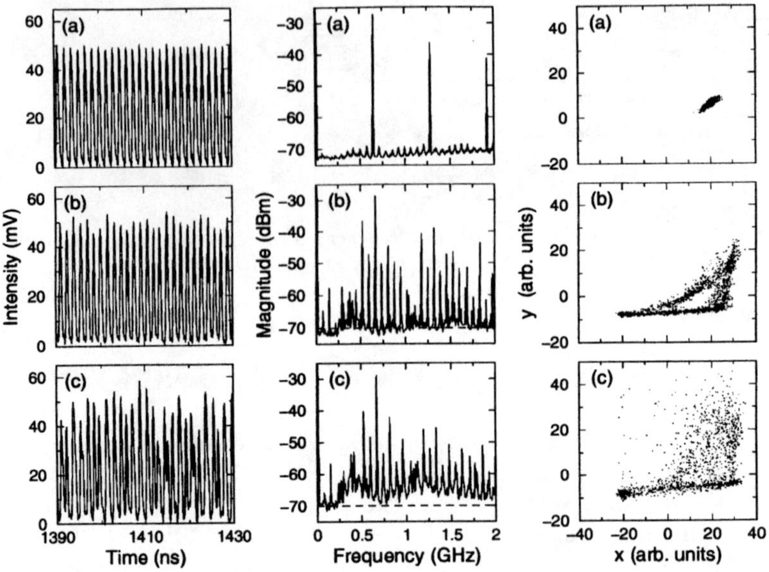

FIGURE 2. Experimental results: time series, power spectra and Poincaré sections at different delay times. From top to bottom: (a) limit cycle oscillations, at $\tau \simeq 7.47$ ns; (b) two-frequency quasi-periodicity, at $\tau \simeq 7.09$ ns; (c) chaos, at $\tau \simeq 6.92$ ns. The broadband background in (c) is much higher than in (b), indicated by the dashed reference lines at -70 dBm.

intensity amplitudes modulated at a certain second frequency, and finally to a chaotic state where the intensities vary irregularly. As limited by the sampling rate of the oscilloscope, the pulses are under-sampled, resulting in large digitizing errors. To avoid the digitizing errors, the power spectra are measured with an RF spectrum analyzer. From the power spectra, it is clear that in Fig. 2(a) there is only one fundamental frequency at $f_1 \simeq 650$ MHz. The small ripples in the spectrum indicate that there is small instability in the laser. Then in Fig. 2(b), a second frequency at $f_2 \simeq 140$ MHz, which corresponds to the inverse of the delay time, is as significant as f_1. The two frequencies f_1 and f_2 are incommensurate and the system is in a two-frequency quasi-periodic oscillating state. Finally in Fig. 2(c), the spectrum is broadened as the system enters a chaotic pulsing state. Though some of the sharp spectral peaks remain in Fig. 2(c), the broadband background, a characteristic of chaos, is much higher in Fig. 2(c) than in Fig. 2(b), as indicated by the dashed reference lines at -70 dBm. The big dot in the Poincaré section in Fig. 2(a) corresponds to regular oscillations. The size of the dot is an indication of the noise in the data. Despite the noise and the under-sampling the toroidal nature of the attractor for $\tau \simeq 7.09$ ns can clearly be seen as a diffused closed curve in Fig. 2(b). In Fig. 2(c) the spread of the Poincaré plot indicates the transition to chaos.

Overall the numerical and experimental results show good agreement. The predicted quasi-periodic route to chaos is experimentally observed.

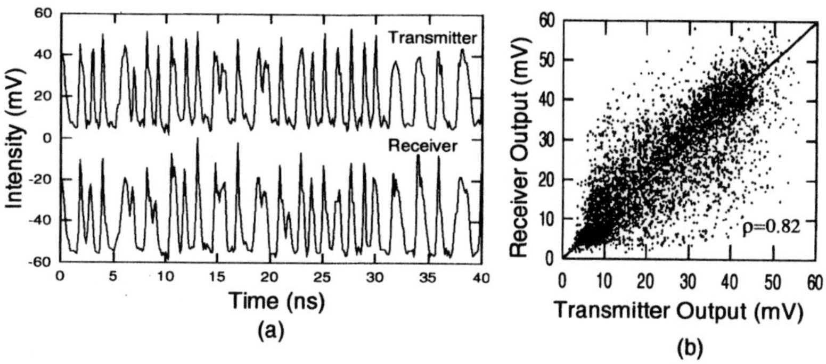

FIGURE 3. Experimental result of chaotic synchronization with c=0.6. (a) Time series of transmitter and receiver. (b) Correlation diagram of the transmitter output vs. the receiver output.

CHAOS SYNCHRONIZATION

The schematic setup of the synchronization experiment is shown in Fig. 1, where a fraction of the transmitter's chaotic output intensity is detected, amplified and coupled into the driving current of the receiver. The receiver is itself a delayed optoelectronic feedback laser. As in our work on the synchronization of doped fiber ring lasers [8, 9] we include the factor c, with $0 \leq c \leq 1$, to indicate the percentage of the total feedback signal in the receiver which is from the transmitter, while $1-c$ corresponds to the fraction from the receiver. When $c = 1$ there is no feedback signal in the receiver and we call this open loop. When $c = 0$ the transmitter and receiver are completely decoupled. We denote by T the transmission time; the time associated with the propagation of the signal thru the open air 'channel' to the receiver system. When the transmitter and receiver are identical, a state where the dynamical variables of both are equal in time is a solution for the coupled system. We found numerically, that for c roughly 0.4 or larger, this solution is stable and synchronization of the electric field and the carrier density in transmitter and receiver is accomplished both with and without spontaneous emission noise [5]. We furthermore showed in [5] that synchronization is most sensitive to parameter mismatch in the delay time, with small synchronization errors for mismatches of up to a few percent for any other parameter.

To achieve synchronization, the laser parameters are matched by carefully choosing a pair of lasers from the same batch with the closest characteristics and then fine-tuning their operating conditions. Fig. 3 shows the experimental results for synchronization with $c = 0.6$, a configuration where the feedback loop in the receiver is closed. A comparison of the time series in Fig. 3(a) shows that they are almost identical. Note that the time series in Fig. 3(a) are shifted in time by the transmission time T (see Fig. 1) to better show the synchronization. The correlation plot Fig. 3(b) is obtained by plotting the transmitter output vs. the receiver output. The data is distributed along the 45° line, indicating identical synchronization. The quality of synchronization is limited by both the intrinsic laser noise and the finite bandwidth and sampling rate of the oscilloscope. We calculate the correlation coefficient ρ of the transmitter and receiver

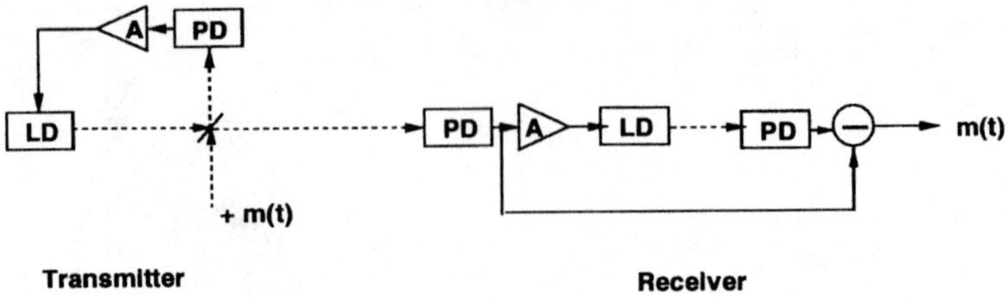

FIGURE 4. Schematic experimental setup for the chaotic communication scheme. LD: laser diode; PD: photodetector; A: amplifier; m(t): message

output to quantify the quality of synchronization. For the data of Fig. 3 the correlation coefficient is $\rho = 0.82$.

In [10] we find that the quality of synchronization increases rapidly for $c > 0.4$ and is largest for the open loop configuration. The quality of synchronization drops fast when the mismatch of the delay time is larger than a few percent. Since the open loop configuration with $c = 1$ does not have the problem of delay time mismatch and shows the highest quality synchronization, $c = 1$ is the preferred configuration for chaotic synchronization and communication with this delayed optoelectronic feedback system.

CHAOS COMMUNICATION

The achievement of synchronization of two chaotic semiconductor lasers with optoelectronic feedback allows us to employ communication schemes in which the message is put into the transmitter dynamics via a reversible operation, e.g. addition or multiplication. The transmitted signal is then a combination of the message and the chaotic output of the transmitter. Since on the receiver side we know the expected chaotic output of the transmitter, due to the identical synchronization, we can use the expected output to extract the message from the received signal. In [5] we evaluated numerically the performance of a multiplicative communication scheme and found favorable bit error rates for a wide range of the c factor. Here we report the result of an experiment in which the reversible function is addition [11].

The schematic experimental setup is shown in Fig. 4, where we use the open loop configuration $c = 1$. It is important to note that the message signal m(t) enters the dynamics of the transmitter via the feedback loop. Therefore the combined signal of output plus message enters symmetrically both the transmitter and receiver, which ensures that synchronization is maintained when we transmit information.

Fig. 5 shows the results of an encoding/decoding experiment of narrow pulses with 100 MHz repetition rate. The time series of the pulse encoding/decoding is shown in Fig. 5. The top trace is the received signal with the message encoded. The second trace is the local receiver laser output, which is due to the synchronization equal to the transmitter output before the addition of the message. The message is therefore recovered by subtracting the receiver output (second trace) from the received signal

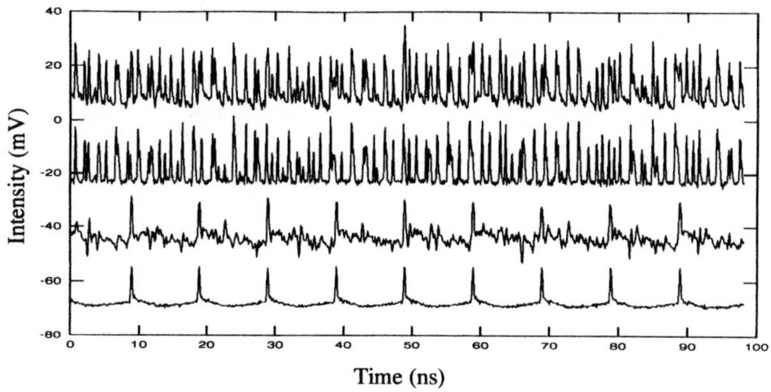

FIGURE 5. Encoding/decoding of digital pulses with 100 MHz repetition rate. Time series of received signal (top), receiver laser output (middle), recovered digital message (third), and original message (bottom).

(top trace). The recovered message shows good quality of decoding as indicated by the recovered pulse train in the third trace, which is to be compared with the original 'message', which is a set of periodic pulses, in the bottom trace.

In conclusion, we demonstrated that optoelectronic feedback lasers exhibit high frequency chaotic oscillations and enter chaos via a quasi-periodic route. We achieved chaos synchronization of two such systems. We furthermore demonstrated that a high-speed pulsing message can be successfully encoded and decoded. Since the transmitter and receiver remain synchronized when a message is transmitted the bit rate of the encoded message can be possibly as high as the repetition rate of the chaotic pulsing, which is about 600 MHz at present time. With even faster components the bit rate can be improved to multigiga bits per second.

REFERENCES

1. Goedgebuer,J. P., Larger, L. and Porte, H., *Phys. Rev. Lett.*, **80**, 2249-2252, (1998).
2. VanWiggeren, G. D., and Roy, R., *Phys. Rev. Lett.*, **81**, 3547-3550, (1998).
3. Abarbanel, H. D. I., Kennel, M. B., Buhl, M., and Lewis, C., *Phys. Rev. A*, **60**, 2360-2374, (1999).
4. Tang, S. and Liu, J. M., *IEEE J. Quantum Electron.*, **37**, 329-36, (2001).
5. Abarbanel, H. D. I., Kennel, M. B., Illing, L., Tang, S., Chen, H. F., and Liu, J. M. "Synchronization and communication using semiconductor lasers with optoelectronic feedback", submitted to *IEEE J. Quantum Electron.* , (2001).
6. Grigorieva, E. V., Haken, H., and Kaschenko, S. A., *Optics Comm.*, **165**, 279-292, (1999).
7. Pieroux, D., Erneux, T., Luzyanina, T., and Engelborghs, K., *Phys. Rev. E*, **63**, 036211-1 to -11, (2001).
8. Abarbanel, H. D. I. and Kennel, M. B., *Phys Rev. Lett.*, **80**, 3153-3156, (1998)
9. Lewis, C., Abarbanel, H. D. I., Kennel, M. B., Buhl, M., and Illing, L., *Phys. Rev. E*, **63**, 016215-1 to -15, (2000)
10. Tang, S. and Liu, J. M., *Optics Lett.*, **26**, 596-598, (2001)
11. Tang, S. and Liu, J. M.,"Message encoding/decoding using chaotic pulsing semiconductor lasers," *OFC 2001 Technical Digest Series*, Anaheim, 2001,paper WDD46, pp. WDD46-1 to -3.

Design and Realisation of Chaotic Encryption Systems

Wolfgang Schwarz and Thomas Falk

Technical University Dresden, Institute for Fundamentals of Electrical Engineering, Mommsenstr.13, 01062 Dresden, Germany

Abstract. Chaotic signal transmission systems are often claimed to be secure by itself. Using a simple example it is shown, that this is not true and that exact design criteria have to be set up before starting the design of a chaotic encryption system.

Then, beginning with statistical design objectives an information encryption system is systematically designed. The structure design leads to a controlled filter structure with overflow nonlinearity, the parameter design has to assure chaotic behaviour and mixing properties of the encoded signal. This defines the limits for the choice of the parameter set representing the key for the encryption. After developing the system structure the system is realized by electronic circuitry. Discrete and IC versions of the solution are presented. In order to prove that the system meets the design requirements experimental results are provided. It can be shown that in a n-th order system the statistical characteristics up to the n-th order of the output signal will not be affected by the input signal. The paper closes with some security estimates for the designed system.

INTRODUCTION

In recent years there has been great interest in secure communication systems based on chaotic systems. Most of the proposed designs can be classified as ad-hoc designs where a structure is more or less evaluated for its security and robustness. In this paper we want to go in the opposite direction: After deriving the objectives for a chaotic decoder a suitable structure will be developed and an implementation is proposed.

There are fundamental differences between a digital transmission system and an analogue one, which is used with most chaotic encryption systems due to the analogue nature of the used chaotic processes. The main difference is that there is always an error in the received signal to be decoded in the analogue system. In digital systems an error free channel can be guaranteed by design, but an analogue encryption system has to be robust against errors caused by a non-ideal channel. This leads to a lower security.

The paper is divided into three parts. In the first one the general scheme for information transmission is reviewed and objectives for the design of its parts are discussed. The second section describes a chaotic encryption system resulting from these objectives. The last section gives an example for the implementation of this system and discusses its performance with a non-ideal transmission channel and its security.

CP622, *Experimental Chaos: 6th Experimental Chaos Conference*, edited by S. Boccaletti et al.

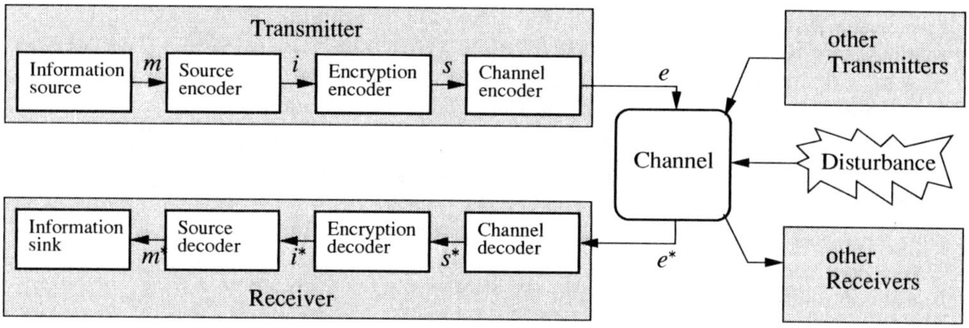

FIGURE 1. General communication scheme

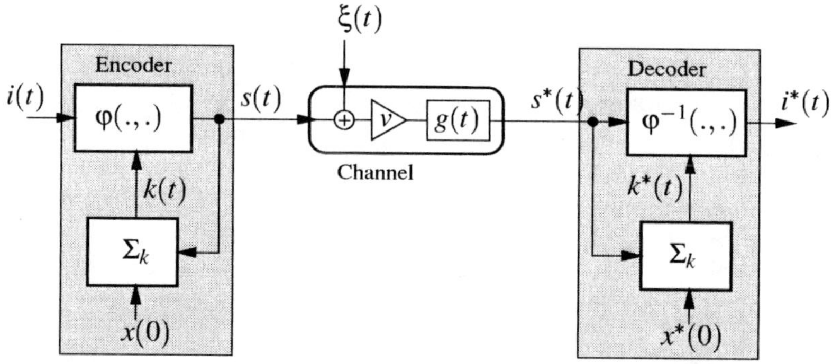

FIGURE 2. Self synchronising encryption system with non-ideal channel

COMMUNICATION SCHEME AND OBJECTIVES FOR SECURE TRANSMISSION

A general structure for information transmission systems is shown in Fig. 1. The *information source* is in general redundant. The reduction of the redundancy is done by the *source coder*. The encryption encoder is able to further increase the entropy of the information signal up to a certain level in order to hide the information. The channel coder adapts the signals to the used channel e.g. by modulation and has to provide a unique signal out of a class of possible signals which can be uniquely decoded by the receiver. The blocks in the receiver have to invert the operation of their counter part in the transmitter. Due to the influences of the channel the received signal e^* will differ from its original e. It is the task of the channel decoder to correct the resulting errors, i.e. to achieve $s^* = s$ as good as possible, at least asymptotically. However this can be done ideal only in the transmission of digital signals. Thus the encryption system used in an analogue transmission system has to be robust against transmission errors.

In the following we restrict our discussion to the encryption system consisting of the encoder and the decoder connected by a non-ideal channel, resulting in the structure shown in Fig. 2. Here the action of the channel is modelled by the addition of a noise signal $\xi(t)$, the attenuation v and a linear filter with the impulse response $g(t)$.

The encoder consists of the map $\varphi(i,k)$ which maps the message i into the encrypted signal s controlled by the key stream k. The key stream generator Σ_k is a dynamic system which generates the key stream depending on the initial condition $x(0)$ and the signal $s(t)$. In the decoder there are the corresponding blocks: the inverse map $\varphi^{-1}(i,k)$ and the an identical key stream generator Σ_k. Both key generators have to generate the same key stream. This can be realised by direct transmission of the key stream to the decoder or –as shown in Fig. 2 – by using a self-synchronising structure.

To design the blocks shown in Fig. 2 the objectives for the design have to be defined. From the demands for a secure communication follows:

1. In order to enable the transmission of information through the system the encoding process has to be invertible indepently from the initial conditions $x(0)$ and $x^*(0)$ in the encoder and the decoder respectively. Here the self synchronising property of the structure is important, but this synchronisation has to work also in the case $s^*(t) \neq s(t)$ which is typical for analogue transmission systems. Note that only asymptotic synchronisation is required.
2. In order to prevent eavesdropping on the channel the transmitted information i and the key parameters k have to be hidden on the channel or at least they have to be hard to detect. This means that the characteristics of the transmitted signal s have to be independent of the message signal i (diffusion) or the mapping of the characteristics of i into those of i has to be very complicated (confusion) [1].

STRUCTURAL DESIGN

The structural design is the process of deriving an actual system from the structure shown in Fig. 2 and the objectives in the previous section. To comply with objective no. 2 the encrypted signal s shall have an n-th order uniform distribution independently from the message signal i. For the map φ we will assume a simple static nonlinearity, while the key generator Σ_k is a dynamic system.

Static Nonlinearity. Because the static nonlinearity maps i and k into the encrypted signal s it has to preserve uniform distributions of k and it has to be invertible for all k regarding i, i.e. $\varphi(.,k)$ has to be bijective for all k. Criteria for distribution preserving maps are well known [3]. From the set of all maps, which meet these requirements on the interval $[-1,1]$ we choose the following subset:

$$\varphi(i,k) = h \circ \mathrm{mod}(i+k) \qquad \text{with} \qquad \mathrm{mod}(x) = x - 2\left\lfloor \frac{x+1}{2} \right\rfloor \tag{1}$$

where h is an arbitrary bijective map on $[-1,1]$. Since $\mathrm{mod}(.,k)$ is invertible on the interval $[-1,1]$ also the concatenation $\varphi = h \circ \mathrm{mod}(.)$ is bijective on the interval $[-1,1]$, see also Fig. 3(a). The resulting map is bijective on the interval $[-1,1]$ and preserves the uniform density distribution. Further examples of h are shown in Fig. 3(b).

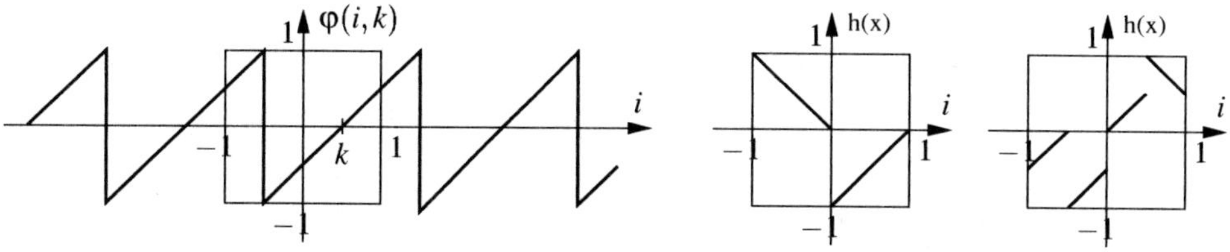

(a) (b)

FIGURE 3. (a) Action of the map $\varphi(i,k)$ on the interval $[-1,1]$ with $h(x) = x$ and (b) further examples of possible maps h

Dynamic Subsystem. The input of the dynamic subsystem Σ_k in the encoder is the encoded signal s with a uniform density which has to be preserved. The key stream k^* generated by the dynamic subsystem in the decoder has to approach at least asymptotically the key stream k in the encoder, i.e. $\lim_{t->\infty}|k^*(t) - k(t)| = 0$ for all initial conditions $x(0), x^*(0)$ and $s^*(0)$. Furthermore the encoder has to be ergodic to generate the required uniform density of s. As a simple example of such a dynamic subsystem is the FIR structure:

$$k(t) = \sum_{i=1}^{n} c_i s(t-i) \tag{2}$$

where the coefficients c_i serve as the encryption key.

IMPLEMENTATION AND MEASUREMENTS

The above derived structure can be implemented by means of analogue circuitry or emulated by a digital signal processor. Both variants have been evaluated. In the following an analogue design using discrete off-the-shelf components will be presented.

The block diagram of the circuit is shown in Fig. 4. Here the dynamic subsystem is implemented as filter structure with adjustable coefficients c_i. For the map φ the characteristics shown in Fig. 3(a) has been used. The whole circuit operates at 40kHz clock frequency. Additional blocks in Fig. principle2 not discussed before are

- the clock regeneration circuit which is realised as a PLL which regenerates the clock from the transmitted signal v_r^* and
- the automatic gain control (AGC) which corrects the attenuation of the channel.

The realised circuit has been tested using a channel model containing an adjustable attenuator and a band pass filter. The results can not all be shown here due to the lack of space but the system proved as quite robust even in this fully analogue realisation. The signals with a sinusoidal transmission and the probability density on the channel are shown in Fig. 5(a). Fig. 6(a) shows the power density spectrum of the transmitted signal and the decoded signal for the same sinusoidal signal. The signal-to-noise ratio of the decoded signal is $\approx$40dB. Note that in the spectrum on the channel the peak of the sinusoidal signal is completely hidden.

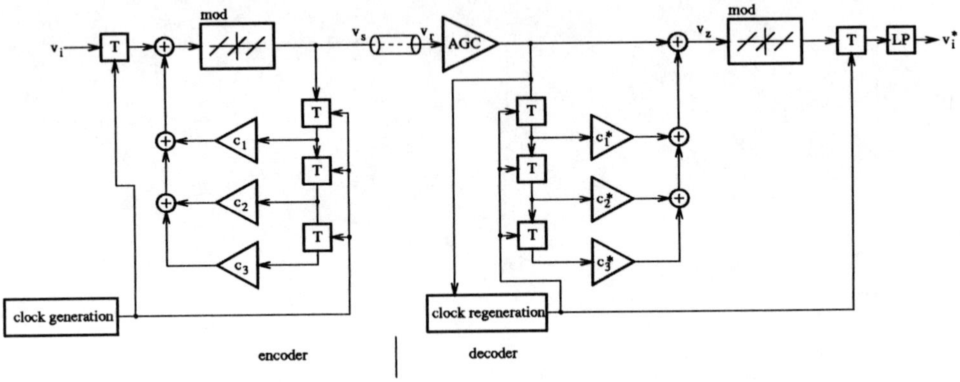

FIGURE 4. Schematic of the realised circuit

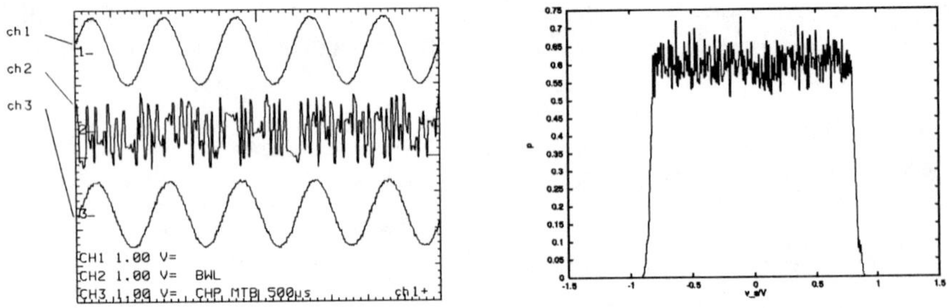

(a) (b)

FIGURE 5. (a) Signals for a sinusoidal transmission (ch1 – v_i, ch2 – v_s, ch3 – v_{i^*} and (b) probability density on the channel

Security

In the proposed structure the coefficients $c_1, \dots, c_n$ of the dynamical system is the key. If the coefficients $c_1^*, \dots, c_n^*$ in the encoder is different, then there is only a noise-like signal at the output with properties similar to these of the transmitted signal s. Fig. 6(b) shows the noise voltage at the output of the decoder as a function of the parameter difference $c = c_1^* - c_1$. As can been seen the optimal parameter $c_1^* = c_1$ can be found simply by minimising the broadband noise component at the output of the decoder. This property is preserved if the system is realised with higher dimensional dynamic systems, but the region in the parameter space where there is a non-zero gradient in the signal-to-noise ratio gets smaller. This behaviour is quite different from that of an digital implemented encryption system, where there is only one key in the key space which decodes the message correctly. This underlines the empirical fact that in analogue systems an enhancement in the security, e.g. by increasing the order of the system, reduces the robustness of the system against the non-ideal channel and vice-versa.

The simple attack using the gradient in the signal-to-noise ratio can be prevented by using an input signal i with the same spectral properties as the signal s on the channel,

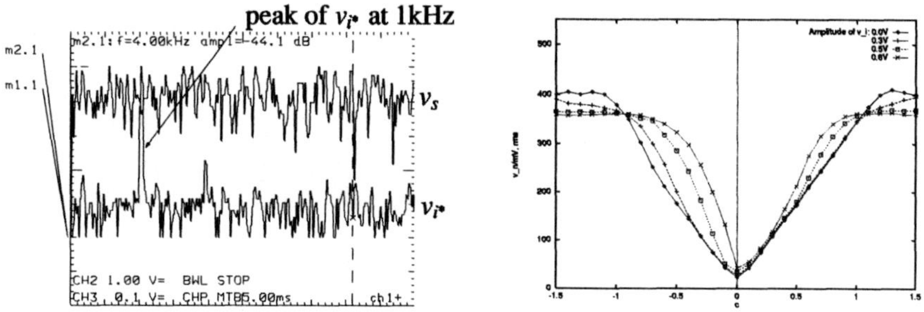

(a) (b)

FIGURE 6. (a) Spectrum of the signal on the channel (upper) and decoded signal v_{i^*} (lower) and (b) noise voltage at the output of the decoder in dependence of parameter difference for a one-dimensional system with the input amplitude as parameter

e.g. by cascading two encoders, however due to the higher order of the resulting system the robustness is reduced.

Other security analyses based on higher order properties of the chaotic signals are presented in [2].

CONCLUSIONS

A systematic approach to the design of chaotic encryption systems has been presented. The resulting structure can be implemented as analogue discrete-time circuit or emulated on a digital signal processor. It seems surprising, that a system designed following the objectives given for a secure communication is in fact not very secure. In our opinion this is based on the analogue realisation where a singular valid key can not be realised due to the required robustness of the whole system. Thus the robustness of the system against a non-ideal transmission channel has to be traded against security.

The authors thank Frank Dachselt for the cooperation and discussions. This work has been supported by the german research agency.

REFERENCES

1. M. Götz, K. Kelber and W. Schwarz. *Discrete-time Chaotic Encryption Systems, Part I: Statistical Design Approach*. In IEEE Trans. Circ. & Systems-I, vol. 44, no. 10, p. 963-970, 1997.
2. F. Dachselt, K. Kelber and W. Schwarz. *Discrete-time Chaotic Encryption Systems, Part III: Cryptographical Analysis*. In IEEE Trans. Circ. & Systems-I, vol. 45, no. 9, p. 883-888, 1998.
3. T. Kilias. *One-dimensional discrete maps*. (in german) PhD thesis, TU Dresden, 1992
4. W. Schwarz, M. Götz, K. Kelber, A. Abel, T. Falk and F. Dachselt. *Statistiacl Analysis and Design of Chaotic Systems*. in *Chaotic Electronics in Telecommunications*. Ed. by M. P. Kennedy, R. Rovatti, G. Setti, CRC-Press, 2000

TRAFFIC

Order and disorder in traffic and self-driven many-particle systems

Dirk Helbing

Institute for Economics and Traffic, Dresden University of Technology, Andreas-Schubert-Str. 23, D-01062 Dresden, helbing@trafficforum.org, www.helbing.org

Abstract. During the last decade, physicists have identified various spatio-temporal patterns of motion in vehicle and pedestrian traffic. Moreover, by applying and extending methods from statistical physics and non-linear dynamics, these have been successfully explained by means of self-driven many-particle models. Some of the questions now understood are the following: Why are vehicles sometimes stopped by so-called "phantom traffic jams", although they all like to drive fast? What are the mechanisms behind stop-and-go traffic? Why are there several different kinds of congestion, and how are they related? Why do most traffic jams occur considerably before the road capacity is reached? Can a temporary reduction of the traffic volume cause a lasting traffic jam? What is the origin of fluctuations in traffic systems and which consequences do they have? Why do pedestrians moving in opposite directions normally organize in lanes, while nervous crowds are "freezing by heating"? Why do panicking pedestrians produce dangerous deadlocks?

One particularly successful approach to modelling vehicle and pedestrian traffic is the treatment as systems of self-driven, motorized, or active "particles" [1, 2, 3, 4, 5, 6]. This requires some slight generalizations of Newton's equation of motion from classical mechanics describing the acceleration $\ddot{\vec{x}}_i(t)$ of a body i of mass m_i subject to pair interactions with other bodies j:

$$m_i\ddot{\vec{x}}_i(t) = \sum_{j(\neq i)} \vec{f}_{ij}(t) . \quad (1)$$

The interaction forces $\vec{f}_{ij}(t)$ are mostly dependent on the locations $\vec{x}_i(t)$ and $\vec{x}_j(t)$ of the interacting bodies i and j at time t. Often, they depend only on the distance vector $\vec{d}_{ij} = (\vec{x}_j - \vec{x}_i)$, but in special cases, they are also functions of the velocities $\vec{v}_i(t) = \dot{\vec{x}}_i(t)$ and $\vec{v}_j(t) = \dot{\vec{x}}_j(t)$. In *driven* many-particle systems such as fluids under the influence of pressure gradients and boundary forces, or vibrated granular media like sand, we need to consider additional interactions with the environment. This concerns (external) driving forces $\vec{f}_0(\vec{x},t)$ due to boundary interactions and gravitational or electrical fields, dissipative friction forces $\vec{f}_i^{\text{fr}}(t) = -\gamma_i\vec{v}_i(t)$ with friction coefficient γ_i, and individual fluctuations $\vec{\zeta}_i(t)$ reflecting thermal interactions with the environment (boundaries, air, etc.) or a variation of the surface structure of the particles:

$$m_i\ddot{\vec{x}}_i(t) = \vec{f}_0(\vec{x}_i(t),t) - \gamma_i\vec{v}_i(t) + \sum_{j(\neq i)} \vec{f}_{ij}(t) + \vec{\zeta}_i(t) . \quad (2)$$

CP622, *Experimental Chaos: 6th Experimental Chaos Conference*, edited by S. Boccaletti et al.

In such systems, the ongoing competition between the driving forces and the dissipative friction forces leads to a spatio-temporal redistribution of energy, which produces a great variety of self-organization phenomena.

In the following, we will generalize Eq. (2) only a little. In contrast to conventional physical systems, we will assume to be confronted with "self-driven", "active", or "motorized" particles with an internal energy reservoir [6]. Therefore, the interaction forces $\vec{f}_{ij}(t)$ do not necessarily fulfill the 3rd Newtonian law $\vec{f}_{ji}(t) = -\vec{f}_{ij}(t)$ *(actio = reactio)*. Additionally, each particle can display an individual driving force $\vec{f}_i^{\,0}(t)$ instead of $\vec{f}_0(\vec{x}_i,t)$. We will see that this will imply various interesting phenomena, which can be observed in living systems. In this connection, it is useful to rewrite Eq. (2) somewhat, where the accelerations $\vec{a}_{ij}(t)$ are often called "generalized forces":

$$\dot{\vec{v}}_i(t) = \frac{v_i^0(t)\vec{e}_i^{\,0}(t) + \vec{\xi}_i(t) - \vec{v}_i(t)}{\tau_i} + \sum_{j(\neq i)} \vec{a}_{ij}(t) . \qquad (3)$$

Here, we have used the notations $\vec{f}_i^{\,0}(t) = \gamma_i v_i^0(t)\vec{e}_i^{\,0}$, $\gamma_i = m_i/\tau_i$, $\vec{f}_{ij}(t) = m_i\vec{a}_{ij}(t)$, and $\vec{\zeta}_i(t) = \gamma_i\vec{\xi}_i(t)$. Equation (3) implies that, while subject to the fluctuations $\vec{\xi}_i(t)$, the velocity $\vec{v}_i(t)$ is adapted to the equilibrium velocity v_i^0 and the direction $\vec{e}_i^{\,0}$ of motion, if particle i is far enough away from other particles j (i.e., $\vec{a}_{ij} = 0$). The adaptation is exponential in time with the relaxation time τ_i. In the following, we will assume that the interaction forces are short-ranged and repulsive.

VEHICLE TRAFFIC ON FREEWAYS

Let us now face freeway traffic as an example for one-dimensional non-equilibrium systems, focussing on the longitudinal dynamics within a single lane. In that case, the directions $\vec{e}_i^{\,0}$ of motion of the vehicles are all parallel. One frequently assumes that the vehicles i mainly react to the respective leading vehicles $(i-1)$, while the response to the following vehicle $(i+1)$ is neglected. In such car-following models, one applies equations of the form

$$\frac{dv_i(t)}{dt} = \frac{v_i^0 - v_i(t)}{\tau_i} + a_{i(i-1)}(t) , \qquad (4)$$

where we drop individual fluctuations for the moment. In simple models, the deceleration $a_{i(i-1)}(t) \le 0$ depends only on the headway (brutto distance) $s_i(t) = [x_{i-1}(t) - x_i(t)]$, but in reality, it is also a function of the own speed $v_i(t)$ and the relative velocity (approaching rate) $\Delta v_i(t) = [v_i(t) - v_{i-1}(t)]$ to the vehicle in front. In the following, we will assume identical driver-vehicle units, so that we have $v_i^0 = v_0$, $\tau_i = \tau$, and $a_{i(i-1)}(s_i, v_i, \Delta v_i) = a(s_i, v_i, \Delta v_i)$. Then, it is often useful to introduce the traffic-dependent "optimal velocity"

$$V(s_i, v_i, \Delta v_i) = v_0 + \tau a(s_i, v_i, \Delta v_i) \le v_0 , \qquad (5)$$

which corresponds to the stationary solution of Eq. (4) with $dv_i/dt = 0$. Driver i tries to adapt his actual velocity $v_i(t)$ to this velocity within the relaxation time τ.

If the "optimal velocity" depends only on the headway $s_i(t)$, we find the following fundamental relation between the traffic flow and the vehicle density $\rho = 1/s$ in homogeneous situations with identical headways $s_i = s$ and velocities $v_i = V$:

$$Q_{\rm f}(\rho) = \rho V(1/\rho) . \tag{6}$$

However, one can show [7] that this homogeneous traffic flow is *linearly unstable,* if

$$\frac{dV(s)}{ds} > \frac{1}{2\tau} . \tag{7}$$

As a consequence, arbitrarily small perturbations, inhomogeneities, or fluctuations of the traffic flow will be amplified, until a congested traffic state with areas of high density and low velocity has evolved.

The *linearly stable* regimes, in which the above condition is not fulfilled and traffic is insensitive to small disturbances, are themselves divided into

- an *absolutely stable* regime, in which even large perturbations fade away,
- a *metastable* regime adjacent to the linearly unstable regime [8].

The metastable regime is characterized by the decay of those perturbations, which are smaller than a certain density-dependent "critical amplitude", while larger perturbations are growing *("nucleation effect")* [8, 9].

Stable traffic flow is only found as long as the function $V(s)$ changes little with the headway s, i.e., when then the vehicle density is low ("free" traffic) or high (since the velocity is anyway small, then). However, for traffic densities around 25 vehicles per kilometer and lane (corresponding to velocities below about 85 km/h), the "optimal velocity" is rapidly falling with decreasing distance. Therefore, small changes are eventually amplified, until a traffic jam is formed, even if there is no accident or bottleneck. One could say that such a "phantom traffic jam" is the consequence of a *chain reaction* based on individual *over-reactions:* The finite adaptation time τ causes a delayed reaction to changes of the traffic situation, so that a vehicle gets too close to its leading vehicle, which requires an over-braking. If the following vehicles already arrive before the correspondingly increased disturbance can be compensated for, the successive vehicles will all brake a little bit more. In the end, traffic will reach a standstill, although all vehicles would like to drive fast.

If the inflow is not significantly reduced, the traffic jam will not dissolve, because accelerating vehicles lose time when leaving the downstream front of congested traffic. The correspondingly increased time gaps imply a considerably reduced outflow from congested traffic, compared to the maximum flow $Q_{\rm max}$ of homogeneous traffic defining the *theoretical* capacity of the road. As a consequence, the outflow $Q_{\rm out}$ from congested traffic is a second, *dynamical* capacity, which is something like a self-organized *"natural constant"* of traffic flow [10].

Most cases of traffic congestion on an n-lane freeway are observed upstream of on-ramps or other bottlenecks. They can be triggered by perturbations significantly below the theoretical capacity, as soon as the sum of the upstream flow $Q_{\rm up}$ and the inflow $\Delta Q = Q_{\rm in}/n$ per lane exceeds the dynamical capacity $Q_{\rm out}$: If a disturbance leads to temporary congestion, the drivers must accelerate again and suffer some time delay, which reduces the capacity to Q_{out}. Therefore, the following vehicles will queue up, and the temporary perturbation grows to form a persistent kind of congestion. The initial perturbation can even be a temporary *reduction* of the traffic flow and/or vehicle density, which can be caused by temporal variations of the traffic volume or even by vehicles leaving the freeway at some off-ramp: When the traffic density has sufficiently increased to reach the metastable regime, the "negative" perturbation will be amplified if it only exceeds the critical amplitude. While it is small, it will move downstream with the vehicles, so one could hope it would pass the bottleneck and leave the system. However, when the density wave grows larger, it will reduce its speed and even change its propagation direction. Once it is fully developed, it moves upstream with constant velocity, since vehicles leave the jam at the downstream front, while new ones join it at the upstream front. Hence, the perturbation returns to the bottleneck like a boomerang and triggers a breakdown of traffic, when it passes the bottleneck in upstream direction, as it thereby reduces the effective capacity to Q_{out}.

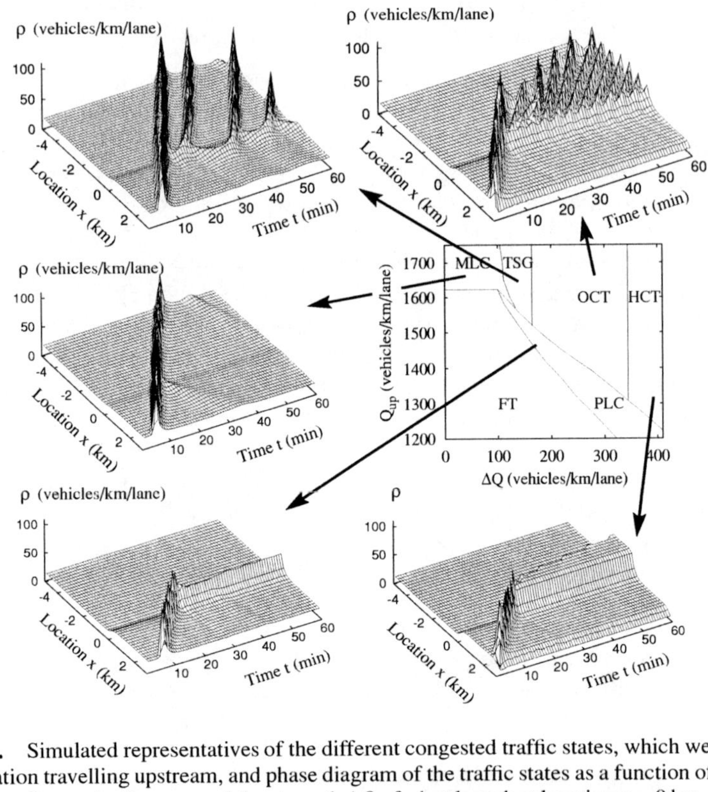

FIGURE 1. Simulated representatives of the different congested traffic states, which were induced by a big perturbation travelling upstream, and phase diagram of the traffic states as a function of the (upstream) traffic volume Q_{up} on the freeway and the strength ΔQ of a bottleneck at location $x = 0$ km, e.g. an on-ramp with inflow ΔQ per freeway lane (center right).

If the total traffic volume $Q_{\text{tot}} = (Q_{\text{up}} + \Delta Q)$ is greater than the dynamic capacity Q_{out}, we will automatically end up with a growing vehicle queue upstream of the on-ramp.

The traffic flow Q_{cong} resulting in the congested area gives, together with the inflow ΔQ, exactly the outflow Q_{out}, i.e. $Q_{\text{cong}} = (Q_{\text{out}} - \Delta Q)$. One can distinguish the following cases [11] (see Fig. 1): If the density ρ_{cong} associated with the flow $Q_{\text{cong}} = Q_{\text{f}}(\rho_{\text{cong}})$ is stable, we find *homogeneous congested traffic* (HCT) corresponding to typical traffic jams during holiday seasons. For a smaller on-ramp flow ΔQ, the congested flow Q_{cong} is linearly unstable, and we either find *oscillating congested traffic* (OCT) or *triggered stop-and-go traffic* (TSG), which often emerges from a spatial sequence of homogeneous and oscillating congested traffic (so-called *"pinch effect"* [12]). In contrast to OCT, stop-and-go traffic is characterized by a sequence of moving jams, between which traffic flows freely. Each traffic jam triggers the next one by inducing a small perturbation at the ramp, which propagates downstream as long as it is small, but turns back when it has grown large enough *("boomerang effect")*. This, however, requires the downstream traffic flow to be linearly unstable. If it is metastable instead (when the traffic volume is further reduced), a traffic jam will usually not trigger a growing perturbation. In that case, one finds either a single *moving localized cluster* (MLC), or a *pinned localized cluster* (PLC) at the location of the ramp. The latter requires the traffic flow in the upstream section to be stable, so that no traffic jam can survive there. Finally, for sufficiently small traffic volumes, we find *free traffic* (FT), as expected.

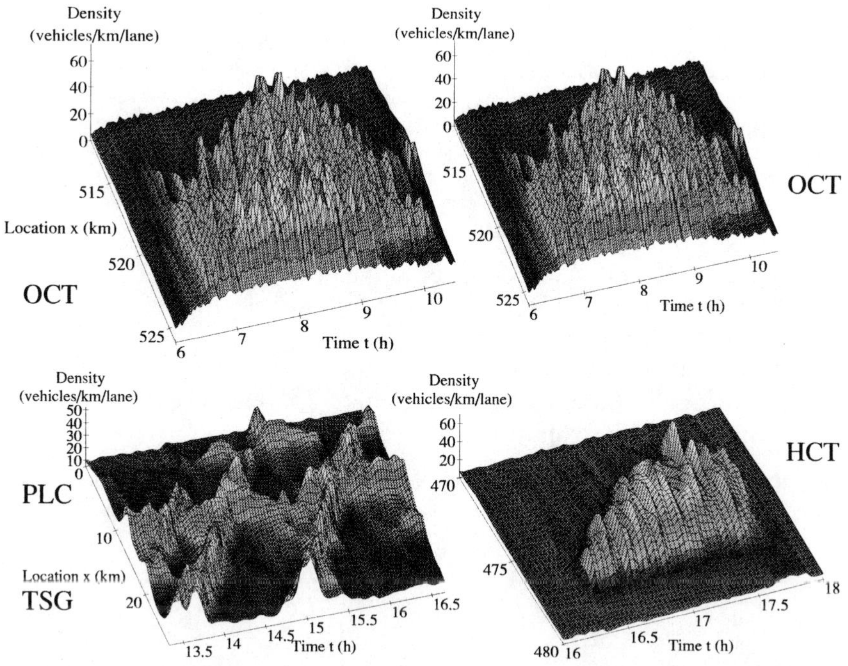

FIGURE 2. Empirical representatives of the different congested traffic states (see text).

The different congested traffic states found in the microsimulations (as displayed in Fig. 1) can all be identified in real traffic data (see Fig. 2). However, the empirically measured patterns look less regular. This is a characteristic feature of traffic, as the flow-density data are *widely scattered* in congested traffic, while the data are quasi-linearly arranged at low densities, see Fig. 3(a). To understand the source of the wide scattering, we have to be aware of several effects:

- Vehicles and driver behavior are not uniform, but heterogeneous. For example, the desired velocities and time headways of vehicles are largely varying. Distinguishing cars and trucks, it is possible to reproduce the empirical features of the flow-density data based on the measured boundary conditions and truck fractions, see Figs. 3(b), (c).

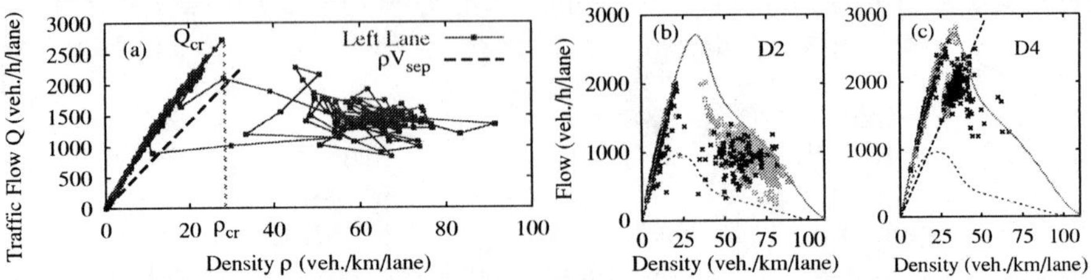

FIGURE 3. (a) Empirical time series of the hysteretic breakdown of traffic flow plotted as a function of the density (from [1]). High-flow states are only observed shortly before the breakdown of traffic. While free traffic is characterized by a quasi one-dimensional flow-density relationship, in the congested regime the flow-density data are widely scattered. The free and congested traffic regimes can be separated by a line ρV_{sep}, where we have chosen $V_{\text{sep}} = 70$ km/h, here. (b), (c) The displayed points in flow-density space correspond to empirical 1-minute data (dark crosses) and related simulation results (grey boxes) at two different cross sections of the freeway. At the section upstream of a bottleneck (b), we find congested flow, while the close downstream section (c) displays homogeneous-in-speed states. The simulations manage to reproduce both, the quasi-linear flow-density relation at small densities and the scattering over a two-dimensional region in the congested regime. For comparison, we have displayed the equilibrium flow-density relations for traffic consisting of 100% cars (—), and 100% trucks (- - -). The respective model parameters were weighted with the measured, time-dependent truck fraction. (From [1], after [13].)

- In congested traffic, the statistical error in the determination of the empirical velocity is particularly high, as is illustrated in Fig. 4(a). This is due to the instability of traffic flow, which is related with oscillations in the velocities and headways of following vehicles [14, 15].
- In congested traffic, we have correlations among several quantities, which affect the determination of the empirical vehicle density.

The latter point is illustrated by the following considerations: If Δt_i denotes the time headway of the ith vehicle and $\langle \dots \rangle$ the arithmetic average over N successive vehicles, we have the relation

$$\frac{1}{Q} = \langle \Delta t_i \rangle = \langle s_i / v_i \rangle = \langle s_i \rangle \langle 1/v_i \rangle + \text{Cov}(s_i, 1/v_i)\,. \tag{8}$$

That is, the fluiddynamic formula $Q = \rho V$ with $1/\rho = \langle s_i \rangle$ and $1/V = \langle 1/v_i \rangle$ is only exact, when the covariance $\text{Cov}(s_i, 1/v_i)$ vanishes. This, however, is not the case in congested traffic, as larger velocities correlate with larger distances, see Fig. 4(b).

A similar problem occurs when the density is determined via the time occupancy of a certain cross section of the road. Considering that $\Delta t_i = T_i + l_i / v_i$, where T_i is the (netto)

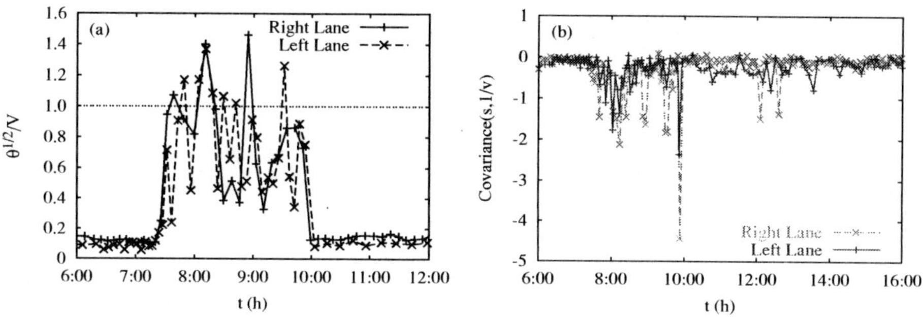

FIGURE 4. (a) Empirical standard deviation $\sqrt{\theta(t)}$ of vehicle velocities divided by the average velocity $V(t)$ as a function of time (from [1], after [16]). The *relative* variation $\sqrt{\theta(t)}/V(t)$ is particularly large during the rush hour, when traffic is congested. This reflects the oscillations in the following behavior due to unstable traffic flow [14, 15] and implies that measurements of average velocities V have a small reliability in the congested traffic regime. (b) The covariance between headways s_i and inverse velocities $1/v_i$ shows significant deviations from zero in congested traffic, while it approximately vanishes in free flow (from [1], after [16]). Even after traffic has recovered, there seem to remain weak correlations between headways and vehicle speeds for a considerable time. These are probably a reminiscence of congestion due to platoons which have not fully dissolved.

time clearance and l_i the length of vehicle i, we have the relation

$$\rho = \rho_{\max} \frac{\sum_i l_i/v_i}{\sum_i \Delta t_i} = \rho_{\max} Q \langle l_i/v_i \rangle = \frac{Q}{\langle l_i \rangle} \langle l_i/v_i \rangle \,, \tag{9}$$

where $\rho_{\max} = 1/\langle l_i \rangle$ is the maximum density and $\langle l_i \rangle$ the average vehicle length. The formula $1/V = \langle l_i/v_i \rangle / \langle l_i \rangle$ for the average velocity results in the correct expression $1/V = \langle 1/v_i \rangle$ only, if the individual vehicle lengths and velocities are not correlated, which is usually not the case.

The fluctuations of the macroscopic traffic variables can be characterized, e.g., by the power spectrum [17, 5]. Despite of signatures of a power-law behavior at low frequencies, white noise is observed at high frequencies. A power law distribution, however, is found for the time durations, during which the flow $Q(x,t)$ at a given cross section x exceeds a given threshold flow Q_{thr} [16]. This may be an effect of platoons of characteristic length distribution forming behind slow vehicles, until these can be passed.

Due to the reduced outflow Q_{out} from congested areas, the breakdown of traffic flow is often triggered by fluctuations. Therefore, most breakdowns can be avoided or delayed by technical optimization measures aiming at flow homogenization. Examples are intelligent speed limits, adaptive on-ramp controls, or driver assistance systems [18].

COLLECTIVE PEDESTRIAN DYNAMICS

Further interesting phenomena are found in two-dimensional pedestrian traffic. As the acceleration time $\tau_i \approx 0.5$ s of pedestrians is very short, the instability phenomena observed in vehicular traffic do not appear.

We describe the *psychological tendency* of two pedestrians i and j to stay away from each other by a repulsive interaction force $A_i \exp[(r_{ij}-d_{ij})/B_i]\vec{n}_{ij}$, where A_i and B_i are constants. $d_{ij} = \|\vec{x}_i - \vec{x}_j\|$ denotes the distance between the pedestrians' centers of mass, and $\vec{n}_{ij} = (n^1_{ij}, n^2_{ij}) = (\vec{x}_i - \vec{x}_j)/d_{ij}$ is the normalized vector pointing from pedestrian j to i. If their distance d_{ij} is smaller than the sum $r_{ij} = (r_i + r_j)$ of their radii r_i and r_j, the pedestrians touch each other. In this case, we assume two additional forces inspired by granular interactions [19, 20], which are essential for understanding the particular effects in panicking crowds: a *"body force"* $k(r_{ij}-d_{ij})\vec{n}_{ij}$ counteracting body compression and a *"sliding friction force"* $\kappa(r_{ij}-d_{ij})\,\Delta v^t_{ji}\vec{t}_{ij}$ impeding *relative* tangential motion, if pedestrian i comes close to j. Herein, $\vec{t}_{ij} = (-n^2_{ij}, n^1_{ij})$ means the tangential direction and $\Delta v^t_{ji} = (\vec{v}_j - \vec{v}_i)\cdot\vec{t}_{ij}$ the tangential velocity difference, while k and κ represent large constants. In summary, we have

$$\vec{f}_{ij} = \left\{A_i \exp[(r_{ij}-d_{ij})/B_i] + k\Theta(r_{ij}-d_{ij})\right\}\vec{n}_{ij} + \kappa\Theta(r_{ij}-d_{ij})\Delta v^t_{ji}\vec{t}_{ij}\,, \qquad (10)$$

where the function $\Theta(x)$ is zero, if the pedestrians do not touch each other ($d_{ij} > r_{ij}$), otherwise equal to the argument x.

The interaction with the boundaries is treated analogously, i.e., if d_{iW} means the distance to wall W, $\vec{n}_{iW}$ denotes the direction perpendicular to it, and $\vec{t}_{iW}$ the direction tangential to it, the corresponding interaction force with the wall reads

$$\vec{f}_{iW} = \{A_i \exp[(r_i - d_{iW})/B_i] + k\Theta(r_i - d_{iW})\}\vec{n}_{iW} - \kappa\Theta(r_i - d_{iW})(\vec{v}_i\cdot\vec{t}_{iW})\vec{t}_{iW}\,. \qquad (11)$$

For normal, "relaxed" situations with small fluctuation amplitudes η_i, our microsimulations of counterflows in corridors reproduce the empirically observed *formation of lanes* consisting of pedestrians with the same desired walking direction, see Fig. 5(a) [21, 22, 1]). If we do not assume periodic boundary conditions, these lanes are dynamically varying (see the Java applet at `www.helbing.org/Pedestrians/Corridor.html`). Their number depends on the width of the street [21], on pedestrian density, and on the noise level. Interestingly, one finds a *noise-induced ordering* [23]: Compared to small noise amplitudes, medium ones result in a more pronounced segregation (i.e., a smaller number of lanes), while large noise amplitudes lead to a *"freezing by heating"* effect, see Fig. 5(b).

The above, symmetric model can explain lane formation without assuming a preference for *any* side [22, 24]. The most relevant point is the higher relative velocity of pedestrians walking in opposite directions. As a consequence, they have more frequent interactions until they have segregated into separate lanes. The resulting collective pattern of motion minimizes the frequency and strength of avoidance maneuvers, if fluctuations are weak. This state is related with a maximum efficiency of motion. One can find such *optimal self-organization* for a large class of driven many-particle systems. That is, if these systems self-organize at all, they simultaneously tend to globally optimize their state [22].

To reflect the effect of getting nervous in panic situations, we assume that the individual level of fluctuations is given by

$$\eta_i = (1-n_i)\eta_0 + n_i\eta_{\max}\,, \qquad (12)$$

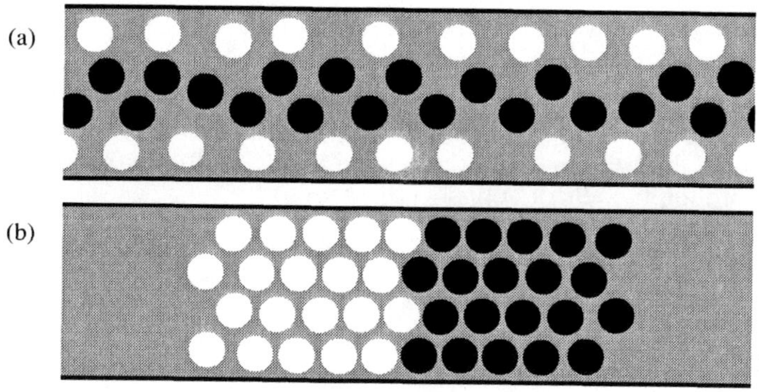

FIGURE 5. (a) Formation of lanes in initially disordered pedestrian crowds with opposite walking directions and small noise amplitudes η_i (after [24, 25, 1]; cf. also [21, 22]). White disks represent pedestrians moving from left to right, black ones move the other way round. (b) For sufficiently high densities and large fluctuations, we observe the noise-induced formation of a crystallized, "frozen" state (after [24, 25, 1]).

where n_i with $0 \leq n_i \leq 1$ measures the nervousness of pedestrian i. The parameter η_0 means the normal and $\eta_{\max}$ the maximum fluctuation strength. It turns out that, at sufficiently high pedestrian densities, lanes are destroyed by an increasing fluctuation strength (which is analogous to the temperature). However, instead of the expected transition from the "fluid" lane state to a disordered, "gaseous" state, a solid state is formed [24]. It is characterized by a blocked situation with a regular (i.e. "crystallized" or "frozen") structure so that we call this paradoxial transition *"freezing by heating"*, see Fig. 5(b).

In simulations of bottlenecks like doors, oscillatory changes of the passing direction are observed, if people do not panic [26, 25, 1] (see Fig. 6). Once a pedestrian is able to pass the narrowing, pedestrians with the same walking direction can easily follow. Hence, the number and "pressure" of waiting and pushing pedestrians becomes less than on the other side of the narrowing where, consequently, the chance to occupy the passage grows. This leads to a deadlock situation which is followed by a change in the passing direction.

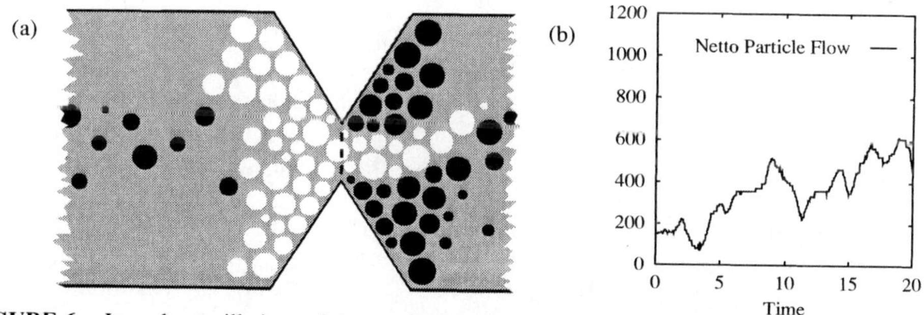

FIGURE 6. Irregular oscillations of the passing direction at a bottleneck (after [1, 25]; cf. also [21]). Dynamic simulations are available at `www.helbing.org/Pedestrians/Door.html`.

Now, we will focus on *uni-directional* flows at bottlenecks. The simulated outflow from a room turns out to be well-coordinated and regular, if the desired velocities $v_i^0 = v_0$ are normal. However, for desired velocities above 1.5 m/s, i.e. for people in a rush, we find an irregular succession of arch-like blockings of the exit and avalanche-like bunches of leaving pedestrians, when the arches break, see Figs. 7(a), (b). This phenomenon agrees with empirical observations and is comparable to intermittent clogging found in granular flows through funnels or hoppers [19, 20].

(a) (b)

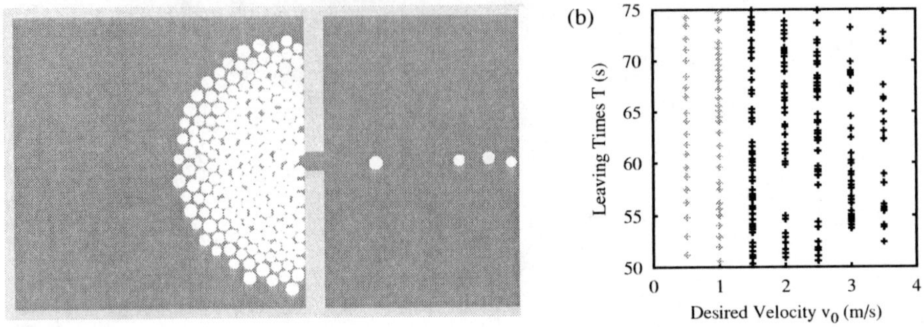

FIGURE 7. When the desired velocity v_0 is too high (e.g. in panic situations), pedestrians come so close to each other, that their physical contacts cause the build up of pressure and obstructing friction effects. This results in temporary arching and clogging related with an inefficient and irregular outflow, while the outflow is regular for small enough desired velocities ($v_0 \leq 1.5$ m/s) [26, 25, 1].

Since clogging is connected with delays, trying to move faster (i.e., increasing v_i^0) can cause a smaller average speed of leaving, if the friction parameter κ is large enough. This *"faster-is-slower effect"* is particularly tragic in the presence of fires, where fleeing people may reduce their own chances of survival.

Since our friction term has, on average, no deceleration effect in the crowd if the walls are sufficiently remote, the arching underlying the clogging effect requires a *combination* of

1. slowing down due to a bottleneck such as a door and
2. strong inter-personal friction, which becomes dominant when pedestrians get too close to each other.

Consequently, the danger of clogging can be minimized by avoiding bottlenecks in the construction of stadia and public buildings. However, jamming can also occur at widenings of escape routes! This surprising result originates from disturbances due to pedestrians, who expand in the wide area because of their repulsive interactions or try to overtake each other. These squeeze into the main stream again at the end of the widening, which acts like a bottleneck and leads to jamming.

Note that panics can be triggered by small pedestrian counterflows [26, 1, 25], which cause delays to the crowd intending to leave. Consequently, stopped pedestrians in the back are getting impatient and pushy. One may describe this by increasing the desired velocity, for example, according to the formula

$$v_i^0(t) = [1 - n_i(t)]v_i^0(0) + n_i(t)v_i^{\max} . \qquad (13)$$

Herein, $v_i^{\max}$ is the maximum desired velocity and $v_i^0(0)$ the initial one, corresponding to the expected velocity of leaving. The time-dependent parameter

$$n_i(t) = 1 - \frac{\overline{v}_i(t)}{v_i^0(0)} \tag{14}$$

reflects the nervousness, where $\overline{v}_i(t)$ denotes the average speed into the desired direction of motion. Altogether, long waiting times increase the desired velocity, which can produce inefficient outflow. This further increases the waiting times, and so on. This tragic feedback can eventually trigger so high pressures that people are crushed or falling and trampled.

Significantly improved (more efficient and safer) outflows can be reached by columns placed asymmetrically in front of the exits, which also prevent the build up of fatal pressures (see the Java applets at `http://angel.elte.hu/~panic/`).

Finally, we discuss a situation in which pedestrians are trying to leave a smoky room, but first have to find one of the invisible exits. Each pedestrian i may either select an individual direction $\vec{e}_i$ or follow the average direction $\langle \vec{e}_j^0(t)\rangle_i$ of his neighbours j in a certain radius R_i (which may be seen as a *dynamical generalization of a Heisenberg spin system*) [27, 1], or try a mixture of both. We assume that both options are weighted with the nervousness n_i:

$$\vec{e}_i^0(t) = \text{Norm}\left[(1-n_i)\,\vec{e}_i + n_i\,\langle \vec{e}_j^0(t)\rangle_i\right] , \tag{15}$$

where $\text{Norm}(\vec{z}) = \vec{z}/\|\vec{z}\|$ denotes normalization of a vector $\vec{z}$. As a consequence, we have individualistic behavior if n_i is low, but herding behavior if n_i is high.

Simulation results for this scenario suggest that neither individualistic nor herding behavior performs well [26]. Pure individualistic behavior means that each pedestrian finds an exit only accidentally, while pure herding behavior implies that the complete crowd is eventually moving into the same and probably blocked direction, so that available exits are not efficiently used, in agreement with observations. Optimal chances of survival are expected for a certain mixture of individualistic and herding behavior, where individualism allows some people to detect the exits and herding guarantees that successful solutions are imitated by the others.

SUMMARY AND OUTLOOK

In recent years, many surprising self-organization phenomena have been discovered in pedestrian and vehicle traffic. They have been successfully simulated and explained with self-driven many-particle models. This includes several instability and breakdown phenomena (leading to "phantom traffic jams", irregular outflows, or the "faster-is-slower effect"), particular segregation and condensation effect effects (such as lane formation, stop-and-go traffic, or "freezing by heating"). Large fluctuations (e.g. in the discharge of vehicle or pedestrian queues at stationary or temporary bottlenecks) are characteristic for these systems, especially, if the vehicles or pedestrians do not behave uniformly, but heterogeneously. Fluctuations are also relevant for empirical measurements, as significant correlations of various quantities arise in dense traffic.

Finally, fluctuations can trigger the breakdown of fluid traffic. This is the background of recent technical optimization measures.

ACKNOWLEDGMENTS

I would like to thank Illés Farkas, Benno Tilch, Martin Treiber, and Tamás Vicsek for inspiring discussions and their help with producing the figures. Moreover, I am grateful to the German Research Foundation (DFG) and the Federal Ministry for Education and Science (BMBF) for financial support through the grants He 2789/1-1, He 2789/2-1, and the project SANDY, grant no. 13N7092.

REFERENCES

This manuscript is partly a translation of D. Helbing: Die wundervolle Welt aktiver Vielteilchensysteme, *Physikalische Blätter* **57**, 27–33 (2001). A more detailed discussion can be found in Ref. [1].

1. Helbing, D., *Rev. Mod. Phys.* **73**, 1067–1141 (2001).
2. Schreckenberg, M., and Wolf, D. E., Eds., *Traffic and Granular Flow '97*, Springer, Singapore, 1998.
3. Helbing, D., Herrmann, H. J., Schreckenberg, M., and Wolf, D. E., Eds., *Traffic and Granular Flow '99: Social, Traffic, and Granular Dynamics*, Springer, Berlin, 2000.
4. Chowdhury, D., Santen, L., and Schadschneider, A., *Phys. Rep.* **329**, 199–329 (2000).
5. Helbing, D., *Verkehrsdynamik*, Springer, Berlin, 1997.
6. Schweitzer, F., *Brownian Agents and Active Particles*, Springer, Berlin, 2001.
7. Bando, M., Hasebe, K., Nakayama, A., Shibata, A., and Sugiyama, Y., *Phys. Rev. E* **51**, 1035–1042 (1995).
8. Kerner, B. S., and Konhäuser, P., *Phys. Rev. E* **50**, 54–83 (1994).
9. Bando, M., Hasebe, K., Nakanishi, K., Nakayama, A., Shibata, A., and Sugiyama, Y., *J. Phys. I France* **5**, 1389–1399 (1995).
10. Kerner, B. S., and Rehborn, H., *Phys. Rev. E* **53**, R1297–R1300 (1996).
11. Helbing, D., Hennecke, A., and Treiber, M., *Phys. Rev. Lett.* **82**, 4360–4363 (1999).
12. Kerner, B. S., *Phys. Rev. Lett.* **81**, 3797–3800 (1998).
13. Treiber, M., and Helbing, D., *J. Phys. A: Math. Gen.* **32**, L17–L23 (1999).
14. Hoefs, D. H., *Untersuchung des Fahrverhaltens in Fahrzeugkolonnen*, Bundesministerium für Verkehr, Abt. Straßenbau, Bonn-Bad Godesberg, 1972.
15. Helbing, D., and Tilch, B., *Phys. Rev. E* **58**, 133–138 (1998).
16. Tilch, B., *Modellierung und Simulation selbstgetriebener Vielteilchensysteme mit Anwendung auf den Straßenverkehr*, Ph.D. thesis, University of Stuttgart, in preparation (2001).
17. Musha, T., and H. Higuchi, *Jpn. J. Appl. Phys.* **15**, 1271–1275 (1976); **17**, 811–816 (1978).
18. Treiber, M., and Helbing, D., *Automatisierungstechnik* **49**, 478–484 (2001).
19. Ristow, G. H., and Herrmann, H. J., *Phys. Rev. E* **50**, R5–R8 (1994).
20. Wolf, D. E., and Grassberger, P., Eds., *Friction, Arching, Contact Dynamics*, World Scientific, Singapore, 1997.
21. Helbing, D., and Molnár, P., *Phys. Rev. E* **51**, 4282–4286 (1995).
22. Helbing, D., and Vicsek, T., *New Journal of Physics* **1**, 13.1–13.17 (1999) (see `www.njp.org/`).
23. Helbing, D., and Platkowski, T., *Int. J. Chaos Theor. Appl.* **5**, 25–39 (2000).
24. Helbing, D., Farkas, I., and Vicsek, T., *Phys. Rev. Lett.* **84**, 1240–1243 (2000).
25. Helbing, D., Farkas, I. J., and Vicsek, T., in *The Science of Disaster: Climate Disruptions, Heart Attacks and Market Crashes*, edited by A. Bunde, J. Kropp, and H.-J. Schellnhuber, Springer, Berlin, 2002, pp. 334–354.
26. Helbing, D., Farkas, I., and Vicsek, T., *Nature* **407**, 487–490 (2000).
27. Vicsek, T., Czirók, A., Ben-Jacob, E., Cohen, I., and Shochet, O., *Phys. Rev. Lett.* **75**, 1226–1229 (1995).

Transportation Networks: Dynamics and Simulation

Stephen G. Eubank

Basic and Applied Simulation Science Group, D-2, MS-M997, LANL, Los Alamos, NM, 87545, USA

Abstract. Analyzing the behavior of socio-technical systems – large urban populations interacting under constraints imposed by technical infrastructure – poses many interesting problems. We describe why simulation is an appropriate experimental method for studying these systems, drawing examples from the development and implementation of the Transportation Analysis and Simulation System (TRANSIMS). TRANSIMS is an urban planning tool that combines census data and household surveys with a description of the transportation network to produce estimates of human mobilty. It iteratively evaluates hundreds of thousands of coupled non-linear models to produce a solution to a million-person game.

INTRODUCTION

TRANSIMS[1] is a machine for turning readily available data into an estimate of a very complicated joint distribution – the location of every individual in an urban region during the course of a day. [2] The input data for TRANSIMS consists of:

marginal distributions in many variables (demographics from census data)

a small sample of some of the joint distributions among those variables (also from census data)

a small sample of activity patterns (surveys, diaries)

details of the transportation infrastructure (maps, transit schedules) in the form of a graph.

The output is the estimated distribution of a collection of demographic variables (in other words, people) at every location at every second.

The estimate respects travel time and cost constraints placed on people's behavior by the transportation infrastructure. It makes an assumption that in general people strive to minimize their cost, but neither assumes that they are perfectly "rational" nor that they all share the same cost function. In particular, they do not necessarily have complete information on which to base decisions.

The heart of the problem to be solved by TRANSIMS is a million-person game. Each player picks a strategy, in this case a route through the transportation network that allows him or her to execute a planned set of activities. A solution is a set of strategies in which nobody is too far from an optimal route. The game is solved iteratively using a traffic simulation to estimate the game's outcome given each player's strategy for that iteration. The solution is self-consistent in the sense that if everyone were allowed to make choices

CP622, *Experimental Chaos: 6th Experimental Chaos Conference*, edited by S. Boccaletti et al.

based on the results of their and everyone else's previous choices, the overall outcome would remain the same.

WHY TRAFFIC?

Traffic, or more generally transportation, has long been an important factor determining how urban environments grow. Traffic modeling provides an experimental testbed for urban planning. It makes possible evaluation of proposed infrastructure changes, including

- construction projects
- changes in the way infrastructure is used, such as HOV lanes
- pricing changes for tolls, mass transit, or parking

Importantly, because of its self-consistency, TRANSIMS takes into acount changes in demand caused by hypothesized infrastructure changes rather than assuming the demand will remain fixed. Evaluation can take the form of total travel times, vehicle-miles, emissions, or many other variables. All these effects can be traced to the demographics of specific individuals, allowing estimates of the differential effect on different, possibly overlapping, subpopulations. For instance, urban planners can determine who will benefit from improvements in local streets (in the form of shorter travel time) and who will be exposed to degradations in air quality.

Important as it is for its own sake, estimating traffic is not the final goal of this simulation process. Human mobility is the basic data needed for understanding the characteristics of social networks. These in turn underlie estimates of mobile telecommunication demand, electric power demand, epidemiology, and many other poorly understood processes. Epidemiology, for instance, is essentially the study of diffusion on a time-dependent graph generated by human mobility. Until now, there has been no good characterization of these graphs. Only ad-hoc hypotheses such as "small-world" models have been studied.

INDIVIDUAL-BASED SIMULATION

Population, Activities, Routes, and Traffic

TRANSIMS synthesizes from census data a population with the correct joint distribution of demographics and locates this population on an actual transportation network. Each household is matched demographically to one of approximately 2000 households whose members filled out activity diaries (the "survey data"). The diaries define a set of activity patterns specifying activity type, start time, and duration – but not location – for each member of a household. Each synthetic household is assigned an activity pattern drawn from those of households with similar demographics. An initial assignment of important anchor locations such as work or school is made based on information such as known employment by location throughout the area.

Based on this set of activity location and time assignmnts, we use a novel shortest-constrained-path technique to find routes through the transportation network.[3] The novel aspects of the route-finder are that it makes use of time dependent travel times on each link of a mutli-modal network and that it easily handles path constraints expressed in a simple language. Examples of constraints that might be appropriate are:

find a path using the modes *walk* followed by *car* followed by *walk*;
find a path using at least one but no more than two bus rides;
find a path that avoids toll roads.

Finding shortest paths through a large network is computationally expensive, but feasible. Finding the shortest path while simultaneously accounting for travel delays induced by everyone else's routes is not feasible. Hence, each person is planned without regard to the rest of the population's plans except in so far as their plans are reflected in the estimated travel times. The first set of paths thus places most drivers on the small set of links with short travel times – everyone takes freeways as much as possible.

The "game" alluded to in the Introduction is to find a shortest path in the context of the choices made by the rest of the population. Here the context provided by other people's paths is summarized in the travel time on each link of the network, which in turn is mostly controlled by the presence or absence of jams.

The traffic simulator provides estimates of travel times for a given set of plans. If it were possible reliably to estimate these directly from the plan set, there would be no need for the traffic simulation. TRANSIMS requires a traffic model that can produce good travel time estimates while updating the entire region in much less than real time. (See Figure 1 for an idea of both the level of detail and the overall scale.) For Portland, Oregon, with population 1.6 million, this requires roughly 10^9 vehicle-seconds per 24 hours, or 10^4 vehicle-seconds per second, of simulated time.

Solving The Game

A perturbative approach to solving this problem seems reasonable, since any one person's travel hardly affects overall travel times, but finding an approximate solution for starting is difficult. Direct calculation is infeasible. TRANSIMS uses iterative approximation. In the first approximation, everyone makes plans as though they were alone on the network. In succeeding approximations, a gradually diminishing fraction of the population is allowed to replan based on the travel times computed from the previous set of plans.

It is tempting to think of each iterate as a new "day" in which people choose a new path based on the traffic they experienced the previous day. This analogy is appropriate in some circumstances, but should not be pushed too far. The iterative process is really a machine learning algorithm for solving a many-player game. For instance, a driver choosing a route in the real world typically does not have complete information about the state of the network on the previous day – only observations about links used on the previous day and only for the specific times at which the driver was present.

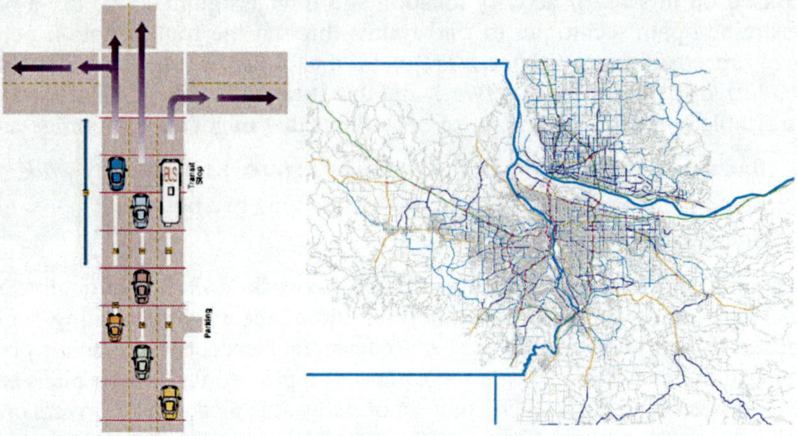

FIGURE 1. Detailed and overall views of the Portland, Oregon network. The simulation includes features such as pocket lanes, traffic signals, transit stops, activity locations, and parking lots. The full network contains about 120,000 edges and 100,000 vertices, all resolved to this level of detail.

The dynamics of this iterative process are more intricate than they at first appear. Many intuitively obvious shortcuts do not work well. For example, allowing the entire population to replan with the most recent travel time estimates can introduce periodic orbits in the state space of all plans. Consider the simplest case of two alternative routes, in which all the traffic oscillates between the two routes. Replanning a diminishing fraction of the population damps these oscillations. The optimal replanning fraction is not known, nor is it even clear what qualities of the network it depends on. Another example of the intricacy of the dynamics is that biasing the selection of those individuals allowed to replan towards those who travel on the most congested links does not seem to improve convergence.

Finally, once the iterative process has found an approximate solution for travel times given activity patterns, travelers can change unrealistic patterns, choose new times for specific activities, or change the locations at which activities are conducted. If the changes at this stage are large, it too can be done iteratively.

An Agent-Based Simulation

It has become almost required in some circles to declare simulations "agent-based". TRANSIMS is agent-based, but it involves an atypical notion of agency. An individual, identified by number, is variously represented as a set of demographics, a set of activities, or a route through the transportation network. Similarly, the route planner and the traffic simulator use radically different representations of the same section of roadway. The agent is not to be found in any of these representations, but rather in the whole collection of them. Moreover, since many of the attributes are the results of simulation, the agents are not predefined, but are endowed with important attributes through their interactions with others.

WHY SIMULATION?

Simulation provides the *only* way to determine the logical entailments of a given model on this scale. In many instances simulation of a particular model is provably more efficient than any other form of calculation. For example, consider the evolution of a cellular automaton. One way to determine the configuration at time t given an initial configuration at time t_0 is to simulate the rules. It can be shown that for a large class of update rules there is no way to determine the configuration at t in fewer steps. Even an analytical technique such as renormalization does not help, since the effective diameter of the rule grows larger and the rule table grows exponentially larger with every attempt to lengthen the time step. Approximation methods may give overall properties of the solution (an invariant distribution, for example), but these are not specific enough for our purposes. It is here that the lack of symmetries in real networks wreaks havoc with attempts to simplify solutions.

Some of these hardness results are most applicable to discrete models. But traffic is inherently discrete, and cannot be forced into a continuous description without losing the identity of the (distinguishable) travelers. Furthermore, determining the maximum flow through a network is the canonical example of a problem that has very different solutions over the integers (discrete flow) or the reals (continuous flow). Since even the lowest-level constituent model in TRANSIMS is best suited to simulation, it is likely that the entire system requires simulation.

MODELS OF TRAFFIC DYNAMICS

Historical Approaches

Because of the social and economic importance of transportation, models for traffic flow have a long history. It would not be surprising to find ideas about how to design efficient transportation networks in ancient civilizations. In the past century, several mathematical approaches have dominated the subject. A detailed review is beyond the scope of this paper. For an entry into the field, see the annotated bibliographies maintained in several places on the world-wide web. For example, `http://www.theo2.physik.uni-stuttgart.de/TrafficForum`.

Several authors have approached traffic from a macroscopic, statistical mechanical point of view. The most well-known among them are Prigogine, Herman, and Montroll. [4] [5][1] This approach develops a master equation for the distribution of speeds.

---

[1] It is interesting to note that one of these authors, R. Herman, was not content with theorizing but also conducted experiments with cars and buses on a racetrack.

Similarly, it has been tempting to use analogies to fluid dynamics, reasoning that smoothly flowing traffic is like laminar flow and jams are a form of turbulence. This approach was primarily codified by Lighthill and Whitham [6] who used an analogue of the Navier-Stokes equation to study the kinematics of density waves in traffic, including phenomena such as the upstream propagation of jams.

Moreover, though not strictly in the category of traffic models, there has been substantial effort invested in direct solution of the "traffic assignment" problem. Typically, these solutions use simple functions for travel time on a link as a function of density. A text illustrating the state of the art has recently been published by Sheffi. [7]

Recently, microscopic models of traffic have become popular.[8] These focus on vehicles as discrete particles rather than on a continuous velocity or density field. They may take the form of ordinary differential equations governing the speed of each vehicle. Nagel and Schreckenberg[9] introduced a model in which both space and time are discrete and the dynamics is represented by a cellular automaton (CA). Because of its simple update scheme, they were able to use the CA model to simulate traffic on the entire German autobahn network. Nagatani has also used a similar approach for studying phase transitions.

Both macroscopic and microscopic results have been important in the design of more efficient transportation systems. For example, an understanding of critical phenomena near the transition from free flow to jammed traffic suggests that designing transportation systems to operate near their critical point is very dangerous. Such designs are likely to produce very long-range (spatial) correlations in fluctuations, which translate into system-wide jams. Detailed models of driver behavior are also important for small-scale studies such as designing the entrance to a heavily used location, or the optimal signal timing for a network of a few intersections.

Nevertheless, these analyses are typically for long, isolated stretches of road – often, indeed, single lane roads. This is in the best reductionist tradition: break the world into its simplest components, find analytic solutions for those components, and let the engineers worry about how to put the pieces back together. A quick glance at the stated goals of the TRANSIMS project shows that we must proceed far beyond this idealized world. We must construct transportation networks consisting of hundreds of thousands of elements, each of which exhibits non-trivial behavior such as phase transitions.

There are three distinct hurdles to overcome in this effort:

1. The first is one of scale. A typical urban population of 10^6 people moves on a network with 10^5 intersections and a similar number of edges. This problem inhabits a kind of no-man's land, neither large enough to treat with statistical mechanics nor small enough to treat analytically. Roughly speaking, neither high-temperature nor low-temperature expansions are appropriate when $1 \ll log(N) \approx 6 \ll log(Avogadro's\ number)$.
2. We expect specific details of the input data to have important consequences. (Else we could not show the effects of changes in the infrastructure.) We cannot make any simplifying assumptions about homogeneity, isotropy, or any other kind of symmetry in the boundary conditions.
3. The particles in this case are not indistinguishable. Even with the drastic simplifications made by TRANSIMS – representing an individual as "economic man"

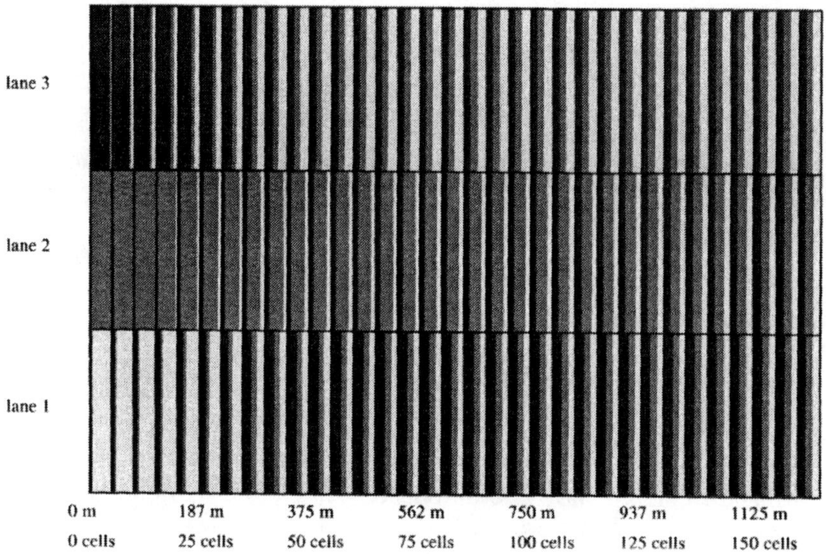

FIGURE 2. A simulated weaving region. Traffic enters from thre right and makes one of three choices at the intersection on the left. In each 37.5 meter segment of the roadway is shown, by lane, the fraction of vehicles in that lane intending to make each turn. The flow rate is such that not all vehicles can merge into the correct lane by the time they reach the intersection. Lane changing logic in the traffic model can be calibrated against observations of actual weaving regions.

with a few demographic variables and providing every driver with the same driving logic – different people can make different choices based on demographics and drivers behave differently in different traffic contexts. Worse yet, not only are these particles distinguishable, they have intentions and need not obey any conservation laws. Indeed, the introduction of drivers following planned routes into a simulation must be done carefully for the simulations to remain consistent with observations. Figure 2 shows the kinds of calibration required in this case.

TRANSIMS uses a CA traffic model modified to represent better multi-lane traffic, network use restrictions (e.g. turning movement or lane use restrictions) and plan-following behavior. Rather than turning bits on and off, a small set of rules emulating the CA rules moves pointers to individual vehicles around a grid.

The driver logic in TRANSIMS is as follows:

Choose a lane
- Pass slow vehicles if there is a gap available
- Move toward desired lane for upcoming turn

Choose speed
- Accelerate until the vehicle is at the speed limit or within range of an obstacle (other vehicles, the end of a lane, traffic signals, etc.).

- Decelerate as required to avoid obstacles.
- Decelerate to provide more opportunity for lane changing if required.
- Decelerate at random occasionally.

This model allows large decelerations in order to remain collision-free. Animation of the traffic may thus look unphysical, but the aspects important for jam dynamics are all correct.

Other elements of the simulation besides drivers must have rules. In particular, traffic signals must provide realistic flow rates. The data collection required to time correctly every signal in the network is too burdensome for most urban planning organizations. Instead, we have implemented adaptive signals that change their timing to accomodate demand. As with animations of the vehicles, close inspection of the traffic signals may reveal behavior that would be unlikely for real signals, but as long as they support the correct rates of flow through an intersection, these details are not crucial.

There are many epicycles introduced on top of these basic rules to handle parking, vehicles that cannot get in the correct lane for a turn because of congestion, priority at intersections, and so on. We have also introduced a few vehicle-specific parameters in circumstances where it seemed necessary to capture important behaviors. For example, trucks and other heavy vehicles have a grade-sensitive acceleration, rather than the uniform 1 $cell/timestep^2$ of other vehicles.

Contact with Experiment and Observation

There is ample observational data for comparison with these models. The quantities of interest are flow, density, and speed. Of these, only two are independent. Density is a natural parameter in most models, and the graph of flow vs. density is often called the "fundamental diagram" of traffic flow. Because of observational difficulties determining density, flow vs. speed is more readily available experimentally. The left side of Figure 3, taken from the Highway Capacity Manual, is representative of the data available. Analogous data from a calibration test of TRANSIMS is shown on the right side of Figure 3.

The modified CA model is up to the challenge of TRANSIMS. It reproduces all the necessary aspects of travel time as a function of demand, and does so quickly and efficiently. Other models are not as well suited to our needs. Analytic models often misplace traffic jams, leading to important differences in travel time estimates. For instance, jams may occur on links where demand is above capacity, but they typically occur at the entrance to those links and for the most part spill back onto the incoming links. Few queueing theory models capture this spillback correctly. Models that try to parameterize driver psychology, or incorporate extremely detailed physical models of car-roadway interaction, tend to run too slowly and in effect only add nuisance parameters.

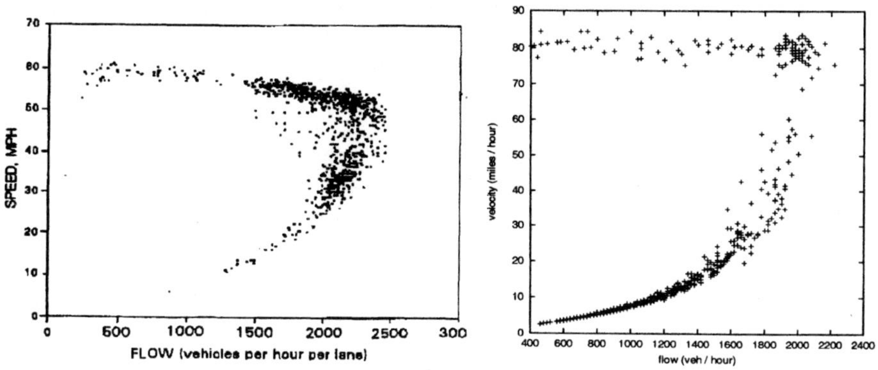

FIGURE 3. Observed (taken from [10]) (left) and simulated (right) traffic flow characteristics. The y-axis scales differ because the speed limits in the two cases were different.

EXTRACTING MEANING

How are we to interpret the results of TRANSIMS? Ironically, although the defining characteristic of the simulation is that it is individual-based, we do not expect the results to be correct for individuals – I cannot find a specific synthetic "me" in the synthetic population. While the demographics of synthetic people living in specific areas are matched with the census, there is no explicit constraint forcing the synthetic people to work or play in the right locations at the right time.[2] Our assumption is that self-consistent travel time estimates together with cost-minimizing behavior supply the necessary constraints.

Of course, with every different set of initial conditions or pseudo-random number sequences, we obtain slightly different results on microscopic scales. This is no different from the world we are trying to model, in which each day brings a slightly different travel experience. Instead, we must consider macroscopic quantities or look at coarse-grained pictures of the outcomes. It is not clear, though, exactly how coarse our pictures must be. We can reliably say that traffic jams are likely to occur in particular locations and times. We can probably estimate the distribution of demographics for those caught in the jams. We probably cannot identify whether a specific individual becomes entangled in a jam. In the case of fluid dynamical simulations, the question does not arise. No one is particularly interested in which molecule of water is part of a shock wave, just where the shock is.

---

[2] This is an important feature of this approach – otherwise, problems of privacy immediately arise.

Not all traffic engineers understand this variability. Some would consider the entire TRANSIMS methodology invalid unless it reproduces traffic counts measured with a pressure hose on a rainy Tuesday, the 12th of May. But even a more generous interpretation of "valid" must provide some comparison of TRANSIMS results with daily variability. Is TRANSIMS invalid if it is outside a 3-sigma probability band on 10% of the available traffic counts? Is daily variability in traffic flow an appropriate determinant of variability to be expected in a simulation? What if the transportation system has multiple basins – sets of microscopically different initial conditions that produce macroscopically similar results? Is there any reason to expect that daily variability would explore the different basins?

It is useful in these circumstances to have an alternative method of determining the equivalence of two dynamical systems. All the results above attempt to compare properties of random samples of state space trajectories. It is possible to compare directly two dynamical systems represented as interaction graphs with local mappings at each vertex together with a schedule for updating the mappings. One can enumerate the possible topologies and multiplicities of state space trajectories from this representation without explicit evaluation of the trajectories themselves.[11] However, this program is not yet mature enough to apply to dynamical systems on this scale.

Sensitivity is intimately related to variability. At this stage of development for socio-techincal simulation, we are interested not so much in quantifying sensitivity to parameters and initial conditions or even classifying it as exponential or linear, but in simply *detecting* it. It is not hard to hypothesize that a certain set of parameters is important but determining its exact influence on a solution is time-consuming. What is really needed is a solution for the inverse problem: given several instances of behavior in a highly nonlinear system, determine the factors responsible.

CONCLUSIONS

Is TRANSIMS worth it? Comparing the total research and development expense for TRANSIMS, on the order of $30 million over 8 years, with the annual federal, state, and local expenditures on transportation, $129 billion in 1995[3], it clearly is. If even one construction project can be designed better using TRANSIMS, it will have paid for itself.

Does traffic depend sensitively on initial conditions? The separation between drivers increases linearly on average on long, isolated sections of roadway. Perhaps the introduction of interactions with the rest of the network introduces more sensitivity. Aggressive drivers' willingness to risk life and limb for a few seconds' gain on nearby traffic implies that they believe it does. Of course, those drivers who often catch up with the aggressive drivers at the next traffic signal probably believe otherwise. In any case, these observations are only relevant at a fine-grained, individual, level of resolution. Macroscopically, daily traffic jams appear in about the same locations and about the same times from day

---

[3] Source: US Department of Transportation, Bureau of Transportation Statistics, as quoted in "Transportation and the Economy".

to day. The problem, once again, is that very few individuals care about the macroscopic view.

TRANSIMS is a good example, and perhaps the enabling technology, of the burgeoning field of socio-technical simulation. Is it Isaac Asimov's psychohistory, which predicts the course of societies through millennia? [12] Of course not, but it may turn out to be a significant first step.

REFERENCES

1. Barrett, C. L., and et al., R. J. B., TRANSIMS: Transportation Analysis Simulation System, Tech. Rep. LA-UR-00-1725, Los Alamos National Laboratory Unclassified Report (2001).
2. Eubank, S. G., "What Makes A Simulation Useful?", in *1999 IEEE International Conference on Systems, Man, and Cybernetics*, IEEE, New York, 1999.
3. Barrett, C. L., Jacob, R., and Marathe, M. V., *SIAM Journal of Computing*, **30**, 809–837 (2001).
4. Prigogine, I., and Herman, R., *Kinetic Theory of Vehicular Traffic*, Elsevier, New York, 1971.
5. Chandler, R. E., Herman, R., and Montroll, E. W., *Operations Research*, **6**, 165–184 (1958).
6. Lighthill, M. J., and Whitham, G. R., *Proc. Roy. Soc A*, **229**, 317 (1955).
7. Sheffi, Y., *Urban Transportation Networks: Equilibrium Analysis with Mathematical Programming Methods*, Prentice-Hall, Englewood Cliffs, New Jersey 07632, USA, 1985, ISBN 0-13-939729-9.
8. Nagatani, T., *J. Phys. A*, **26**, L781–L787 (1993).
9. Nagel, K., and Schreckenberg, M., *Journal de Physique I France*, **2**, 2221 (1992).
10. Banks, J.H., "Freeway Speed-Flow-Concentration Relationships: More Evidence and Interpretation" in *Transportation Research Record* **1225**. Transportation Research Board, National Research Council, Washington, D.C., 1989.
11. Mortveit, H. S., and Reidys, C. M., *Discrete Mathematics*, **226**, 281–295 (2001).
12. Asimov, I., *The Foundation Trilogy*, Doubleday, Garden City, New York, 1951.

COMBUSTION

Controlling Cyclic Combustion Variations in Lean-Fueled Spark-Ignition Engines

C. S. Daw*, J. B. Green, Jr.*, R. M. Wagner*, C. E. A. Finney*, L. I. Davis**, Jr., L. A. Feldkamp**, J. W. Hoard**, F. Yuan**, and F. T. Connolly**

**Oak Ridge National Laboratory, National Transportation Research Center, 2360 Cherahala Boulevard, Knoxville, Tennessee 37932-6472*

***Ford Research Laboratory, Dearborn, Michigan 48121*

Abstract. Under conditions of lean fueling or high exhaust gas recirculation, interactions between residual cylinder gas and freshly injected fuel and air produce undesirable combustion instabilities in spark-ignition engines. The resulting dynamics can be described as noisy bifurcations of a nonlinear map and are complicated by cylinder-to-cylinder coupling. We discuss the key dynamic features relevant to control and demonstrate simple feedback control of a multi-cylinder test vehicle.

INTRODUCTION

Cycle-by-cycle combustion variations (CV) can be particularly severe in spark-ignited, internal combustion engines for lean air-fuel mixtures (i.e., when the ratio of air to fuel is greater than required by chemical stoichiometry) or for high levels of exhaust gas recirculation (EGR). CV is a problem because it increases engine roughness, decreases fuel efficiency, and produces excessive emissions of unburned hydrocarbons. Up to now, the approach for minimizing CV has been to simply avoid operating conditions where it occurs. In the current regulatory climate, however, lean fueling and high EGR are needed to reduce nitrogen oxide emissions. Thus there is strong interest in understanding and controlling the CV phenomenon.

Grunefeld et al. [1] discuss deterministic fluid dynamic effects during engine intake and exhaust strokes and conclude that they are dominant contributors to CV. The importance of the residual cylinder gas (both composition and amount) has also been recognized and is generally regarded as the cause of the frequently observed alternating pattern of high and low work output cycles, although other mechanisms have been proposed [2]. Stevens, Shayler, and Ma [3] considered CV from the standpoint of understanding the mechanism well enough to effect a reduction in the variation by imposing control. They found that significant correlation exists between consecutive firings of a particular cylinder, and that various measurable quantities of combustion intensity, such as indicated mean effective pressure, are subject to reasonable prediction one cycle in advance. These authors also considered various means of imposing control, such as through changes of spark timing and fuel delivery.

CP622, *Experimental Chaos: 6th Experimental Chaos Conference*, edited by S. Boccaletti et al.

In previous publications (see for example, [4-6]), we presented evidence for a dominant (but noisy) period-doubling bifurcation that develops during lean fueling. The visibility of low-dimensional period doubling is remarkable when one considers the complexity of internal combustion engines and the high-dimensional inputs (e.g., turbulent mixing in the cylinder). In this paper we focus on improved characterization of the low-dimensional dynamics and linking those dynamics to engine control.

EXPERIMENTAL EQUIPMENT AND PROCEDURES

The experiments reported here were done with three different types of engines: 1) a single-cylinder Cooperative Fuel Research (CFR) engine; 2) a four-cylinder 2.3-liter General Motors (GM) "Quad-4" engine; and 3) a 4.6-liter V8 Ford engine used on both the Expedition and Grand Marquis vehicles. Common features of all of these engines include throttled intake, spark ignition, and individual port fuel injection.

For the first two engines, lean-fueling combustion dynamics were studied with the engines mounted on test stands and connected to dynamometers. Fuel-to-air ratio was adjusted only with injected fuel pulse width, and throttle position was held constant to minimize changes to the air intake dynamics. Spark timing was set at stoichiometric fueling to achieve minimum ignition advance for maximum brake torque and minimum variance and was then held fixed at all fueling conditions investigated. The CFR dynamometer had both absorbing and motoring capability, allowing us to maintain a constant speed. The Quad-4 dynamometer had absorbing capability only, so that speed could only be maintained by reducing frictional load as the amount of fuel relative to inlet air was reduced. Nominal operating conditions for the CFR engine were 1000 rpm and a spark advance of 27° before top dead center (BTDC). Nominal operating conditions for the GM engine were 1200 rpm and a spark advance of 23° BTDC. The CFR residual cylinder gas fraction was varied by throttling the exhaust (i.e., more exhaust back pressure decreases the pressure ratio of intake to exhaust and increases the residual gas fraction). This allowed us to directly observe the effects of residual gas on the bifurcation dynamics.

Two different V8 engine configurations were used, one test-stand-mounted and one vehicle-mounted. Each of these engines was of the same basic type but not completely identical. Specifically, the test-stand engine was configured for a recent edition Ford Expedition (a sport utility vehicle), while the vehicle engine was installed in a 1994 Grand Marquis (a large sedan). The test-stand configuration included an absorbing/motoring dynamometer to maintain constant speed, and fueling adjustments were made by manipulating both throttle position and fuel injection to maintain constant torque. Nominal test-stand conditions for all fueling levels were 1200 rpm, 25 N-m torque, and 20° BDTC spark timing. The vehicle experiments were conducted with the engine idling 400-700 rpm (automatic transmission set in park position) at constant throttle and a constant nominal fuel-to-air equivalence ratio of 0.7 (with no EGR). While the objective of the test-stand experiments was to study the dynamic behavior of eight cylinders in detail, the vehicle experiment focused on demonstrating simultaneous feedback control implementation in eight cylinders using existing sensors and reprogramming of the electronic engine control (EEC) computer.

In all test-stand experiments, pressures in each cylinder were recorded at the rate of once per crank angle degree. Pressure data were later converted to integrated combustion parameters including net heat release (HR) and indicated mean effective pressure (IMEP) [7]. Whenever engine settings were changed (e.g., fuel adjustments made), data recording was delayed until the engine stabilized (i.e., reached "steady state"). All engine feedback controllers were shut off to minimize non-combustion artifacts, and operation was entirely open loop except for the dynamometer speed controllers. Data set sizes for one operating condition on the CFR and Quad-4 engines typically included 10,000 and 3,000 engine cycles, respectively. Data sets for the test-stand V8 were typically limited to 350 engine cycles (all eight cylinders recorded simultaneously) by the memory of the data acquisition system. Because of this limitation, multiple 350-cycle blocks were collected at each operating condition.

Measurements of the vehicle-mounted engine were made using standard sensors available on the car. Specifically, we used existing measurements from the crankshaft position encoder that produces a timing interval (the so-called PIP interval) to estimate changes in crankshaft acceleration. With the proper care, it is possible to correlate crankshaft acceleration to the corresponding energy release from each combustion event. This correlation allowed us to use estimated crankshaft speed fluctuations as a surrogate for direct combustion measurements. Based on the surrogate combustion measurements, cycle-by-cycle adjustments were then made to the fuel injection pulse to offset combustion instabilities. As described later, the successful implementation of control (by means of reprogramming the EEC computer) required overcoming complex interactions among the cylinders as well as constraining the net effect of all the control actions to leave the total mean fueling level unchanged. The only non-standard equipment in the vehicle experiments was a second interfacing computer, which was used to program and record output from the on-board EEC computer. As with the test stand, computer memory only allowed recording of approximately 350 contiguous cycles from all cylinders. Thus, repeated blocks of 350 cycles were recorded to improve statistics and gain information about long-term behavior.

EXPERIMENTAL RESULTS AND DISCUSSION

The primary experimental data for all test-stand experiments consisted of heat release time series for each cylinder at each operating condition. For the vehicle experiments, the primary data were estimated crankshaft acceleration time series. In the following discussion we describe the test-stand experimental results and then describe how feedback control was implemented in the vehicle.

CFR and Quad-4 Test Stand Experiments

Typical return maps for combustion heat release in the single-cylinder CFR engine are shown at four different fuel-to-air ratios in the top row of panels in Fig. 1. As has been reported previously, the heat release pattern near stoichiometric fueling appears as a noisy fixed point on the diagonal, and a characteristic bowed pattern of scattered points develops as the fuel-to-air ratio decreases. This transformation progresses through a series of modified bow shapes with increasingly lean fueling until finally, another noisy fixed point is reached and combustion is shut down.

For purposes of predictive control, we assume there is an underlying nonlinear map dominating the observed cycle-to-cycle heat release variations similar to our previously proposed physical model [5]. We are inspired by numerical experiments with this model, which demonstrate that the noise-free dynamics can be reconstructed by appropriately averaging noisy realizations of the return map. We want to increase the generality of our method, however, by not assuming *a priori* that our experimental data will be completely consistent with this specific model. Instead we want to use empirically derived maps that minimize model assumptions.

Based on the success of our model studies, we attempted to reconstruct approximations of the experimental maps by two methods: 1) least-square fitting of the observed heat release pairs to simple polynomial functions (i.e., $HR(i+1) = f[HR(i)] = a + b*HR(i) + c*HR(i)^2 \ldots m*HR(i)^n$) and 2) dividing the values of HR(i) into discrete bins and determining the average corresponding value of HR(i+1). In the first method, the appropriate polynomial order is determined by fitting the data with increasingly higher-order polynomials until the goodness of fit reaches a plateau. Typically, we find that the best polynomial order is 4-7, which is consistent with our numerical model studies. From the second method, we also obtain an estimate of map uncertainty based on the dispersion of the HR(i+1) values around the mean value.

Example fitting results are illustrated in the bottom two rows of panels in Fig. 1. The second row depicts results from polynomial fitting of the data in the top row, while the third row illustrates results from bin averaging. Both methods appear to give consistent results. Even though there is considerable noise in the original data, the general features of these estimated maps are similar to the noise-free version of our physical model [5]. Two interesting characteristics are shown in the two lower rows. First, stable fixed points for the noise-free maps are plotted as dark circles in the second row, illustrating a transition from period-1 to period-2 and then back toward period-1 at the leanest condition. Second, the estimated maps show a peak in HR(i+1) at low values of HR(i) and a sharp decline in HR(i+1) at lower HR(i). This latter feature is more prominent for experimental data than for our physical model.

Similar plots are shown in Fig. 2 for conditions where the residual in-cylinder gas fraction is increased. Comparing Fig. 2 with Fig.1, we observe, as expected, that higher residual fraction increases the degree of bifurcation at a given fuel-to-air ratio. Likewise, it is apparent that noise-free dynamics with higher-order periodicities and possibly chaos are possible. The latter can be seen more clearly in Fig. 3(b), which is an estimate of the low-dimensional component of the experimental bifurcation sequence shown in Fig. 3(a). The noise-free version of the bifurcation sequence was produced by estimating the noise-free map at each fueling level from 500 pairs of observed heat release values. Compared with Fig. 3(c), which is the noise-free bifurcation sequence predicted by the physical model, we note that the relative fueling range over which higher-order periodicities and chaos occur experimentally is larger than predicted by the model. This difference suggests that control strategies should be able to handle more complex behavior than simple periodic oscillations.

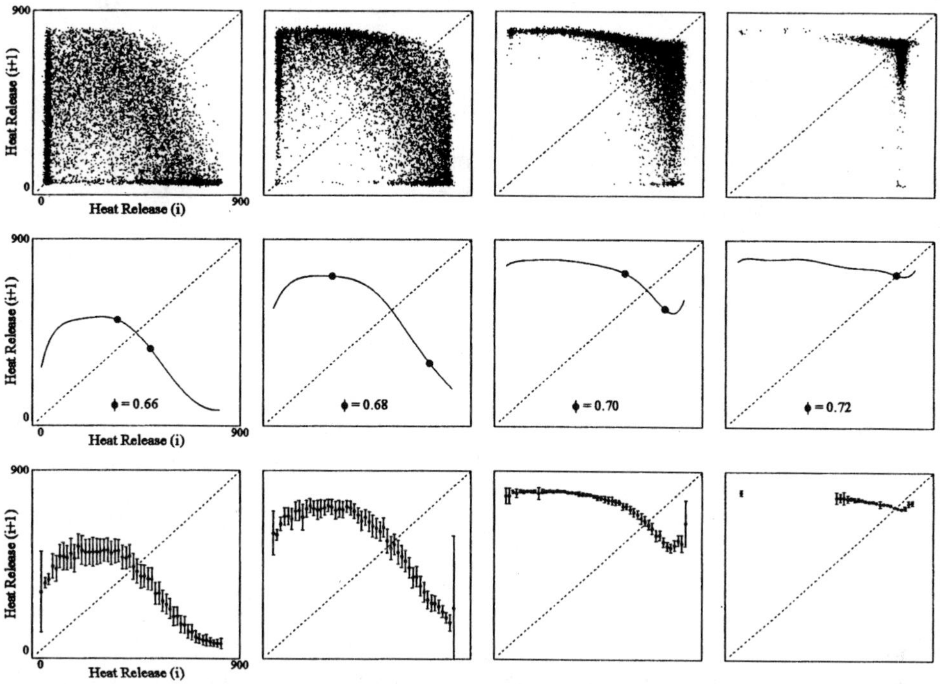

FIGURE 1. Heat release return maps for the single-cylinder CFR engine operating with a residual gas fraction of 0.09. Top row is raw data. Middle row is polynomial fit with stable fixed points indicated as solid circles. Bottom row is binned fit with 95% confidence limits.

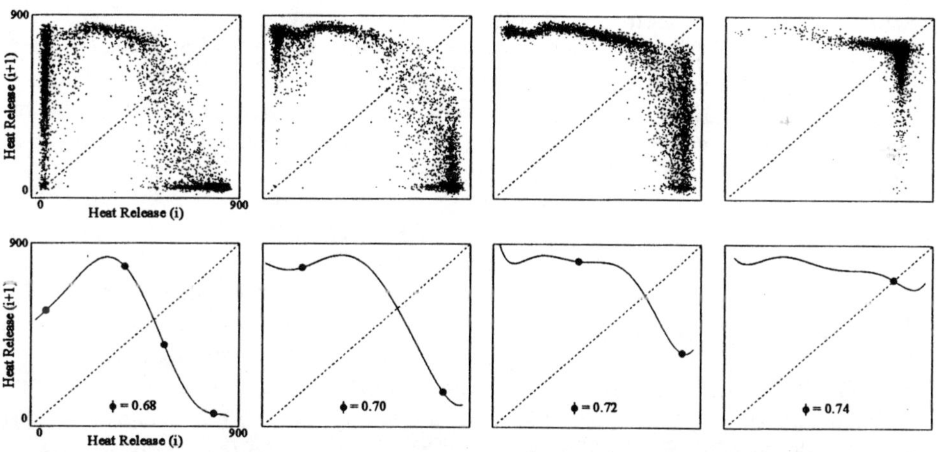

FIGURE 2. Heat release return maps for the CFR engine operating with a residual gas fraction of 0.16. Top row is raw data. Solid circles indicate fixed points of fitted map.

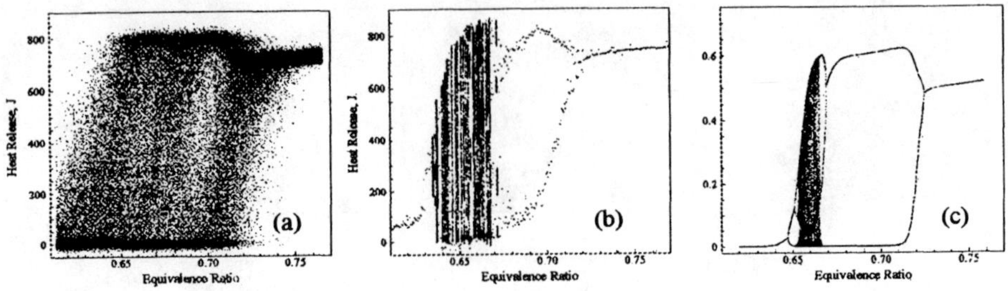

FIGURE 3. Bifurcation diagrams for the single-cylinder CFR engine operating with a residual gas fraction of 0.16. Part (a) is original experimental data; part (b) is reconstruction from polynomial map approximations; part (c) is physical model prediction without parametric noise.

Further analysis of the model and map differences has confirmed that the model assumptions are inadequate to account for the effect of reduced in-cylinder temperature on subsequent combustion events. Interestingly, this effect is only apparent in the map shape in the lower 10-15% of the heat release range, and yet this small difference in shape seems to be the major cause of more rapid experimental bifurcation. While an improved physical model can provide additional fundamental insight, it is also clear that empirically derived maps should be suitable for control.

Figure 4 illustrates experimental return maps and fitted functions for multi-cylinder data from the Quad-4 engine. Here we observe that, although the same nominal amount of fuel is fed to each cylinder, there is clearly a varying degree of bifurcation. This cylinder-to-cylinder difference reflects the intake air imbalance problem common to all multi-cylinder engines. Due to inevitable differences in the path between each cylinder and the air intake manifold, there is always some air imbalance at any given operating condition. As engine speed and load change, complex flow dynamics and acoustic interactions shift air distribution in unpredictable ways. Because fuel injection occurs in the intake port to each cylinder, air-fuel mixing is also uneven. Thus it may not generally be possible to predict *a priori* how individual cylinders will differ. For the condition in Fig. 4, it appears that the cylinders farthest from air intake (cylinders 3 and 4) receive the least amount of air. There also appear to be differences in the level of parametric noise experienced by each cylinder.

V8 Test-Stand Experiments

Typical return maps for lean fueling in the test-stand-mounted V8 engine are illustrated in Fig. 5. Note again the considerable variation from cylinder to cylinder. Simultaneous measurements of all cylinders also permitted investigation of possible cylinder-to-cylinder interactions (i.e., synchronization). One approach we used for synchronization detection was cross correlations between cylinder pairs as illustrated in Fig. 6(a). For one pair (cylinders 1 and 5), a positive correlation developed when the as-injected fuel-to-air equivalence ratio was decreased to 0.66. Likewise, a strong anti-correlation developed for the cylinder pair 1-3. There was little or no correlation for the pairs 1-6 and 2-6. When all cylinders are bifurcated to some degree, one expects occasional apparent correlations or anti-correlations due to chance.

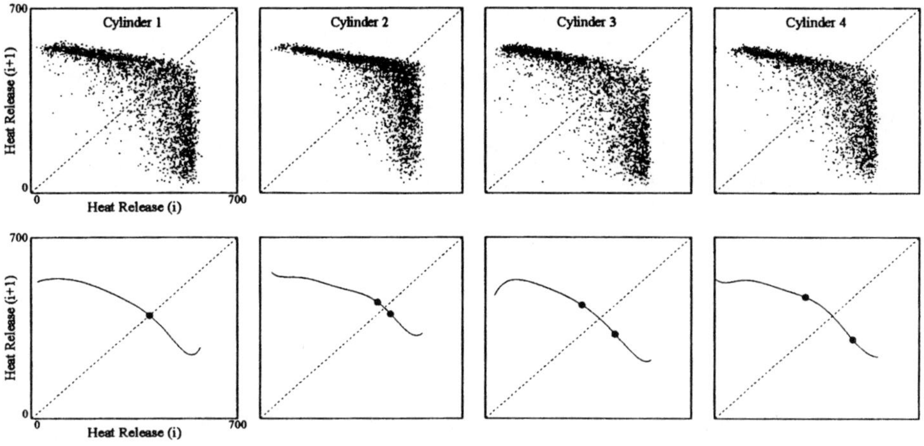

FIGURE 4. Heat release return maps for the GM Quad-4 engine with a fuel-to-air ratio of 0.586. Top is raw data. Bottom is polynomial approximation. Solid circles indicate fixed points of fitted map.

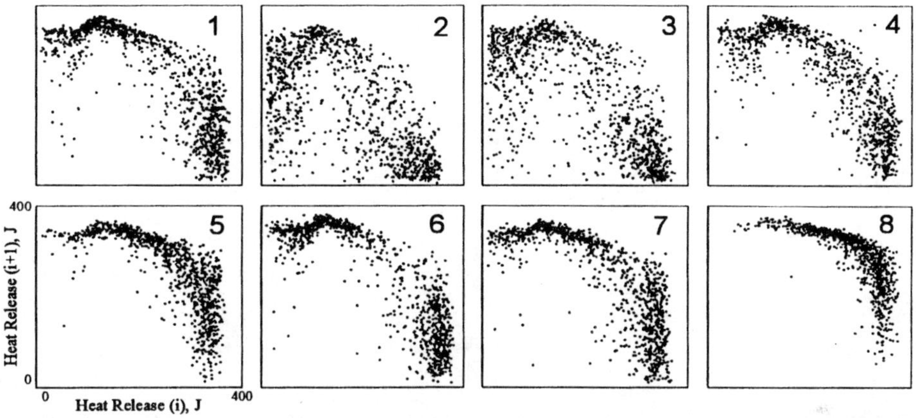

FIGURE 5. Heat release return maps for the test-stand-mounted V8 engine at a fuel-to-air ratio of ϕ= 0.66. Numbers in panels refer to physical location of cylinder. Firing order was 1-3-7-2-6-5-4-8.

In this case, the cross-correlation plot was averaged for three discontinuous segments of 350 cycles, making the likelihood of a chance artifact remote. Additional analyses of many similar simultaneous data sets compared to truly independent heat release series provide convincing evidence for persistent positive and negative correlations between cylinders.

We find that symbolic methods are also useful for discerning synchronization. In one version of this approach, heat release values from each cylinder are converted into 1-bit symbols (i.e., 0 or 1) based on whether each value falls above or below the median for that cylinder. Symbol "words" (i.e., short symbol sequences loosely analogous to time-delay embedding vectors) are then constructed according to:

$$S2(j) = [X(j-1), Y(j-1), X(j), Y(j), X(j+1), Y(j+1)] \quad (1)$$

where S2(j) is the symbolic "state" of the cylinder pair at engine cycle j, and X(j) and Y(j) are the symbolic heat release values at cycle j for the first and second cylinder pair members, respectively. For convenience, we refer to each particular symbol sequence according to a unique index number between 0 and 63 and defined by:

$$Index(S2) = X(j-1)*32 + Y(j-1)*16 + X(j)*8 + Y(j)*4 + X(j+1)*2 + Y(j+1) \quad (2)$$

Figure 6(b) illustrates a "synchrogram" from above scheme for two selected cylinder pairs. Engine cycle number is on the horizontal axis, while the current sequence index number is on the vertical axis. When combustion oscillations develop in both cylinders, peaks typically appear at sequence indices 12, 25, 38, and 51, respectively corresponding to symbol sequences **[0,0,1,1,0,0]**, **[0,1,1,0,0,1]**, **[1,0,0,1,1,0]**, and **[1,1,0,0,1,1]**. When the cylinders remain unsynchronized, the heights of these target peaks occur with equal probability and there is no persistent relationship between the cylinders. However, when the cylinders become synchronized (Fig. 6(b)), peaks 12 and 51 (for correlated behavior) or peaks 25 and 38 (for uncorrelated behavior) become persistently dominant. As seen in Fig. 6(b), patterns of correlation or anti-correlation can persist for long periods.

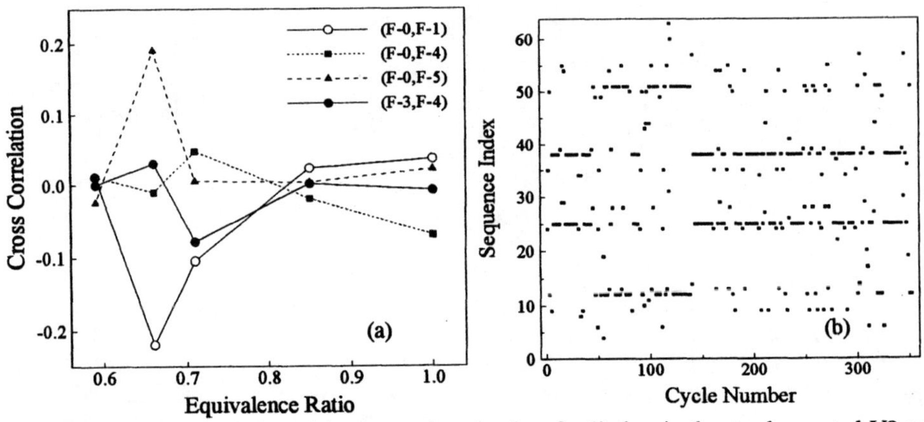

FIGURE 6. Cross correlation (a) between selected pairs of cylinders in the stand-mounted V8 engine as fuel-to-air ratio is changed. Symbol synchrogram (b) illustrating a predominantly anti-correlated cylinder pair in the test-stand-mounted V8 engine at $\phi = 0.66$. Persistent dark bands indicate synchronization.

From observations over many experiments, we have observed that synchronization occurs simultaneously with the onset of bifurcations in two or more cylinders. When one cylinder becomes strongly bifurcated, it can act as a "driver", stimulating other cylinders to bifurcate. The most easily recognizable form of synchronization is when pairs of cylinders develop persistent correlation or anti-correlation with each other. Synchronization appears to occur episodically, so that a certain patterns can persist for long periods and then suddenly shift to a different pattern. The mechanisms for cylinder-to-cylinder coupling are not well understood, but it appears that acoustic and flow effects in the intake and exhaust air manifolds and fuel "spillover" (described later) are all involved.

V8 Vehicle Experiments

Figure 7 illustrates typical uncontrolled crankshaft acceleration return maps for all cylinders in the Grand Marquis when the engine was idling at a fuel-to-air equivalence ratio of 0.77. The labels "Sync 0", "Sync 1", etc. refer to the temporal firing order of the cylinders. The average acceleration due to all cylinders is depicted in the center return map and the time series at the bottom. Even though crankshaft acceleration measurements are less accurate than direct combustion measurements, they appear to be quite adequate for detecting combustion bifurcations. As with previous multi-cylinder observations, there is considerable variation from cylinder to cylinder. It appears that some of the cylinders (e.g., 5, 6, and 7) are fully bifurcated into a noisy period-2 condition. One or two of the cylinders (e.g., 1 and 4) appear to be even further bifurcated, possibly into higher-order periodicities or chaos.

A simple (but naïve) approach for controlling combustion oscillations is to perturb the fueling for each cylinder to be out of phase with dominant high-low combustion fluctuations. Thus if the current acceleration is low, one anticipates that the next acceleration will be high and reduces the fueling by an appropriate amount. If one targets the period-1 fixed point as the control objective, the simplest strategy would be to make the fueling adjustments proportional to the current deviation. The only major constraint needed would be to make sure that the overall fueling perturbations sum to zero so that the average fuel-to-air ratio remains unchanged. Our experience shows that, unfortunately, such a simple approach is not effective because of cylinder-to-cylinder interactions. Typical experimental results are illustrated in Fig.8 where the variation in rms crankshaft speed for the Grand Marquis is plotted against the magnitude of the applied (anti-correlated) feedback gain. As the gain is increased from zero, the rms magnitude continues to increase, indicating that this control increases combustion instability.

We now understand that a more sophisticated approach is needed for multi-cylinder control. It is well known that fuel injection to one cylinder can influence other cylinders through the so-called "spillover" effect. Spillover is due to the considerable fraction of injected fuel that wets the walls of the intake port and does not become completely vaporized and inducted into the intended cylinder. This residual fuel then becomes available for induction into successively fired cylinders and creates an unintended fuel spike in those cylinders. Also, because the cylinders tend to synchronize, one must be careful that the control actions themselves don't become globally correlated. For an improved control strategy, we developed a modified prediction for the next acceleration event for each cylinder based on consideration of the current acceleration, the previous fueling perturbation, and the previous fueling perturbations made to all the other cylinders. The form of this prediction is:

$$A_i(k+1) = a*A_i(k) + b_i*\Delta f_i(k+1) + b_{i-1}*\Delta f_{i-1}(k) + b_{i-2}*\Delta f_{i-2}(k)\ldots\ b_{i-7}*\Delta f_{i-7}(k) \quad (3)$$

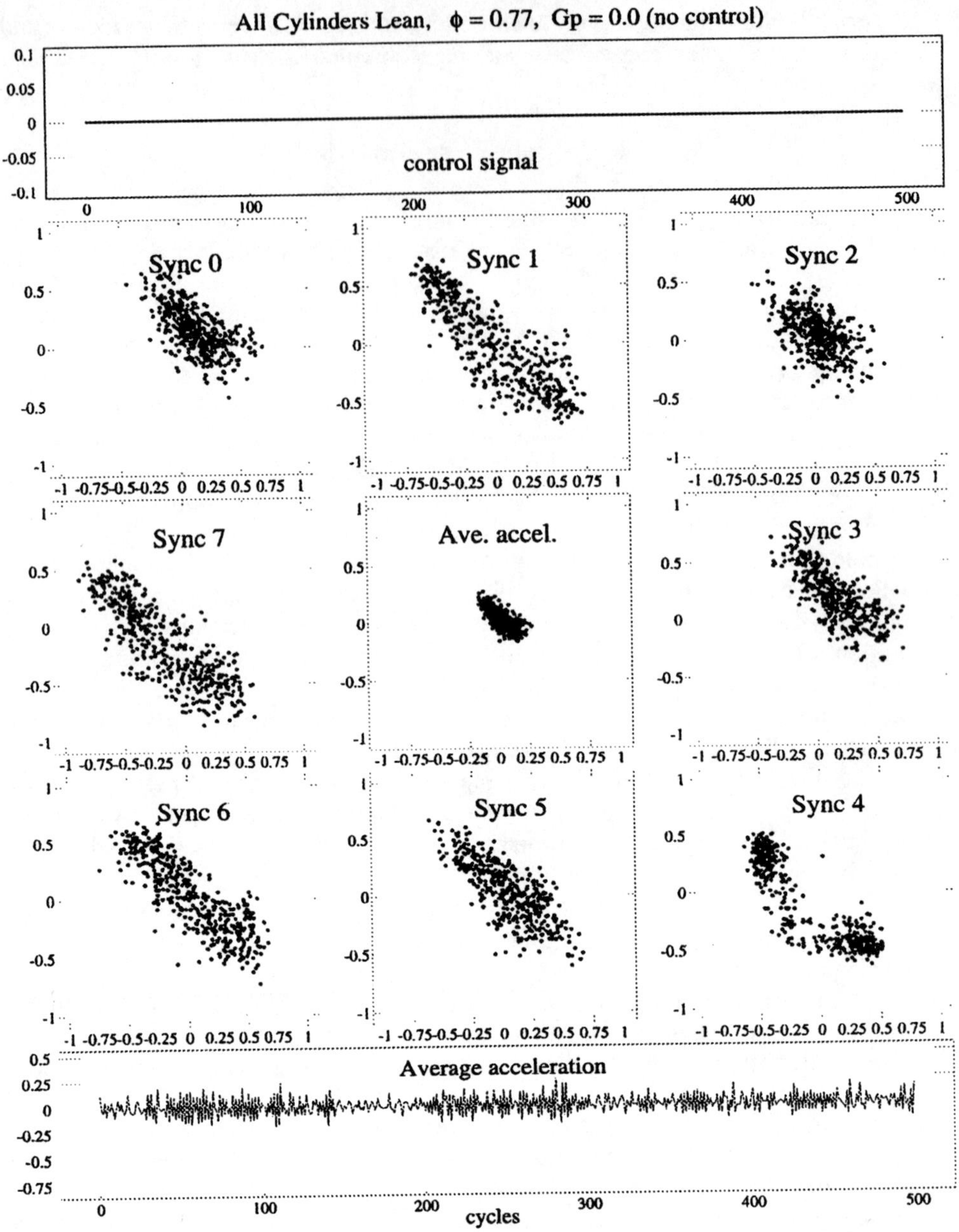

FIGURE 7. Return maps for crankshaft acceleration in uncontrolled vehicle-mounted V8 engine.

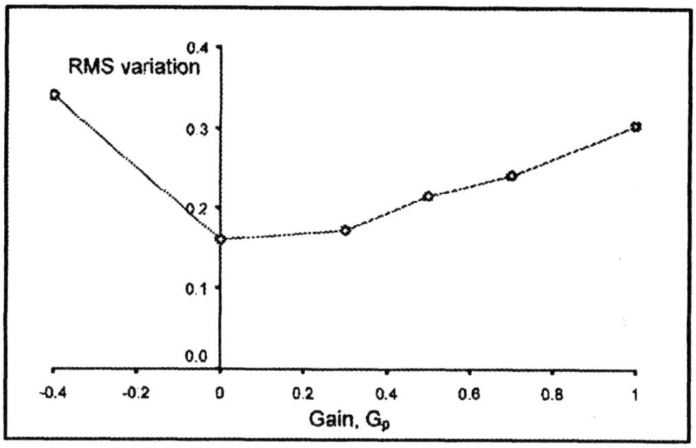

FIGURE 8. Effect of simple proportional feedback on overall engine speed variations in the test vehicle. Minimum fluctuations occur for zero feedback gain (i.e., no control).

where $A_i(k)$ is the acceleration in cylinder i for cycle k, $\Delta f_i(k)$ is the fueling perturbation made to cylinder i during cycle k, and the a and $b_i \ldots b_{i-7}$ are empirically determined coefficients reflecting the single-cylinder map and spillover response, respectively. When we wish to determine $\Delta f_i(k+1)$ for any given cylinder, we have the previous acceleration in that cylinder ($A_i(k)$), all the previous fuel perturbations made up to that moment ($\Delta f_{i-x}(k)$), and the target acceleration for the unstable period-1 fixed point ($A_i(k+1)$). The above equation can then be solved directly to obtain $\Delta f_i(k+1)$.

To guard against control synchronization, we monitor the net acceleration phase (determined from the difference between the current acceleration and a running average of accelerations) averaged over the past eight events. An additional negative feedback correction is then applied to the initial fueling perturbation determined from Eqn. (3) (i.e., a further adjustment is made to offset global system oscillations). This correction also helps keep the average fueling level constant. We have confirmed that, without this negative feedback, large global oscillations inevitably result.

The effectiveness of the improved feedback scheme is illustrated in Fig. 9. Now we observe a marked decrease in the combustion oscillations and engine speed fluctuations. We believe this degree of improvement is impressive considering that a single feedback gain has been used for all cylinders and no attempt has been made to explicitly account for the cylinder-to-cylinder differences. Figure 10 further illustrates the effect of the feedback gain on engine speed. Unlike the previous strategy (Fig. 8), there is now a clear minimum rms achieved for a finite feedback. (Note that a global rms minimum occurs for Gp =1.0, but this is not a desirable condition because some of the cylinders are completely shutdown). We expect that greater improvements are possible with adjustable gains for each cylinder and/or additional control actions that remove time-average cylinder imbalance.

These results demonstrate that active control is clearly possible for the combustion instabilities in lean-fueled spark-ignition engines. Additional issues need to be addressed for commercial implementation including the large cylinder-to-cylinder differences that occur in multi-cylinder engines, the highly transient conditions under which most engines actually operate, and the use of other sensor measurements as control inputs (e.g., measurements of cylinder coupling and parametric noise).

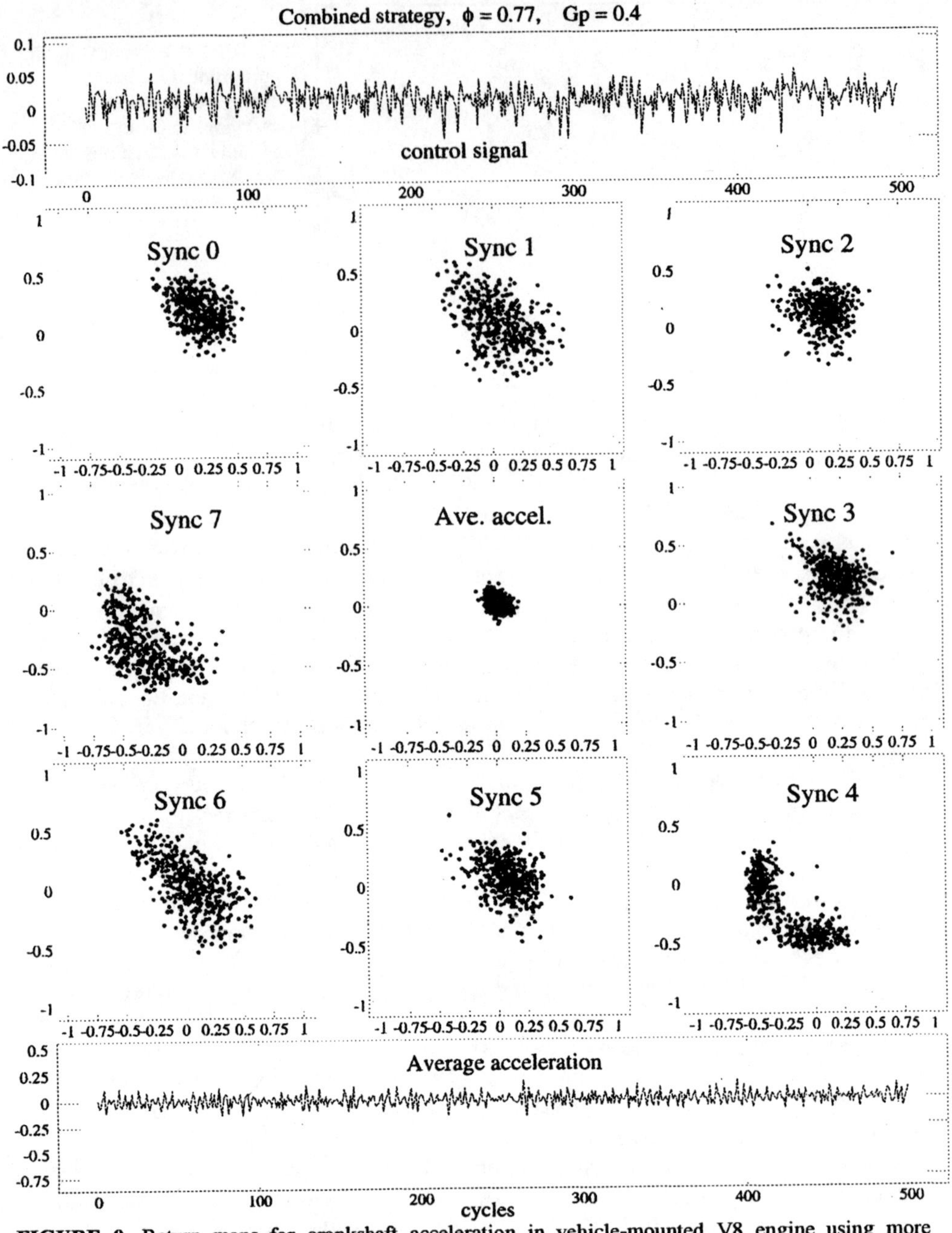

FIGURE 9. Return maps for crankshaft acceleration in vehicle-mounted V8 engine using more sophisticated feedback control strategy.

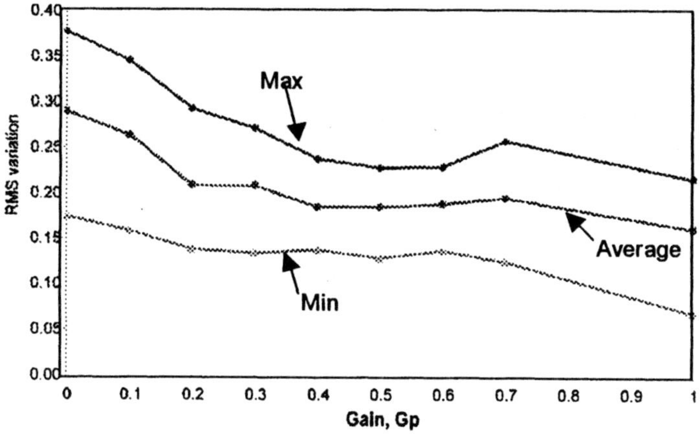

FIGURE 10. Effect of more sophisticated feedback strategy on engine speed variations in the test vehicle. The bottom and top lines correspond to the best and worst cylinders, respectively. Minimum variations occur at a feedback gain of 0.5- 0.6.

ACKNOWLEDGMENTS

This work was funded by the U.S. Department of Energy and Ford Motor Company under Cooperative Research and Development Agreement ORNL-95-0337.

DISCLAIMER

REFERENCES

1. Grunefeld, G., Beushausen, V., Andresen, P., and Hentschel, W., "A major origin of cyclic energy conversion variations in SI engines," Society of Automotive Engineers (SAE) Technical Paper No. 941880 (1994).
2. Hancock, M.S., Buckingham, D.J., and Belmont, M.R., "The influence of arc parameters on combustion in a spark-ignition engine," SAE Technical Paper No. 860321 (1986).
3. Stevens, S.P., Shayler, J., and Ma T.H., "The impact of combustion phasing on cycle-to-cycle performance of a spark ignition engine," SAE Technical Paper No. 950687 (1995).
4. Daw C.S., Finney, C.E.A., Kennel, and Connolly, F.T., "Cycle-by-Cycle Combustion Variations in Spark-Ignited Engines," in Proceedings of the 4th Experimental Chaos Conference, edited by M. Ding, W. Ditto, L. Pecora, M. Spano, and S. Vohra, World Scientific, Singapore, 1998, pp 329-340.
5. Daw, C.S., Kennel, M.B., Finney, C.E.A., and Connolly, F.T., *Phys. Rev. E*, **57:3**, pp 2811-2819 (1998).
6. Wagner, R.M., Drallmeier, J.A., and Daw, C.S., *Int J Engine Research*, 1:4, pp 301-320 (2000).
7. Heywood, J.B., *Internal Combustion Engine Fundamentals*, McGraw-Hill, New York (1988).

Criticality on Flame Spreading Over Liquid Fuels

Eugenio Degroote*and Pedro L. Garcia-Ybarra†
Universidad Politécnica de Madrid - E.U.I.T.Agrícolas
C/Ciudad Universitaria, s/n, Madrid-28040, Spain

Abstract Flame spreading over liquid fuels has been experimentally characterized for four different alcohols in a complete range of temperatures. Five different spreading regimes (and four critical temperatures T_1, T_2, T_3, T_4 have been observed. The critical temperature T_1 has been described as steady state bifurcation, T_2 resembles a transcritical bifurcation, while T_3 corresponds to a Hopf bifurcation, and T_4 is a homoclinic connection.

INTRODUCTION

Flame spreading over liquid fuels is accompanied by liquid motions near the flame tip. The temperature gradient on the liquid fuel surface produces surface tension gradients that generate liquid fuel motions in front of the flame. Flame propagation, which is controlled by heat diffusion in the condensed phase (as for solid fuels) is, for liquid fuels, *assisted* by heat convection in the condensed phase. This phenomenon can bring new flame spreading regimes, depending on the coupling of both phases (liquid and gas). This coupling can even be unsteady, producing then an oscilatory behavior. The goal of this work is to bring some light to this problem by characterizing the different regimes of flame spreading over liquid fuels experimentally observed, as well as all the critical transition temperatures.

*e-mail: edegroote@agricolas.upm.es
†e-mail:pg.ybarra@ciemat.es

CP622, *Experimental Chaos: 6th Experimental Chaos Conference*, edited by S. Boccaletti et al.

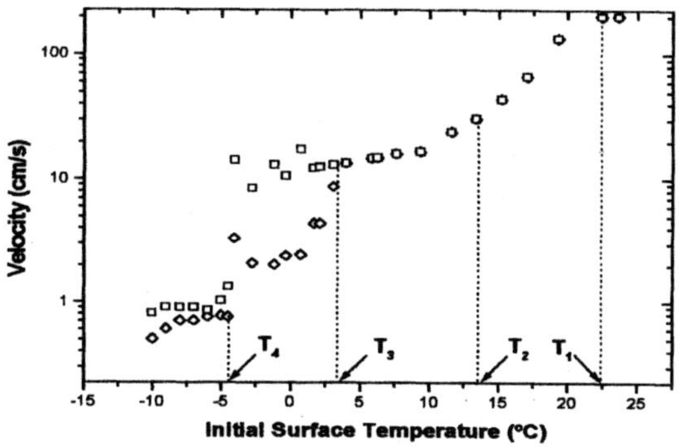

Figure 1: Bifurcation diagram for the flame spreading over liquid ethanol for the 40cm long channel. Squares are the maxima spreading velocities (v_{fmax}); diamonds are the minima spreading velocities (v_{fmin})

EXPERIMENTAL SETUP

To analyze the characteristics of flame spreading over liquid fuels a long channel, that contains the fuel, has been prepared. The initial fuel surface temperature could be kept uniform along the horizontal with a refrigerant circuit. The gas mixture could be ignited at one end of the channel and the spreading observed until the flame reached the other end. Two different duralumin channels and four alcohols were used (methanol, ethanol, 1-butanol and 2-propanol). The first channel was 2.5 cm wide, 40 cm long and 4 cm deep; the second channel was 3.4 cm wide, 100 cm long and 1.5 cm deep. Two techniques were set up to analyze the flame spreading:

- An array of eight Cr-Al thermocouples ($\phi \approx 25\mu m$) disposed on the fuel surface at regular intervals along the channel measure surface liquid temperature.
- A video camera recorded the flame front evolution in the central part of the channel.

The voltage generated by the temperature differences between the two welds of each thermocouple was amplified and collected on a computer through an AD/DA acquisition card. The arrival of the flame to each thermocouple could be easily detected. Also, from the video recording, a composite image of the flame front

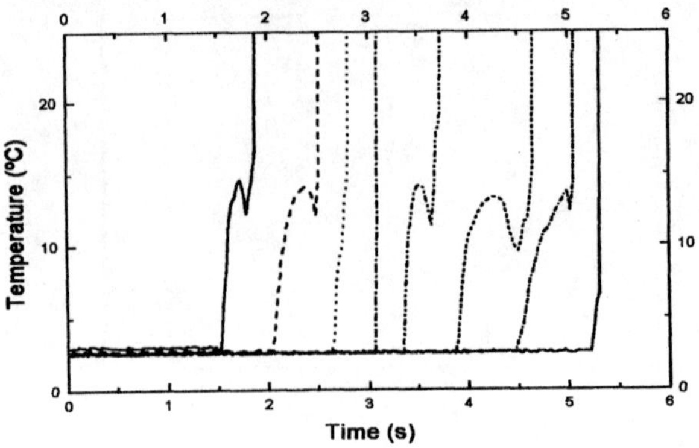

Figure 2: Evolution of the thermocouples readings(propanol $T_\infty = 2.4^0C$

position $x_f(t)$ could be obtained. This line can be used to derive the instantaneous flame spreading velocity profile $v_f(t)$ (more details of the experimental setup can be found in [1]).

RESULTS

Using this experimental setup, the maximum and the minimum spreading velocity of the flame can be represented as a function of the initial liquid surface temperature T_∞. The plot of these extreme spreading velocities for methanol is shown in figure 1. Five different spreading regimes are clearly identified:

- For temperatures $T_\infty > T_1$, flame spreading is uniform and nearly constant, with flame propagation velocities of order $10^2\ cm/s$. In this case, , the fuel vapor pressure is so high that it leads to a flame spreading regime purely controlled by the gas phase, where the presence of the liquid fuel is ignored by the flame.
- For $T_2 < T_\infty < T_1$, flame spreading is uniform. T_2 is a value close to the flashpoint temperature of the liquid fuel. A *solid fuel-like regime* of uniform flame spreading velocity v_f controlled by heat diffusion in the condensed phase appears. Flame velocities vary from order $10^2\ cm/s$ to approximately $10\ cm/s$, decreasing almost linearly with decreasing temperatures, with a slope in the $T_\infty - v_f$ diagram of the order of $10\ cm/s \cdot^0 C$.

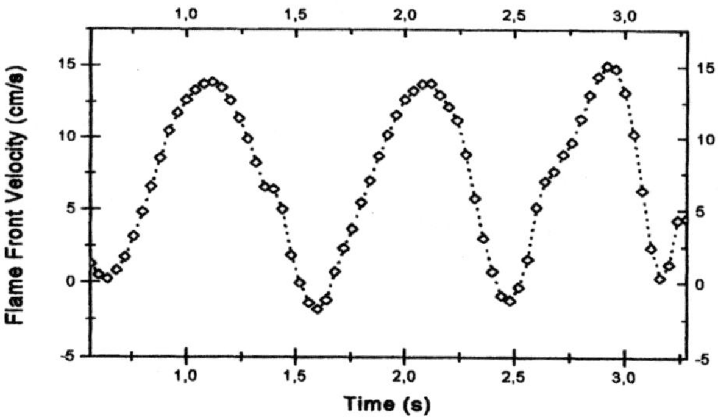

Figure 3: Temporal evolution of the flame spreading velocity in the pulsating regime(ethanol $T_\infty = 2.0^0C$

- For temperatures $T_3 < T_\infty < T_2$, , flame spreading is still uniform, but the slope is in this case of order $1\,cm/s \cdot {}^0C$, then showing a sharp transition with respect to the preceding regime. The temperature T_2 corresponds to the point where flame spreading, controlled by heat diffusion in the condensed phase, turns to flame propagation *assisted by heat convection* in the condensed phase. (assisted regime).
- For temperatures $T_4 < T_\infty < T_3$, , an oscillatory regime is observed for v_f (pulsating regime). For a given temperature T_∞, flame spreading velocity oscillates between the values v_{fmin} and v_{fmax} with a characteristic period P.
- For temperatures $T_\infty < T_4$, flame spreading is almost constant, with flame velocities close to $1\,cm/s$ (pseudouniform regime).

Typical plots of the pulsating regime are shown in figures 2,3; figure 2 corresponds to the temporal evolution of the thermocouples for propanol; figure 3 is the flame front velocity. The appearance of the pulsating regime is related with the existence of a vortex structure in the liquid fuel ahead of the flame (a complete description of the pulsating process is given in [1]). Phase diagram is depicted in figure 4.

CRITICAL TEMPERATURES T_1, T_2, T_3, T_4

The critical temperatures T_1 resembles a steady state bifurcation point . The fuel temperature T_2 can be considered as a bifurcation point between two steady

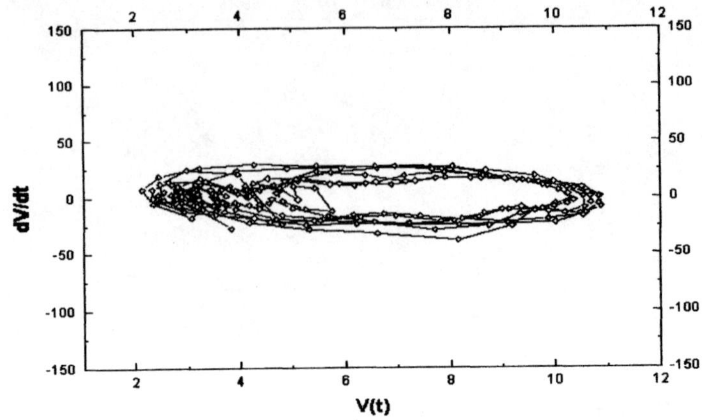

Figure 4: Phase diagram corresponding to ethanol (100 cm long channel).

states that (within the the experimental accuracy of our system) show a linear behavior with different slopes. Therefore, T_2 should be considered a *transcritical bifurcation*. For temperatures ($T_2 < T_\infty < T_1$) flame spreading proceeds similarly to the flame spreading on a solid fuel, and thermocapillary convection is negligible. For $T_3 < T_\infty < T_2$ a slight temperature augmentation preceding the flame arrival is detected, indicating the existence of a *preheated region* of warm liquid fuel ahead of the flame. The characteristic length L vanishes close to T_2, whereas, near T_3, it is close to 1 cm. Therefore, the warming of liquid fuel ahead of the flame reduces heat losses, increasing flame propagation velocity. This mechanism results now to be the driving one, producing then an abrupt variation in the slope (of one order of magnitude) when the temperature T_2 is crossed.
Finally, for fuel temperatures values $T_\infty < T_3$, the energy balance just described in the liquid fuel becomes unstable and flame spreading pulsation occurs for $T_4 < T_\infty < T_3$. A limit cycle appears at $T_\infty = T_3$, as a *Hopf bifurcation* from the thermocapillary assisted regime, corresponding to $T_3 < T_\infty < T_2$ (more details can be found in [1]).

The evolution of the pulsation period can be seen in figure 5. The data correspond to all the alcohols used in our experiments. For every fuel used in our experiments, there is a logaritmic dependence between the oscillation period P and T_∞:

$$\frac{P-a}{b} = ln\left(\frac{T_\infty - T_4}{T_4}\right)$$

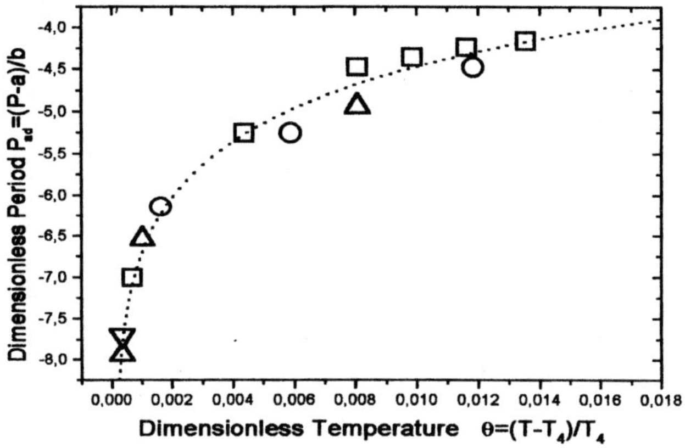

Figure 5: Period of flame spreading oscillation *vs* Initial Surface Temperature. The data corresponds to *all* the alcohols used in our experiments.

that presents a divergence for $T_\infty = T_4$. For lower temperatures ($T_\infty < T_4$) a new steady state regime appears; both states collision for $T_\infty = T_4$. Therefore, T_4 should be considered as *homoclinic orbit.*

CONCLUSION

Five diferent spreading regimes have been experimentally observed. Four transition points T_1, T_2, T_3, T_4 have been characterized. The study of the channel dimensions, can be the subject of future work.

References

[1] Degroote, E., Garcia-Ybarra, P.L., *Eur. Phys. J. B*, **13**, 381-386, (2000).

[2] Akita, K., *Fourteenth Symposium (International) on Combustion (The Combustion Institute, Pittsburgh, 1973)*, 1075-1083, (1973).

[3] Mackinven, R., Hansel, J., Glassman, I., *Comb. Sci. Tech.*, **1**, 293, (1970).

[4] Ross, H.D., *Prog. Energy and Combust. Sci.*, **20**, 17-63, (1994).

Dynamics of Pulsating and Cellular Flames Using a High-Speed, High Sensitivity Camera

Michael Gorman
Department of Physics
University of Houston
Houston, TX 77204-5506

Abstract. A high-speed, high sensitivity camera has been assembled to record the spatiotemporal dynamics of pulsating and cellular flames at frequencies above 15 Hz, the Nyquist frequency of standard videotape. A high-speed CCD camera has been equipped with an image intensifier to record the dynamics of low-intensity, high frequency pulsating flames and to combine both electronic and video data on each frame. The display of stability boundary diagrams of both pulsating and cellular flames will also be discussed.

INTRODUCTION

This paper will discuss technical issues related to the study and characterization of physical systems that exhibit spatiotemporal dynamics in which emitted electromagnetic radiation is a dynamical variable. Such systems include semiconductor plasmas[1], combustion systems[2], the pulsating aurora, lightning, gas discharges, NMR imaging, and coated biological systems. In these cases the experimenter is limited only by the resolution of the sensor in determining the spatial characteristics of the dynamics.

DYNAMICS OF PREMIXED FLAMES

Our experiments[4] on flat, premixed flames stabilized on a porous plug burner have found a large (>70) number of states of pulsating and cellular flames, each with its own distinctive spatiotemporal dynamics. These states form a set of examples that can be used to develop and test techniques to analyze such dynamics.

CP622, *Experimental Chaos: 6th Experimental Chaos Conference*, edited by S. Boccaletti et al.

An instructive example of pulsating dynamics is seen in the operation of a small airplane engine. After take-off and achieving cruising altitude, the pilot reduces the fuel until the engine exhibits a put-put-put sound. He/she then adds fuel until these oscillations stop. The operating point of the engine has been set just outside the region of instability that precedes the extinction limit. This point is close to the point of maximum operating efficiency of the engine. The efficiency of jet engines for propulsion and of gas turbine combustors for electrical power generation could benefit if the operating point could be pushed to leaner mixtures by controlling the combustion front in dynamical regimes. Another benefit of operating as close as possible to the lean-burn limit is the decrease in pollution, which is a strong function of the temperature.

A premixed flame stabilized on a circular porous plug burner exhibits both pattern formation (cellular flames) and oscillations (pulsating flames). Cellular flames are observed in rich, heavy hydrocarbon-air mixtures in which mass diffusivity of the deficient species (the oxydizer) dominates. In dynamics states of cellular flames[5] a uniform, flat, circular flame front becomes unstable to the formation of one or more concentric rings of cells, with brighter, hotter regions of excess oxydizer bounded by darker, cooler regions with excess fuel. At certain values of the control parameters the ordered states bifurcate to dynamic states in which the cells move in collective motion. Most of this dynamics has frequencies below 6 Hz.

In dynamic states of pulsating flames[6] the flat flame front executes a complex motion in which the entire flame front moves in collective motion. Pulsating flames are observed in light hydrocarbon-air mixtures and in hydrocarbon-oxygen mixtures in which thermal diffusivity (effectively, the heat lost to the burner) dominates. Two kinds of states are observed: standing wave states in which the motion resembles the modes of a vibrating drumhead and traveling wave states, in which the the flame front exhibits rotating spiral motion or which breaks up into a number of bright regions that rotate. Most of this dynamics occurs at frequencies greater than 15 Hz.

Experimental Considerations

Standard video in the US (Europe) operates at the line frequency of 60 Hz (50 Hz). Each frame is composed of two interlaced fields resulting in a frame rate of 30 Hz (25 Hz). This situation yields an Nyquist frequency of 15 Hz (12.5 Hz), which is the highest frequency physical phenomenon that can be properly recorded using such video. Phenomena that occur at higher frequencies are aliased. The most familiar example of this effect is the apparent rotation of the wheels of a car (or any other moving vehicle) in a direction opposite to its translational motion due to the "folding back" of the dynamics at the sampling frequency.

The flicker frequency of the eye is typically 12 Hz. Uniform motion above that frequency can appear continuous; nonuniform motion typically appears too complex to understand. One way to observe dynamics at higher frequencies is to view the flame front through the wheels of a rotating chopper blade, effectively stroboscopic viewing. As the frequency of the blade is tuned to a characteristic frequency of the motion, the spatial characteristics of that single mode can be identified. However, it is quite difficult to obtain a record of the motion or to display that motion so that others can view it.

The conditions of our experiments necessitate the use of an image intensifier a in addition to a high-speed camera.. Three factors contribute to the low intensity available to observe the dynamics. First, as the frequency of the dynamics increases, the frame rate of the camera must also increase, thereby reducing the amount of light available per frame. Second, our experiments are conducted at low pressure, which further reduces the emitted light intensity because the number of chemical reactions is directly proportional to the number of molecules. (This low pressure operation has the beneficial effect of eliminating the ambient oxygen that causes a diffusion flame so that the premixed flame front can be studied as an isolated dynamical system A standard

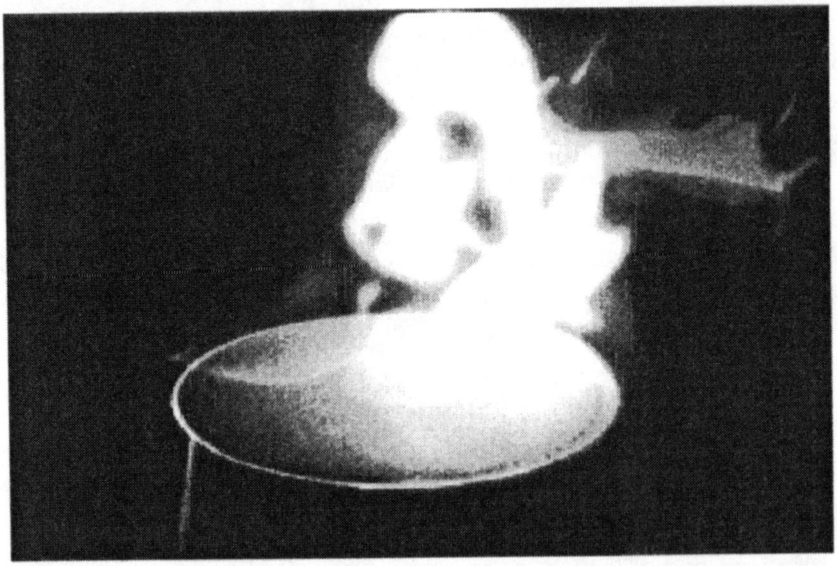

FIGURE 1. Initial stages of ignition for a circular porous plug burner.

Bunsen burner has an inner, hotter, conical premixed flame front and an outer, cooler diffusion flame.. The latter is caused by the excess fuel in the rich mixtures, which travels through the premixed front without reacting with the premixed oxydizer. It then combines with the ambient oxygen.) Third, pulsating dynamics is observed near the extinction limits, where either the fuel or oxydizer is deficient in the premixed gas, still further reducing the number of reactions relative to those of stochiometric mixtures.

The CCD camera is a Redlake PCI Motionscope capable of 1000 frames/second. A Roper image intensifier, which is coupled to the CCD through a lens system, has an adjustable gain up to 100,000. The interface between the camera and the computer is constructed so that a National Instruments Data Acquisition System can record up to 16 channels of electronic data and sychronize this data with the video frames.

Figure 1 shows a frame of videotape, taken at 500 frames/second, of the porous plug burner shortly after ignition by a heating element. The flat, premixed flame that sits approximately 2 mm above the burner surface, has spread over half the burner surface and the plume-shaped diffusion flame, which appears brighter and ribbon-like, extends above the burner.

Display of Electronic Stability Boundaries

A main theme of our experimental investigations is the observation of many examples of spatiotemporal dynamics, which can be used to develop and implement techniques to analyze complex dynamics. To that end, we areinvestigating different burner geometries, different fuel mixtures, and different burner orientations. We have also constructed a linear burner, a ring burner (width smaller than the size of a cell), an annular burner (width greater than the size of a cell). We also have designed a spherical burner and a cylindrical burner that are currently under construction.

Most of our experiments have used isobutane-air and propane-air mixtures to study cellular flames. However, mixtures using butane, pentane and ethane fuels will also be used. Substituting oxygen for air results in a dramatic change in the dynamics. cellular flames, which require the lighter oxydizer to diffuse relative to the heavier fuel, cannot form in oxygen mixtures.

The advent of electronic journals, such as the SIAM Journal of Applied Dynamical Systems and the Journal of Nonlinear Science, make it possible to display the complexity of this dynamics. Stability boundary diagrams[7] can now be constructed in which the dynamics at each point in parameter space is represented by a video clip. Figure 2 shows a small part of such a diagram for a propane-air cellular flame at 1/2 atmosphere. Three dynamical regions are portrayed: a stable 10/4 state with ten cells in the outer ring and four cells in the inner ring, an intermittently ordered state[8] just below the 10/4 ordered state; and an intermittently ordered state just above a stable 10/3 ordered state (not shown). Each point on the stability boundary diagram is clickable, which results in an MPEG movie to appear.

The operation of our experiments generates so much data that it is prohibitive for the experimenter to write down the control parameters corresponding to each video clip and correlate the electronic and video data after the experiment. The control parameters are recorded in two ways: in long runs on videotape, a split screen is used to combine the video output from the flow controller with the video of the flame dynamics; and in (short) runs with the high speed camera (and computer), the National Instruments board creates a companion file for the flow rates and any other electronic data--such as pressure, light intensity, or applied voltage—that is correlated with each electronic frame.

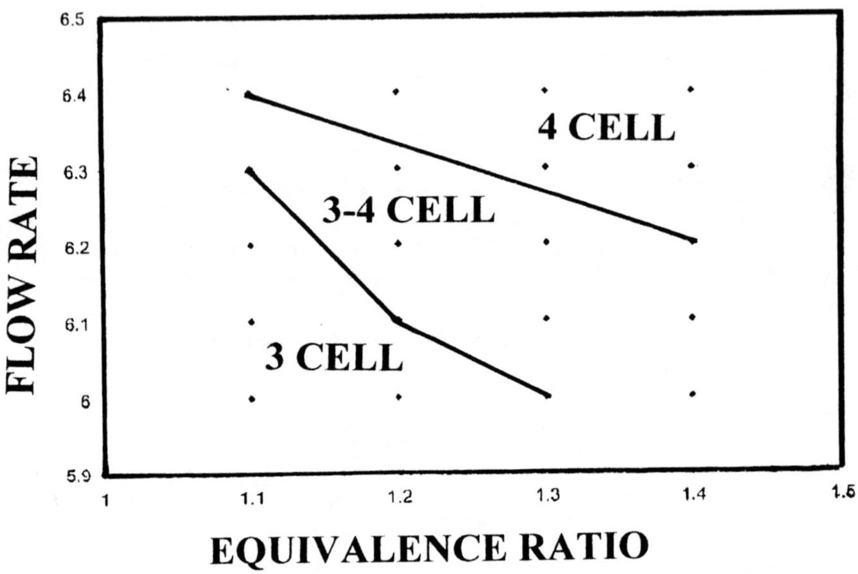

Figure 2. A small section of a stability boundary diagram of propane-air cellular flames. The Flow Rate is given in units of liters/min. The Equivalence Ratio is the ratio of fuel to oxydizer relative to that of stochiometry. Each dot is clickable and produces a mpeg sequence characterizing the dynamics at that point in paramter space.

ACKNOWLEDGEMENTS

The author wishes to thank Kay Robbins for her long-time assistance in conducting this research. Earl Taylor and Robert Brockman helped in rebuilding the experiment. Eric van Grevenyghe took the data on ignition used in this paper.

REFERENCES

1. Zakrzewski, Z., Moisan, M. and Suave, G, in *Microwave Excited Plasmas*, edited by M. Moison and J. Pelletier, Elsevier, 1992, Ch. 4.
2. M. el-Hamdi, M. M. Gorman, M., and. Robbins, K. A., Deterministic Chaos in Laminar Premixed Flames: Experimental Classification of Chaotic Dynamics., in *Proceedings of the International Conference on Pulsating Combustion*, Comb. Sci. and Technol. **94**, 87-101 (1993)
3. Senseman, D. M. Correspondence between visually evoked voltage-sensitive dye signals and synoptic activity recorded in cortical pyramidal cells with intracellular microelectrodes. Vis. Neurosci 13, 963-977 (1996).
4. Gorman, M. el-Hamdi and K. A. Robbins, Spatiotemporal Chaotic Dynamics of Premixed Flames, 1st Experimental Chaos Conference, edited by L. Pecora and W. Ditto, Wiley, New York, 1992, pp. 76-89.
5. Gorman, M., el-Hamdi, M., and Robbins, K. A. Experimental Observation of Ordered States in Cellular Flames, Comb. Sci. and Technol. **98**, 37-45 (1994).
6. Blackshear, J. I., Mapp, J. W., and Gorman, M. An Experimental Study of Pulsating Low Pressure Flames, Comb. Sci. Technol. **35**, 311, 1984.
7. Mapp, J., Blackshear, J. I., and Gorman, M., Nonplanar, Nonsteady Modes of Propagation in Premixed Burner-Stabilized Flames, Comb. Sci. Technol. **43**, 217 (1985).
8. Stone, E., Gorman, M., el-Hamdi M., and Robbins, K. A, Identification of Intermittent Ordered Patterns as Heteroclinic Connections, Phys. Rev. Lett. **76**, 2061-64 (1996).

OPTICS

Dynamics Of Coupled Nd:YVO_4 Microchip Lasers

B. Forsmann, A. Kornfeld, N. Benkert, J. Schüttler, M. Schülke, and M. Möller*

Institut für Angewandte Physik, Westfälische Wilhelms-Universität, Corrensstr. 2-4, 48149 Münster, Germany
**now at Physikalisch-Technische Bundesanstalt, Labor 8.12, Abbestr. 2-12, 10587 Berlin, Germany*

Abstract. We report investigations of the dynamical behavior of a system of spatially coupled microchip lasers, operating under stationary conditions near the threshold of phase locking. We observe a number of different types of intensity and phase correlation instabilities. These include synchronized intensity oscillations of two or three lasers, as well as other types. Furthermore the transient behavior at the switch-on of pump power is investigated.

INTRODUCTION

Experimental and theoretical research on intensity and phase dynamics of arrays of spatially coupled lasers, i.e., nonlinear oscillators, has produced some fascinating results. In particular arrangements with a small number of lasers allow to gain basic insights into the rich dynamical behavior these systems exhibit [1-4].

The experimental system under investigation comprises two or three spatially coupled Nd:YVO_4 lasers in a semi-monolithic microchip resonator, generated by a Titanium-Sapphire pump laser. Control parameters are the mutual detunings and the spatial separation, determining the coupling strength.

EXPERIMENTAL SETUP

The experimental setup consisting of two main parts –the pump section and the analysis section– is shown schematically in figure 1. The purpose of the pump section is to provide up to three parallel pump laser beams with the ability to adjust the spacing in the laser crystal in the range of 400 to 700 μm. Furthermore, with expected Nd:YVO_4 laser mode radii of about 120 μm, the pump beam foci should be considerably smaller than this, thus ensuring coupling between the lasers is entirely due to the overlap of the laser fields.

Three single-mode fibres with individual output collimating optics form three pump light sources with beams waists of 160 μm. The two-dimensional arrangement of the collimators can be arbitrarily adjusted with a minimal distance of 3 mm. This 'source plane' is imaged onto the Nd:YVO_4 crystal using a telescope with a 10:1 reduction

CP622, *Experimental Chaos: 6th Experimental Chaos Conference,* edited by S. Boccaletti et al.

and thus producing pump foci of 16 μm at a minimal distance of 0.3 mm. The actual source of the pump light is a Ti:sapphire laser at 808 nm. The distribution of the laser power into the three fibres is controlled by a combination of polarizing beam splitters and half-wave plates. Polarization controllers in each fibre ensure optimal alignment to the preferred Nd:YVO_4 crystal axis.

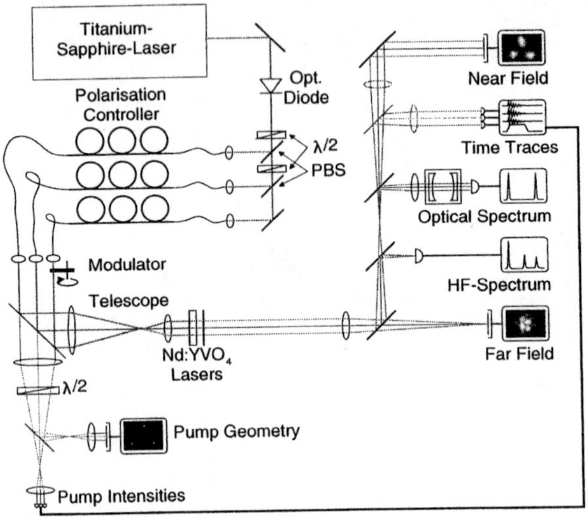

FIGURE 1. Schematic diagram of the experimental setup, PBS denotes polarizing beam splitters, λ/2 denotes half-wave plates

For switch-on, respectively, switch-off experiments an acousto-optical modulator in one of the optical fibre arms allows to attentuate the pump power to 30% of the stationary value. Additionally the pump geometry is imaged onto a CCD camera and the pump intensities are recorded via a quadrant photo diode.

The lasing material consists of a 1.2 mm thick, 3×3 mm^2 Nd:YVO_4 crystal (2 at.% Nd), that is mounted on a copper heat sink. One of the crystal facets carries a dielectric coating with a high transmittance for the 808 nm pump light and a high reflectance for the 1064 nm laser light, thus serving as a resonator mirror. The resonator is completed by a planar output coupler with 98% reflectivity at a distance of about 0.2 mm from the laser crystal. Tilting the output coupler using PZT actuators allows adjusting the frequency detunings between the lasers. Thermal lensing induced in the Nd:YVO_4 crystal creates individual and stable cavities emitting TEM_{00} infrared laser beams with waist radii of about 120 μm. The coupling strength can be continuously changed by varying the spacing of the pump laser.

Intensity time traces are recorded with fast amplified InGaAs photodetectors (Thorlabs PDA255) and a four-channel digitizing oscilloscope (LeCroy LT344). A scanning Fabry-Perot interferometer is used to check that the lasers oscillate on a single longitudinal mode each. The intensity distribution in the plane of the output coupler, i.e., the 'near-field' is imaged onto a CCD camera. Another CCD camera is situated in the focal plane of the first imaging lens, giving access to the 'far-field' intensity distribution. The optical frequency differences and the relaxation oscillation frequencies are recorded with a photodetector in the far-field plane and measured with a radio-frequency spectrum analyser.

MODEL

The coupled Nd:YVO_4 lasers are modeled as single transverse and longitudinal mode class-B Lasers that are coupled to each other through spatial overlap of their electrical fields. Writing the complex amplitudes as $\mathrm{E_i} \equiv E_i \exp(\mathrm{i}\varphi_i)$ leads to a system of coupled differential equations for the laser amplitudes E_i and phases φ_i and the gains G_i [1]:

$$\begin{aligned}
\frac{\mathrm{d}E_i}{\mathrm{d}t} &= \frac{1}{\tau_c}(G_i-\alpha)E_i - \sum_{i\neq j}\kappa_{ij}E_j\cos(\varphi_j-\varphi_i)\\
\frac{\mathrm{d}G_i}{\mathrm{d}t} &= \frac{1}{\tau_f}\left(p-G_i-G_iE_i^2\right)\\
\frac{\mathrm{d}\varphi_i}{\mathrm{d}t} &= \omega_i - \sum_{i\neq j}\frac{\kappa_{ij}}{\tau_c}\frac{E_j}{E_i}\sin(\varphi_j-\varphi_i)
\end{aligned} \tag{1}$$

In these equations, τ_c is the cavity round-trip time, τ_f is the fluorescence time of the upper lasing level of the Nd^{3+} ion in YVO_4 (1064 nm), p is the pump coefficient, α is the cavity loss coefficient and the ω_i are the detunings of the lasers from a common cavity mode, respectively. The lasers are coupled linearly to each other with strengths κ_{ij}. The coupling strength for each pair of laser evaluates to the term $\kappa = \exp(-d^2/2w^2)$, as determined by the overlap integral of two Gaussian beams with $1/\mathrm{e}^2$ radius w, separated by the distance d.

A single uncoupled laser possesses a stable focus point as an attractor in phase space. When perturbed, the laser will show intensity relaxation oscillations with a characteristic frequency ω_{rel}. In the coupled laser system, destabilization can occur through a resonance effect if one of the mutual detunings ω_j - ω_i is in the range of ω_{rel}.

EXPERIMENTAL RESULTS

The lasers are operated at pump powers of about 55 mW each, corresponding to 1.7 times the threshold. Relaxation frequencies are in the range of 1 to 2 MHz.

Two Laser Case

Depending on the control parameters –the mutual detuning and the spacing of the lasers– several types of intensity instabilities can be observed: synchronous harmonic oscillations and pulsing, irregular pulses and the type of localized oscillations In the latter case one of the lasers exhibit a pulsing behavior; the other one oscillates relatively weakly, typically with a phase shift between corresponding maxima in the two lasers. The two lasers can switch roles, thus indicating a spontaneous symmetry breaking of the system [5]. More complex nonlinear phenomena are observed as well, like period doubled synchronous oscillations. These types of behavior and their numerical reproduction are presented in detail in [3].

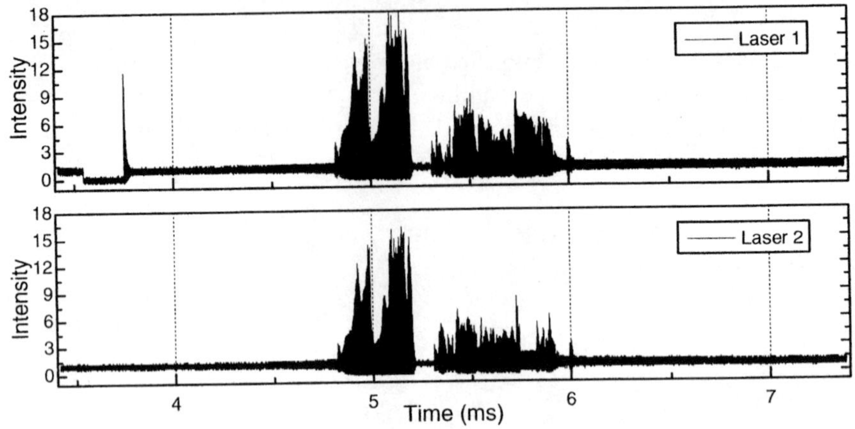

FIGURE 2. Intensity time traces for a switch-on of pump power of one laser

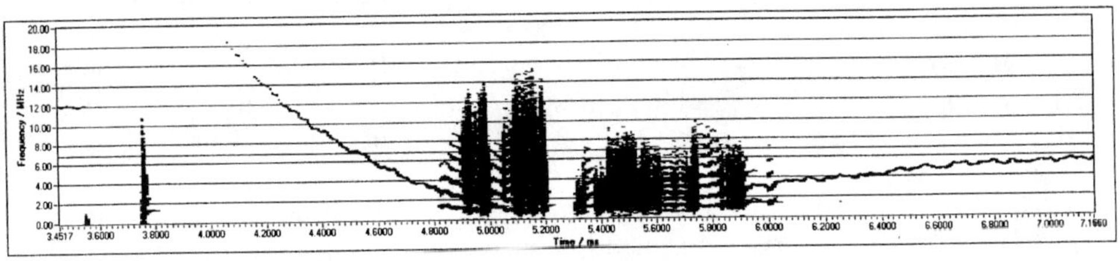

FIGURE 3. JTFA diagram for the intensity traces in figure 2.

Recent experiments are concerned with the question, how the switch-on of pump power for one laser during stationary emission of the other laser influences the dynamics of the two laser system. An experimental result is shown in fig. 2: The pump power of laser 1 is switched off and then switched on again after about 200 μs. Laser 1 exhibits relaxation pulses before reaching a stationary state; after a while a period of intensity dynamics in both lasers can be seen.

This behavior is caused by a perturbation of the thermal load balance of the laser crystal through the switch-off of pump power of one laser. This leads to a change of the thermal lens of this laser cavity and hence to a drift of the optical oscillation frequency. This in turn influences the mutual detuning of the two lasers. With the help of a joint time-frequency analysis (JTFA) of both intensity time traces (see fig. 3) it is possible to make the the variation of the detuning visible and interpret the dynamical scenario: Several hundred microseconds after switch-on a slowly decreasing common frequency in both lasers, that we identify with the mutual detuning, can be seen. When reaching a critical value of about three times the relaxation frequency, strong pulsing is triggered. A further decrease of the detuning leads to a phase-locked state. After, presumably, changing its sign, the detuning causes further intensity instabilities of a different type, compared to first region of dynamics. Because of the fact that the model equations (1) are symmetric with respect to the sign of the detuning, this behavior indicates multistability. A comparison with numerical integration of equations (1) is shown in figure 4. In both diagrams the intensity maxima of the lasers are mapped with the mutual detuning $\Delta = \omega_j - \omega_i$ adiabatically scanned in a range from zero to three times the relaxation frequency ω_{rel}. The two scans, which differ in their scan direction, reveal a hysteresis; indicating multistability.

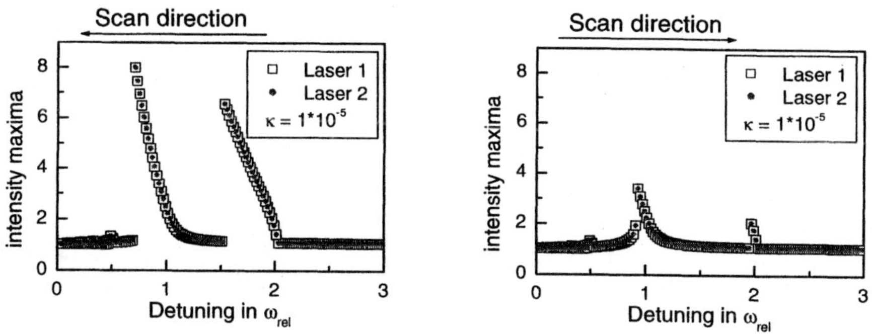

FIGURE 4. Intensity maxima diagrams for an adiabatic scan of the detuning ω_2-ω_1. The two diagrams differ in their scan direction. Numerical result for integration of equation (1) with control parameter $\kappa=1\times10^{-5}$ and system parameters $\alpha=0.01$, $p/\alpha=2.01$, $\tau_c=20$ps, $\tau_f=10\mu$s; $I_i=|E_i|^2$.

Three Laser Case

Generic types of intensity dynamics and their numerical reproduction in a three laser system can be found in detail in [4].

A new type of oscillations that cannot be observed in the two-laser system are 'modulated pulses', i.e., pulses with a base frequency in the range of the relaxation frequency ω_{rel} and an additional supermodulation at a frequency of about seven times ω_{rel} (see fig. 5 (left part)). We observe these oscillations with the three lasers arranged in an equilateral triangle in a regime of relatively strong coupling ($\kappa = 1\times10^{-4}$). This type of instability is numerically reproduced as shown in fig. 5 (right part) and interpreted as a situation where the coupled laser system is driven by two detunings of different order of magnitude. This is confirmed by a JTFA analysis as well.

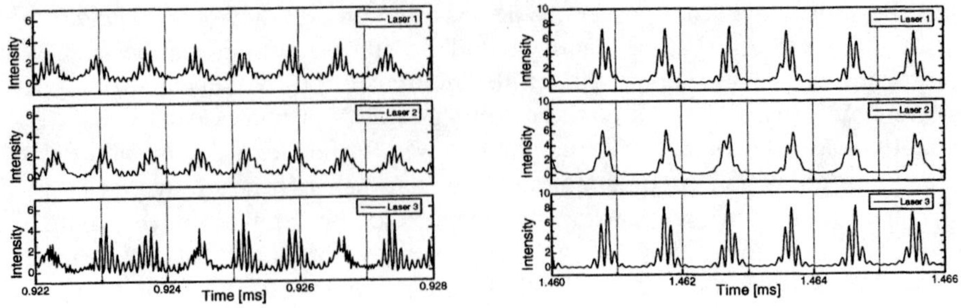

FIGURE 5. Modulated Pulses: <u>left</u>: Experiment; <u>right</u>: Numerical result for integration of equation (1) with control parameters $\kappa=5\times10^{-4}$, $\omega_2-\omega_1=3.5\times\omega_{rel}$, $\omega_3-\omega_1=12\times\omega_{rel}$ and system parameters $\alpha=0.015$, $p/\alpha=1.5$, $\tau_c=11$ps, $\tau_f=23\mu$s; $I_i=|E_i|^2$.

Higher transverse modes

In experimentally generated two laser situations, where a relatively weak higher transverse mode is present in at least one laser, a strongly enhanced occurence of localized oscillations is observed. This correlation between the laser mode structure and localized oscillations leads to the question, if the rather simple model of coupled (fundamental) modes has to be extended or if the onset of a higher transverse mode only represents a peculiar kind of perturbation for the coupled fundamental modes.

CONCLUSIONS

In a system of two coupled Nd:YVO_4 lasers various stationary as well as transient dynamical states were investigated. The joint time-frequency analysis (JTFA) of laser intensity time series reveals the influence of rapid detuning variations. Effects like hysteresis could be explained numerically, using a coupled-mode model.

Futhermore, a correlation between the presence of higher transverse modes and the frequency of localized oscillations was noticed.

In the three laser system a new type of behavior, 'modulated pulses', were observed. From the JTFA analysis as well as from numerical investigations it is deduced that here the system is driven by two detunings of different order of magnitude.

REFERENCES

1. Fabini, L., Colet, P., Roy, R., Lenstra, D., *Phys. Rev. A* **47**, 4287 (1993)
2. Thornburg, K.S. Jr., Möller, M., Roy, R., Carr, T.W., Li, R.-D., and Erneux, T., *Phys. Rev. E* **55**, 3865 (1997)
3. Möller, M., Forsmann, B. and Lange, W., *Quantum Semiclass. Opt.* **10**, 839 (1998)
4. Möller, M., Forsmann, B. and Lange, W., *J. Opt. B: Quantum Semiclass. Opt.* **2,** 371 (2000)
5. Erneux, T., Kuske, R., *Optics Communications* **139**, 125-131 (1997)

Localized structures in nonlinear optics: spatial features and interactions

P.L. Ramazza*, E. Benkler†, S. Boccaletti*, U. Bortolozzo*, S. Ducci** and F.T. Arecchi*

*Istituto Nazionale di Ottica Applicata, Largo E. Fermi, 6 50125 Firenze, Italy
†Institute of Applied Physics, Darmstadt, Germany
**Ecole Normale Supérieure de Cachan, 94235 Cachan Cedex, France

Abstract.
We study the localization of spatial dissipative structures in a nonlinear system formed by a liquid crystal light valve closed in an optical feedback loop. After reviewing the transition from patterns extended across the whole wavefront to localized structures, we investigate how stability and shape of the observed patterns depend on experimental parameters. Interactions between these "dissipative solitons" lead to the formation of several molecule-like bound states, formed by pairs of localized structures. This fact is explored experimentally and justified theoretically.

INTRODUCTION

After the intense investigation on pattern formation and competition in nonlinear dissipative dynamical systems that took place over the last two decades [1], and the studies about formation of solitons in Hamiltonian systems [2], a great attention has recently been paid to characterize the emergence of localized structures (LS) in extended dissipative systems. In this framework, several analytical and numerical efforts to identifying the mechanisms leading to structure localization have been presented [3], while observation of LS has been experimentally offered e.g. in fluid dynamics [4], chemistry [5], granular materials [6] and nonlinear optics [7, 8, 9].

Here we report an experimental investigation of localized structures (LS) in a nonlinear optical system, consisting of a Liquid Crystal Light Valve (LCLV) with optical feedback. The conditions for the existence of LS are identified, and the analogies between our system and general models for pattern formation in dissipative media are discussed.

We investigate the role of the relevant experimental parameters in determining both the stability and the spatial profile of the localized structures, and discuss how the presence of spatial oscillations on the LS tails gives rise to interactions between pairs of LS's, resulting in the existence of a discrete set of bound states.

CP622, *Experimental Chaos: 6th Experimental Chaos Conference*, edited by S. Boccaletti et al.

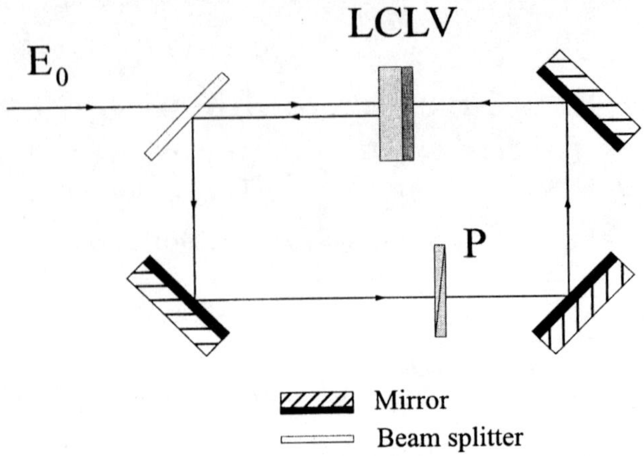

FIGURE 1. Schematic drawing of the experimental setup. A LCLV operating in reflection is closed in an optical feedback and enlighted by an incident laser. Diffraction along the feedback propagation path is present. Interference between the extraordinary polarized component of the light, affected by the phase retardation φ, and the ordinary one, that is a constant reference wave, is introduced by means of polarizer P. For technical details see [9].

LOCALIZATION OF PATTERNS IN THE LCLV

The nonlinear optical system used in our experiments consists of a LCLV with feedback [9]. A schematic drawing of the setup is shown in Fig. 1.

The mechanism of pattern formation in such system can be explained as follows. An initial plane wave sent into the interferometer has a phase $\varphi(\vec{r},t)$ that evolves according to

$$\tau\frac{\partial\varphi}{\partial t} = -(\varphi-\varphi_0)+l_d^2\nabla_\perp^2\varphi+\alpha I_0 \mid e^{\frac{-il\nabla^2}{2k_0}}(Be^{-i\varphi}+C)\mid^2, \qquad (1)$$

where $\varphi_0 = \pi$ is the phase working point, τ the relaxation time and l_d the diffusion length of the LCLV. k_0 and I_0 are respectively the incident laser wavenumber and intensity, l the interferometer length, α the Kerr-like response constant of the LCLV. $B = \cos^2\theta$ and $C = \sin^2\theta$ are two constants measuring the relative weight of diffractive vs. interferential effects, θ being the angle between the director of the nematic liquid crystals of the LCLV and the transmissive axis of a polarizer oriented along the polarization direction of the incident light.

By inspection of Eq. (1) and of the relations for B and C, it is seen that for $\theta = 0$ interference is absent, and a purely diffractive case is recovered. This corresponds to the well known case of a Kerr slice with optical feedback, in which typically hexagons or other delocalized patterns are formed. By increasing θ, the interplay between diffraction and interference leads to a bifurcation diagram of the kind shown in Fig. 2. We observe here the coexistence of a uniform solutions displaying bistability, with a patterned solution. The lower branch of the uniform solution and the patterned solution are stable, while the upper branch of the uniform solution is unstable.

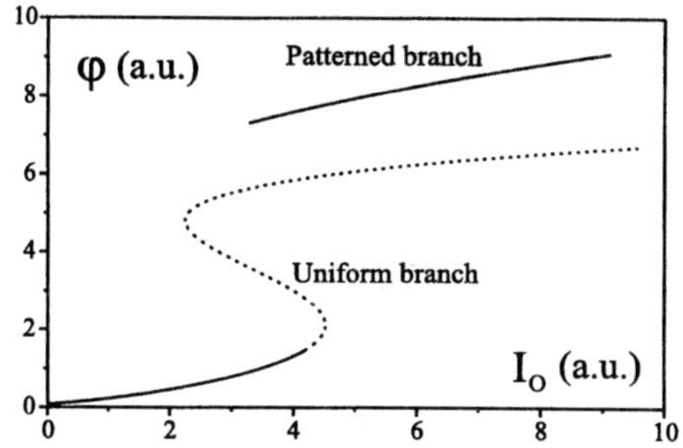

FIGURE 2. Typical bifurcation diagram of the system, showing the coexistence of a uniform and a patterned branch.

In the presence of a subcritical bifurcation connecting a lower uniform and an upper patterned branch, formation of localized structures is typical [10, 11, 12]. In a one-dimensional system, the reason for localization has been clearly identified in the pinning of two fronts connecting the lower and upper states moving along opposite directions. By *pinning* it is meant the interaction between spatial variations of the dynamical variables occurring at different scales. These scales are the pattern wavelength and the width of a front connecting it to the uniform solution. It has been shown that, in a scenario similar to the one here considered [13], the velocity of such a front is given by $v(x) = v_0(\lambda) - \alpha\cos(k_0 x)$, where λ denotes the control parameter, and k_0 is the pattern wavenumber. If a parameter range for which $| v_0(\lambda) | < \alpha$ exists, it follows that the front velocity vanishes at an infinite number of positions along x. A LS can then be interpreted as a patterned part of the system, limited at left and right by two stationary fronts connecting it with the uniform solution. Since successive positions at which $v(x)$ vanishes are separated by a pattern wavelength, it is expected an infinite set of LS, each one containing an integer number of elementary cells of the pattern.

In two or three-dimensions the situation is complicated by the presence of curvature effects, though numerical simulations have extensively shown that the pinning mechanism of pattern localization still works [10, 14].

EXISTENCE AND FEATURES OF LOCALIZED STRUCTURES

As we have seen, the angle θ plays a central role in determining the LS existence range. Other relevant parameters are the intensity I_0 of the input light beam, and the interferometer spatial frequency bandwidth q_b. This last one is controlled by means of a variable aperture put in a Fourier plane (not shown in Fig. 1). Fig. 3 reports the state diagram of the system in the (I_0, θ) plane, for three different values of q_b (normalized to the diffractive wavenumber $\pi\sqrt{\lambda l/2}^{-1}$.

The feature common to all the three curves of Fig. 3 is that LS exist for a broad range of θ values approximately centered at 45°, corresponding to the value at which the interferential term in the feedback ($\sim BC = \cos^2\theta\sin^2\theta$) is maximum. In these conditions the S-shape of the bifurcation diagram shown in Fig. 2 is maximally pronounced, i.e.

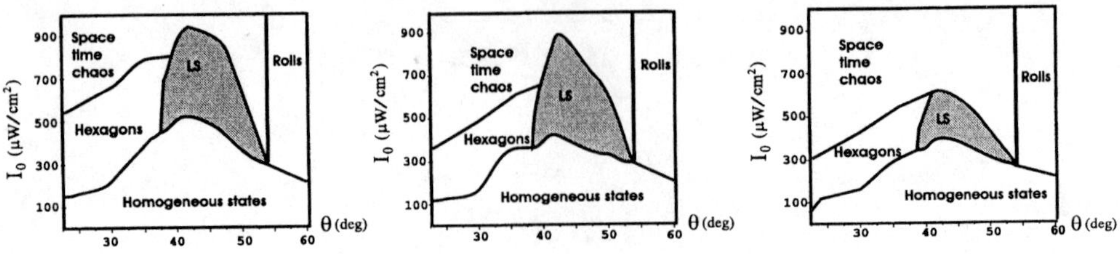

(a) (b) (c)

FIGURE 3. State diagram of the system in the (I_0, θ) parameter plane for $q_b = 1.4$ (left plot), $q_b = 2.1$ (center plot) and $q_b = 3.0$ (right plot).

the range of I_0 at which bistability between the uniform and patterned branches exists is maximum. This explains why the intensity range for LS'existence is maximum at $\theta = 45^o$ and decreases when departing from this value.

The delocalized hexagonal patterns observed at small values of θ do not necessarily correspond to a disappearance of bistability properties. We recall, indeed, that even for the purely diffractive case ($\theta = 0$), the system displays a subcritical bifurcation to hexagons, though the width of the bistable range is considerably smaller than the ones sketched in Fig. 2. We are then led to the conjecture that at small values of θ localization is not observed due to a quantitative rather than qualitative reason. With reference to our previous discussion about the pinning of fronts, it is likely that , for these values of θ there is no parameter range in which the velocity of the fronts connecting the patterns with the uniform solution vanishes.

At high values of θ localization of patterns is lost again, this time in favor of stripes. This is related to the vanishing of the quadratic nonlinearity that occurs at $\theta \sim 60^o$. Both hexagons and stripes occur as multiscale patterns close to threshold in the cases of $q_b = 2.1$ and $q_b = 3.0$. This is due to the fact that the spatial frequency response of the LCLV is very flat, so that more than one diffractive bands can be excited already at low input intensities.

At higher input intensities, the occurrence of space time chaos is generic. However, the quantitative values of parameters at which this regime occurs varies considerably with θ and q_b. Also, the qualitative features of these chaotic regimes are strongly dependent on the parameters, so that considering it as a unique dynamical state is a rough approximation.

In Fig. 4 we show some pictures of LS observed at $\theta = 42^o$, for different values of system bandwidth and input intensities. We see here that LS are formed by a central peak, (whose width is insensitive to q_b), plus a set of concentric rings forming a tail whose spatial oscillations have a decreasing amplitude for increasing distances from the LS center. A remarkable feature is that the spatial scale of the oscillations in the tail does depend on q_b, and decreases for increasing q_b until $q_b \simeq 3$, where it saturates to a constant value.

The consequence is that LS have a natural oscillation spatial frequency, that is in fact tunable by constraining the system to operate within a suitable spatial frequency bandwidth. The occurrence of LS oscillatory tails has been reported in other physical

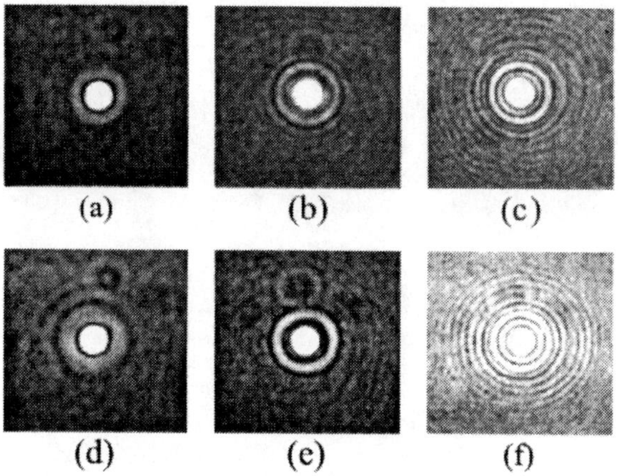

FIGURE 4. Snapshots of the observed LS for $\theta = 42^o$. (a) $I_0 = 540\ \mu W/cm^2, q_b = 1.4$; (b) $I_0 = 410\ \mu W/cm^2, q_b = 2.1$; (c) $I_0 = 380\ \mu W/cm^2, q_b = 3.0$; (d) $I_0 = 810\ \mu W/cm^2, q_b = 1.4$; (e) $I_0 = 700\ \mu W/cm^2, q_b = 2.1$; (f) $I_0 = 540\ \mu W/cm^2, q_b = 3.0$.

systems [10, 15, 16], and is considered to be a typical signature of the formation of LS via pinning of the fronts connecting the uniform and the patterned state.

Variations in the input intensity produce an increase of the modulation depth of the oscillating tails. This is expected to be of relevance with regards to LS interactions, that are indeed strongly dependent on the tail properties.

INTERACTIONS AND BOUND STATES

LS's interactions have been predicted in general theoretical models, and experimentally observed in some systems. A general expectation is that the forces between different LS are determined by the features of each single LS's tail [10, 16]. Namely, while for monotonically decreasing tails only attractive or repulsive forces are expected, when the tails are oscillating in space, both attractive and repulsive forces occur, depending on the distance between the centers of the interacting LS.

These results have been predicted in the context of general models of pattern forming systems, though in the case of optics an intuitive argument can be used to gain a better insight into the physical origin of these phenomena. Let us consider a localized structure in a nonlinear optical medium; to be specific, we focus the attention on a Kerr medium, and on a LS with monotonically decaying tails. The nonlinearity induces a refractive index perturbation proportional to the local light intensity. Hence, on the tails of the LS there will be a spatial distribution of refractive index that acts, roughly speaking, like a wedged (and curved) piece of glass. A second LS that travels in this spatial region is deflected by the spatial distribution of refractive index induced by the first LS. As a net result, an attractive or repulsive force between the LS's is observed, depending on the sign of the medium nonlinearity. The argument can be extended to the case of

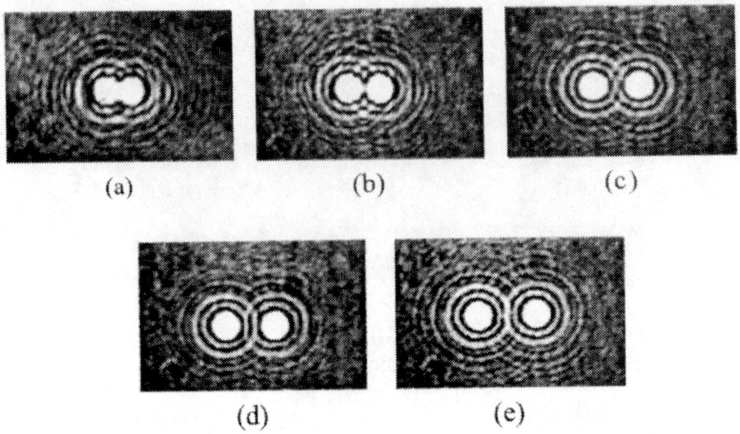

FIGURE 5. Snapshots of different bound states observed at $q_b = 3.5$. The experimental trial used to induce the appearance of these states is discussed in the text.

oscillating LS's tails, where attractive or repulsive forces are expected, depending on the mutual position of the two structures.

To provide experimental evidence of these forces, we adopted the following simple procedure. We first induce a pair of LS at fixed positions on the wavefront. This is achieved by sending weak laser pulses on the rear side of the LCLV, at the chosen locations. When these addressing pulses are switched off, the positions of the LS's evolve due to the mutual interactions, till eventually a stable bound state is reached. However, the motion induced by these two effects display different properties. Namely, while the system inhomogeneities induce some slow, regular drift, the LS interaction manifests in rapid variations of the mutual orientation and distance of the two structures. Our observations pointed to the fact that the evolution of the distance between the centers of two interacting LS *is not* monotonic in time, but present features that arise from the existence of forces of alternate sign.

In Fig. 5 we show a set of bound states observed for q_b = 3.5, at a value of I_0 close to the threshold for pattern localization. These states form an ordered set, that follows a given rule. Observing Fig. 5 (a) to (e), it is seen that each bound state is related to the previous one by the addition of a wavelength of the tails in the region between the two LS maxima. This feature is observed along the whole set of considered q_b values, even though at small bandwidth only few bound states appear, due to the fact that the binding energy of the state varies with q_b, and in some cases it is not sufficient to keep the LS pair actually bound in the presence of unavoidable system inhomogeneities and fluctuations.

A theory of the formation of bound states is given in Ref. [10], in a study of a subcritical Swift-Hohenberg equation. In that approach, the interactions between a pair of LS's with oscillating frequency ν and decaying with decay rate μ produce a long time evolution for the distance R between the LS centers:

$$\frac{dR}{dt} = \frac{1}{R}\frac{d}{dR}(e^{-\mu R}cos(\nu R)) \equiv F(R,\mu,\nu), \tag{2}$$

thus inducing a countable infinite number of stable bound states, whose separation correspond approximately to ν^{-1}. These results have been obtained in the limit in which the amplitude of one LS is small in a region where the amplitude of the other is large, which is the case for all bound states observed in our experiment.

We conclude that the selection rule for the formation of LS bound states observed in our experiment is in agreement with the one obtained by setting $F = 0$, $\frac{dF}{dR} < 0$ in Eq. (2). This is not surprising, since our experiment displays a subcritical bifurcation of a real order parameter to a patterned state, and therefore is appropriately modeled by an order parameter equation of that kind.

REFERENCES

1. M. Cross and P.C. Hohenberg, Rev. Mod. Phys. **65**, 851 (1993).
2. See e.g. E. Infeld and G. Rowlands, *Nonlinear waves, solitons and chaos*, (2nd ed., Cambridge University Press, 2000).
3. H. Riecke, "Localized structures in Pattern-Forming Systems" in *Pattern Formation in Continuous and Coupled Systems*, ed. by M. Golubitsky, D. Luss and S. Strogatz (IMA Volume 115, Springer, 1999), p. 215.
4. E. Moses, J. Fineberg and V. Steinberg, Phys. Rev. **A35**, 2757 (1987).
5. H. H. Rotermund, S. Jakubith, A. Von Oertzen and G. Ertl, Phys. Rev. Lett. **66**, 3083 (1991).
6. P. Umbanhowar, F. Melo and H. Swinney, Nature **382**, 793 (1996).
7. M. Saffman, D. Montgomery and D.Z. Anderson, Opt. Lett. **19**, 518 (1994); V. B. Taranenko, K. Staliunas and C.O. Weiss, Phys. Rev. Lett. **81**, 2236 (1998).
8. A. Schreiber, B. Thuring, M. Kreuzer and T. Tschudi, Opt. Comm. **136**, 415 (1997).
9. P.L. Ramazza, S. Ducci, S. Boccaletti and F.T. Arecchi, J. of Optics B: Quantum and Semiclassical Optics **2**, 399 (2000).
10. I.S. Aranson, K.A. Gorshkov, A.S. Lomov and M.I. Rabinovich, Physica **D43**, 435 (1990).
11. M. Tlidi, P. Mandel and R. Lefever, Phys. Rev. Lett. **73**, 1328 (1994).
12. P. Coullet, C. Riera and C. Tresser, Phys. Rev. Lett **84**, 3069 (2000).
13. D. Bensimon, B.I. Shraiman and V. Croquette, Phys. Rev. **A38**, 5461 (1988).
14. H. Sakaguchi and H.R. Brand, Physica **D97**, 274 (1996).
15. W. J. Firth and A.J. Scroggie, Phys. Rev. Lett. **76**, 1623 (1996).
16. C. Schenk, P. Schultz, M. Bode and H.G. Purwins, Phys. Rev. **E57**, 6480 (1998).

Fractals in Microcavities: Giant Coupled, Multiplicative Enhancement of Optical Responses

Won-Tae Kim[†], Vladimir P. Safonov*, Vladimir M. Shalaev[†], and Robert L. Armstrong[†]

†Department of Physics, New Mexico State University, Las Cruces, NM 88003
**Institute of Automation and Electrometry, Siberian Branch of the Russian Academy of Science, Novosibirsk 630090, Russia*

Abstract. Lasing at extremely low pump intensities, below 1 mW, and dramatically enhanced Raman scattering were observed for a novel class of optical materials, microcavities doped with nano-structured fractal aggregates. Nanostructured fractal aggregates lead to giant local-fields concentrated in "hot spots" much smaller in size than the size of the fractal and the wavelength. Coupling the localized plasmon modes in fractal aggregates with microcavity resonances further increases the local fields. These microcomposites possess unique optical characteristics, including extremely large probabilities of radiative processes. An extremely pronounced time-dependent lasing effect of dye molecules adsorbed on silver particles was also observed in our experiment. This time dependence is likely due to photo-stimulated aggregation of silver colloid particles and the gradient forces pulled fractal aggregates into the high-field area associated with the morphology-dependent resonance mode.

INTRODUCTION

The giant enhancement of the optical response in metal nanocomposites and thin metallic films containing nanoscale surface features has been intensively studied in recent years [1-4]. This enhancement is associated with the excitation of surface plasmons, collective electromagnetic modes whose characteristics are strongly dependent on the geometrical structure of the metallic component of the medium.

Nanocomposites often look similar in different scales, i.e., they are scale-invariant and their structure is characterized by the fractal geometry. Since fractal objects do not possess translational invariance, they cannot transmit the running waves that are characteristic of homogeneous media. Rather, collective optical excitations in fractals result from the near-field multiple scattering within sub-wavelength regions of the fractal medium; therefore, the optical modes, such as surface plasmons, are spatially localized in fractals [2]. This localization leads to the presence of nanometer-scale spatial regions of high local electric fields, "hot" spots, and accordingly, to significant enhancement for a variety of optical processes, such as Raman scattering, four-wave mixing, quadratic electro-optical effect, and nonlinear absorption and refraction [2-4].

The "hot" spots associated with localized optical modes in fractals represent the sharp peaks in the spatial distribution of the local field and can be viewed as giant

CP622, *Experimental Chaos: 6th Experimental Chaos Conference*, edited by S. Boccaletti et al.

field fluctuations [2]. Since the nonlinear optical signal is mostly generated from the "hot" spots the local enhancement in these areas may exceed the average one by many orders of magnitude; e.g., whereas the average enhancement for an optical process on a fractal is, say, 10^6 or so, the local enhancement in the "hot" spots can be as large as 10^{15} [2]. This opens a fascinating possibility for nonlinear local spectroscopy of single molecules and nanoparticles in fractals [2-4]. The position of these "hot" spots providing enormous enhancement, as well as the peaks' height and their spatial separations, strongly depend on the wavelength of the applied field [2,4].

An alternative approach for achieving large enhancement of the optical response involves the excitation of morphology-dependent resonances (MDR's) in dielectric microcavities [5]. These resonances, which have high quality factors ($Q = 10^5 \sim 10^9$), result from confinement of the radiation within the microcavity by the total internal reflection. Light emitted or scattered in the microcavity may couple to the high-Q MDR's lying within its spectral bandwidth, leading to enhancement of both spontaneous and stimulated optical emissions. For example, enhanced fluorescence emission from an organic dye-doped cylindrical or spherical microcavity occurs when either the laser pump or the fluorescence (or both) couple to microcavity MDR's [6]. Moreover, the increased feedback produced by MDR's is sufficient to obtain laser emission from a dye-doped microdroplet under both cw [7] and pulsed [8] laser excitation, with the threshold cw pump intensity three orders of magnitude lower than that of a conventional dye laser in an external cavity.

The existence of high-Q microcavity modes is also responsible for numerous stimulated nonlinear effects including: (1) Stimulated Raman and Rayleigh wing scattering and four-wave parametric oscillation under moderate intensity cw excitation [9]. (2) A large (>100) quantum electrodynamics (QED) enhancement [7, 9] arising from MDR-induced changes in the density of states. (3) Optical bistability connected with the thermal nonlinearity of fused silica microspheres under low power (10^{-5} W) cw excitation [10]. (4) Finally, experiments in which a microcavity was doped with two fluorescent species, one a dye-doped sol, the other a liquid dye, exhibited lasing due to both enhanced radiative (MDR) and non-radiative (Förster-type) energy transfer mechanisms [11,12].

As discussed above, strong existing evidence suggests that fractal nanocomposites and microcavity resonators individually result in large enhancements of optical emissions. In the present paper, we demonstrate that huge, multiplicative enhancement factors are obtained under the simultaneous, combined action of these two resonant processes when the emitting species is adsorbed onto metal fractal aggregates contained within high Q microcavities.

EXPERIMENT

Sample Preparation

Silver colloid solutions were prepared by reduction of silver nitrate by sodium citrate in an aqueous solution [13]. Boiling this solution results in the formation of

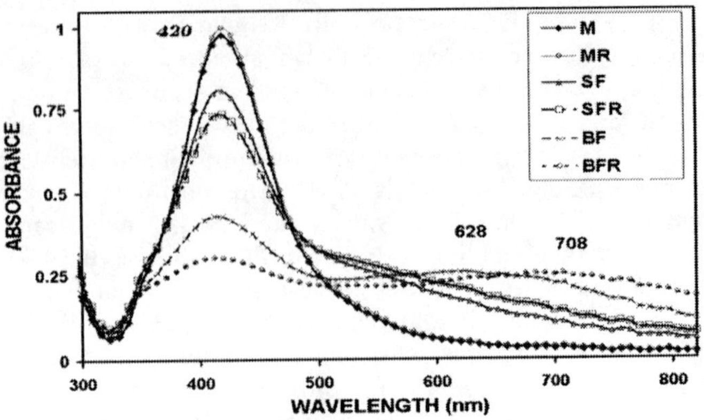

FIGURE 1. Absorption spectra of silver colloidal particles (M, MR) and their fractal aggregates (SF, SFR, BF, and BFR) in 1 mm path length cell. M stands for monomer (fresh silver colloidal solution comprising non-aggregated particles). F represents fractal aggregates, which were grown by adding approximately 0.03 M (1/50 volume ratio) of aqueous fumaric acid is added to the fresh silver colloid solution, resulting in the large long-wavelength wing in the absorption spectrum of large fractal aggregates solution. This wing is due to the aggregation of colloidal particles into fractals that provide absorption in the red and near-infrared parts of the spectrum. S and B stand for small and big (large) fractal aggregates. R in the legend represents addition of 10^{-4} M of R6G solution with 1/200 volume of the solution of M, SF, or BF (the resulting concentration of R6G in the solution is 5×10^{-7} M). R6G in methanol is added to fractal aggregates solutions leading to a further increase of the aggregation initiated first by the acid. Note that adding small amount of R6G only (without fumaric acid) to the fresh colloidal solution doesn't initiate itself the aggregation (see MR), but it does promote the aggregation if the acid was initially added.

silver nanoparticles (monomers) with an average diameter of 25 nm. Addition of an organic acid (0.03 M saturated fumaric acid) to the monomer solution in volume proportions from 1:50 to 1:100 promotes the aggregation of colloidal nanoparticles into fractal clusters, containing, typically, $\sim 10^3$ monomers. Electron microscopic analysis of these aggregates demonstrate that they possess a fractal structure with the fractal dimension $D \approx 1.77$ that characterizes the cluster-cluster aggregation of particles [14].

Extinction spectra of non-aggregated silver colloids in the visible region of the spectrum exhibit a single resonance feature centered at 420 nm, with the width 80 nm; this peak is due to the surface plasmon excitation in single silver nanoparticles. Aggregation leads to the appearance of a broad wing extending toward the long-wavelength part of the spectrum. Enhanced extinction (absorption and scattering) in the long- wavelength region (see Figure. 1) is a consequence of induced high-Q optical modes in metal fractal aggregates [2].

Lasing experiments were done with Rhodamine 6G (R6G) dye. A small amount of a parent solution of 10^{-4} M R6G in methanol was added to the silver colloid solution; the resulting dye concentration in the samples studied ranged from 10^{-8} to 10^{-5} M. Cylindrical microcavities were fabricated from cylindrical quartz tubes (inner diameter is 0.7 mm and outer diameter is 1.0 mm), with the dye/colloidal solution placed within the tube. Addition of R6G to the colloid solution resulted in increased aggregation (see spectra SFR and BFR in Figure 1).

Experimental Set-up

A 10 mW cw Argon-ion laser (λ = 514.5 nm) and a 0.75 mW cw green He-Ne laser (λ = 543.5 nm) were used as pumping sources. The pump beam (approximately 2 mm in diameter) was focused into the tube by a 75 mm focal length lens; focal plane beam diameters were 70 μm and 35 μm for an Argon and a He-Ne lasers, respectively. Pump beam polarization was vertical (along the axis of the tube), and output radiation was collected at 90 degrees to the incident radiation as shown in Figure 2; configuration 2a (b) was used when we did (did not) wish to excite MDR's in the cylindrical microcavity. Spectroscopic measurements were performed using a CCD camera mounted to a SpectraPro 300i spectrograph; an 1800 groove per millimeter grating provided a spectral resolution of 0.04 nm (full width at half maxima, FWHM).

The image of the emitting surface of the lasing tube was transferred to the entrance slit of a spectrograph, facilitating study of the spatial structure of the emission along the axis of the tube (which was parallel to the vertical entrance slit). It was found that the fluorescence emission has a structure in the vertical direction with a characteristic size 50 μm, and the location of the resonance peaks occur in various spatial locations; these features are characteristic of the mode structure of microcavity MDR's. The lasing spectral peaks are also observed to change in time during cw excitation; specifically, when the detector integration time was greater than one second, the lasing spectrum became smoothed. We suggest that this is primarily due to laser-heating-induced changes in the refractive index and, hence, in the MDR wavelength.

Elastic scattering of a laser beam passed through the outer edge of the empty cylindrical tube exhibited well-defined, MDR angular structure. When the beam is passed through the inner edge of an empty tube (see Figure 2a) the MDR intensity is significantly less. Nonetheless, most of our experiments were performed using this illumination geometry because it provides the effective optical excitation of fractals within a cavity and their coupling to the MDR's. Filling the tube with a colloidal solution resulted in strong elastic scattering with a clearly resolved MDR angular structure; this observation suggests that light scattering by colloidal particles facilitates trapping of the radiation in the MDR cavity modes. Elastic scattering by fractal

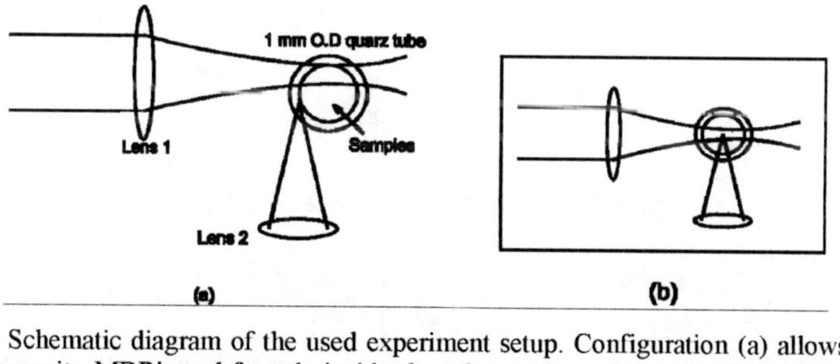

(a) (b)

FIGURE 2. Schematic diagram of the used experiment setup. Configuration (a) allows excitation of both the microcavity MDR's and fractals inside the microcavity, whereas in the configuration (b) the MDR's are almost not excited by the incident beam. The lens 1 has the focal length 75 mm; the lens 2 is used to collect the output radiation at 90 degree. The inner diameter of the quartz cylindrical cavity is 700 μm.

aggregates and monomers also contributes to output-coupling of radiation from microcavity MDR's. Scattering, together with absorption, decreases the quality-factor Q of the cavity modes according to $Q^{-1}= Q_a^{-1} + Q_{sv}^{-1} + Q_{ss}^{-1}$, where Q_a^{-1}, Q_{sv}^{-1}, and Q_{ss}^{-1} are losses due to absorption, volume scattering, and surface scattering, respectively [5]. (Diffraction losses are negligible in our case.) If colloidal aggregates are present in the microcavity, the volume absorption is the most important loss mechanism [2].

RESULTS AND DISCUSSIONS

Morphology-Dependent Resonances in a Fractal-Microcavity Composite

The luminescence spectrum of a dye solution in a microcavity without colloidal particles has a maximum at $\lambda = 560$ nm and a FWHM of 30 nm for the $\lambda = 514.5$ nm excitation. The presence of fractal aggregates of silver colloidal particles results in large peaks in the spectrum with the mode spacing $\Delta\lambda \approx 0.066$ nm, for the spectral interval shown in Figure 3. This spacing is close to, but somewhat less than the calculated inter-mode spacing, $\Delta\lambda = (\lambda^2/2\pi a)[(n^2 - 1)^{-1/2}\tan^{-1}(n^2 - 1)^{1/2}] \approx 0.076$ nm, for a quartz ($n = 1.46$) cylinder of radius $a = 0.5$ mm and water as host medium ($n_h = 1.54$). Note that, in our experiments, the effective path length of a single traversal of the cavity may be effectively increased because of the scattering in aggregated colloidal solution, resulting in the observed decrease in $\Delta\lambda$. The spectral widths of the peaks are limited by our instrumental resolution, and they are approximately 0.04 nm.

The measured absorption coefficient at 543.5 nm is $\alpha = 5$ cm^{-1}, so that $Q_a = 2\pi n/\alpha\lambda = 3.4 \times 10^4$. The measured scattering loss, $Q_{sv}^{-1} + Q_{ss}^{-1}$, is smaller than Q_a^{-1} by, at least, one order of magnitude, implying that scattering may be regarded in our experiments simply as an output-coupling mechanism for micro cavity radiation. An alternative

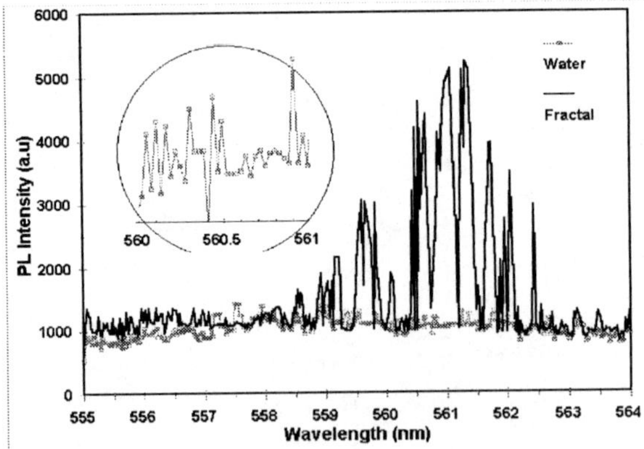

FIGURE 3. Luminescence spectrum of dye solution (the dye concentration in solution is 5×10^{-7} M) with and without fractals in a microcavity, for the $\lambda = 514.5$ nm excitation by Ar-ion laser. The volume ratio of water and methanol is 200:1 in both cases. The inserted figure is to show clearly typical mode widths and intermode spacings.

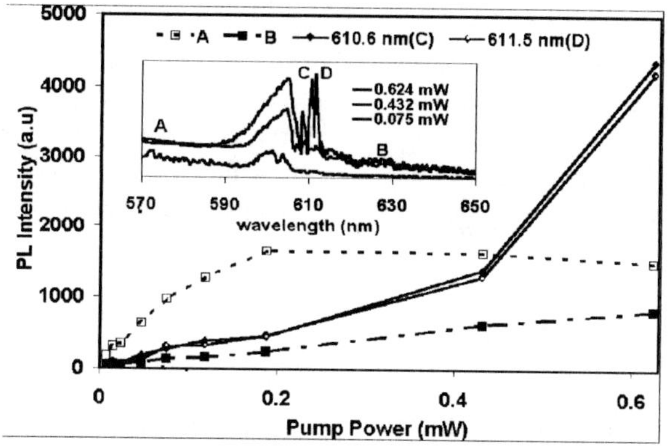

FIGURE 4. Lasing threshold dependence of solution with dye-doped fractal aggregates of silver colloid particles in a microcavity. Inserted figure represents photoluminescence spectra at different intensities of the pump beam (He-Ne laser; λ = 543.5 nm). The threshold effect for lasing for denoted spectral components is shown in the insert. Note that the luminescence signal from non-aggregated monomers is multiplied by a large factor, 10^2 to 10^3

output coupling mechanism, which is based on a resonator with a deformed cross section [15], is more appropriate for active cavity media with high reflectivity. The emission spectrum (with Argon laser pumping) of an aggregated silver colloid solution with a R6G concentration of 5 x 10^{-7} M, dramatically differs from that of the pure R6G solution; Figure 3 illustrates the huge increase in peak intensities in a narrow spectral region centered near λ = 561 nm with a bandwidth of approximately 3 nm. Note, that the mode structure in this spectral range is approximately the same as for pure dye solution in a cavity with no fractals. This indicates that the presence of fractals in a microcavity doesn't affect much the cavity MDR's.

Lasing and Photo-Stimulated Aggregation of Colloid Particles

Lasing emission from R6G dye molecules adsorbed onto silver colloidal aggregates inside a cylindrical microcavity was obtained for dye concentrations approximately 3 orders of magnitude lower than for the corresponding microcavity dye laser in the absence of colloidal aggregates, and for a threshold pump intensity approximately 3 orders of magnitude less than for a conventional dye laser.

We also studied emission intensity of different spectral components as a function of the pump intensity. It was found that this dependence is linear for low excitation intensities for all spectral components. However, when the pump intensity exceeds some critical value in the range between 20 and 50 W/cm^2, some peaks grow drastically, exhibiting a lasing threshold dependence (Figure. 4). The threshold power for the λ = 543.5 nm He-Ne pump laser is as small as 2 x 10^{-4} W. It is noteworthy that the R6G concentration was only 5 x 10^{-4} M in these experiments, three orders of magnitude lower than that for conventional dye lasers with an external cavity and three orders of magnitude lower than that for a microdroplet laser without silver fractal aggregates [7] (the minimum R6G concentration that results in lasing in our experiments is as low as 10^{-8} M). These findings suggest that the lasing effect is due to dye molecules adsorbed on the surface of silver aggregates. This conclusion is also supported by the fact that increasing the R6G concentration up to 10^{-5} M does not result in additional growth of the lasing peak intensities. This dye concentration is apparently too large to be adsorbed onto the silver particles, and the additional dye, remaining in solution as free molecules, does not effectively contribute to lasing.

Our experiments show that placing R6G dye in a microcavity leads to 10^3 to 10^5 enhancement of the dye photoluminescence. By adding non-aggregated silver colloidal particles to the dye solution in a microcavity, further (multiplicative) enhancement is obtained, varying between 10^2 and 10^3. Finally, aggregation of the colloidal particles in the microcavity into fractals results in more multiplicative enhancement, which also can be as large as ten thousands. Thus, the overall emission enhancement provided by fractals in a microcavity can be enormously large, up to 10^{12}.

As previously reported, the spectra of both the R6G cw microlaser [7] and the pulsed microlaser [16] are shifted to the long-wavelength wing of the luminescence band. The authors of both [7,16] suggested that this effect was caused by the reduction in R6G absorption with increasing the wavelength. In our experiments, there is one additional important reason for the observed shift. Namely, the fractal resonance-induced enhancement factor for the local electric field, $G = 2(\varepsilon^2/3\varepsilon_h\varepsilon')/\lambda^2$, is larger for longer wavelengths [2] (where $\varepsilon = \varepsilon' + \varepsilon''$ and ε_h are the dielectric functions for silver nanoparticles and the host medium (water), respectively).

An extremely pronounced time-dependent effect was also observed when a fresh silver colloid solution was irradiated with a cw, 543.5 nm green He-Ne pump laser at a power of 0.75 mW (Figure 5). It is known that irradiation of a colloidal solution causes the aggregation of sols [17,18]. As seen in Figure 6, R6G luminescence increases with time until several peaks near 610 nm are driven above the lasing threshold. Note that the aggregation process is limited to only a small volume of the

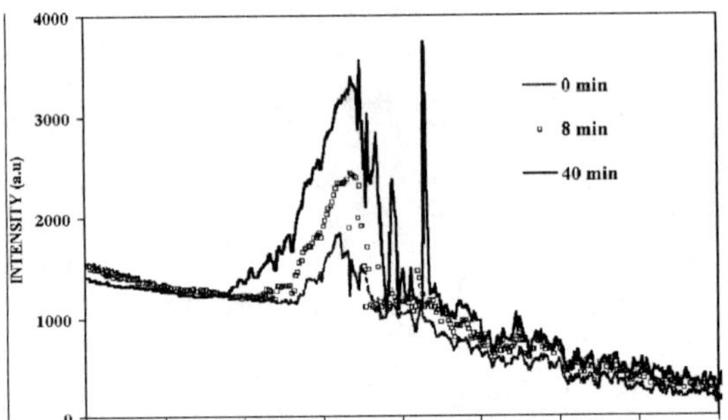

FIGURE 5. Spectra of the photo-stimulated aggregates of colloidal particles for different irradiation times. The maximum pump intensity of He-Ne laser is 0.75 mW (λ = 543.5 nm); the laser beam is focused to a cylindrical microcavity resulting in the 35 μm irradiation spot. After 40 minutes of continuous irradiation, the drastic increasing in the lasing intensity is observed. Note that the aggregation process is limited to only a small volume of the sample where the pump beam is focused. Photo-stimulated aggregation of silver colloid solutions with a 0.75 mW green He-Ne laser: After 40 minutes illumination, lasing peaks start to appear

sample where the pump beam is focused. A plausible explanation of the strong time-dependent interaction is that, as time progresses, so does the degree of aggregation, and when the aggregation reaches a certain level, fractal resonant enhancement lowers the lasing threshold to the level which is, in fact, being provided by the green He-Ne pump laser. This result provides direct evidence of the effect of fractal resonance enhancement on lasing emission. This is a very important finding for micro-technology, since it opens the possibility of activating lasing emission by remote, non-invasive techniques.

A plausible explanation for this is that photo-stimulated aggregation is induced by surface enhanced (two-photon) ionization of silver sols [18]. The beam radius at the sample is about 35 μm in order to get sufficient optical signal from the morphology-dependent enhancement of the microcavity; the pump irradiance in this case is at most 50 W/cm^2. After approximately 30 minutes of irradiation to the solution contained non-aggregated silver particles doped by Rhodamine 6G dye molecules, strong lasing emission spontaneously appeared from the microcavity. The overall concentration of R6G dye in the solution is 5×10^{-7} M, which is to generate extremely weak luminescence in a non-aggregated colloid solution. Photo-stimulated aggregation was determined from spectroscopic analysis and transmission electron micrographs.

We conjecture that this time dependent lasing effect is likely due to photo-stimulated aggregation, whereby initially non-aggregated silver colloidal particles undergo *in situ* aggregation in response to irradiation by a He-Ne pump laser. Lasing experiments demonstrate that large optical signals only occur from large fractals. Several experiments were performed using fresh colloid and small fractal solutions, but no strong signals were observed. A secondary conjecture, which follows from the spatial inhomogeneity of the microcavity MDR fields, is that the resulting fractal aggregates are “pulled” into high-field regions of the MDR's as a result of electromagnetic gradient forces, which may be caused by light-induced diffusive flow of electrons resulting from non-uniform spatial distributions of the light intensity in the sample as well as light-induced currents and/or potential differences [19]. Light-induced currents and diffusive forces move the nanocomposites, including individual particles and fractal aggregates, into the most intense regions of the electromagnetic fields, where self-trapping of the pump beam can occur. This process may initiate and continue the growth of fractal aggregates. Light-induced drifts in the solution may be more effective in the fractal/microcavity composites because both pump light and emitted light are trapped inside a cavity. Several experimental trials confirm that nanocomposites migrate and aggregate themselves due to field gradient or diffusive field induced by the pump light.

Surface Enhanced Raman Scattering

Surface-enhanced Raman scattering (SERS) spectra from sodium citrate molecules adsorbed on silver fractal aggregates were obtained under two experimental geometries where MDR's either were, or were not, excited (see Figure 2). We found that SERS is 10^3 to 10^5 times more intense when the MDR's are excited. Apparently, this strong enhancement occurs despite the fact that the presence of fractals in microcavity somehow decreases its quality-factor because of the absorption (mainly) and scattering.

Figure 6 shows fragments of this spectrum between λ =575 nm and λ =615 nm. Analogously to the case of Argon-ion laser excitation, we have a set of peaks whose minimal spacing is approximately equal to the intermode spacing of our cylindrical microcavity. We have identified the largest (4 times more intensive than emission at 600 nm) peak B at 612.6 nm with the combination Raman mode (1310+780) cm^{-1} of sodium citrate, used to grow silver monomers and present in minute (3 x 10^{-4} M) concentrations in our monomer solutions; the other peak A at 580 nm corresponds to 1160 cm¯1 fundamental Raman mode of the citrate molecule. These peaks exhibit a nonlinear dependence on pump intensity and may be regarded as surface-enhanced stimulated Raman scattering in a microcavity, notably, excited by a 0.75 mW cw pump laser. These peaks A, and B in Figure 6 are coincident with spectral features of the luminescence or inelastic light scattering from silver colloid solutions in a microcavity without R6G. We suggest that these peaks coincide with the highest-Q MDR's. As noted previously, the observed width of the peaks essentially corresponds to our instrumental width, so that the real peak width is smaller than the measured value $\delta\lambda = 0.04$ nm. This allows us to estimate a lower bound for the quality factors, Q > $\lambda/\delta\lambda$ = 1.5 x 10^4. Note that this lower bound value is consistent with the previous estimate of Q $\cong$ 3.4 x 10^4.

By comparing Raman signals from sodium citrate adsorbed on silver colloid aggregates and from a high-concentration sodium citrate solution without colloidal particles, we obtained that the SERS enhancement due to fractal aggregates only is 10^5 to 10^6, which is consistent with the data of Ref. 20. With the further enhancement provided by a microcavity, the resultant average enhancement by fractals in a microcavities is estimated to be in the range 10^8 to 10^{11}.

It is important that the local enhancement in the "hot" spots of fractals in a microcavity can be larger than the above average by many orders of magnitude. It was recently shown theoretically and experimentally that local SERS enhancements on

FIGURE 6. Inelastic scattering spectra of silver colloid solutions in a cylindrical microcavity obtained by irradiation of a sample with a cw He-Ne laser (the maximum power is 0.75 mW; λ = 543.5 nm). The top solid-lines represent spectra of the dye-doped fractal aggregates of silver colloid particles; the bottom dashed lines show spectra of non-aggregated colloidal particles.

fractal aggregates of metal nanoparticles may be as high as 10^{12} to 10^{15} which makes possible observation of single-molecule Raman scattering [21-23]. Using the factor 10^{14} for the fractal hot-spot enhancement and taking into account the multiplicative enhancement factor provided by a microcavity, we find that the local enhancement in the hot spots of fractals placed in a microcavity can be as large as 10^{18}. This opens new fascinating possibilities for optical micro-analysis, including Raman spectroscopy of single molecules and nanoparticles.

To summarize, results reported in the present paper promise an advance in the design of micro/nano-lasers, operating on single molecules adsorbed on metal nanostructures within a micro/nano-cavity, as well as offering the possibility of combining surface enhanced radiative processes and high-Q morphology dependent resonances in microcavities.

ACKNOWLEDGMENTS

Authors acknowledge useful discussions with Prof. Richard K. Chang. This work was supported by NSF under Grants DMR-9623663 and DMR-9810183, and by ARO under grant DAAG55-98-1-0425.

REFERENCES

1 *Electron Transport and Optical Properties of Inhomogeneous Media*, edited by W. L. Mochan and R. G. Barrera, North-Holland, Amsterdam, 1994; Bergman, D. J. and Stroud, D., *Solid State Phys.* **46**, 147 (1992); *Surface Enhanced Raman Scattering*, edited. by R. K. Chang and T. E. Furtak Plenum, NY, 1982; Moskovits, M., *Rev. Mod. Phys.* **57**, 783 (1985); Boyd, R. W., Gehr, R. J., Fisher, G. L. and Sipe, J. E., *Pure Appl. Opt.* **5**, 505 (1996); Flytzanis, C., *Prog. Opt.* **29**, 2539 (1992).

2 Shalaev, V. M., *Phys. Reports* **272**, 61 (1996); Markel, V. A., Shalaev, V. M., Stechel, E. B., Kim, W., and Armstrong, R. L., *Phys. Rev.* **B53**, 2425 (1996); Shalaev, V. M., Poliakov, E. Y., and Markel, V. A., *Phys. Rev.* **B53**, 2437 (1996);. Poliakov, E. Y , Markel, V. A., Shalaev, V. M., Botet, R., *Phys. Rev.* **B57**, 14901 (1998); Shalaev, V. M., Sarychev, A. K., *Phys. Rev.* **B57**, 13265 (1998).

3 Rautian, S. G., Safonov, V. P., Chubakov, P. A., Shalaev, V. M., Stockman, M. J., *JETP Lett.* **47**, 243 (1988); Safonov, V. P., Shalaev, V. M., Markel, V. A., Danilova, Yu. E., Lepeshkin, N. N., Kim, W., Rautian, S. G., and Armstrong, R. L., *Phys. Rev. Lett.* **80**, 1102 (1998).

4 Shalaev, V. M., Markel, V. A., Poliakov, E. Y., Armstrong, R. L., Safonov, V. P., and Sarychev, A. K., *J. Nonlinear Optical Physics and Materials*, **7**, 131 (1998).

5 *Optical Processes in Microcavities*, edited by R. K. Chang and A. J. Campillo, World Scientific, Singapore, New Jersey, London, Hong Kong, 1996.

6 Owen, J. F., Barber, P. W., Dorain, P. B., and Chang, R. K., *Phys. Rev. Lett.* **47**, 1075 (1981).

7 Tzeng, H.-M., Wall, K. F., Long M. B., and Chang, R. K., *Opt. Lett.***9**, 499 (1984).

8 Biswas, A., Latifi, H., Armstrong, R. L., and Pinnick, R. G., *Opt. Lett.* **14**, 214 (1988).

9 Lin, M. B., and Campillo, A., J., *Phys. Rev. Lett.* **73**, 2440 (1994).

10 Braginsky, V. B., Gorodetsky, M. L., and Ilchenko, V. S., *Phys. Lett.* **A137**, 393 (1989).

11 Armstrong, R. L., p.257 in Ref. 5.

12 Armstrong, R. L., Xie, J. G., Ruekgauer, T. E., and Pinnick, R. G., *Opt. Lett.***17**, 943 (1992).

13 Lee P. C., and Meisel, D., *J. Phys. Chem.* **86**, 3391 (1982).

14 Weitz, D., and Oliveria, M., *Phys. Rev. Lett.* **52**, 1433 (1984); Jullien, R., and Botet, R., *Aggregation and Fractal Aggregates*, World Scientific, Singapore, 1987.

15 Gmachi, C., Capasso, F., Narimanov, E. E., Nockel, J. U., Stone, A. D., Faist, J., Sivko, D. L., and Cho, A. Y., *Science* **280**, 1556 (1998).
16 Armstrong, R. L., Xie, J. G., Ruekgauer, T. E., Gu, J. and Pinnick, R. G., *Opt. Lett.* **18**, 119 (1993).
17 Kreibig, U., and Vollmer, M., *Optical Properties of Metal Clusters*, Springer-Verlag, Berlin Heidelberg, 1995.
18 Karpov, S. V., Popov, A. K., and Slabko, V. V., *JETP Lett.* **66**, 106 (1997).
19 Shalaev, V. M., Douketis, C., Stuckles, J. T., and Moskovits, M., *Phys. Rev.* **B53**, 1138 (1996).
20 Siiman, O., Bumm, L. A., Callaghan, R., Blatchford, C. G., and Kerker, M., *J. Phys. Chem.* 87, 1014 (1983).
21 Nie, S. and Emory, S. R., *Science* **275**, 1102 (1997).
22 Kneipp, K., Yang, Y., Kneipp, H., Perelman, L. T., Itzkan, I., Dasari, R. R., and Feld, M. S., *Phys. Rev. Lett.* **78**, 1667 (1997).
23 Gadenne, P., Brouers, F., Shalaev, V. M., and Sarychev, A. K., *J. Opt. Soc. Am.* **B 15**, 68 (1998).

Chaotic on-off keying in laser systems for optical secure communications

A. Uchida[*], M. Shinozuka[¶], S. Kinugawa[*], S. Yoshimori[*], and F. Kannari[¶]

[*] *Department of Electronics and Computer Systems, Takushoku University, 815-1, Tatemachi, Hachioji, Tokyo, 193-0985, Japan*

[¶] *Department of Electronics and Electrical Engineering, Keio University, 3-14-1 Hiyoshi, Kohoku-ku, Yokohama, 223-8522, Japan*

Abstract. A digital message is experimentally encoded in the chaotic carrier by turning on and off the external modulation of an acousto-optic modulator. Two binary states can be distinguished in the receiver by using the difference of accuracy of chaos-synchronization between turning on and off the modulation.

INTRODUCTION

Optical communications using chaos synchronization have attracted increasing interest as novel hardware-oriented security systems [1-3]. Optical chaos has a potential to guarantee privacy for the signal transmission because the chaotic temporal waveforms of the laser output cannot be reproduced without synchronizing two lasers or without knowing the matching parameters between the transmitter and the receiver. These hardware-oriented security systems could guarantee higher privacy than that obtained with conventional software-oriented cryptograph systems.

The methods for achieving chaos communication using lasers can be classified into three types: chaotic masking [4-6], chaotic modulation [1-3], and chaos shift keying [7,8]. For the chaotic masking method, a message is externally added on a chaotic carrier outside the chaotic oscillator, whereas for the chaotic modulation method a message is added in the chaotic oscillator and strongly affect to the generation of a new chaotic oscillation. In both methods, chaos is used for concealing the message components. On the other hand, chaos is used for representing the message signal itself for chaos shift keying method. In this method, two different chaotic attractors are generated as binary signals of 0 and 1 when one of the system parameters in the transmitter is slightly switched between two parameter values. In the receiver there are two chaotic generators which are set to one of the corresponding parameter values. Since synchronization of chaos can be obtained between the transmitter and one of the chaotic generators in the receiver with the same parameters, one can determine the transmitted binary bits by knowing which generator is synchronized. As a simpler case of the chaos shift keying, chaotic on-off keying method has been proposed in electrical circuit systems [9]. The chaotic on-off keying method requires only one chaotic generator in a receiver, whose parameter is matched to that of one of the transmitters.

CP622, *Experimental Chaos: 6th Experimental Chaos Conference*, edited by S. Boccaletti et al.

One can determine the binary bits by checking if synchronization is achieved in the receiver.

In this study, we experimentally demosntrate a revised chaotic on-off keying method in laser systems. We use one chaotic oscillation state and another oscillation state that is slightly shifted from the original chaotic state by application of an additional periodic oscillation, and switch between these two states. The accuracy of chaos synchronization in the receiver is always degraded in the shifted state. The transmitted binary signals can be determined by knowing how accurate the synchronization is. Our method has some advantages compared with the conventional chaos shift keying method. Since we add a small external modulation to degrade the accuracy of chaos synchronization instead of the parameter switching, the laser in the transmitter always remains stable without change in the temperature of the crystal or the amount of laser power. In addition, it is difficult to distinguish the two chaotic states created by an external modulation only from the transmission data due to its small modulation depth.

EXPERIMENTAL SETUP

Figure 1 shows our experimental setup for signal transmission. We used two Nd:YVO_4 microchip lasers which were pumped by laser diodes [10]. Chaotic outputs of the microchip lasers were easily obtained by modulating the injection current of the laser diode at a frequency of 0.95 MHz . The chaotic output of the microchip laser was externally modulated with an acousto-optic modulator (AOM) (IntraAction Corp., AOM-40R) with a sinusoidal periodic wave at a frequency of 4.0 MHz and a depth of less than 0.2%. The modulation of the AOM was turned on and off at a frequency of 100 kHz to encode a sequence of digital bits in the laser output as a message signal. The two oscillation states created by turning on and off the AOM can be regarded as binary signals of 1 and 0, respectively. The laser output with a message was transmitted to the receiver (no chaos in the absence of transmitted light) in a free space, and injected into the laser cavity in the receiver for chaos synchronization [10]. The output powers of the microchip lasers at the transmitter and receiver were 1.50 and 0.20 mW, respectively (measured in front of the receiver). The chaotic laser outputs of the transmitter (chaos + modulation) and the receiver (synchronized chaos) were measured simultaneously with a digital oscilloscope and a RF spectrum analyzer in the receiver.

EXPERIMENTAL RESULTS

Figure 2 shows the correlation plots between waveforms of the two laser outputs (a) without and (b) with the external modulation using an AOM. Accurate synchronization is achieved only when the modulation is switched off, as shown in Fig. 2(a). This state corresponds to binary signal 0. On the other hand, synchronization is achieved with a slightly degraded accuracy in the presence of the external modulation due to a change of the optical intensity through the AOM, as shown in Fig. 2(b). This state corresponds to binary signal 1. One can easily distinguish two different chaotic states by knowing how accurate the synchronization is.

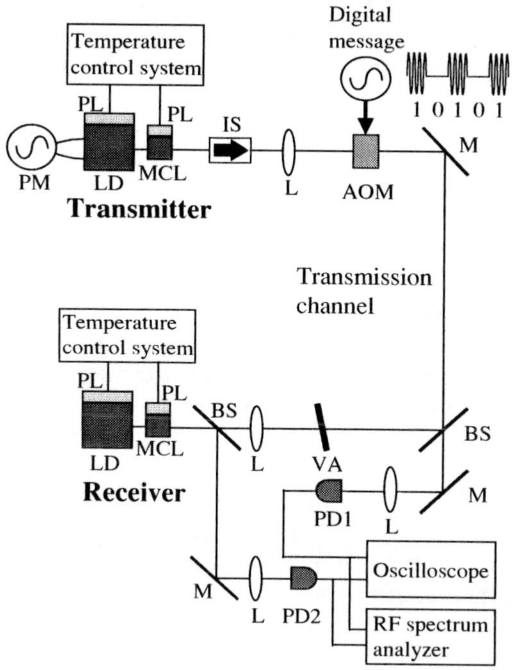

FIGURE 1. Experimental setup for signal transmission based on the chaotic on-off keying method: AOM, acousto-optic modulator; BS's, beam splitters; L's, lenses; M's, mirrors; VA, variable attenuator (neutral density filter); LD's, laser diodes; MCL's, Nd:YVO$_4$ microchip lasers; PL's, Peltier devices; IS, optical isolator; PD's, photodiodes; PM, pump modulation.

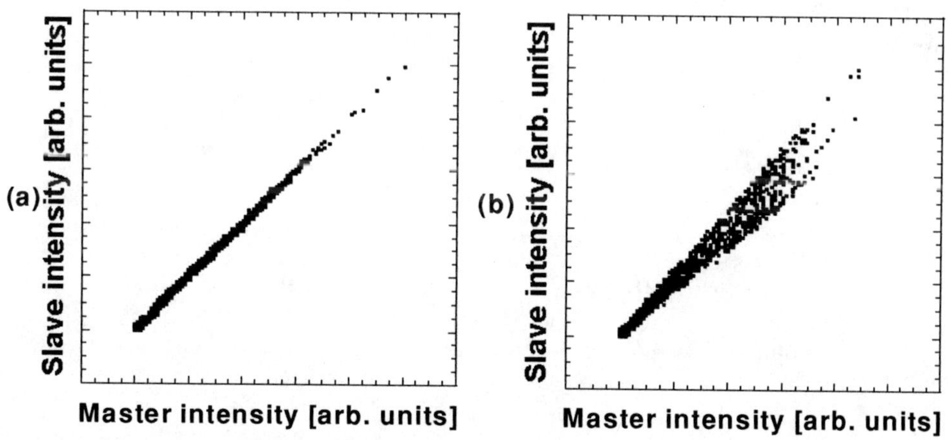

FIGURE 2. Correlation plots between waveforms of the two laser outputs (a) without and (b) with the external modulation at a frequency of 4.0 MHz by using an AOM.

To investigate the possibility of distinguishing the two binary states from only the transmission signal, we measured the RF power spectra of the two laser outputs when the modulation was turned on and off. Figure 3 shows the RF spectra of the transmitter and receiver outputs (a) without and (b) with the modulation. The modulation signal cannot be found in the RF spectrum of the transmitter in Fig. 3(b) due to its small modulation. Thus, it is very difficult to distinguish the binary bits only from the transmission data. In the receiver, the spectral peak corresponding to the external modulation frequency is clearly observed in the RF spectrum of the receiver only in Fig. 3(b). The existence of the modulation peak corresponds to the degeneration of accuracy of synchronization as shown in Fig. 2(b).

We have experimentally demonstrated transmission of a digital message encoded in a chaotic laser output. Figure 4 shows (a) the temporal waveform of the transmitter output with a digital message, (b) the temporal waveform of the synchronized receiver output, (c) the encoded digital message, (d) the absolute value of the difference between the two normalized laser outputs of (a) and (b), and (e) the decoded temporal waveform with a low-pass filter from the signal (d). Message components cannot be observed in the output of the transmitter and the receiver in Fig. 4(a) and (b). When the two laser outputs are normalized and subtracted to each other, the messages can be obtained as an envelope of chaotic oscillation as shown in Fig. 4(d). We can recover the original messages by filtering the difference of the two outputs with a low-pass filter. The digital sequence clearly appears in Fig. 4(e) and the binary bits can be judged according to a certain threshold value. Therefore, we have demonstrated a chaotic on-off keying method in laser systems.

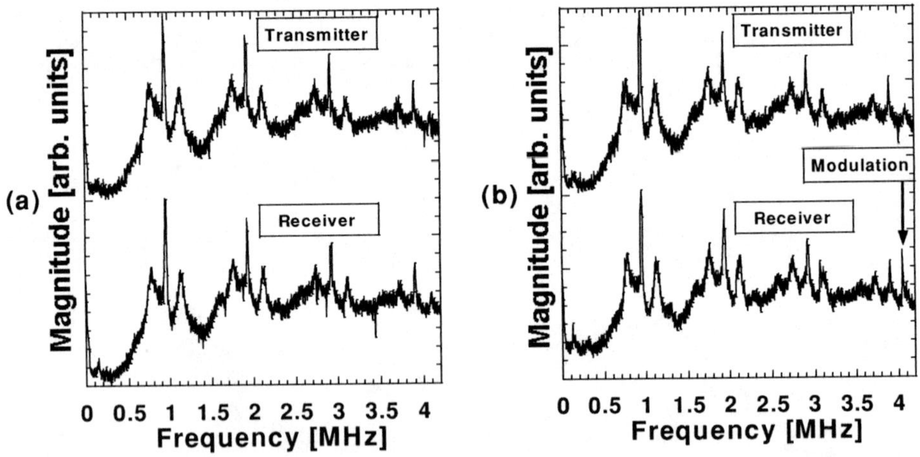

FIGURE 3. Radio-frequency spectra of the transmitter and receiver outputs (a) without and (b) with the external modulation.

Since this method is not dependent on the type of laser medium, it could be applied to semiconductor lasers in which faster chaotic oscillations can be generated at a frequency of a few GHz. One of the further works necessary for developing this system is on the transient time of chaos synchronization when the modulation is turned on and off [8]. The transient time requires more than 10 times the reciprocal of the modulation frequency of the AOM. It should be shortened for higher transmission rates.

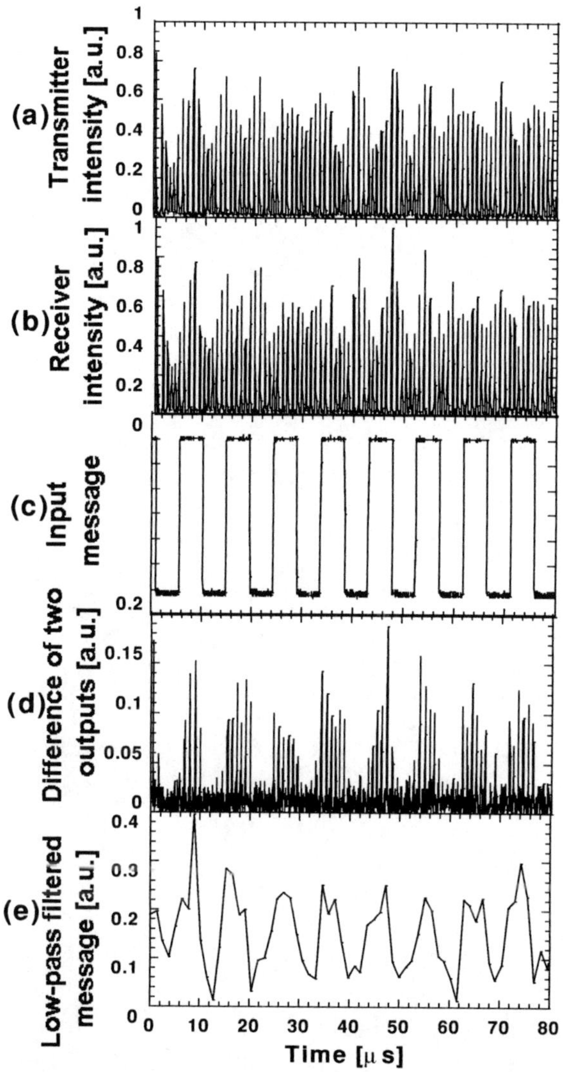

FIGURE 4. (a) Temporal waveform of the transmitter output with a digital message, (b) temporal waveform of the synchronized receiver output, (c) encoded digital message, (d) absolute values of the difference between two normalized laser outputs of (a) and (b), (e) decoded temporal waveform with a low-pass filter from the signal (d). Note the different range of the vertical axes in (a), (b) and (d).

CONCLUSION

We experimentally demonstrate a chaotic on-off keying method for secure communications by using chaos synchronization in two microchip lasers. The output of the microchip laser in the transmitter is externally modulated with an acousto-optic modulator at around 4 MHz. A digital message is encoded in the chaotic carrier by turning on and off the modulation at 100 kHz. Since the accuracy of synchronization for the slave laser in the receiver tends to be degraded in the presence of the external modulation on the injection laser signal, one can distinguish two binary states. The digital message can be recovered as an envelope of the chaotic oscillation when the difference between the two laser outputs of the transmitter and the receiver is calculated.

ACKNOWLEDGMENTS

This work was financially supported by the Sumitomo Foundation, the Sasagawa Scientific Research Grant from The Japan Science Society, the Telecommunications Advancement Foundation, Grant-in-Aid for Encouragement of Young Scientists from the Japan Society for the Promotion of Science, and the promotion and mutual aid corporation for private schools of Japan.

REFERENCES

1. VanWiggeren, G. D., and Roy, R., *Science* **279,** 1198 (1998).
2. Goedgebuer, J.- P., Larger, L., and Porte, H., *Phys. Rev. Lett.* **80,** 2249 (1998).
3. Luo, L. G., Chu, P. L., and Liu, H. F., *Photon. Tech. Lett.* **12,** 269 (2000).
4. Sivaprakasam S., and Shore, K. A., *Opt. Lett.* **24,** 1200 (1999).
5. Fischer, I., Liu, Y., and Davis, P., *Phys. Rev. A* **62,** 011801 (2000).
6. Colet, P., and Roy, R., *Opt. Lett.* **19,** 2056 (1994).
7. Annovazzi-Lodi, V., Donati, S., and Scire, A., *IEEE J. Quantum Electron.* **32,** 953 (1996).
8. Parlitz, U., Chua, L. O., Kocarev, L., Halle, K. S., and Shang, A., *Int. J. Bifur. Chaos* **2,** 973 (1992).
9. Kennedy, M. P., Kolumban, G., Kis, G., and Jako, Z., *Proceedings of the 1998 IEEE International Symposium on Circuits and Systems (ISCAS'98)* **4,** 461 (1998).
10. Uchida, A., Ogawa, T., Shinozuka, M., and Kannari, F., *Phys. Rev. E* **62,** 1960 (2000).

GHZ BANDWIDTH CHAOTIC OPTICAL DATA ENCRYPTION USING EXTERNAL CAVITY DIODE LASERS

J. Paul, S. Sivaprakasam, P. Spencer, P. Rees and K. A. Shore

University of Wales, Bangor, School of Informatics, Dean Street, Bangor LL57 1UT Wales, UK

Abstract. Synchronized external-cavity diode lasers are used for chaotic optical communication. A 1 GHz message is encoded, transmitted and decoded.

INTRODUCTION

The dynamics of various non-linear systems have been studied in the past for both fundamental understanding and practical application purposes. The possibility of controlling chaos in such systems [1-7] has been a powerful stimulant for research work and its possible application in communication systems [8]. Synchronisation of chaos has, in particular, been studied in various systems such as electronic circuits [9,10], and lasers [11-28]. An experimental demonstration of message encryption and decryption has been shown using wavelength chaos [19]. Because of the ease of operation of semiconductor lasers, it is of interest to develop a chaotic communication system utilising these lasers as the source of optical chaos [20-24]. The underlying concept for this work is that the transmitted message should be encoded within the noise like output of a chaotic transmitter. Extraction of the message requires a receiver in which the same chaos is generated as in the transmitter, which can be achieved by synchronisation of the chaos of the transmitter and the receiver. Significant interest in this field has been directed at the development of chaotic data encryption techniques suitable for use in optical communication systems which would utilise chaotic lasers as transmitters and receivers. Recent demonstration of synchronisation of chaotic diode lasers [20,25] has led to the encoding and decoding of kHz messages using chaotic diode lasers [21]. However, for

CP622, *Experimental Chaos: 6th Experimental Chaos Conference*, edited by S. Boccaletti et al.

applications in optical communications it is of great importance to effect encoding and decoding of GHz bandwidth signals. Such GHz bandwidth message transmission has been shown experimentally using erbium doped fibre lasers [26] and has been analysed theoretically for semiconductor lasers [27]. A recent work [28] on chaotic diode lasers discusses transmission of message at 561 MHz, however no explicit decoding of the message is shown.

In this paper we present experimental work showing encoding and decoding of a 1 GHz bandwidth message using chaotic diode lasers. This demonstrates the opportunity for fully exploiting the modulation bandwidth of laser diodes in chaotic optical data transmission.

EXPERIMENTAL SETUP

The experimental arrangement is shown schematically in figure 1. We have used two single mode Fabry Parot lasers emitting at 825 nm, with a line width of 200 MHz (Access Pacific, model No. APL-830-40) for our experiments. The side mode suppression ratio is –30dB. The lasers are driven by ultra low noise current sources (ILX-Lightwave, LDX-3620, noise level: $2pA/Hz^{1/2}$) and are temperature controlled using thermo-electric controllers (ILX Lightwave LDT-5412) to a precision of 0.01 K. The laser output is collimated using AR coated diode objective (Newport – FLA17). The master and slave lasers are subject to optical feedback from external mirrors (M1 and M2), the feedback strength being controlled by continuously variable neutral density filters (NDF1 and NDF2). The cavity length is 50cm in both cases. The optical isolators (OFR-IO-5-NIR-HP) ensure that the lasers are free from back reflection and the typical isolation is –41dB per isolator. Isolator IO1 and IO2 ensure that the transmitter is isolated from the receiver. The plane of polarisation of the laser is parallel to the plane of the table. Each isolator introduces a change in the plane of polarisation by 45^0. A half wave plate (HWP) corrects this polarisation change introduced by IO1 and IO2. Two identical fast photo detectors (Newport, model no. Ad-70xr) of bandwidth 6GHz are used for detection. The output of the transmitter is coupled to photodetector PD1 via beam splitter BS1 and BS2. Beam splitter BS3 acts as a coupling element between the transmitter and receiver. Beam splitter BS4 couples the receiver output to photodetector PD2. The photodetector outputs are stored in a digital storage oscilloscope (LeCroy LC564A 1GHz) and then acquired by a PC. The message is derived from a signal generator (Marconi Instruments - Model 2022) introduced by direct amplitude modulation of the laser drive current via a bias T connector (Mini-Circuits, model ZFBT-6GW) to the transmitter laser.

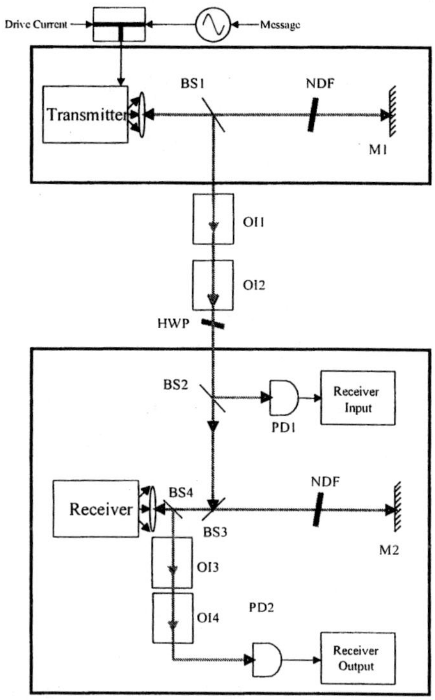

Figure 1. Schematic representation of the experimental arrangement. BS1-BS4, beam splitters; PD1, PD2, photodetector; OI1-OI4, optical isolators; M1, M2, mirrors; NDF's, neutral density filters; HWP, half wave plate.

SYNCHRONIZATION AND ENCODING

The transmitter and receiver lasers are driven to chaos by application of appropriate feedback from the external-cavity mirrors. The effective external cavity reflectivities for transmitter and receiver lasers are 1.05×10^{-4} and 1.1×10^{-4} respectively. The free-running threshold current (I_{th}) values are 27mA for both the lasers. The temperature and current of the master (27.53 ^{0}C, 30.51mA) and slave (25.82 ^{0}C, 27.23mA) are so adjusted to ensure that they operate at the same wavelength. A 1GHz sine wave from the signal generator is used as the message and is coupled to the laser drive current using a bias T connector, achieving direct amplitude modulation of depth 0.1% of the chaotic transmitter laser output. A fraction (4.2%) of the master optical output is fed to the receiver laser, which leads to synchronization of the lasers. The receiver laser set-up is as identical to the transmitter laser set-up as possible, especially in terms of cavity length, feedback strength and operating current. The time traces of the transmitter and receiver output are shown in figure 2(a) and (b) respectively. The power spectra of the transmitter and receiver output are shown in figure 2(c) and 2(d) respectively.

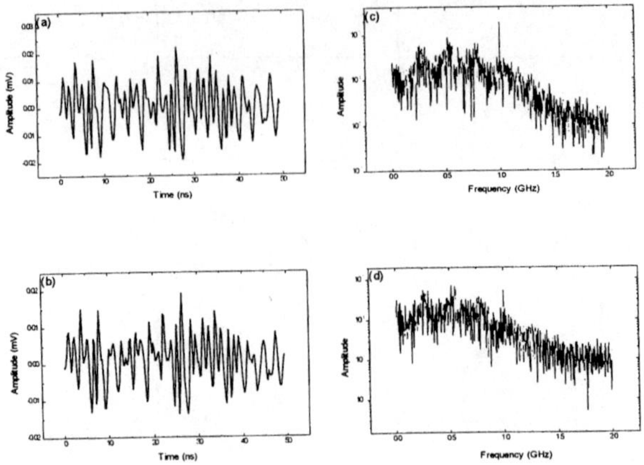

Figure 2. (a) transmitter output, (b) receiver output, (c) power spectrum of the transmitter output, the strong peak shows the message component and (d) power spectrum of the receiver output, the strong peak shows the message component.

For the present experiment deliberate use has been made of a relatively strong message signal whose presence is clearly seen in figures 2(c) and 2(d). Figures 2(a) and 2(b) show that the receiver laser output follows the output of the master as is confirmed by plotting the synchronization diagram shown in figure (3).

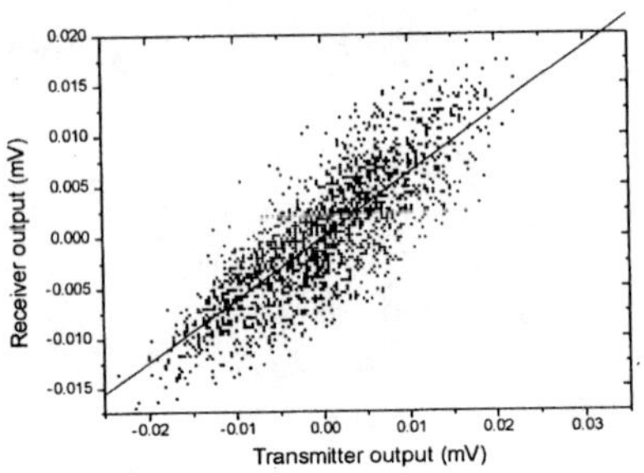

Figure 3. Synchronization plot. Receiver output intensity is plotted against the transmitter intensity. The straight line shows the fitting of the synchronization plot.

DECODING

A similar technique to that followed by Van Wiggeren and Roy [18] is used to decode the message. The message is recovered by taking the difference between the output intensities of the transmitter and receiver. The recovered message signal is shown in figure 4(a). The power spectrum of the decoded message is shown in figure 4(b), which confirms the recovery of the 1GHz message and the suppression of other frequency components.

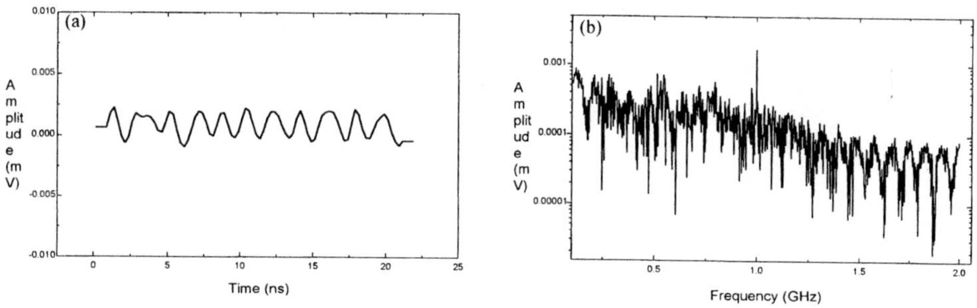

Figure 4. (a) Recovered message and (b) its power spectrum.

It has thus been shown experimentally that it is possible to encode, transmit and decode a 1 GHz bandwidth signal using chaotic external-cavity diode lasers. In future work attention will be given to the minimum signal strength which is consistent with message decoding thus demonstrating the security of the chaotic message transmission. The demonstration of this capability provides the basis for the use of chaotic optical data encryption in practical optical communication systems.

Acknowledgements

This work is partly supported by the UK EPSRC under grant GR/N63093/01 and by EPSRC Industrial CASE Studentship support by Nortel Networks, Harlow, UK for J.Paul

References

1. E. Ott, C. Greboggi and J. A. Yorke, "Controlling Chaos," *Phys. Rev. Lett.*, **64**, pp. 1196-1199, 1990.
2. J. Mφrk, B. Tromborg and J. Mark, "Chaos in semiconductor lasers with optical feedback: Theory and Experiment", *IEEE. J. Quant. Electron.*, **28**, pp. 93-107, 1992.
3. S. Bielawski, M. Bouazaoui, D. Derosier and P. Gloireux, "Controlling unstable periodic orbits by delayed continuous feedback," *Phys. Rev. E*, **49**, pp. 971-973, 1994.
4. V. Annovazzi-Lodi, S. Donati and M. Manna, "Chaos and locking in a semiconductor laser due to external injection", *IEEE. J. Quant. Electron.*, **30**, pp. 1537-1541, 1994.

5. Y. Liu and J. Ohtsubo, "Chaos control and noise suppression in external-cavity semiconductor lasers", *IEEE. J. Quant. Electron.*, **33**, pp. 56-65, 1997.
6. A. V. Naumenko, N. A., Loiko, S. I. Turovets, P. S. Spencer and K. A. Shore, "Controlling dynamics in external-cavity laser diodes with electronic implusive delayed feedback", *J. Opt. Soc. Am. B*, **15**, pp. 551-561, 1998.
7. A. Uchida, T. Sato, T. Ogawa and F. Kannari, "Nonfeedback control of chaos in a microchip solid-state laser by internal frequency resonance", *Phys. Rev. E*, **58**, pp. 7249- 7255, 1998.
8. S. Hayes, C. Greboggi and E. Ott, " Communicating with chaos", *Phys. Rev. Lett.*, **70**, pp. 3031-3034, 1993.
9. L. M. Pecora and T. L. Carroll, "Synchronisation in chaotic systems", *Phys. Rev. Lett.*, **64**, pp. 821-824, 1990.
10. K. M. Cuomo and A. V. Oppenheim, "Circuit implementation of synchronised chaos with applications to communication" *Phys. Rev. Lett.*, **71**, pp. 65-68, 1993.
11. H. G. Winful and L. Rahman, "Synchronised chaos and spatiotemporal chaos in arrays of coupled lasers", *Phys, Rev. Lett.*, **65**, pp. 1575-1578, 1990.
12. K. A. Shore and D. T. Wright, "Improved synchronisation algorithm for chaotic communications system", *Electron. Lett.*, **30**, pp. 1203-1204, 1994.
13. R. Roy and K. S. Thornburg, Jr., "Experimental synchronisation of chaotic lasers", *Phys. Rev., Lett.*, **72**, pp. 2009-2012, 1994.
14. P. Colet and R. Roy, "Digital communication with synchronised chaotic lasers", *Opt. Lett.*, **19**, pp. 2056-2058, 1994.
15. T. Sugawara, M. Tachikawa, T. Tsukamoto and T. Shimisu, "Observation of synchronisation in laser chaos", *Phys. Rev. Lett.*, **72**, pp. 3502-3506, 1994.
16. C.R. Mirasso, P. Colet and P. Garcia-Fernandez, "Synchronisation of chaotic semiconductor lasers: application to encoded communications", *IEEE. J. Phot. Tech. Lett.*, **8**, pp. 299-301, 1996.
17. D. Y. Tang, R. Dykstra, M. W. Hamilton, and N. R. Heckenberg, "Observation of generalised synchronisation of chaos in a driven chaotic system", *Phys. Rev. E*, **57**, pp. 5247-5251, 1998.
18. G. D. VanWiggeren and R. Roy, "Communication with chaotic lasers", *Science*, **279**, pp. 1198-1200, 1998.
19. L. Larger, J. P. Goedgebuer and F. Delorme, "Optical encryption system using hyperchaos generated by an optoelectronic wavelength oscillator", *Phys. Rev. E*, **57**, pp. 6618 -6624, 1998.
20. S. Sivaprakasam and K. A. Shore, "Demonstration of optical synchronisation of chaotic external-cavity laser diodes", *Opt. Lett.*, **24**, pp. 466-468, 1999.
21. S.Sivaprakasam and K.A.Shore, "Signal masking for chaotic optical communication using external-cavity diode lasers", *Opt. Lett.*, **24**, pp. 1200-1202, 1999.
22. S. Sivaprakasam and K. A. Shore, "Message encoding and decoding using chaotic external-cavity diode lasers", *IEEE. J. Quant. Electron.*, **36**, pp. 35-39, 2000.
23. S.Sivaprakasam, E.M.Shahverdiev and K.A.Shore, "Experimental verification of the synchronisation condition for chaotic external cavity diode lasers", *Phy. Rev. E*, **62**, pp. 7505-7507, 2000.
24. S.Sivaprakasam and K.A.Shore, "Critical signal strength for effective decoding in diode laser chaotic optical communications", *Phys. Rev. E*, **61**, pp. 5997-5999, 2000.
25. H. Fujino and J. Ohtsubo, "Experimental synchronisation of chaotic oscillations in external-cavity semiconductor lasers", *Opt. Lett.* **25**, pp. 625-657, 2000.
26. L.G.Luo, P.L.Chu and H.F.Lui," 1-GHz Optical Communication System Using Chaos in Erbium-Doped Fiber Lasers", IEEE.J.Photo.Tech.Lett.,12,pp269
27. H.F.Chen and J.M.Liu," Open-Loop Chaotic Synchronization of Injection-Locked Semiconductor Lasers with Gigahertz Range Modulation",IEEE.J.Quant.Electron.,**36**,27,2000
28. I.Fischer,Y.Lui and P.Davis,"Synchronization of chaotic semiconductor laser dynamics on subnanosecond time scales and its potential for chaos communication",Phys.Rev.E,**61**,5997,2000

Coherence Resonance in Stimulated Brillouin Scattering

Jennifer D. Simonotto, Valeri Kovalev*, and Robert G. Harrison

Nonlinear Dynamics Group, Department of Physics, Heriot Watt University, Edinburgh EH14 4AS Great Britain

**Also with: Lebedev Physics Institute, Moscow, Russia*

Abstract. We examine data from a stimulated Brillouin scattering experiment with very small feedback (<<1%), from a single-mode optical fiber with a CW YAG laser, and find evidence of Coherence Resonance. For low laser pump levels, the signal is stochastic in nature, as shown by analysis with local dispersion and surrogate data. As the pump level increases, a periodic signal emerges, corresponding to the round trip time of light in the fiber; at higher pump levels this periodic signal is lost, and the data once again becomes stochastic. These findings raise fundamental issues on the emergence of deterministic signals from seemingly stochastic behaviour in nonlinear optical interactions.

INTRODUCTION

An important issue of current interest in the field of non-linear optics concerns whether a signal growing from noise can evolve to possess deterministic dynamical behaviour through a non-linear interaction and, if so, what the influence of the noise on the dynamics is. In non-linear optics there are many interactions that are initiated by stochastic processes, the sources of which are either thermal fluctuations or quantum noise. A paradigm of such interactions is Brillouin scattering, which is a spontaneous light scattering process generated in a medium from acoustic waves; these waves are generated from thermal fluctuations in the medium. If the incident light field is increased to a certain level, the process undergoes exponential gain through stimulated Brillouin Scattering (SBS), in which up to 100% of the incident light may be scattered back. In this system, the process converts input pump power into mainly noise as previously described [1].

SBS without feedback is thought to be stochastic [2] (though the evidence used to suggest this is based on looking at the shape of the autocorrelation function, and should be reexamined), while with sufficient feedback (~3.5%), SBS is deterministic/chaotic [3]. But in such a system, can there be a crossover between stochasticity and determinism? In this report we consider SBS with very weak feedback (<<1%), that is, between the two limiting cases described above.

We report experimental evidence of Coherence Resonance (CR), a phenomenon well known in the nonlinear dynamics community [4-10], in which a non-zero amount of noise added to a system causes the system to produce a periodic output. In our

CP622, *Experimental Chaos: 6th Experimental Chaos Conference*, edited by S. Boccaletti et al.

experiment, through increase of the pump intensity into the system only, we see seemingly stochastic output evolve to have periodic modulation, which decays back to stochasticity upon further increase of the pump intensity.

DATA

All experimental data used was obtained for conditions of very weak feedback (~10^{-8} %) and was far below the feedback at which the dynamical behavior of the Stokes signal is mainly deterministic [3,11,12]. The feedback strength, determined by the angle of cleave of the fiber ends, was varied by variation of the angle of cleave of the exit end of the fiber only, with the input end being set at ~80°, corresponding to a reflectivity $R_1 \approx 10^{-8}$ % [13]. A representative example of data used in analysis is shown in Fig.1 which shows time recordings of the AC component of the Stokes signal power for three pump levels, taken using a fiber of length L = 240 meters. For a pump power near the SBS threshold the Stokes signal displays irregular behavior (Fig.1a), the corresponding power spectrum showing an essentially broad band distribution (see Fig.2b). On increasing the pump power, the signal shows evidence of underlying periodicity which grows to become the dominant feature in the Stokes signal (Fig.1b) and then decreases for further increase of pump power to finally revert to aperiodicity (Fig.1c). The period of the oscillatory emission is measured to be ~ 2.3 μsec, which corresponds to the round trip time for light in the fiber,

$$t_r = \frac{2nL}{c}; \tag{1}$$

the corresponding frequency f_0 = 430 kHz and harmonics are clearly evident in the power spectra in Fig.2a.

DETERMINISM VS STOCHASTICITY

In order to quantify our findings, that of periodicity emerging and subsequently submerging into irregular behavior, it is necessary to establish the nature of such aperiodicity, whether stochastic or (deterministic) chaos.

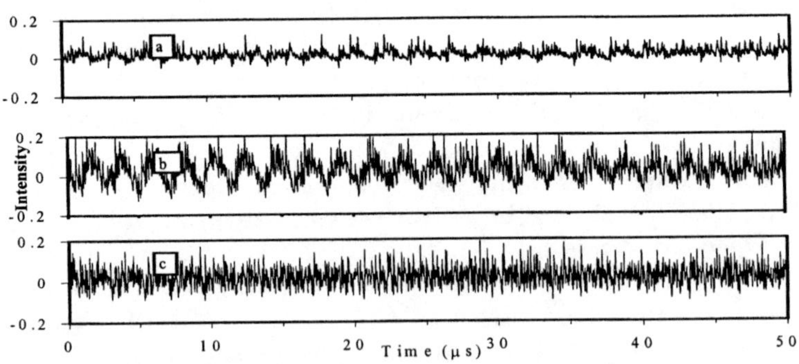

FIGURE 1. AC component of the Stokes signal at pump levels $2P_{th}$ (a), $4P_{th}$ (b), and $7P_{th}$ (c). The vertical axis is Stokes intensity for each pump level in relative units.

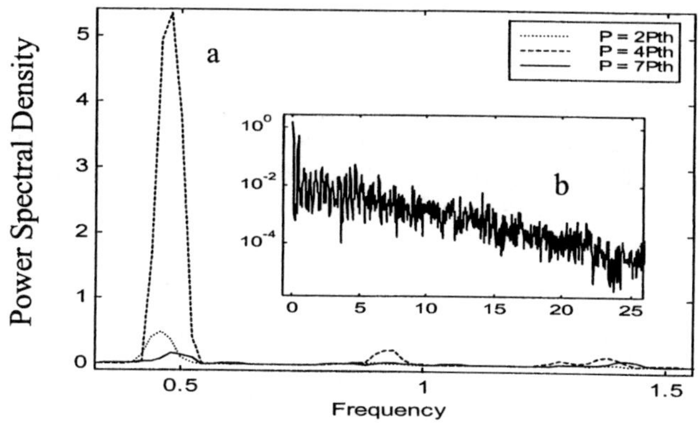

FIGURE 2. Power spectral density for three pump levels: $2P_{th}$ (dotted line), $4P_{th}$ (dashed line), and $7P_{th}$ (solid line). Main figure (a) shows the linear-linear plot of the power spectrum around the characteristic period (a). The inset shows a longer-range power spectrum of $2P_{th}$ in a log-linear plot (b). Both frequency scales are in MHz.

Surrogate data and Local Dispersion

In order to tell if data exhibits deterministic characteristics, we should distinguish it clearly from a random process. One way to do this is to create surrogate data, data that is stochastic in nature, but has certain properties in common with our data. [14] There are several different types of surrogates in use today; we use mainly real data with randomized Fourier phases (surrogate type 1) to test the hypothesis that the data is simply linearly autocorrelated gaussian noise, and Amplitude-Adjusted Fourier Transform surrogate (surrogate type 2) to test whether the data is a static nonlinear transform of linear gaussian noise [14].

We can then compare the data and surrogates in a statistical test (usually some sort of test that demonstrates clearly if data is deterministic or not) to find whether our data has the same properties as the surrogate data. If test results from our real data are not sufficiently distinguishable from those of surrogate data, then the data is stochastic in nature, depending on the type of surrogate used [14]. Many different measurements have been used: correlation dimension, Lyapunov exponent, forecasting error, local flow, local dispersion, to name a few [14-16]. We use local dispersion to do our hypothesis testing, since it remains fairly robust for up to 25% signal noise [16], gives robust results for small data sets [16], and allows the use of surrogates [15-17].

Local Dispersion is a statistical quantification of false nearest neighbors [18]. In order to calculate the local dispersion of a dataset, we first embed the data into an m-dimensional space [19], and then trace the paths of points very close to each other. If a cluster of points diverge as the embedding dimension increases, we consider these to be false nearest neighbors. This idea can be quantified as $y_n = x_n(t+\tau) - x_n(t)$, where $x_n = x_0, x_1, \ldots, x_k$; the point x_0 and its k nearest neighbors [15]. We can measure y_n at many points on the embedded space, thus making a more statistically significant measure, called the translation error [15], defined by:

$$e_{trans} = \frac{1}{k+1}\sum_{j=0}^{k}\frac{\left\|y_j - \langle y\rangle\right\|^2}{\left\|\langle y\rangle\right\|^2}. \tag{2}$$

Calculating the translation error for a range of embedding dimensions gives a series of numbers that can then be compared to the surrogate data; if a system is deterministic, it should have a translation error less than that of the surrogate data [15,16].

Characterizing the behavior of non-resonance data

We used less highly sampled data taken from a 100 meter fiber to measure whether or not the system displayed stochastic behavior. The sampling of the new system was approximately ¼ the sampling rate of the previous data, thus we were able to get data sets that were 4 times as long. We broke the data into sets of length 4000, a length at which the change in variance from one set of data to another was not large. We then ran the local dispersion test on data and surrogates (segmented in the same way), with a maximum embedding dimension of 15, a time lag of 2, number of nearest neighbors 4, exclusion time 1, and translation horizon of 1. We generated surrogates of type 1 and 2, (so testing the hypothesis that the data was different from linearly autocorrelated gaussian noise and whether the data is a static nonlinear transform of linear gaussian noise). We ran the test over 25 data sets and both types of surrogates for each pump level. The error bars shown in Fig. 3 are all one standard deviation. We tested six pump levels: 75 mW, 90 mW, 110 mW, 120 mW, 140 mW and 170 mW. One pump level (140 mW) showed obvious periodicity in the data, another example of Coherence Resonance. Looking at the Results, we see that in the 140 mW case (the periodic one, see fig 3c) the data is clearly distinguishable from the data, and stochasticity in all the rest (75 mW, 120 mW, and 170 mW are shown in fig 3a, 3b, and 3d respectively. The other pump levels, though not shown, also seemed to be stochastic), meaning that the data could not be distinguished from the surrogate data.

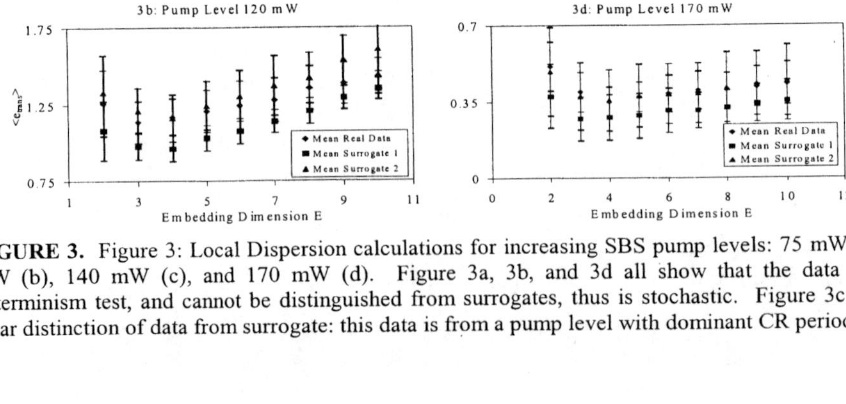

FIGURE 3. Figure 3: Local Dispersion calculations for increasing SBS pump levels: 75 mW (a), 120 mW (b), 140 mW (c), and 170 mW (d). Figure 3a, 3b, and 3d all show that the data fails the determinism test, and cannot be distinguished from surrogates, thus is stochastic. Figure 3c shows a clear distinction of data from surrogate: this data is from a pump level with dominant CR periodicity.

COHERENCE RESONANCE

To quantify CR, the time series data were taken in digital form at a sampling rate of 0.5 nanoseconds, corresponding to frequency of 2 GHz. Five sets of data were taken at each pump level, and each data set comprised 25000 data points. We analyzed the AC component of the intensity of the back-scattered Stokes radiation. All of the data sets used were taken with the laser pump intensity above the SBS threshold but below the SBS lasing threshold. The SBS threshold value of pump power was 220 mW for the fiber sample of length 240 m. For the 240 meter fiber, the highest available pump power was about seven times the threshold value for SBS.

As shown in Fig.2a and previously mentioned in the data section, the power spectral density shows definite sharp peaks associated with this periodic oscillation; the peak is highest for some medium pump value. If we normalize the power spectral densities to the background noise, we can compare the heights of the peaks more carefully, by comparing heights at

$$h_1 = e^{-1/2} h_0 , \tag{3}$$

where h_0 is the peak height of each of the power spectra normalized to the noise background at the frequency f_0. We see that indeed the spectrum for pump level 4Pth has the highest height h_l (triangles in Fig.4); the general distribution of these heights is similar to some results obtained by Wang et. al. [8] and also Gang et.al. [6].

In order to quantify the coherency of the system, we use the measure of coherence β, which has also been interpreted as similar to the signal-to-noise ratio [6]. β is a convenient measure to use as it is robust enough for statistical measures with the amount of data that we have, and can be calculated as:

$$\beta = h_0 \left(\frac{\Delta f}{f_0} \right)^{-1} , \tag{4}$$

where $\Delta f/f_o$ is the familiar quality factor Q; Δf is the peak width at h_l, and f_o is the frequency of resonance [6]. The β factor for the backscattered intensity was calculated for each dataset taken at seven different pump levels (see dots in Fig. 4). Calculating the mean and standard deviation of the Stokes signal at each pump level, we see a

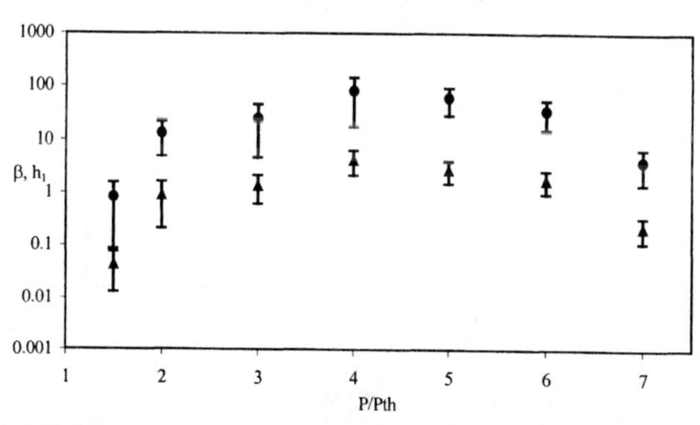

FIGURE 4. β and h_l versus pump power. Dots are mean β; triangles are the mean of the peak height at $h_l = e^{-1/2}h_0$ of the normalized power spectral density. The error bars are one standard deviation for both.

robust increase and then decrease in β at around $4P_{th}$ as the laser power is increased, with the entire range of coherency ranging over 3 orders of magnitude. We note similarities in our results in order of magnitude and shape to work by Wang et.al. [8], and similar in shape to Lee et.al. [7], Gang et.al. [6], and Potsnov et.al. [9,10].

CONCLUSIONS

In summary, we find experimental evidence of Coherence Resonance in an autonomous noise generating system subject to weak feedback. The feedback seems to serve the role of an ordering parameter in this system by which coherency emerges from and submerges into stochasticity, as shown by local dispersion testing. We do make the assumption that the system does not have more than 25% noise, above which the local dispersion test has been shown to be unreliable [16], but further testing is under way. We emphasize that our observations are for a feedback parameter value well below that for which the threshold for laser oscillation occurs, at and beyond which the system displays deterministic dynamics. The appearance of CR in SBS is made possible by the long interaction length (which gives a sufficiently high and variable power of noise), and sensitive control of feedback, both allowed by the use of a fiber medium. It is reasonable to expect CR to be found in other stimulated scattering phenomena and in fiber lasers where the noise is of amplified spontaneous emission and may be large.

REFERENCES

1. Simonotto, JD, Kovalev, V, Harrison, RG, submitted to Phys. Rev. Lett., (2001).
2. Gaeta, AL and Boyd, RW, Phys. Rev. A, **44**, 3205-3209 (1991).
3. Harrison, RG, Uppal, JS, Johnston, A, and Moloney, JV, Phys. Rev. Lett., **65**, 167-170 (1990).
4. For a general review of nonlinear ordering and Stochastic Resonance, see: Landa, PS, Zaikin, AA, Ushakov, VG, Kurths, J, Phys. Rev. E, **61**, 4809-4820 (2000); Landa, PS, McClintock, PVE, Physics Reports, **323**, 1-80 (2000); Anishchenko, VS, Neiman, AB, Moss, F, Schimansky-Geier, L, Physics-Uspekhi, **42**, 7-36 (1999); and Gammaitoni, L, Hänggi, P, Jung, P, Marchesoni, F, Reviews of Modern Physics, **70** 223-287 (1998).
5. Pikovsky, A, Kurths, J, Phys. Rev. Lett., **78**, 775-778 (1997); Cartwright, JHE, Phys. Rev. E, **62**, 1149-1154 (2000); Lindner B, and Schimansky-Geier, L, Phys. Rev. E, **60**, 7270-7276 (1999); and Giacomelli, G, Giudici, M, Balle, S, and Tredicce, J, Phys. Rev. Lett., **84**, 3298-3301 (2000).
6. Gang, H, Ditzinger, T, Ning, CZ, Haken, H, Phys. Rev. Lett., **71**, 807-810 (1993).
7. Lee, SG, Neiman, A, Kim, S, Phys. Rev. E, **57**, 3292-3297 (1998).
8. Wang, Y, Chik, DTW, Wang, ZD, Phys. Rev. E, **61**, 740-746 (2000).
9. Potsnov, DE, Han, SK, Yim, TG, Sosnovtseva, OV, Phys. Rev. E, **59**, R3791-R3794 (1999).
10. Han, SK, Yim, TG, Potsnov, DE, Sosnovtseva, OV, Phys. Rev. Lett., **83**, 1771-1774 (1999).
11. Gaeta, L and Boyd, RW, Int. J. of Nonlinear Optical Physics, **1**, 581-594 (1992).
12. Damming, M, Zinner, G, Mitschke, F, and Welling, H, Phys. Rev. A, **48**, 3301-3309 (1993).
13. Marcuse D and Stone, J, J. Lightwave Technol. **4**, 377-381 (1986).
14. Theiler, J, Eubank, S, Longtin, A, Galdrikian, B, and Farmer, JD, Physica D, **58**, 124-141 (1992).
15. Wayland, R, Bromley, D, Pickett, D, and Passamante, A, Phys. Rev. Lett. **70,** 580-582 (1993).
16. Chang, T, Sauer, T, and Schiff, SJ, Chaos **5**, 118-126 (1995).
17. Schiff, SJ, Jerger, K, Chang, T, Sauer, T, Aitken, PG, Biophysics Journal, **67**, 684-691 (1994).
18. Kennel, MB, Brown, R, Arbarbanel, HDI, Phys. Rev. A, **45** 3403-3411 (1992).
19. Sauer, T, Yorke, JA, Casdagli, M, J. Stat. Phys., **65** 579-616 (1991).

Experimental observation of synchronization and anti-synchronization of chaotic low-frequency-fluctuations in external cavity semiconductor lasers

Immo Wedekind* and Ulrich Parlitz*

*Drittes Physikalisches Institut, Universität Göttingen
Bürgerstraße 42-44, D-37073 Göttingen, Germany

Abstract. Experimental observations of synchronization and anti-synchronization of low-frequency-fluctuations of external cavity semiconductor lasers are presented. Antisynchronization occures not only for the well known power drop-out dynamic but also for the power jump-up dynamic. Detailed experimental investigations of the parameter space reveal several seperate areas of (anti-)synchronization. Numerical simulations with the Lang-Kobayashi model equations are presented for the antisynchronized power drop-out case.

Chaotic fluctuations in semiconductor lasers can occure in different situations: due to periodic modulation of the pump current [1], due to electro-optical feedback where the pump current is modulated by the emitted light intensity [2, 3], or due to an external cavity (optical feedback) [4].

External cavity semiconductor lasers (ECSLs) has been a subject of extensive research during the last 15 years, because of its importance in technical applications like optical data storage or optical fiber communications [7]. In most of these cases, effects due to optical feedback are unwanted and engineers try to avoid them.

A typical phenomenon occurring in ECSLs are low frequency fluctuations (LFFs), where the intensity of the emitted light oscillates in the 5-10 GHz range with an envelope that shows relatively slow, irregular fluctuations (about 10 MHz). These slow fluctuations are either power drop-outs or power jump-ups as illustrated in Fig.1.

Power drop-outs have been studied very much in detail during the last 15 years [8] and are usually meant when the phenomenon of LFFs is addressed. In particular, the question whether the underlying dynamics is (mainly) a stochastic process or governed by a chaotic attractor has been discussed controversely [9, 10, 11, 12]. Numerical simulations based on a deterministic model [13] indicate that LFF dynamics is governed by a very high-dimensional chaotic attractor and is thus very difficult to distinguish from a stochastic signal, a feature that makes them interesting candidates for an optical encryption system.

For high pump currents and high reflectivity of the mirror of the external cavity power jump-ups occur (Fig.1b) that have first been reported by Pan et al. [14], but have not been investigated further since then.

Synchronization of low frequency power drop-outs has been shown first numerically

CP622, Experimental Chaos: 6th Experimental Chaos Conference, edited by S. Boccaletti et al.

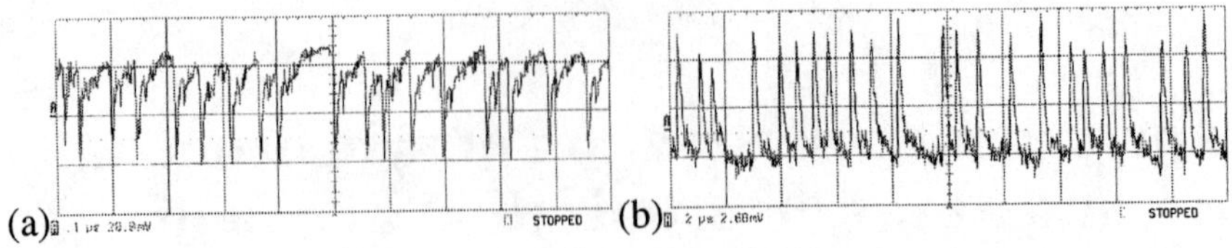

FIGURE 1. Intensity vs. time for an ECSL showing low frequency fluctuations: (a) power drop-outs (b) power jump-ups.

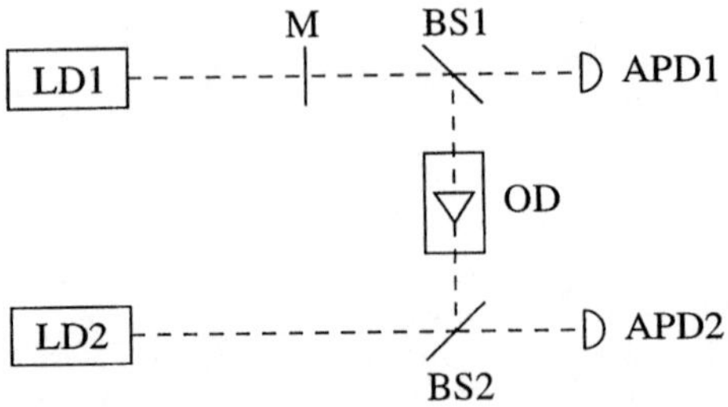

FIGURE 2. Setup of uni-directionally injection coupled external cavity semiconductor lasers studied in this report. The size of the external cavity is about 40cm. (LD, laser diode; M, mirror; BS, beam splitter; OD, optical diode; APD, avalanche photo diode.)

by Ahlers et al. [13] simulating a pair of unidirectionally injection coupled semiconductor lasers as shown in Fig. 2. Soon after this successful simulation synchronization was also demonstrated experimentally by Takiguchi et al. [15] and Sivaprakasam and Shore [16]. Figure 3a shows an experimental example for synchronization of this kind of LFFs.

Depending on the coupling the phenonmenon of "anticipation" of chaos synchronization [17] may occur where the intensity fluctuations of the response laser run ahead the output of the drive [18].

Only recently it was demonstrated experimentally that not only synchronization of power drop-outs can be achieved but also anti-synchronization where the intensity of the response laser jumps up when the drive intensity drops [19, 20] as shown in Fig. 3b. Switching between both types of synchronization is possible, for example, by changing slightly the pump current of the drive system [19]. This feature may be of importance for potential applications in communication systems.

In this article we present first experimental observations of (anti-)synchronization of low frequency power jump-ups as shown in Figs. 3c,d. The drive laser produces irregular jump-ups and the response laser follows these jump-ups (synchronization, Fig. 3c) or shows synchronized power drop-outs (Fig. 3d) that we call anti-synchrony.

In order to answer the question where in parameter space (anti-)synchronization occurs we have systematically varied the pump currents and temperatures of both unidirectionally coupled lasers. To monitor synchronization, intensity time series of both the drive and the response laser have been sampled by means of an avalanche photo diode (see Fig. 2). For each data set a *synchronization index S* was computed as the product of the peak-to-peak-amplitude of the response laser intensity and the largest cross correlation of both signals. Vanishing values of the synchronization index indicate lack of any

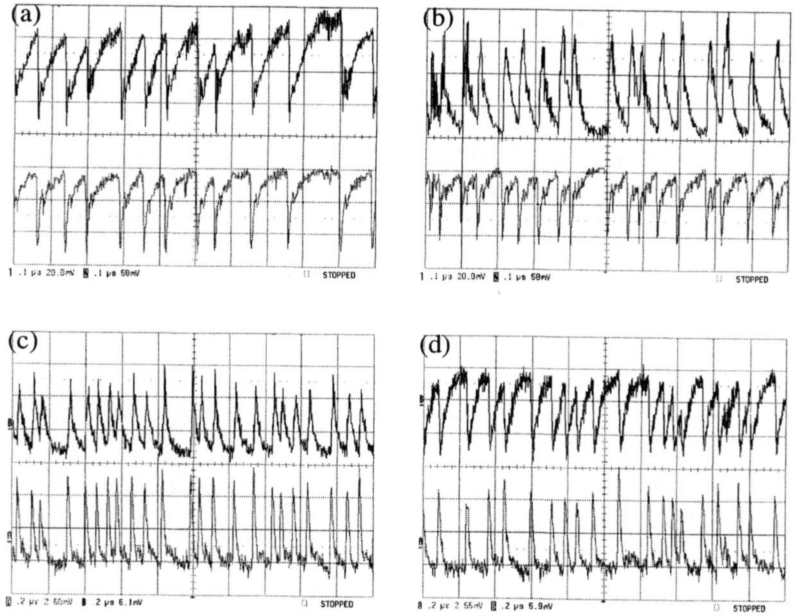

FIGURE 3. Experimental observation of synchronization and anti-synchronization of low frequency fluctuations. The lower and upper traces in each subfigure correspond to drive and response, respectively. (a) power drop-outs, synchronization; (b) power drop-outs: anti-synchronization; (c) power jump-ups: synchronization; (d) power jump-ups: anti-synchronization.

synchronization, large positive values correspond to synchronization and negative values of large magnitude occur for anti-synchronization. Weighting the linear correlation by the response amplitude turned out to be necessary to avoid spurious results, because outside the synchronization regime the intensity of the response laser is so weak that random fluctuations due to discretization errors may indicate artificial correlations. Figure 5 shows four phase diagrams where the synchronization index S is plotted grey-scaled vs. pump current and temperature of one of the lasers (the pump current and temperature of the other laser are kept fixed). Figure 5a shows the synchronization diagram for the case where the drive laser displays low frequency power drop-outs and the pump current I_2 and temperature T_2 of the response laser are varied. The white and black stripes correspond to regions where synchronization or anti-synchronization occurs, respectively. A similar pattern is found when varying the pump current I_1 and the temperature T_1 of the drive laser while keeping the parameters of the response laser fixed (Fig. 4b).

Figures 4c and 4d show corresponding results for low frequency jump-up dynamics of the driving laser.

The stripe structure depends on both temperature and pump current. This leads to the conclusion that the responsible parameter for synchronization and anti-synchronization is whether pump current nor temperatur directly but the tuning of both laser light frequencies.

In order to model the observed synchronization phenomena we used the well-known

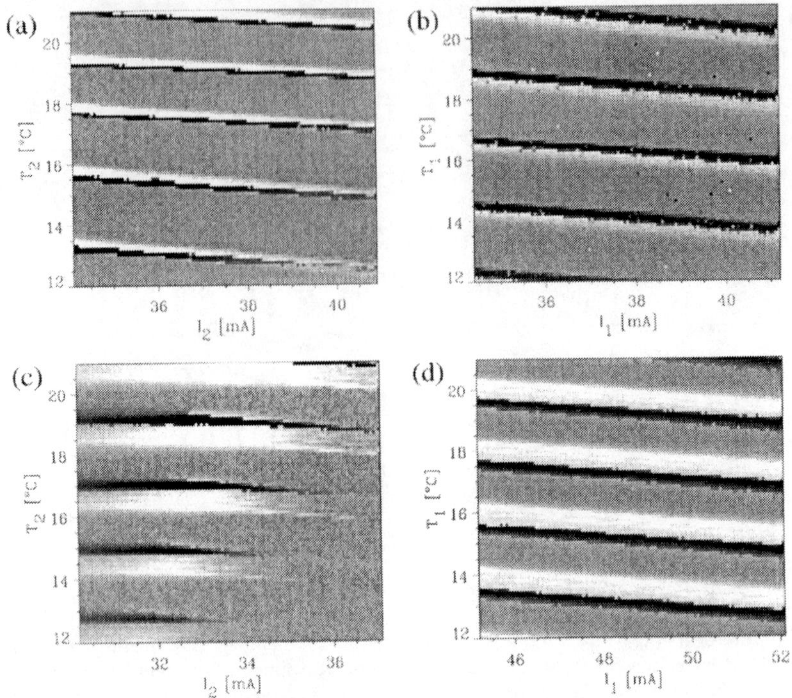

FIGURE 4. Synchronization index S (grey scaled) vs. pump current I and temperature T. Light grey indicates synchronization and dark grey anti-synchronization. (a) drive laser: power drop-outs, S vs. pump current I_2 and temperature T_2 of the response. (b) drive laser: power drop-outs, S vs. pump current I_1 and temperature T_1 of the drive. (c) drive laser: power jump-ups, S vs. pump current I_2 and temperature T_2 of the response. (d) drive laser: power jump-ups, S vs. pump current I_1 and temperature T_1 of the drive.

Lang-Kobayashi equations [21] for the electron population N and the slowly varying complex amplitude $E(t)$ of the intracavity optical field oscillating at the angular frequency ω_0. These equations are generally considered to give a valid approximation of a single mode SL with weak to moderate optical feedback from an external resonator.

$$\begin{aligned}
\frac{d}{dt}E(t) &= \frac{1}{2}\Big(-i\alpha G_N[N(t)-N_{sol}]+[G(N)-\Gamma]\Big)E(t) \\
&\quad +\kappa E(t-\tau)e^{i\omega_0\tau} \\
\frac{d}{dt}N(t) &= pJ_{th}-\gamma N(t)-G(N)|E(t)|^2
\end{aligned}$$

$$\begin{aligned}
\frac{d}{dt}\tilde{E}(t) &= \frac{1}{2}\Big(-i\alpha G_N[\tilde{N}(t)-N_{sol}]+[G(\tilde{N})-\Gamma]\Big)\tilde{E}(t) \\
&\quad +\rho\langle E(t-\tau_c)e^{i(\omega_0\tau_c+\Delta\omega_0 t)},e^{i\varphi}\rangle e^{i\varphi} \qquad (1)
\end{aligned}$$

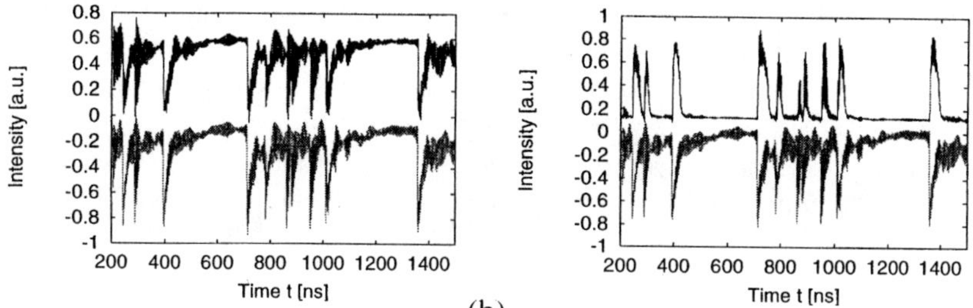

(a) (b)

FIGURE 5. Numerical simulation of (a) synchronization and (b) anti-synchronization of low frequency power drop-outs. For comparison with the experimentally measured intensities the signals are low-pass filtered (500 MHz). The lower and upper traces in each subfigure correspond to drive and response, respectively.

$$\frac{d}{dt}\tilde{N}(t) \quad = \quad \tilde{p}J_{th} - \gamma\tilde{N}(t) - G(\tilde{N})|\tilde{E}(t)|^2$$

The delay τ_c due to coupling depends on the distance between both lasers and was chosen to equal zero in the simulations. The meaning and values of the other parameters are given in Table 1.

The second laser system possesses no external cavity. The difference between the emitted light frequencies of both lasers $\Delta\omega_0$ leads to a modulation $e^{i\Delta\omega_0 t}$ of the coupling term in equation (1). The polarization due to the Faraday isolator used for uni-directional coupling (see Fig. 2) is modelled by a scalar product $\langle\cdot,\cdot\rangle$ of the transmitted electrical field with $e^{i\varphi}$ in the coupling term where φ stands for the angle of the output of the polarizer. Numerical simulations showed that it is necessary to take the effect of the polarizer into account in order to obtain anti-synchronization.

TABLE 1. Parameters used in the simulations, values taken from Ref. [14].

solitary laser carrier number	$N_{\rm sol}$	5.9×10^8
differential optical gain	G_N	$3.5 \times 10^3\,{\rm s}^{-1}$
external cavity roundtrip time	τ	4 ns
linewidth enhancement factor	α	3.0
stimulated emission rate	G	$\approx G_N \cdot [N - N_{sol}]$
carrier decay rate	γ	$5.29 \times 10^8\,{\rm s}^{-1}$
cavity decay rate	Γ	$7.14 \times 10^{11}, {\rm s}^{-1}$
reflection coefficient	κ	1.75×10^{11}
coupling coefficient	σ	5×10^{11}
laser threshold current	$J_{\rm th}$	$3.12 \times 10^{17}{\rm s}^{-1}$
wavelength	$2\pi c/\omega_0$	635 nm

Figure 5 shows simulated synchronization (Fig. 5a) and anti-synchronization (Fig. 5b) of low-frequency drop-outs. Similar results for power jumps-ups have not been observed yet due to severe numerical difficulties upon solving the Lang-Kobayashi delay differential equations for (very) high reflection coefficients. Another challenge is the proper explanation of the physical mechanisms underlying the phenomena of synchronization and anti-synchronization. The occurrence of "stripes" of (anti-)synchronization in Fig. 3

may be explained by changes of the wavelength when varying pump current or temperature, but a complete model still lacks.

We thank C. Massoller, R. Roy, H.D.I. Abarbanel, C. Mirasso and the members of our nonlinear dynamics group for stimulating discussions on chaos synchronization in semiconductor lasers. Furthermore, we thank K. Lautscham, W. Ebrecht and H. Hohmann for technical advice and assistence and the BMBF (13N7979) for financial support.

REFERENCES

1. J. Sacher et. al, *Phys. Rev. A* **45**(3), 1893 (1992).
2. C.-H. Lee and S.-Y. Shin, *Appl. Phys. Lett.* **62**(9), 922 (1993)
3. S. I. Turovets, J. Dellunde, and K. A. Shore, *J. Opt. Soc. Am. B* **14**(1), 200 (1997)
4. J. Mørk, B. Tromborg, and J. Mark, *IEEE J. Quantum Electron.* **28**, 93 (1992).
5. J.-P. Goedgebuer, L. Larger, and H. Porte, *Phys. Rev. Lett.* **80**, 2249 (1998).
6. H.F. Chen and J.M. Liu, *IEEE J. Quantum Electron* **36**, 27 (2000).
7. I. Fischer, G. H. M. van Tartwijk, A. M. Levine, W. Elsäßer, E. Göbel, and D. Lenstra, *Phys. Rev. Lett.* **76**, 220 (1996); I. Fischer, T. Heil, M. Münkel, and W. Elsäßer, *SPIE* **3283**, 63 (1998).
8. H. Tempin, N. A. Olsson, J. H. Abeles, R. A. Logan, and M. B. Panish, *IEEE J. Quantum Electron.* **22**, 286 (1986)
9. T. Sano, *Phys. Rev. A* **50**(3), 2719 (1994)
10. J. Sacher, W. Elsäßer, and E. O. Göbel, *Phys. Rev. Lett.* **63**, 2224 (1989).
11. G. H. M. van Tartwijk, A. M. Levine, and D. Lenstra, *IEEE J. Sel. Topics Quantum Electron.* **1**, 466 (1995).
12. C. H. Henry and R. F. Kazarinov, *IEEE J. Quantum Electron.* **QE-22**, 294 (1986); A. Hohl, H. J. C. van der Linden, and R. Roy, *Opt. Lett.* **20**, 2396 (1995); M. Giudici, C. Green, G. Giacomelli, U. Nespolo, and J. R. Tredicce, *Phys. Rev. E* **55**, 6414 (1997); D. W. Sukow, J. R. Gardner, and D. J. Gauthier, *Phys. Rev. A* **56**, R3370 (1997).
13. V. Ahlers, U. Parlitz and W. Lauterborn, *Phys. Rev. E* **58**(6), 7208 (1999).
14. M.-W. Pan, B.-P. Shi and G.R. Gray, *Optics Letters* **22**(3) 166 (1997).
15. Y. Takiguchi, H. Fujino and J. Ohtsubo, *Optics Letters* **24**(22), 1570 (1999).
16. S. Sivaprakasam and K.A. Shore, *Optics Letters* **24**(7), 466 (1999); S. Sivaprakasam, E.M. Shahverdiev, and K.A. Shore *Phys. Rev. E* 62(5) 7505 (2000).
17. H.U. Voss, *Phys. Rev. E* **61**(5) 5115 (2000).
18. C. Masoller, *Phys. Rev. Lett.* **86**(13) 2782 (2001).
19. I. Wedekind and U. Parlitz, *Int. J. Bif. Chaos* **11**(4), 1141 (2001).
20. S. Sivaprakasam, I. Pierce, P. Ress, P.S. Spencer, and K.A. Shore, *Phys. Rev. A.* **64**, 013805 (2001).
21. R. Lang and K. Kobayashi, *IEEE J. Quantum Electron.* **QE-16**, 347 (1980).

HYDRODYNAMICS

Lagrangian measurement in fully developed turbulence

N. Mordant, P. Metz, J.-F. Pinton* and O. Michel†

*Laboratoire de Physique, École Normale Supérieure, 46 allée d'Italie, F-69007 Lyon, France
†Laboratoire d'Astrophysique, Université de Nice, Parc Valrose, F-06108, Nice, France

Abstract. We have developed a new experimental technique to measure the Lagrangian velocity of tracer particles in a turbulent flow, based on ultrasonic Doppler tracking. This method yields a direct access to the velocity of a single particule at a turbulent Reynolds number $R_\lambda = 1200$. Its dynamics is analyzed with two decades of time resolution, below the Lagrangian correlation time. We observe that the Lagrangian velocity spectrum has a Lorentz form $E^L(\omega) = u_{\rm rms}^2 T_L/(1+(T_L\omega)^2)$, in agreement with a Kolmogorov-like scaling in the inertial range. The probability density function (PDF) of the velocity time increments displays a change of shape from quasi-Gaussian a integral time scale to stretched exponential tails at the smallest time increments. This intermittency, when measured from relative scaling exponents of structure functions, is more pronounced than in the Eulerian framework.

INTRODUCTION

Lagrangian characteristics of fluid motion are of fundamental importance in the understanding of transport and mixing. It is a natural approach for reacting flows or pollutant contamination problems to analyze the motion of individual fluid particles [1]. Another characteristic of mixing flows is their high degree of turbulence. For practical reasons, most of the experimental work concerning high Reynolds number flows has been obtained in the Eulerian framework. Lagrangian measurements are challenging because they involve the tracking of particle trajectories: enough time resolution, both at small and large scales, is required to describe the turbulent fluctuations.

Early Lagrangian information have been extracted from the dispersion of particles, following Taylor's approach. Recently numerical and experimental studies have focused on resolving the motion of individual fluid or tracer particles. The emerging picture is as follows. The one-component velocity auto-correlation function is quasi-exponential with a characteristic time of the order of the energy injection scale [2, 3, 4]. The velocity power spectrum is expected to have a scaling $E^L(\omega) \propto \omega^{-2}$, as recently reported [5, 6] and expected from a Kolmogorov similarity arguments. In the same spirit, the second order structure function should scale as $D_2^L(\tau) = C_0\varepsilon\tau$, where ε is the the power dissipation. Measurements of atmospheric balloons [7] have given $C_0 = 4 \pm 2$ and a limit $C_0 \to 7$ has been suggested in stochastic models [8]. Recent experiments [9, 10] using high speed optical techniques have shown that the statistics of the Lagrangian acceleration are strongly non-Gaussian.

We have developed a new experimental method, based on sonar techniques [11], in

CP622, *Experimental Chaos: 6th Experimental Chaos Conference*, edited by S. Boccaletti et al.

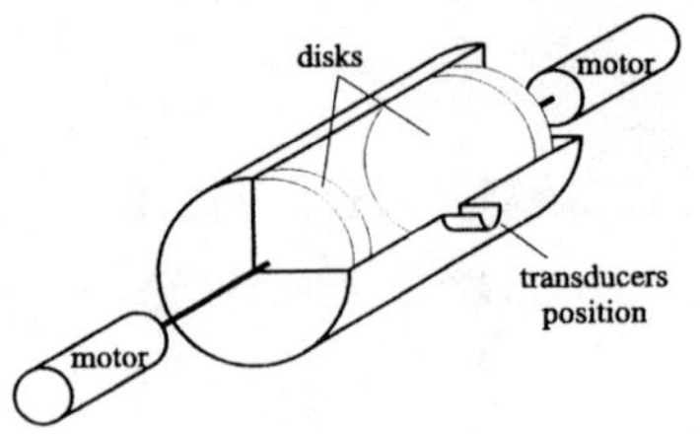

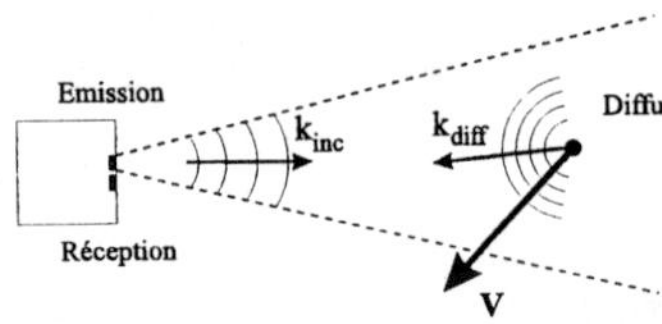

FIGURE 1. Setup & Measurement principle. The flow is produced in a cyminder in the gap of two counter rotating dics. It is seeded with one 250 μm neutrally buoyant sphere. As it is continuously insonified by a monochrmatic acoustic wave, it scatters a Doppler shifted sound which is detected by several arrays of transducers. A numerical tracking of the varying Doppler shift yields the velocity of the particle.

order to study in a laboratory experiment the Lagrangian velocity across the inertial range of time scales. We obtain the first measurement of single particle velocity for times up to the flow large scale turnover time, at high Reynolds number. In this Letter, we report the results of this measurements and compare with previous observations and numerical predictions.

SETUP AND METHOD

The turbulent flow is produced in the gap between two counter-rotating discs [12]. This setup has the advantage to generate a strong turbulence in a compact region of space, with no mean advection. In this way, particles can be tracked during times comparable to the large eddy turnover time. Discs of radius $R = 9.5$ cm are used to set water into motion inside a cylindrical vessel of height $H = 18$ cm – cf. figure 1. To ensure inertial entrainment, the discs are fitted with 8 blades with height $h_b = 5$ mm. In the measurement reported here, the power input is $\varepsilon = 25$ W/kg. It is measured on the experiment cooling system, from the injection-dissipation balance. The integral Reynolds number is $Re = R^2\Omega/\nu = 6.5\ 10^4$, where Ω is the rotation frequency of the discs (7.2 Hz), and $\nu = 10^{-6}$ m^2/s is the kinematic viscosity of water. A conventional turbulent Reynolds number can be computed from the measured *rms* amplitude of velocity fluctuations ($u_{rms} = 0.98$ m/s) and an estimate of the Taylor microscale ($\lambda = \sqrt{15\nu u_{rms}^2/\varepsilon} = 0.88$ mm); we obtain $R_\lambda = 740$. This value is consistent with earlier studies in the same geometry; it corresponds to the range of turbulent Reynolds numbers where measurements of particle acceleration have been reported [10].

Our technique is based on the principle of a continuous Doppler sonar. A small

(2mm×2mm) emitter continuously insonifies the flow with a pure sine wave, at frequency $f_0 = 2.5$ MHz (in water). A moving particle backscatters the ultrasound towards an array of receiving transducers, with a Doppler frequency shift related to the velocity of the particle: $2\pi\Delta f = \mathbf{q}.\mathbf{v}$. The scattering wavevector $\mathbf{q}$ is equal to the difference between the incident and scattered directions. The intersection of the incident sound emitting cone and the of the receiver dtetection cone define the measurement volume. In our setup it is located in the center of the cylinder where the flow is quasi homogeneous and isotropic. A very small number of particles (about 5) seed the flow so that at any given time there is only one scatterer in the measurement volume. This scheme enables us to track single particles velocity for times comparable to the flow forcing time.

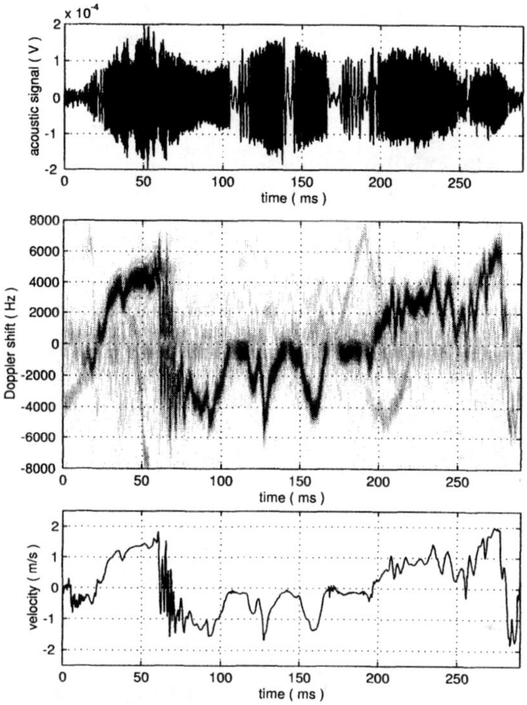

FIGURE 2. Signal processing. The Doppler shifted sound scattered by the moving particle which is first heterodyned at low frequency and receorded digitally (top): low frequencies correpond to slow velocity of the particle whereas high frequencies correspond to large velocities (along the scattering direction). This shown in the middle figure in the spectrogram of the signal. The actual tracking of the Doppler shift is performed using a high resolution parametric method and yields the final velocity profile shown in the bottom figure.

The velocity of the particle is measured as the Doppler shift of the incident sound. On the signal recorded at the receiving transducer array it appears as a frequency modulation. The demodulation is done numerically, to obtain the component of the particle velocity along the scattering wavevector $\mathbf{q}$. The study reported here is made with a single array of transducers so that only one Lagrangian velocity component is measured. To obtain a fast detection of the Doppler shift, we have developed a high

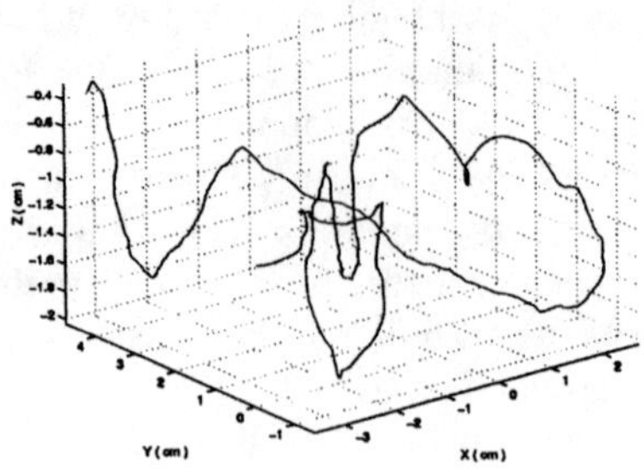

FIGURE 3. Particle trajectory: the position of the particle results from the tracking of the direction of emmission of the scattered sound by the array of receiving transducers. In this example the tracer particle is followed during 200 ms.

resolution parametric method which relies on an Approximated Maximum Likelihood scheme coupled with a generalized Kalman filter [11].

The flow is seeded with a small number of neutrally buoyant (density 1.06) polystyrene spheres with diameter $d = 250\ \mu$m . It is expected that the particles follow the fluid motion up to characteristic times of the order of the turbulence eddy turnover time, at a scale corresponding to their diameter, i.e. $\tau_{\min} \sim d/u_d \sim \varepsilon^{-1/3} d^{2/3}$, using standard Kolmogorov phenomenology. For beads of diameter 250 μm, one estimates $\tau_{\min} \sim 1.3$ ms. This value is within the resolution of the demodulation algorithm whose cut-off frequency is at 3 kHz. Note that the Kolmogorov dissipative time ($\tau_\eta = \sqrt{\nu/\varepsilon} = 0.2$ ms) is smaller, so that we do not expect to resolve the dissipative region. The statistical quantities are calculated from 3×10^6 velocity data points, taken at a sampling frequency equal to 6500 Hz.

CHARACTERISTICS OF PARTICLE MOTION

Trajectory

As the acoustic receivers are made of two crossed arrays of transducers, it is possible to decode the the angle from which the moving particles scatters the incoming sound. Each line array gives an angle and so the two line arrays in each receiver define one line of emmission. With two receivers at 90 degrees, one gets two lines of emmission whose shortest distance is found to be less than the diamter of the moving particle. This method thus yields the position of the particle. This technique is not used to compute the velocity, which is obtained directly from the Doppler shift tracking, but it gives an idea of the performance of the method and of the particle tracking. A typical trajectory is shown in figure 3: the particle is tracked for 200 ms, a time that is longer that the disc rotation period. Its motion spans a volume of (5 cm $\times$ 5 cm $\times$ 2.5 cm). This exmaple confirms that our techniques allows a measurement the motion of tracer particles during time and length scales comparable to the turbulent flow integral characteristic scales.

Trajectory

At small scales, the motio of the particle is very irregular. As noted in the study of La Porta *et al.* [10], the acceleration of the particle can take very larges values. In our measurement although the acceleration of the tracer particle is lower than that of a fluid particles (size effect), we have measured values larger than 2000 m/s^2 – see figure 4.

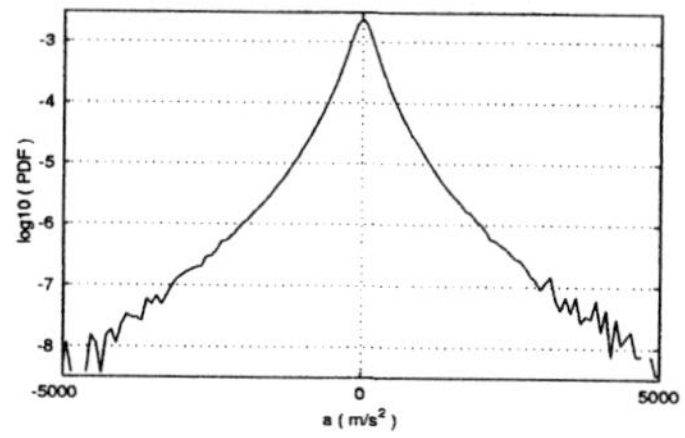

FIGURE 4. Particle acceleration : Probability Density Function (PDF) of the acceleration of the tracer particle. Note the extremely large values reached ($> 500g$!) and the strongly non-Gaussian shape of the distribution.

This value is considerably higher that what is expected by a smooth K41 cascading process; indeed, in that framework, the acceleration at scale ℓ should be of the order of $\varepsilon^{2/3}\ell^{-1/3}$. Its largest value is at the Kolmorov scale, it is equal to $a_\eta = \varepsilon^{2/3}\eta^{-1/3} \sim$ 300 m/s^2 and its value at λ (the size of the tracer particle) is $a_\lambda = \varepsilon^{2/3}\lambda^{-1/3} \sim 90$ m/s^2. Our measurements show particle accelerations that are over one order of magnitude larger. The values measured are however consistent with velocities of the order of u_{rms} being realized over distances of the order of the Taylor microscale (the particle size): $u_{\mathrm{rms}}^2/\lambda \sim 1000$ m/s^2. Such events are usually associated with the formation of vortex sheets and vortex tubes (filaments), which have been studied swirling flows [13]. Exemple of occurence of this type are shown in figure 5.

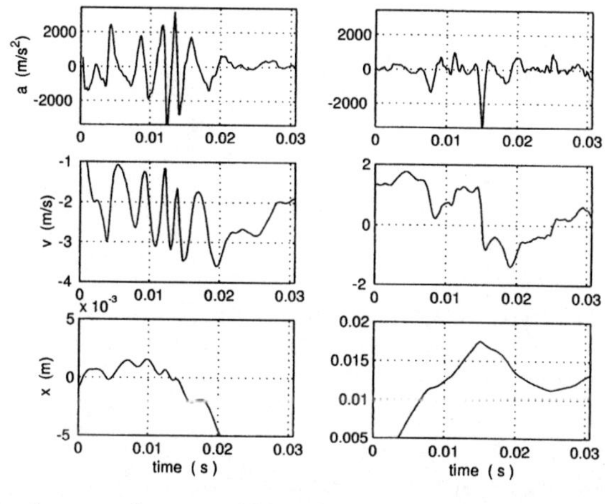

FIGURE 5. Example of events where very hign particle acceleration are reached. (left): the position, velocity and acceleration of the particle oscillates as if it is trapped into a spiraling vortex motion; (right): the velocity of the particle undergoes a sudden change as would if crossing a mixing layer between two large eddies (vortex sheet).

STATISTICS OF THE LAGRANGIAN VELOCITY

Auto-correlation function and spectrum

We now turn to statistical characteristics of the Lagrangian velocity and first consider the Lagrangian velocity auto-correlation function:

$$R^L(\tau) = \frac{\langle v(t)v(t+\tau)\rangle_t}{\langle v^2\rangle} \quad . \tag{1}$$

We observe – figure 6 – that it has a slow decrease which can be modeled by an exponential function $\rho_v(\tau) \sim e^{-\tau/T_L}$. This expression defines an integral Lagrangian time scale $T_L = 22$ ms. For comparison, the period of rotation of the discs is 140 ms and the sweeping period of the blades is 17 ms. The measured Lagrangian time scale thus appears as a time characteristic of the energy injection. The exponential reproduces extremely well the variation of the auto-correlation function, from about $5\tau_\eta$ at small scales to $4T_L$. These limits coincide with the upper and lower resolution of the technique, so that we observe an exponential decay over the entire range of our measurement. However, as the variance of the acceleration must be finite [10] there has to be some lower cut-off to this behavior, at times of order τ_η. These observations extend and confirm previous numerical and experimental studies at moderate Reynolds numbers [2, 3, 6]. Note that the exponential decay of the Lagrangian velocity auto-correlation is a key feature of stochastic models of dispersion since it appears as a linear drift term in a Langevin model of particle dynamics [1, 8].

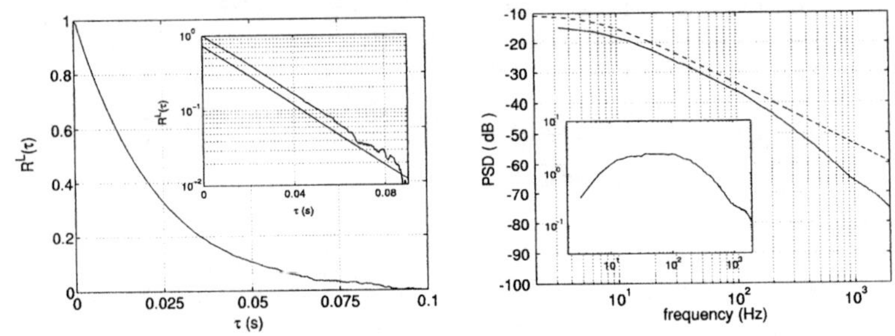

FIGURE 6. (a) Velocity autocorrelation function. A best exponential fit is $\rho_v^L(\tau) = 1.03e^{-45.7\tau}$. It is shown, slightly shifted for clarity, as the linear curve in the inset. (b) corresponding power spectrum; the upper curve is the Lorentzian function calculated from the exponential fit of the auto-correlation function (shifted for clarity).

We also show in Figure 6 the velocity power spectrum, computed both from the data and as the Fourier transform of the exponential decay of the auto-correlation function:

$$E^L_{\text{fit}}(\omega) = \frac{u^2_{rms}T_L}{1+(T_L\omega)^2} \quad . \tag{2}$$

We observe a clear range of power law scaling $E^L(\omega) \propto \omega^{-2}$. This is in agreement with a Kolmogorov K41 picture in which the spectral density at a frequency ω is a dimensional function of ω and ε: $E^L(\omega) \propto \varepsilon\omega^{-2}$. To our knowledge, this is the first time that it is directly observed at high Reynolds number and in a laboratory experiment, although it has been reported in oceanic studies [5] and in lower Reynolds number direct numerical simulations [6]. Departure from the Kolmogorov behavior is observed at low frequencies in agreement with the exponential decay of the auto-correlation. At high frequencies, the spectrum deviates from the Lorentzian form due to the particle response. Note on the spectrum plot that the measurement is made over a dynamical range of about 50 dB.

Second order structure function

We now consider the second order structure function of the velocity increment

$$D_2^L(\tau) = \langle (v(t+\tau) - v(t))^2 \rangle_t = \langle (\Delta_\tau v)^2 \rangle \ . \tag{3}$$

We emphasize that these are time increments, and not space increments as in the Eulerian studies. The profile $D_2^L(\tau)$ is shown in the inset of figure 7. It is linked to the auto-correlation by $D_2^L(\tau) = 2u_{\rm rms}^2\left(1 - R^L(\tau)\right)$: at small times one observes the trivial scaling $D_2^L(\tau) \propto \tau^2$ and at large times $D_2^L(\tau)$ saturates at $2u_{\rm rms}^2$ (as $v(t)$ and $v(t+\tau)$ are uncorrelated).

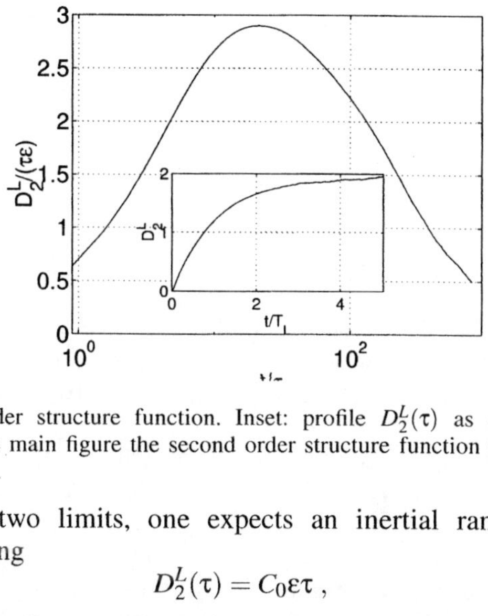

FIGURE 7. Second order structure function. Inset: profile $D_2^L(\tau)$ as a function of time, non-dimensionalized T_L. In the main figure the second order structure function is non-dimensionalized by the Kolmogorov scaling $\varepsilon\tau$.

In between these two limits, one expects an inertial range of scales with a Kolmogorov-like scaling

$$D_2^L(\tau) = C_0\varepsilon\tau \ , \tag{4}$$

where C_0 is a 'universal' constant. Such a behavior is consistent with dimensional analysis and with an ω^{-2} scaling range in the velocity power spectrum. Figure 7 shows $D_2^L(\tau)/\varepsilon\tau$; a plateau with a constant C_0 is not observed. The function reaches a maximum at $20\tau_\eta$, for which $C_0 \sim 2.9$. This value is in agreement with the estimation $C_0 = 4 \pm 2$ in [7] and in the range of values (between 3 and 7) used in stochastic models for particle dispersion [14]. In our set of measurements between $R_\lambda = 100$ and $R_\lambda = 1100$, we have observed an increase of C_0 (defined in the same way) from 0.5 to 4. We point out that in the absence of an equivalent of the Kármán-Howarth relationship for the Lagrangian time increments, a limit value of C_0 is not *a priori* fixed. Dimensional analysis yields $D_2^L(\tau) = C_0(Re)\varepsilon\tau$ and similarity arguments give $C_0(Re) \to$ const. or $C_0(Re) \to Re^{\alpha}$ in the limit of infinite Reynolds numbers.

Intermittency

To further describe the statistics of the Langrangian velocity fluctuations, we have analyzed the statistics of the velocity increments $\Delta_\tau v$. Their PDF Π_τ for τ covering the accessible range of time scales is shown in figure 8.

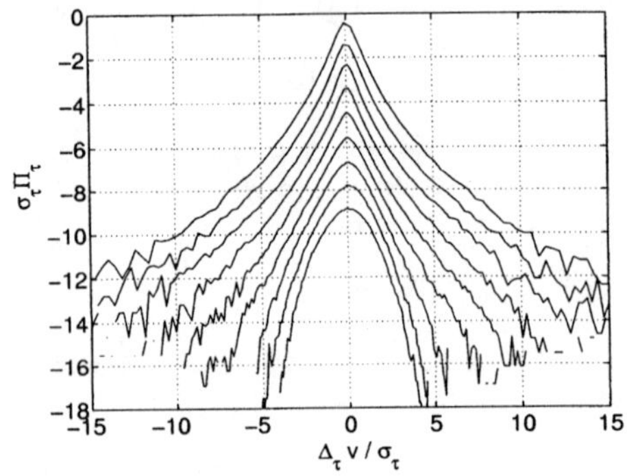

FIGURE 8. PDF $\sigma_\tau\Pi_\tau$ of the normalized increment $\Delta v_\tau/\sigma_\tau$. The curves are shifted for clarity. From top to bottom: $\tau = [0.15, 0.3, 0.6, 1.2, 2.5, 5, 10, 20, 40]$ ms.

To emphasize the functionnal form, the velocity increments have been normalized by their standard deviation so that all PDFs have unit variance. A first observation is that the PDFs are symmetric, in agreement with the local symmetries this flow. Another is that the PDFs almost Gaussian at integral time scales and progressively develop stretched exponential tails for small time increments. At the smallest increment, the stretched exponential shape is in agreement with measurements of the PDF of Lagrangian acceleration at identical Reynolds numbers [10]. In our case, the limit form of the velocity increments PDF is not as wide as that of the acceleration because the Kolmogorov scale is not resolved. Note that in regards of the evolution of the PDF, the intermittency is at least as developed in the Lagrangian frame as it is in the Eulerian one [15].

The continuous evolution with scale can be quantified using the flatness factor. We show in figure 9 the variation excess kurtosis $K(\tau) = \langle(\Delta_\tau v)^4\rangle / \langle(\Delta_\tau v)^2\rangle^2 - 3$. It is null at integral scale as expected from the Gaussian shape of the PDF and increases steeply at small scales. Below about $5\tau_\eta$, the increase is limited by the cut-off of the particle; an extrapolation of the trend to τ_η yields $K(\tau_\eta) \sim 40$ in agreement with acceleration measurements in [10].

More generally, one can choose to describe the evolution of the PDFs by the behavior of their moments (or 'structure functions') $D_q^L(\tau) = \langle|\delta_\tau v|^q\rangle$. Indeed, a consequence of the change of shape of the PDFs with scale is that their moments, as the flatness factor

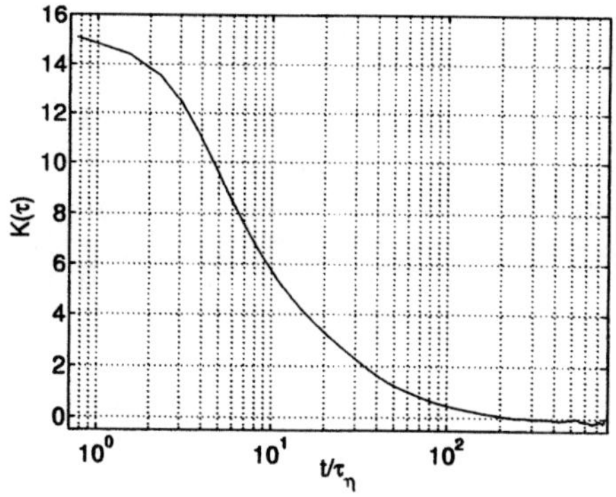

FIGURE 9. Evolution of the excess kurtosis factor $K(\tau) = \langle(\Delta_\tau v)^4\rangle / \langle(\Delta_\tau v)^2\rangle^2 - 3$ for the PDFs of the time velocity increments.

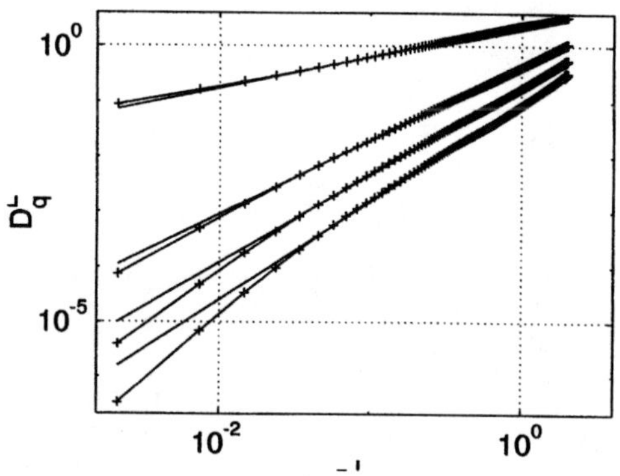

FIGURE 10. ESS plots of the structure function variation (in double log coordinates). The solid curves are best linear fits with slopes equal to $\xi_q^L = 0.56 \pm 0.01, 1.34 \pm 0.02, 1.56 \pm 0.06, 1.8 \pm 0.2$ for $p = 1, 3, 4, 5$ from top to bottom.

above, vary with scale. Classicaly in the Eulerian picture, one expects scaling in the inertial range, $D_q^E(r) \propto r^{\zeta_q}$, at least in the limit of very large Reynolds numbers. At the finite Reynolds number where most experiments are made, the lack of a true inertial range is usually compensated by studying the relative scaling of the structure functions – the ESS ansatz [16]. We use the second order structure function as a reference. Indeed the dimensionnal extimation of D_2^L (as that of D_3^E) depends linearly on the increment and on the dissipation. Figure 10 shows that, as in the Eulerian frame, a relative scaling is observed for the Lagrangian structure functions of orders 1 to 5, $D_q^L(\tau) \propto D_2^L(\tau)^{\xi_q}$. We observe that the relative exponents follow a sequence close to, but more intermittent than the corresponding Eulerian quantity. Indeed, we obtain: $\xi_1^L/\xi_3^L = 0.42$, $\xi_2^L/\xi_3^L = 0.75$, $\xi_4^L/\xi_3^L = 1.17$, $\xi_5^L/\xi_3^L = 1.28$ to be compared to the commonly accepted Eulerian values [17] $\xi_1^E/\xi_3^E = 0.36$, $\xi_2^E/\xi_3^E = 0.70$, $\xi_4^E/\xi_3^E = 1.28$, $\xi_5^E/\xi_3^E = 1.53$.

CONCLUSION

In conclusion, using a new experimental technique, we have obtained a Lagrangian velocity measurement that covers the inertial range of scales. Our results are consistent with Kolmogorov-like dimensionnal predictions for second order statistical quantities. At higher orders, the observed intermittency is very strong. How the Lagrangian intermittency is related to the statistical properties of the energy transfers is an open question. From a dynamical point of view, the Navier-Stokes equation in Lagrangian coordinates could be modeled using stochastic equations. Work is currently underway to compare the dynamics of the Lagrangian velocity to predictions of Langevin-like models.

acknowledgements: We thank Bernard Castaing for interesting discussions and Vermon Corporation for the design of the ultrasonic transducers. This work is supported by grant ACI No.2226 from the French Ministère de la Recherche.

REFERENCES

1. Pope S.B., *Annu. Rev. Fluid Mech.*, **26**, 23, (1994).
2. Virant M., Dracos T., *Meas. Sci. Technol.*, **8**, 1539, (1997).
3. Sato Y., Yamamoto K., *J. Fluid Mech.*, **175**, 183, (1987).
4. Yeung P.K., Pope S.B., *J. Fluid Mech.*, **207**, 531, (1989).
5. Lien R-C., D'Asaro E.A., Dairiki G.T., *J. Fluid. Mech.*, **362**, 177, (1998).
6. Yeung P.K. , *J. Fluid Mech.*, **427**, 241, (2001).
7. Hanna S.R., *J. Appl. Meteorol.*, **20**, 242, (1981).
8. Sawford B.L., *Phys. Fluids*, **A3**, 1577, (1991).
9. Voth G.A., Satyanarayan K., Bodenschatz E., *Phys. Fluids*, **10**, 2268, (1998).
10. La Porta A., Voth G.A., Crawford A., Alexender J., Bodenschatz E., *Nature*, **409**, 1017, (2001).
11. Mordant N., Michel O., Pinton J.-F., *submitted to JASA*, (2000) and ArXiv:physics/0103083.
12. Mordant N., Pinton J.-F., Chillà F., *J. Phys. II France*, **7**, 1729-1742, (1997).
13. B. Dernoncourt, J.-F. Pinton, S. Fauve, *Physica D*, **117**,181-190, (1998).
14. Du S., Sawford B.L., Wilson J.D., Wilson D.J., *Phys. Fluids*, **7**, 3083, (1995).
15. Anselmet F., Gagne Y., Hopfinger E.J., Antonia R.A. *J. Fluid Mech.*, **140**, 63, (1984).
16. Benzi R., Ciliberto S., Baudet C., Ruiz-Chavarria G., Tripiccione C., *Europhys. Lett*, **24**, 275, (1993).
17. Arncodo A. et al., *Europhys. Lett*, **34**, 411, (1996).

CIRCLE MAP DYNAMICS IN THE BUBBLE GUN EXPERIMENT

Alberto Tufaile[1] and J. C. Sartorelli

Instituto de Física, Universidade de São Paulo, Caixa Postal 66318, CEP 05315-970 São Paulo, SP, Brazil.

Abstract. We have studied the air bubble formation in a submerged nozzle in a water/glycerol solution inside a cylindrical tube submitted to a sound wave perturbation, the bubble gun experiment. Using the sound wave amplitude as the control parameter, we have observed transition from quasiperiodicity to chaos and transition to chaos via period doubling as the same way as the two-dimensional circle map. We compared experimental reconstructed attractors with attractors obtained with two-dimensional circle map dynamics.

INTRODUCTION

Combining the developments in nonlinear dynamics with the interest in engineering processes of bubbling systems, some authors looked for the chaotic bubbling during the nineties. Studying the bubbling from only one orifice, Tritton and Edgell [1] have observed some attractors by the signal obtained from a hot-film anemometer placed close to the bubble formation using the method of reconstructing a phase-space from a single time series. Mittoni *et al* [2] have given further evidence that bubbling can display chaotic behavior by the characterization of positive Lyapunov exponents from the experiment with non-invasive technique. Rejecting the premise that bubble formation in dense fluid resemble an inverted dripping faucet with period-doubling to chaos, Nguyen *et al* [3] have shown spatio-temporal dynamics in a train of rising bubbles is the dominant source of instability in the bubbling dynamics. On the other hand, studying bubbling from perforated plates with two orifices, Ruzicka *et al* [4] have reported complex behavior in the transition between synchronous and asynchronous regimes.

We have studied the bubble gun experiment [5], which consists of the air bubble formation from only one nozzle submerged in a water/glycerol solution inside a cylindrical tube submitted to a sound wave above the fluid. This experiment exhibits transition from quasiperiodicity to chaos, period doubling cascade and Hénon-like dynamics [6] as the same way as in the circle map. In this letter, we present the comparison between reconstructed attractors from the experiment and the circle map.

[1] E-mail: tufaile@if.usp.br

CP622, *Experimental Chaos: 6th Experimental Chaos Conference*, edited by S. Boccaletti et al.

THE CIRCLE MAP DYNAMICS

The circle map dynamics has been used as a paradigm of chaotic dynamics in systems with two competing frequencies, for example in the impulsively excited oscillator [7], the dynamics of a bouncing ball [8], to model human heartbeats [9], and the dynamics of a particle in a rotating liquid [10]. In addition to temporal evolution, the circle map dynamics has been formulated in terms of spatial arrangement to describe an adsorbed atom layer on a crystal surface, as in the Frenkel-Kontorova model [11]. The two-dimensional circle map is given by:

$$\begin{aligned} x_{n+1} &= x_n + \Omega - \frac{K}{2\pi} sin(2\pi x_n) + \beta y_n \quad \mathrm{mod}\,(1), \\ y_{n+1} &= \beta y_n - \frac{K}{2\pi} sin(2\pi x_n), \end{aligned} \tag{1}$$

where x_n and y_n are the dynamical variables, β regulates the damping, Ω is the ratio between two independent frequencies, and K is the intensity of non-linear external impulse. When $\beta << 1$, the system is strongly dissipative and reduces to an one-dimensional map. When $\beta = 1$, the system is nothing but the standard map which has been widely used to model conservative Hamiltonian systems [11].

The determinant of the Jacobean of the map of eq. (1), det **J** is:

$$\det \mathbf{J} = \frac{\partial(x_{n+1}, y_{n+1})}{\partial(x_n, y_n)} = \beta\,, \tag{2}$$

which shows that the phase space shrinks uniformly and, in the extremely dissipative limit $\beta = 0$.

EXPERIMENTAL APPARATUS

The experimental apparatus is shown in Fig. 1. The bubbles are issued by injecting air through a metallic nozzle submerged in a viscous fluid column (20 % water/ 80%glycerol). The inner diameter of the cylindrical container is 53 mm and 70 cm in height. The inner diameter of the nozzle is 0.78 mm. The nozzle is attached to a capacitive reservoir, and a solenoid valve controlled by a PID controller (BTC-220) sets the air flux. The flow rate is measured by a flowmeter Aalborg (GFM47).

The detection system is the same as the one of the dripping faucet experiment, (see Refs. [5,12] for details). The delay times between successive bubbles are measured by a time circuitry inserted in a PC slot, with a time resolution equals to 1 μs. The input signals are voltage pulses, induced in a resistor, defined by the beginning

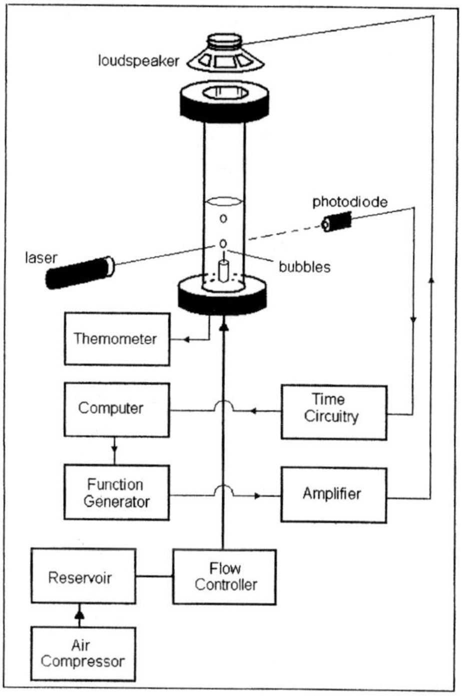

FIGURE 1. Diagram of the bubble gun experiment.

(ending) of scattering of a laser beam. The pulse width is the time interval t_n (n is the bubble number), and the time delay between two pulses defines the crossing time (dt_n) of a bubble through the laser beam, so that the total time interval is $T_n = t_n + dt_n$. Series of time intervals between bubbles {T} were obtained, and then the bubble mean frequency was calculated as $f_b = 1/\langle T \rangle$. We have estimated the total experimental noise as ~100μs in the period 1 behavior.

Whenever an air bubble crosses an observation point above the needle, it displaces an amount of fluid, and after the bubble passage through the fluid, the initial configuration is restored. Hence, when a bubble train crosses the observation point, it displaces the fluid with the bubbling frequency f_b. This movement is an oscillatory system while the sound wave (frequency f_s) represents the other one (see Fig. 2) that drives the bubbling. In this way, we have constructed a forced oscillator analogous to that occurring in the circle map model, where the sound wave amplitude A corresponds to the coupling strength K, and the bubbling time T_n is equivalent to the variable y_n. The Ω value is defined by $(f_s/f_b)_0$, when the oscillators are uncoupled (A = 0), and the winding number is $W = f_s/f_b$.

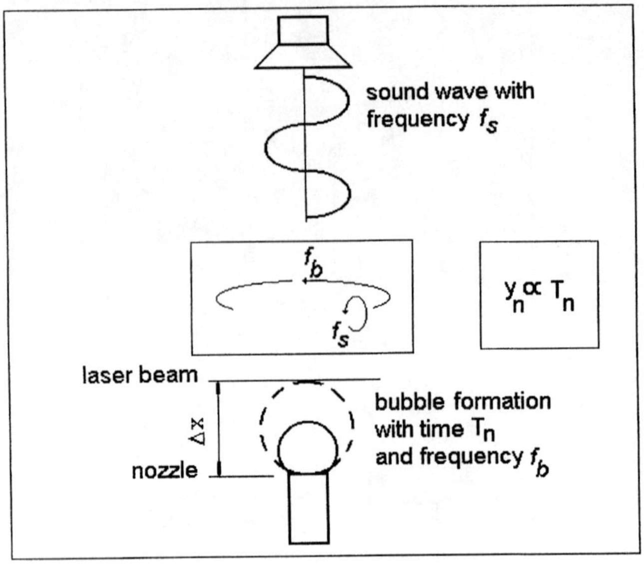

FIGURE 2. Bubble formation driven by the sound wave.

RESULTS AND DISCUSSION

The air flow rate and the sound wave frequency were initially chosen at ~ 39.5 bubble/s and 150 Hz respectively. We changed the bubble formation dynamics increasing the driven voltage in the loudspeaker as is shown in Fig. 3 with the reconstructed attractors (T_{n+1} versus T_n). We started at a fixed point at f_b = 39.5 bubbles/s as shown in Fig. 3(a). By applying the sound wave, the fixed point looses its stability and a limit cycle appears, as shown in Fig. 3(b). In Fig. 3(c) the limit cycle looses its stability and a fixed point, near (26.7, 26.7 ms), is visited in an intermittent behavior. After that, another amplitude increasing stabilizes the bubbling system in a period-1, as is shown in the reconstructed attractor of Fig. 3(d). The bubbling system evolves with another step in the amplitude in Fig. 3(e), and the Fourier transform applied to the data has indicated a noisy period-2 behavior, without any bubbling rate variation. With further amplitude increasing, we have observed the appearance of a two-band attractor, in Fig. 3(f). The last three attractors, which are shown in Figs. 3(g-i), present chaotic behavior.

In Fig. 4 we have obtained a similar sequence using the two-dimensional circle map with β = 0.1. These reconstructed attractors were obtained increasing the parameter K, with the first return map y_{n+1} versus y_n. The Ω value used was obtained from the experimental frequency rate of Fig. 3(a), and $\Omega = (f_s/f_b)_0 = 3.8$. The similarity

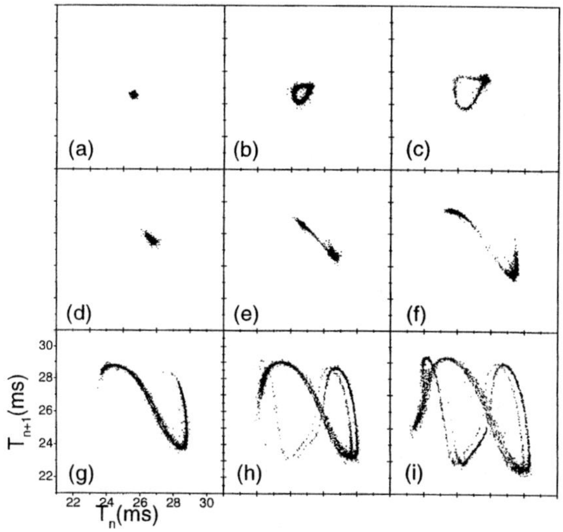

FIGURE 3. Reconstructed attractor from the bubble gun experiment increasing the sound wave amplitude.

between each map suggests that we are observing the same evolution in the both dynamical systems. The data from Fig. 4(a) to 4(c) represent a transition to chaos via quasiperiodicity. In Fig. 4(d), the period 1 is a resonant state (W = 4:1). A period doubling occurs in the circle map as is shown in Fig. 4(e), followed by two-band chaotic attractor in Fig. 4(f). The occurrence of Hénon-like attractor [6] due to stretching and folding mechanism is represented in Fig 4(g). The last two chaotic attractors appear due to a boundary crisis in the two-dimensional circle map.

CONCLUSIONS

These comparisons between the experimental results and the data obtained from the circle map enable us to realize that the bubbling system is a self-sustained oscillator with frequency f_b, that can be driven by an external sound wave with frequency f_s. We have observed features present in the experimental system similar to the circle map, such as quasiperiodic and intermittent behaviors, subharmonic synchronization, period doubling, Hénon-like attractors and boundary crises.

ACKNOWLEDGMENTS

This work has been supported by the Brazilian agencies: CNPq and FAPESP.

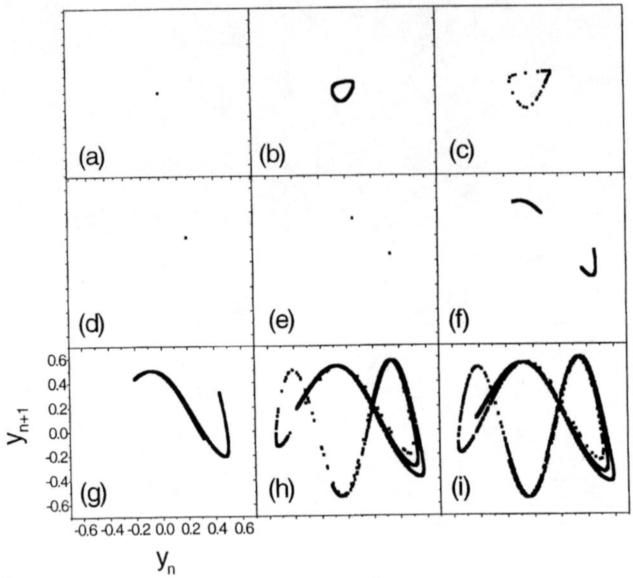

FIGURE 4. The reconstructed attractors (y_{n+1} *vs.* y_n) from the two-dimensional circle map dynamics, with $\beta = 0.1$, $\Omega = 3.8$, and K: (a) 0.0, (b) 0.5, (c) 1.0, (d) 2.0, (e) 2.8, (f) 3.1, (g) 3.15, (h) 3.39, and (i) 3.5.

REFERENCES

1. Tritton, D. J., and Egdell, C., *Phys. Fluids* **5**, 503-505 (1993).
2. Mittoni, L. J., Schwarz, M. P., and La Nauze, R. D., *Phys. Fluids* **7**, 891-893 (1955).
3. Nguyen, K., Daw, C. S., Cheng, M., Bruns, D. D., Finey, C. E. A., and Kennel, M. B., *Chem. Eng. J.* **64**, 191-197 (1996).
4. Ruzicka, M., Drahos, J., Zahradnik, J., and Thomas, N.H., *Chem. Eng. Sci.* **55**, 421-429 (2000).
5. Tufaile, A., and Sartorelli, J. C., *Physica A* **275**, 336-346 (2000).
6. Tufaile, A., and Sartorelli, J. C., *Phys. Let. A* **275**, 211-217 (2000).
7. Argyris, J., Faust, G., Haase, M., "Routes to turbulence" *in An Exploration of Chaos*, edited by J. Agyris, Elsevier Science, Amsterdam, 1994, 476- 509.
8. Everson, R. M., *Physica D* **19**, 355-383 (1986).
9. Haken, H., Peper, C. E., Beek, P. J., and Daffershofer, A., *Physica D* **90**, 176 (1996).
10. Piva, M., Calvo, A., and Wesfreid, J. E., *Appl. Phys.* **33**, 1431 (2000).
11. Bai-Lin, H., "Circle and Two-dimensional Mappings", in *Elementary Symbolic Dynamics and Chaos in Dissipative Systems*, World Scientific, Singapore, 1989, 250-252.
12. Tufaile, A., Pinto, R. D., Gonçalves, W. M., and Sartorelli, J. C., *Phys. Let. A* **255**, 58-64 (1999).

The Dynamics of Bound States of Topological Defects in Extended Spatially Periodic Structures

A.B. Ezersky*, S.V. Kiyashko* and A.V. Nazarovsky*

**Institute of Applied Physics, Russian Academy of Science, 46 Uljanov Street, 603950 Nyzhny Novgorod, Russia*

Abstract. Comparative analysis of characteristics of defects, bound states of topological defects and spatio-temporal chaos caused by their interaction is presented for the structures arising at a Benard-Marangoni convection and at a parametric excitation of capillary waves on the free surface of liquid. Analogy between dynamics of defects in spatially periodic hydrodynamical systems and dynamics of defects in crystals is presented.

INTRODUCTION

Disturbances of order in spatially periodic structures caused by instability in large aspect ratio hydrodynamical systems have much in common with disturbances of order in crystal lattices. For example, a topological defect in a roll structure is a two-dimensional analog of a edge dislocation in a lattice of atoms [1, 2], and a penta-hepto defect in a hexagonal lattice arising in thermoconvection is analogous to the structures formed in carbon nanotubes whose bending is due to the fact that, in a hexagonal lattice of carbon atoms forming the wall of the tube, one atom has five neighbours, and the other seven neighbours [3]. Edge dislocations in crystal lattice are known to interact effectively to produce the states of topological defects of complex organization [1]. We report results of research on dynamics of defects in spatially periodic systems generated by hydrodynamic instability and show that the interaction of defects may result in either regular pattern, such as bound states, domain wall, etc., or spatio-temporal chaos of defects.

BOUND STATES OF TOPOLOGICAL DEFECTS

Formation of bound states of topological defects is typical of systems with quadratic nonlinearity that leads to resonance three-mode interactions. In this section we will consider results of experiments on dynamics of the bound states arising at thermoconvection and at parametric excitation of capillary waves.

Three modes having wave vectors at an angle of 120 degrees are resonantly coupled at thermoconvection. A bound state arises as a result of interaction of two oppositely charged topological charges belonging to different modes forming a hexagonal lattice. Experiment [4] and numerical computation [5] revealed that such charges approach each other along the corridor inside which there is a large gradient of phases of the modes

CP622, *Experimental Chaos: 6th Experimental Chaos Conference*, edited by S. Boccaletti et al.

forming the structure. Parametric excitation of surface waves by spatially homogeneous monochromatic oscillations occurs as a result of three-waves interactions, with the pump of frequency Ω and the counter propagating waves with frequencies $\Omega/2$ and wave vectors k and $-k$ being resonantly coupled. Such nonlinear interaction may give rise to bound states of topological defects that were investigated in detail in [6].

The phase field of the bound state formed as a result of interaction of two topological defects has quadrupole components. These components have been calculated theoretically and observed in experiments for both the penta-hepto defects in hexagonal lattice [4, 7] and the bound state at parametric excitation of waves by a homogeneous field [6]. Unlike the case of a solitary topological charge whose phase field is proportional to azimuthal angle θ, the interaction leads to appearance of a component proportional to $\cos(2\theta)$.

Besides pair interactions, topological defects can form linear chains. Example of such a chain of dislocations for parametrically excited ripples is given in Fig.1a [8]. We call

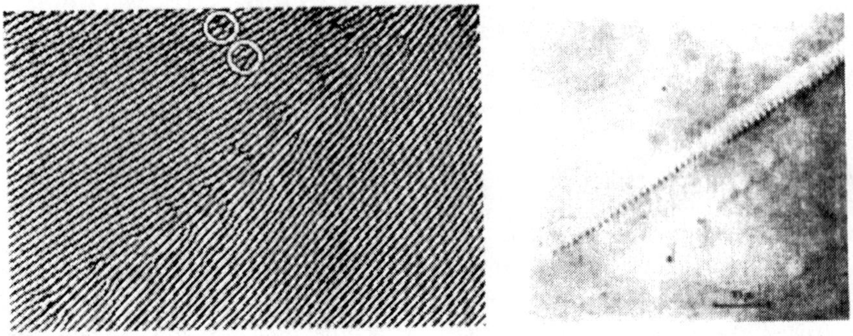

a b

FIGURE 1. Chain of dislocations in the structure formed at parametric excitation of capillary ripples and in a germanium crystal.

such chains domain walls, bearing in mind that they separate regions having different orientations of wave vectors of the modes forming the pattern. Similar linear chains exist in crystal lattices too. The dislocations arranged at the boundary between the grains in germanium are depicted in Fig.1b. The orientations of the neighbouring grains differ by about 0.015 degrees.

TRANSIENT CHAOS

When the parameter governing instability of the system changes, there may arise transient chaos of defects that exists during a finite time interval. Characteristics of such chaos were investigated, for example, in a system with double diffusion [9]. In this section we will report results of investigations into characteristics of the transient chaos of topological defects in a system of parametrically excited capillary waves and will focus on the analogy with characteristics of defects in crystals.

The main idea of the experiment is the following. It is known that, at the instant a source of vertical vibrations is switched on abruptly, initial perturbation is produced on the vertical walls of a couette near its boundaries due to the presence of a viscous boundary layer. If the experimental couette is cylindrical, then the perturbation propagates to the centre in the form of circular waves and, simultaneously, the amplitude grows in time due to parametric pumping [10]. The amplitude of circular standing waves increases up to the value close to a stationary one, after which it ceases to grow. However, if experimental parameters are chosen such that two mutually perpendicular standing waves are a stable state, then the cylindrical wave starts to break down. A structure of the target type exists only for finite time. A square lattice in which mutually orthogonal modes compete weakest of all is more favourable for the system in terms of energy. Breaking down of a target starts from appearance in some regions of the couette of perturbations with wave vectors locally parallel to the front of the cylindrical wave. Topological defects arise at the boundaries of such regions. As the transition from the target to a square lattice occurs as a result of growing amplitude of random perturbations, the arrangement of the defects in the space of the couette is also random.

In our experiments at fixed supercriticality a source of vibrations of a couette was switched on abruptly and the process of arising of defects was recorded by a video camera. The number of defects on the surface of the liquid in the couette was calculated by the recorded images for different values of frequency of force. It was found that the number of the defects N in the couette was proportional to the wave number of the excited mode (see Fig.2a), and the number of the defects in the rings of one width did not depend on the radius R of the rings (Fig.2b).

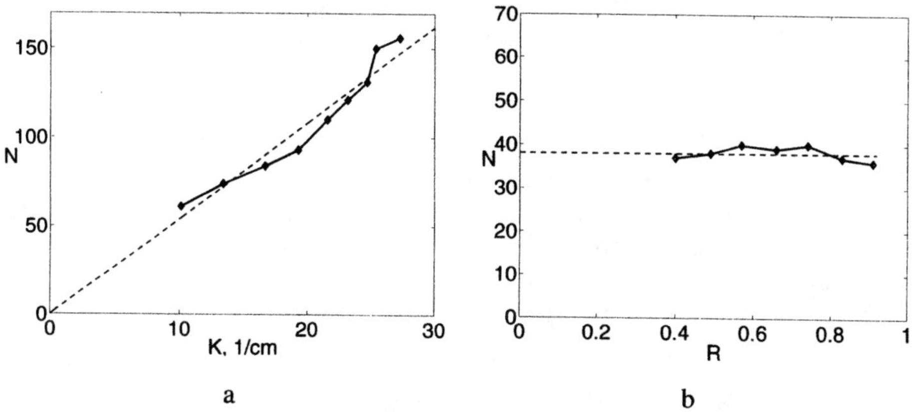

FIGURE 2. (a) Dependence of number of defects N on the wave number of excited mode; (b) dependence of N on radius of the ring.

These results can be understood from the simplest geometrical considerations. Suppose that the total number of defects, N, in the couette of radius R is proportional to the length of the side boundary of the couette measured in wavelengths of perturbations, λ, with wave vectors orthogonal to the wave fronts of the target. Then, we obtain $N \sim \frac{2\pi R}{\lambda}$ or the density of defects $\rho = \frac{N}{\pi R^2} \sim \frac{1}{R\lambda}$. Analogous dependence is observed for defects in a bent crystal [2]. The density of defects, ρ, is proportional to $\rho \sim \kappa b^{-1}$, where κ

is the crystal curvature and b is the modulus of the Burgers vector. For parametrically excited waves in a circular couette, κ is the curvature of the wave front in the target type structure, and the modulus of the Burgers vector is the wavelength of perturbations.

STATIONARY CHAOS

The topological defects in spatially periodic structures grow in number with the increase of supercriticality. Interaction of the defects can give rise to spatio-temporal chaos. Example of such chaos in depicted in Fig.3 for parametric excitation of capillary waves.

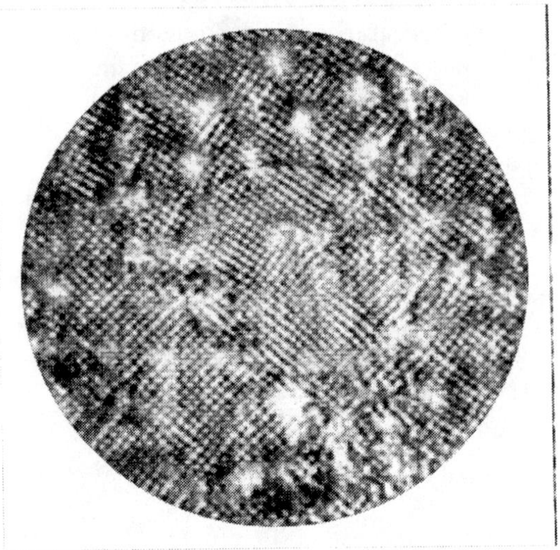

FIGURE 3. Chaos of defects in parametrically excited ripples: image of pattern, comprising the image of two mutually orthogonal standing waves with defects.

A square structure formed by two standing waves contains an ensemble of interacting dislocations which, having formed randomly near the side walls of the couette, move in it, scatter at each other, annihilate, and form bound states. Position of the topological defects was determined by the phase field of the envelope of one of the mutually orthogonal standing waves. Details of this procedure were described in [8]. Here we only note that the phase field had the size of 512 x 512 and we used for processing the modes whose wave vectors were almost parallel to the OY vertical axis.

For the characteristic of the spatio-temporal chaos we calculated the space correlation function $C(x,y)$ of the defect position and the dependence of the number of defects, N, on supercriticality $N = N(\varepsilon)$. The correlation function is defined by

$$C(x,y) = \sum_i \sum_j g(i,j)g(i-x,j-y),$$

where $g(x,y) = 1$ if there is a defect at the point (x,y), and $g(x,y) = 0$ if there is no defect. The correlation function averaged over eight images $C(x,y)$ as well as $C_x =$

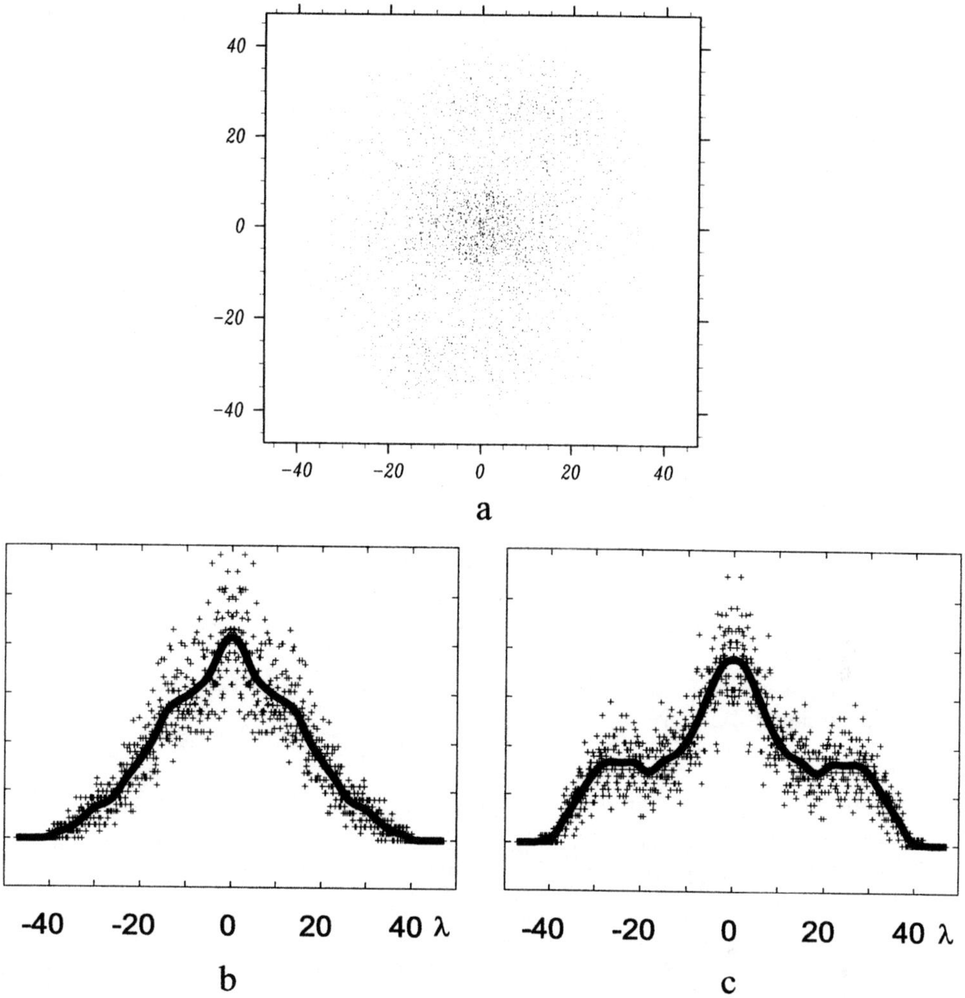

FIGURE 4. (a) Two-dimensional correlation function $C(x,y)$; (b) function $C_x(x)$; (c) function $C_y(y)$.

$\Sigma_y C(x,y)$ and $C_y = \Sigma_x C(x,y)$ are shown in Fig.4 for supercriticality $\varepsilon = 1.33$. One can see from Fig.4 that functions C_x and C_y differ markedly. Functions C_y has a plateau at $x \in (-30\lambda, 30\lambda)$, which is evidently explained by the fact that the defects in this experiment move primarily along the fronts (climb motion) and the number of defects formed per unit time near the side wall is maximal at an angle close to 45 degrees (if the front is parallel or perpendicular to the side boundaries, then the boundary conditions sustain a perfect structure).

The number of topological defects, N, increase with increasing supercrituality ε. When plotting the dependence $N(\varepsilon)$ we calculated the number of defects averaged over several images. The growth of the number of defects can be approximated by the Arrhenius formula $N(\varepsilon) = D\exp(-B/\varepsilon)$, where D and B are constants. An analogous dependence of the number of defects on supercriticality was obtained in the study of transitions to spatial disorder in the Marangoni-Benard convection [11]. The authors of [11] investigated the formation of the penta-hepto defects in hexagonal structures arising in thin layers of a liquid heated from below.

CONCLUSION

To conclude we note that the topological defects observed in the spatially periodic structures produced by hydrodynamic instabilities possess similar properties irrespective of the system in which they are realized. We have shown that the defects can form bound states and domain walls both in tetragonal and hexagonal lattices. The average number of defects increases with increasing supercriticality by the same law. Evidently, at the stage when the spatially periodic structure has been formed, dynamics of the individual cells forming the structure is more essential than dynamics of linearly unstable collective modes of the dissipative medium. Consequently, the analogy with crystals, where atoms play the part of cells, is fairly profound. Manifestations of this analogy can be found in that bound states of defects and domain walls, exist in both crystals and hydrodynamic patterns, and bending of wave fronts, like bending of crystals, leads to the formation of defects.

The research was done under financial support of the Russian Foundation for Basic Research (grant 99-02-16493).

REFERENCES

1. Read, W.T., *Dislocations in cristals*, McGraw - Hill publishing Company LTD, N-Y - London - Toronto.
2. Cottrell, A.H, *Theory of crystal dislocations*, Blackie and Son LTD, London - Glasgow, 1964.
3. Dekker, C., *Physics Today*, N5, 22-28, (1999).
4. Afenchenko, V.O., Ezersky, A.B., Nazarovsky, A.V., Velarde, M.G., *Int. J. Chaos and Bifurcation*, (2001), accepted for publication.
5. Rabinovich M.I., Tsimring, L.Sh., *Phys. Rev.E*, **49**, R35, (1994).
6. Ezersky, A.B., Nazarovsky, A.V., Kiyashko, S.V., *Physica D*, **152-153**, 310-324, (2001).
7. Pismen, L., Nepomnyashchy, A., *Europhys. Lett.*, **24**, 461, (1993).
8. Ezersky, A.B., Ermoshin, D.A., Kiyashko, S.V., *Phys.Rev. E*, **51**, 4411, (1995).
9. La Porta, A., Surko, C.M., *Phys.Rev. Lett.*, **77**, 2678, (1996).
10. Matusov, P.A., Tsimring, L.Sh., "The propogation of the front of parametrically excited capillary ripples", in *Non linear waves III*, ed.by A.V.Gaponov-Grekhov, Springer - Verlag, Berlin, 1990, pp.98-105.
11. Perz-Garcia, C., Cerisier, P., Ocelli, R, In *Propogation in the systems far from equilibrium*, ed by J.E.Wesfreid et. al., Springer - Verlag, Berlin, 1988, p.232.

COMPLEXITY AND SELF-ORGANIZATION IN LARGE ARRAYS OF ELASTIC RODS IN AN AIR CROSS-FLOW

Masaharu Kuroda

Applied Complexity Engineering Group, AIST, Japan

Francis C. Moon

Sibley School of Mechanical and Aerospace Engineering, Cornell University, USA

Abstract. Experiments on spatio-temporal dynamics in large arrays of cantilevered elastic-rods in uniform and stationary wind-tunnel cross-flow have been carried out. From 90 to 1000 steel and polycarbonate rods with gap ratios ranging from 1.0 to 2.5 are used. As the Reynolds number (based on the rod diameter) increases, a pattern with the characteristics of spatio-temporal chaos emerges in the global behavior of the array of elastic rods. There are local and global patterns. The local patterns are composed of transient rest, linear motions and elliptical motions. In the 90-rod experiments, a cluster-pattern entropy measure is introduced based on these three patterns as a quantitative measure of the local complexity. Below a threshold wind velocity, no significant dynamics appears. Video images reveal that at first each rod moves individually, then clusters consisting of several rods emerge and finally global wave-like motion takes place with higher flow velocities. Spatial patterns in the rod-density distribution appear as more rods suffer impacts with nearest neighbors. Furthermore, these collective nonlinear motions of rods are observed and categorized into several global modes. Using accelerometer data, the rod impact rate versus flow velocity shows a power-law scaling relation. This phenomenon may have application to plant-wind dynamics and damage as well as fluid-structure heat exchange systems. This experiment may also be a two dimensional analog of the impact dynamics of granular materials in a flow.

INTRODUCTION

Fluid-elastic vibrations occur in heat exchanger systems, agricultural crop-wind interactions [(1)-(6)] and bio-mechanical problems such as a ciliated inside-throat. Classic problems include the flow of fluid through tube structures and around tubes and have generally focused on linear dynamics. In this study we discuss experimental nonlinear dynamics of a large array of up to 1000 in a cross flow. The rod-like structures are cantilevered at a base and free to vibrate at the top. (Fig. 1) The coupling between rods consists of fluid forces and contact when the vibration amplitude becomes too large. The fluid forces are of two types; fluid-elastic nearest neighbor forces equivalent to springs and dampers, and non-nearest neighbor forces produced by vortices leaving

CP622, *Experimental Chaos: 6th Experimental Chaos Conference*, edited by S. Boccaletti et al.

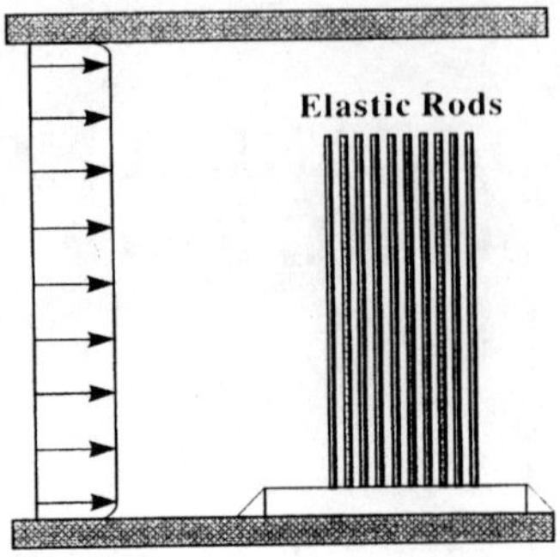

FIGURE 1. Sketch of Experimental Set-Up

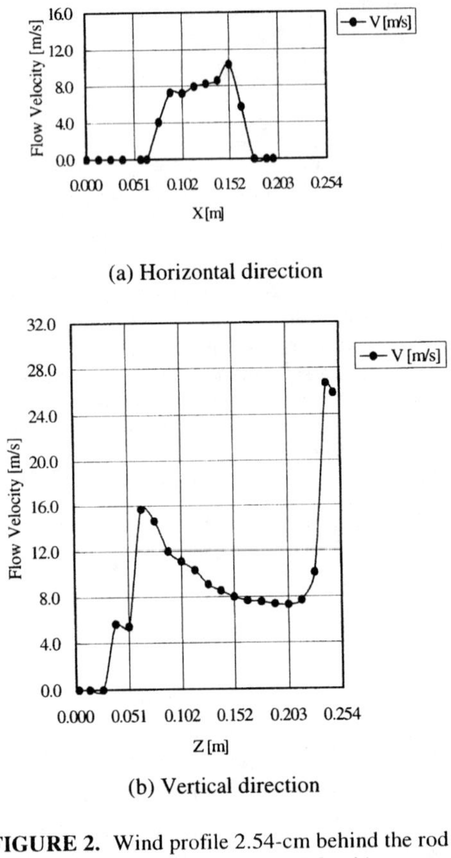

FIGURE 2. Wind profile 2.54-cm behind the rod array (Upstream Flow Velocity = 7.0 m/s)

the forward rows of cylinders and effecting the dynamics of the rearward rows of cylinders [7].

Observations of the tip vibration dynamics reveal complex patterns of motion of the rods, some of which appear nearly stationary, others vibrate in a straight-line motion in-line and at an angle to the upstream flow and others vibrate in elliptical patterns sometimes associated with rod to rod contact. While the rod frequencies lie close to their natural frequencies, the phases and types of motion, i.e., stationary, straight-line and elliptic show no regular pattern and change in time. It is important to emphasize that the linear theory would predict 2xNxM eigenmodes for an NxM array of rods. However, no clear modal pattern emerges as the wind tunnel velocity is increased. Thus, we have sought to apply entropy measures of complexity to describe the dynamics. Using a 90 steel rod array, we have already reported the complex dynamics of the local behavior of vibrating rods and introduced 2-dimensional spatial entropy measure [8], [9] named cluster-pattern entropy to quantify the complexity of the global behavior of the vibrating rods [10], [11]. In this paper, wind-tunnel experiments are extended from 90-rod array to 300-and 1000-rod arrays.

DESCRIPTION OF EXPERIMENT

As shown in Fig. 1, the rod-like structures are cantilevered at a base and free to vibrate at the top. The 17.1-cm long rods were steel with a diameter of 1.59 mm and a gap equal to the rod diameter. The first eigen frequency of a steel rod is 26 Hz. The coupling between rods consists of fluid forces and contact when the vibration amplitude becomes too large. The wind tunnel is a low turbulence system with a 25.6 cm x 25.6-cm cross section. The wind speeds ranged from 0-12 m/s. The Reynolds number based on the rod diameter ranges from 200-900. Dynamic data are obtained by instrumenting one or two rods as well as video taping of the motion of the ends of the rods of the entire array. Also photographic images with a shutter speed on the order of the vibration period for 10 Hz were taken.

The 90-rod experiments were extended to 300 rods of the same diameter and spacing with 10 rows by 30 steel rods. In these experiments, the flow behind the 10-row array in a cross-flow was measured with a pitot tube. Horizontal-directional and vertical-directional flow speed distributions are shown after cross section of the observatory part of the wind tunnel is narrowed, respectively, in Fig. 2 (a) and in Fig. 2 (b).

PATTERN FORMATION OF SPATIO-TEMPORAL DYNAMICS

This figure shows contour maps of rod-density distribution as a result of an image processing of videotape frames capturing the 300 steel-rod array behavior from above. The array has 25 columns and 40 rows with the gap ratio 1.0. The lower side is upstream and the upper side is downstream in each map.

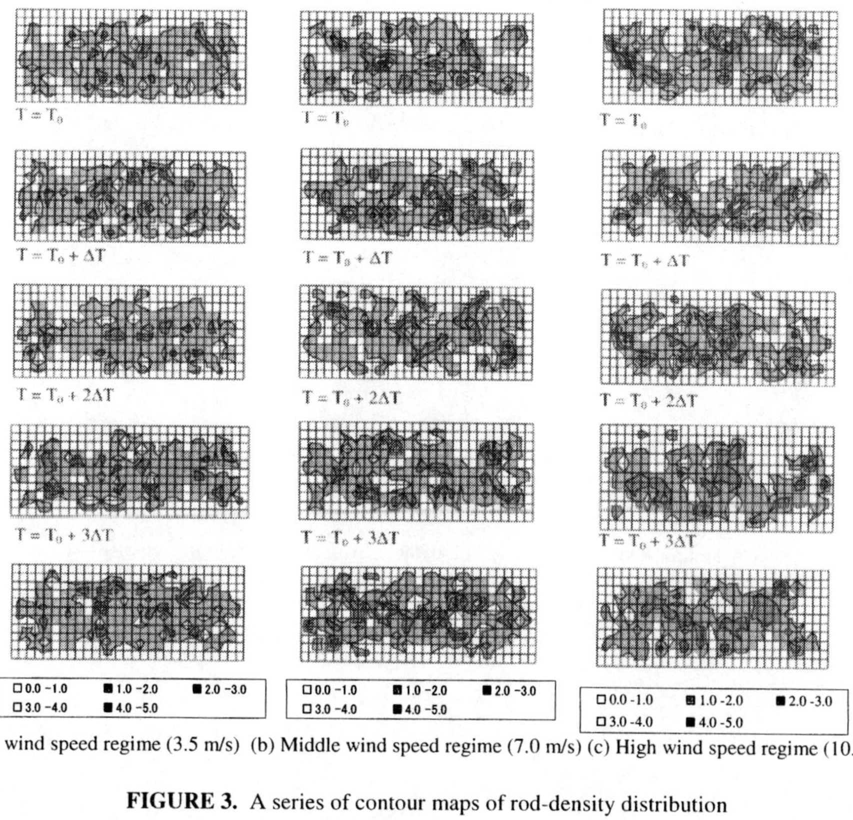

(a) Low wind speed regime (3.5 m/s) (b) Middle wind speed regime (7.0 m/s) (c) High wind speed regime (10.5 m/s)

FIGURE 3. A series of contour maps of rod-density distribution

In the low wind-speed regime, observed from the top, rod tops seem to oscillate individually and they do not show so significant difference in density distribution as time passes. Understood from the contour maps of the rod density, white area around the circumference is small; the rod-density shading is just bunched over whole area.

In the middle wind-speed regime, the white area on the contour maps spread wider than before; the light-gray area appears as though it has torn itself and dense and scattered areas alternates remarkably as time elapses. Finally, clusters, i.e., collective movements of some rods, emerge in the rod-density distribution. It is confirmed that these clusters do not move along a specific pattern at this stage, but rather just move and freely collide with one another.

In the high wind-speed regime, the white area on the contour maps continues to widen; rifts and gaps appear on the light-gray area and the number of clusters displays continued increase. Furthermore, it is most characteristic and important that those clusters are linked together as a chain and that those chains were borne at the both corners on the front most row; they moved coherently in a diagonal direction to the center part of the backmost row repeatedly, as waves do.

DISCUSSION ON TEMPORAL DYNAMICS

On the other hand, time-series data on one rod shown in Fig. 4 suggests that bursting phenomena are probably related to impact between rods as flow velocity is increased. These time-series data were obtained from a small accelerometer attached 1cm-up from the foot of the rod in the middle of the front row. Judging from Fig. 4, we can infer that increased collision frequency among rods alters spatio-temporal patterns of vibrating-rods array shown from those patterns previously.

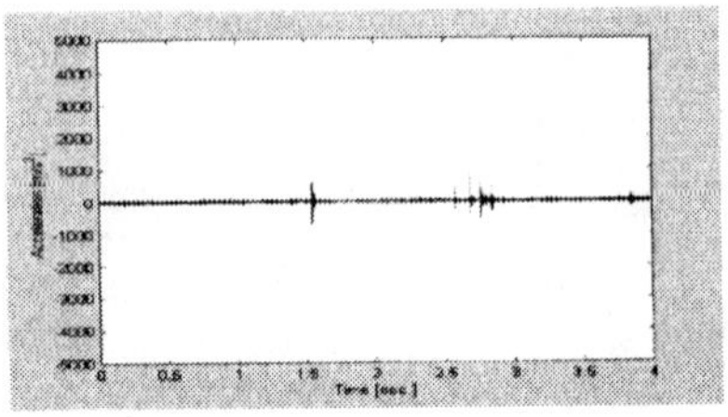

(a)

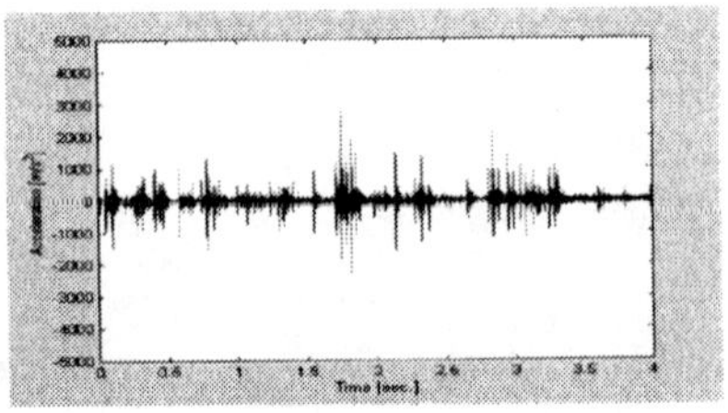

(b)

(a) Temporal dynamics measured by accelerometers at 7.0 m/s wind speed
(b) Temporal dynamics measured by accelerometers at 11.2 m/s wind speed

FIGURE 4. Time-series data of bursting phenomena

Scaling Law

In addition, the number of bursting waveforms for 24 seconds as a function of wind speed is plotted in Fig. 5. Bursting phenomena take place due to rod collisions. Bursting peaks were counted by the program coded on MATLAB.

Finally, we confirm that a scaling law exists in cluster generation and that the scaling law is predicted to be described by a fractal dimension. The basic relationship between the number of bursts and wind velocity is

$$V^2 = c\,N^{\alpha}, \tag{1}$$

where V represents wind velocity, N shows the number of busts and c is a constant. The term V^2 is in proportion to input energy into the rod array. By taking the logarithm of both sides of this equation, we obtain a log N-log V plot.

We obtain the graph slope; Fig. 6 indicates that α is almost 0.25. Therefore, the energy inputted into the rod array by wind and the number of bursts follow the power law with a fractal dimension of 0.25.

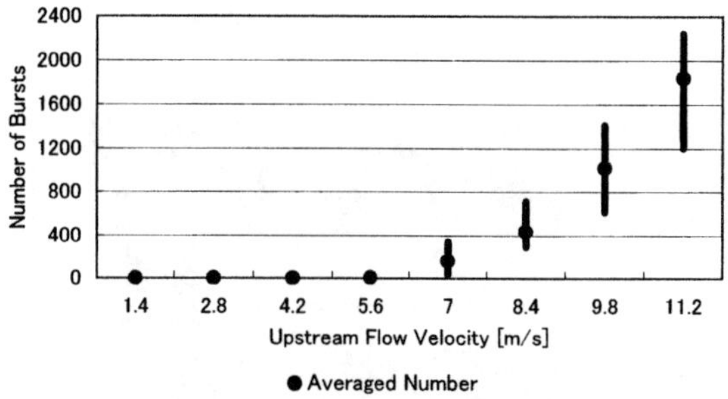

FIGURE 5. Number of bursting waveforms as a function of wind velocity (threshold = 0.5 x 10^4 m/s^2)

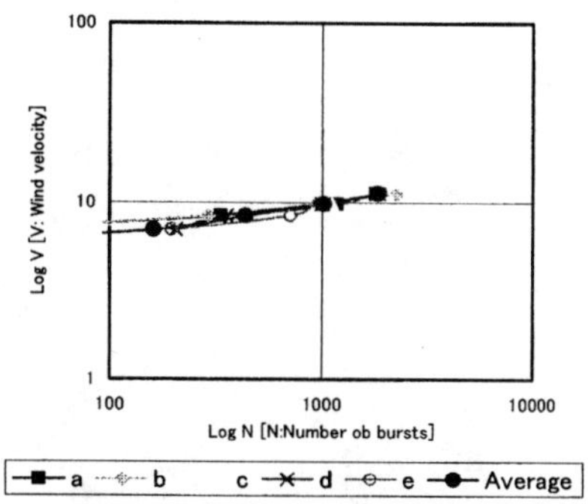

FIGURE 6. Power law between the number of bursts and wind velocity (threshold = 0.5 x 10^4 m/s^2, V^2=cN^{α}, α = 0.242)

COLLECTIVE NONLINEAR MODES OF THE OSCILLATORS

Finally, collective nonlinear modes of the entire rod array are introduced. In large arrays of elastic rod oscillators, global nonlinear modes were discovered while local complex behavior was revealed. Furthermore, it was clarified that observed nonlinear modes are limited to two types; a symmetrical mode and an asymmetrical mode. Also, the overall vibrating rod shape switches one mode to another repeatedly as time elapses.

25 column x 12 row Array of Polycarbonate rods

Theses figures show videotape frames capturing the behavior of a 300 polycarbonate rod array from above at 11.20 m/s wind speed. Rods are 20.0-cm long polycarbonate with 3.20-mm diameter. The first eigen frequency of the polycarbonate rod is 18 Hz. The array has 25 columns and 12 rows with the gap ratio 1.5. The upper side is upstream and the lower side is downstream in each picture. Each two-picture set shows successive images; the rod-array spatial pattern oscillates from that in the upper picture to that of the lower picture.

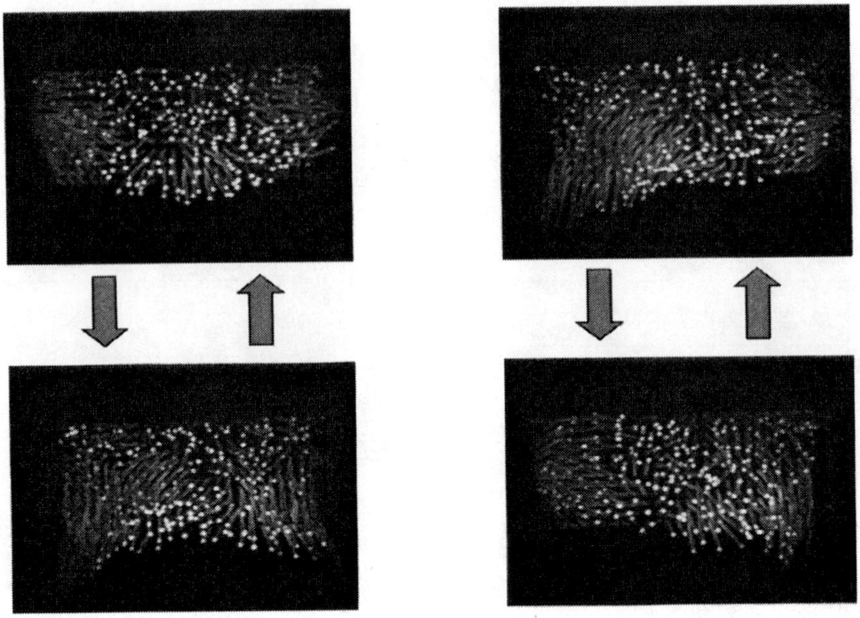

FIGURE 7. Symmetrical nonlinear mode and asymmetrical nonlinear mode of a 300-rod array

25 column x 40 row Array of Polycarbonate rods

These figures show videotape frames capturing behavior of a 1000-polycarbonate-rod array from above at 8.40 m/s wind speed. Rods are 20.0-cm long polycarbonate with 3.20-mm diameter. The first eigen frequency of the polycarbonate rod is 18 Hz. The array has 25 columns and 40 rows with the gap ratio 1.5. The right-hand side is upstream and the left-hand side is downstream in each picture. Each two-picture set shows successive images, and the rod array spatial pattern oscillates from that in the upper picture to that of the lower picture. By examining videotaped rod motions, we found that several characteristic frequencies of motion exist: the frequency of the switch between symmetrical mode and asymmetrical mode, those of waves proceeding on the boundary edges of the array, and those of waves passing through the inner area of the array.

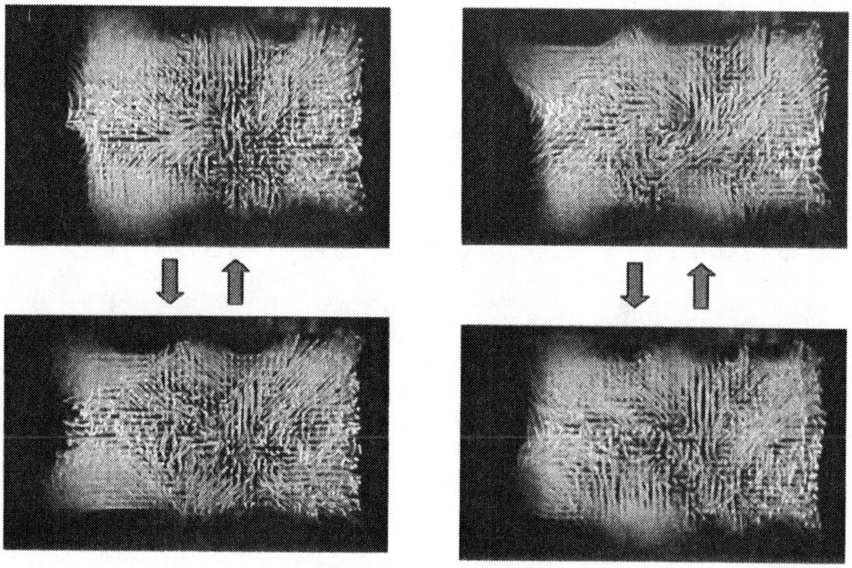

FIGURE 8. Symmetrical nonlinear mode and asymmetrical nonlinear mode of a 1000-rod array

CONCLUSIONS

Concluding remarks obtained by experiments using 300 rods and 1000 rods inside the wind tunnel are described below.

(1) Non-stationary complex phenomena occurring in the array of 300 vibrating rods and that of 1000 vibrating rods in the wind tunnel were observed and recorded.

(2) Hierarchical spatio-temporal pattern formations such as clusters of rods and wave-like motion of clusters were demonstrated.

(3) Power-law expressed by fractal dimension 0.25 exists in the scaling relation between the rod impact rate and the flow velocity.

(4) It was clarified that rod-array collective behaviors can be categorized into two types of global nonlinear modes: symmetrical mode and asymmetrical mode.

Future studies are summarized as follows.

(1) A theoretical model must be constructed to explain experimental results.

(2) A control method of spatio-temporal patterns appearing in a large array of elastic rods in a cross-flow must be developed.

REFERENCES

(1) Coutts, M.P. & Grace, J. Eds., *Wind and Trees*, Cambridge University Press, Cambridge, UK, (1995).
(2) Finnigan, J.J. "Turbulence in Waving Wheat," *Boundary-Layer Meteorology*, 16 (1979), 181-211.
(3) Finnigan, J.J. & Mulhearn, P.J., "Modeling Waving Crops in a Wind Tunnel," *Boundary-Layer Meteorology*, 14 (1978), 253-277.
(4) Grace, J., *Plant Response to Wind*, Academic Press, London, (1977).
(5) Inoue, E., "Studies of Phenomena of Waving Plants ("HONAMI") Caused by Wind," *Agric. Meterol.* (Japan), 11 (1955), 18-22.
(6) Niklas, K.J., *Plant Biomechanics*, Chapters 7 & 9, University of Chicago Press, (1922).
(7) Thothadri, M. & Moon, F.C., "Helical Wave Oscillations in a Row of Cylinders in a Cross-Flow," *J. Fluids and Structures*, 12 (1998), 591-613.
(8) Williams, G.P., *Chaos Theory Tamed*, Joseph Henry Press, Washington D.C., (1999).
(9) Moon, F.C., *Chaotic and Fractal Dynamics*, J. Wiley & Sons, NY, (1992).
(10) Kuroda, M and Moon, F. C., "Complexity Appearing in Large Arrays of Vibrating Rods in an Air Cross-Flow," (in English), *in Proc. of JSME Dynamics & Design Conference 2000*, (2000), Paper No. 238.
(11) Moon, F. C. and Kuroda, M., "Spatio-Temporal Dynamics in Large Arrays of Fluid-Elastic Toda-Type Oscillators," *Physics Letters A*, 287 (2001), 379-384.

Stability And Transition To Chaos In The Spherical Couette Flow With Counter Rotating Boundaries.

S.Gertsenshtein, D.Jilenko, O.Krivonosova

Institute of mechanics of Moscow State University

Abstract. Transition to chaos in Spherical Couette flow was experimentally investigated by velocity pulsation measurements. The stability curve was obtained in a wide range of control parameters: $0 < Re_i < 1000$, $-1200 < Re_o < 800$. The laminar-turbulent transitions have been studied near the local maximum at the stability curve at $Re_o = -810$ ($-700 < Re_2 < -950$). This transition was found to occur throw four bifurcations at the right of the maximum and throw five at the left. Spectra of all supercritical flow regimes are formed from the frequencies, corresponding to three revealed flow space structures. The transition always occur from one-frequency regime, which structure is symmetric with respect to the equator and include three vortex tubes in each hemisphere. The symmetry-breaking bifurcation has a considerable hysteresis. The coordinates were found, where the relative frequencies of the supercritical regimes (i.e. the frequencies divided by the frequency of rotation of the outer boundary) don't depend on the Re_o.

INTRODUCTION

By analogy with the plane and cylinder Couette flow the flow of viscous incompressible fluid between two concentric coaxially rotating spherical boundaries is called spherical Couette flow (SCF).

Three dimensionless parameters determine the structure of SCF: the Reynolds numbers for inner ($Re_i = \Omega_i r_i^2 / \nu$) and outer ($Re_o = \Omega_o r_o^2 / \nu$) spheres and relative gap size $\delta = (r_o - r_i)/ r_i$. Here r_i, r_o are the radii and Ω_i, Ω_o the angular velocities of inner (index i) and outer (index o) spheres, ν is the cinematic viscosity of the fluid in the gap.

Basic and supercritical flows and primary instability in different layers were experimentally investigated by Belyaev *et al* [1], Munson and Menguturk [2], Nakabayashi and Tsuchida [3]. The basic flow in SCF consists of the rotation in azimuthal direction and complex motion in meridional plane. The basic flow is symmetric about the equator plane and the rotational axis. Secondary flows (the flows which formed after primary instability) loose one or both symmetries and usually are periodical ones. Naturally supercritical flows are more complicated.

Transition to chaos were experimentally investigated by Nakabayashi and Tsuchida [4] in the layer 0.14 , Belyaev *et al* [5] in the layer 1.006, Egbers *et al* [6] in the layers 0.33 and 0.5, all with fixed outer sphere. The extended investigations of the supercritical flows with both rotating boundaries were carried out in cylinder Couette

CP622, *Experimental Chaos: 6th Experimental Chaos Conference*, edited by S. Boccaletti et al.

flow [7]. This paper presents the results of the experimental investigation in the layer $\delta = 1.006$ when both spheres rotate [8,9].

EXPERIMENTAL PROCEDURES

The experiments were carried out at the set-up «Sphere» at the Institute of Mechanics of Moscow State University [10]. The spherical layer with $\delta = 1.006$ is formed by two transparent spheres (the radius of the outer sphere Re_o=150mm). Controlled conditions, such as temperature and angular velocities of the spherical boundaries, allow us to fix the Reynolds numbers with accuracy within 0.05%. The experimental technique is based on the simultaneous visualization of the flow structure in meridional plane and measurement of the velocity pulsation. For the visualization a small amount (less then 0.002%) of aluminum flake was added to the silicon oil, used as working fluid. Velocity fluctuations measurements were carried out by TSI and DISA LDA in combination, with dual beam forward scatter receiving optic, in two points of the layer (distance from equator was 45 and 75 mm, distance from the rotation axis was 120 mm). Spectra of velocity pulsation were calculated by FFT method for discrete analog-signal samples consisting of 16384 or 32768 points, separated by 0.06 s intervals. For all flow regimes presented here we have followed the next procedure: first the outer sphere is slowly accelerated from rest to chosen value with the inner sphere at rest, then the inner sphere is quasi-statically accelerated (with a step $\Delta Re_i/Re_i < 0.2\%$) from rest to the chosen value.

PRIMARY INSTABILITY

The stability curve was obtained for $0 < Re_i < 1000$, $-1200 < Re_o < 800$. The negative values of the Re_o corresponds to the counter rotation of the boundaries. Within the experimental accuracy the loose of stability along the curve demonstrates no hysteresis. The stability curve $a - g$ and the schemes of the basic (6 – 9) and secondary (10) flows in meridional plane are presented at Fig.1.

At the Fig.1 scheme 6 the structure of basic flow in point O (with fixed outer sphere) is presented: the boundary between the flows in upper and lower hemispheres coincides with the equator plane, meridional motion consists of one circulation in each hemisphere. The loose of stability in point O was discovered by Belayev *et al* [10], the flow becomes periodical with four vortexes in the upper and four vortexes in the lower hemispheres, all propagating in the azimuthal direction. The vortexes in upper and lower hemispheres are displaced for one half of period ones above the others. The boundary between the flows in upper and lower hemispheres is no more coincides with the equator plane, it hesitates with the amplitude, arising from zero near the inner sphere to its maximum value near the outer sphere. For Reynolds numbers $|Re_2| < 100$ (segment $d - e$) the secondary flow is somewhat like the described above secondary flow in point O. The secondary flow near $c - d$ segment is also analogues to one in point O, but all processes are much slower and the amplitudes are greater.

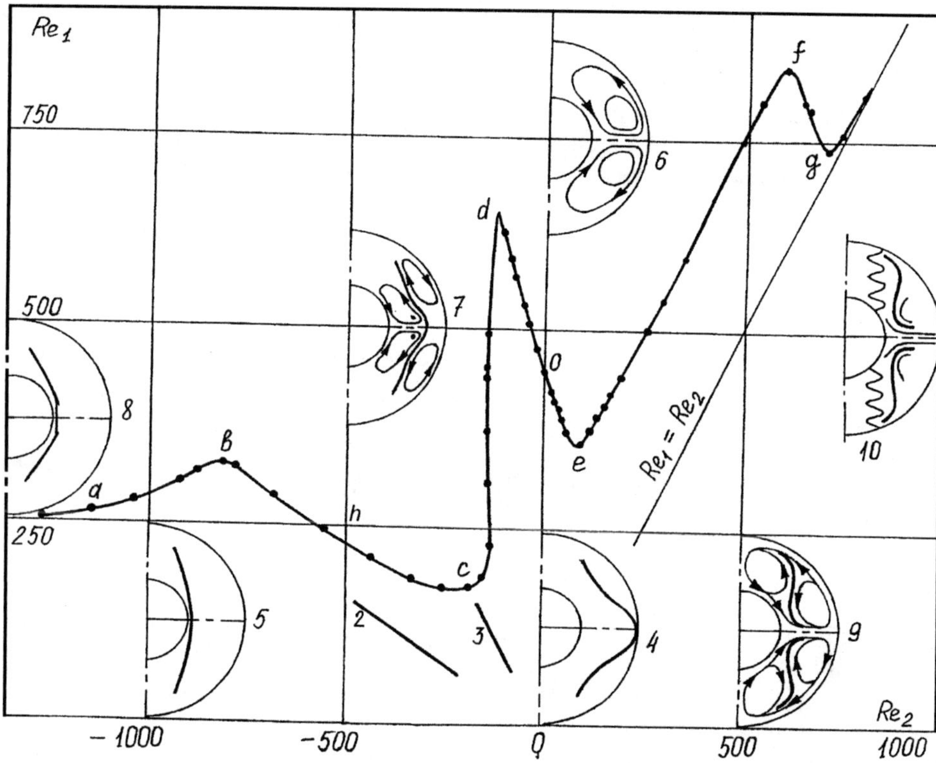

FIGURE 1. Stability curve ($a-g$) and flow structure.

The structure the of the basic flow with co-rotating boundaries is rather complex and depends on the relative influence of boundaries. Fig.1, scheme 9 shows the structure of the basic flow before the primary instability at the segment *e-f*, the S-shaped layer divides two circulation of one direction, combined into one common circulation of the same direction. The circulation which is nearer (inner circulation) to the rotational axis is much weaker than the other one. In this case the instability appears in destroying of the outer steady circulation and generating a number of vortices, propagating in azimuthal direction. The secondary flow holds the equatorial symmetry. At the segment *f-g* the structure of the basic flow is qualitatively the same, but the way of the transition to instability and the structure of the secondary flow are differ from described above. The secondary flow is represented in Fig.1, scheme 10.

Three types of the meridional motion of basic flow were observed while counter-rotating boundaries. Lines 2, 3 in Fig.1 represents the regions with different types of the meridional motion. As was established everywhere the meridional motion has two steady circulation in each hemisphere. The distinction is in the location of the line, dividing these two circulation. Between lines 2 and 3 this line doesn't touch neither inner nor outer spherical boundary. At the right of line3 this line touches the outer spherical boundary (Fig.1, scheme 4), at the left of line 2 – the inner spherical boundary (Fig.1, scheme 5).

At Fig.1, the scheme 7 the structure of the basic flow near the $b – c$ segment of the stability curve is shown. The line dividing two circulation extended near the equator to the outer sphere, centers of both circulation approach to the equator. But the secondary flow is different for segments b - h and $h – c$. The $h – c$ secondary flow holds its equator symmetry and represents two vortices, propagating in the direction of rotation of inner sphere. The $b – h$ secondary flow has no symmetry and represents so-called spiral waves, propagating both along the rotational axis and in the azimuth direction. These spiral waves restricted by the line dividing circulation near the equator and die out moving away from the equator.

The secondary flow at $a – b$ segment, called circumpolar motion, is like spirals propagating from pole to equator.

TRANSITION TO CHAOS

Transition to chaos was investigated in the vicinity of the local maximum on the stability curve at Re_o = -810 (point b). Because of different secondary flows we can treat point b as an intersection of two neutral curves, corresponding to different types of instability.

Let us consider the sequence of supercritical flows for Re_o = -700. The first bifurcation (primary instability) at $Re^{(1)}$ leads to the appearance of spiral waves and a single frequency f_s in the spectrum. The second bifurcation at $Re^{(2)}$ – to the quasi-periodical flow regime, in the structure of the flow the slowly circumpolar motion is added to the spiral waves and in the spectrum frequency f_p is added to f_s. After the third bifurcation (at $Re^{(3)}$) the new flow regime is formed, which we have called the regime of localized vortices. This regime has one frequency f_t and its harmonics in the spectrum and following spatial structure: in both upper and lower hemispheres three equidistant vortices, somewhat like broken Taylor rings, propagate in the direction of rotation of outer sphere, the vortices inclined with respect to equatorial plane. The forth bifurcation (at $Re^{(4)}$) leads to chaotic flow regime. In contrast to the first two regimes the last ones are of hard excitation. Fig.2a shows the dependence of relative frequencies (normalized to the rotation frequency of the outer sphere f_o) of supercritical flow regimes on the Re_i.

For any Re_o in the range $-810 < Re_o < -700$ and any location of the mentioned above measurement points, a quasi-static increase of Re_i results in the same order of supercritical flow regimes. Curves *1* – *3* in Fig.3a shows the boundaries of onset for the quasi-periodical (*1*), localized-vortex (*2*) and stochastic (*3*) flow regimes at $-810 < Re_o < -700$.

For the segment $a – b$ let us consider the sequence of supercritical regimes for Re_o = -900. Fig.2b shows the relative frequencies in this case. The differences from the previously described sequence are 1) in the order of arising of two first regimes, namely spiral waves and circumpolar motion, and 2) in three frequency flow regime, which arises at $Re^{(3)}$ after quasi-periodical one and changes in localized-vortex at

$Re^{(4)}$. The new frequency in three frequency regime is f_g , its amplitude in the spectrum exceeds those for f_s and f_p . The values of f_g and f_t are very close to each other near $Re^{(4)}$, this allows us to assume, that the motion at the frequency f_g is somewhat like localized-vortex.

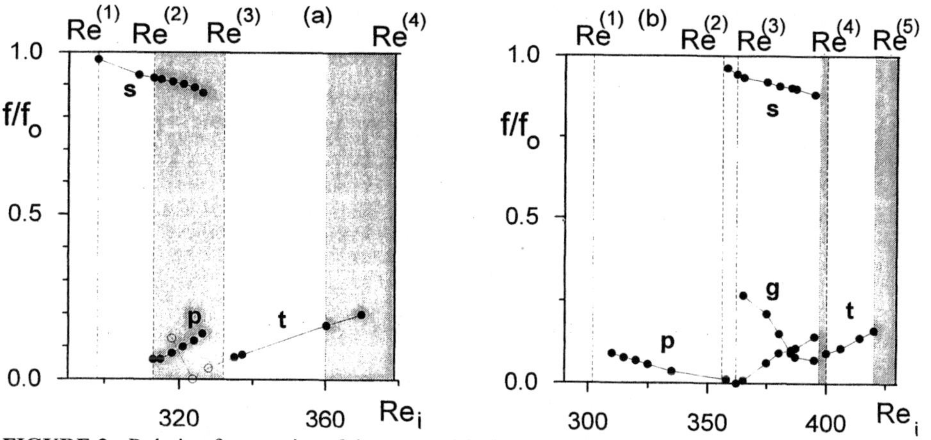

FIGURE 2. Relative frequencies of the supercritical flow regimes at (a) Re_2 =-700 and (b) Re_2 =-900. **s** – spiral waves, **p** – circumpolar motion, **t** – localized vortexes, **g** - the third frequency in three frequency regime. Hysteresis zones are shaded. Black and white symbols correspond to the increasing and decreasing values of Re_i.

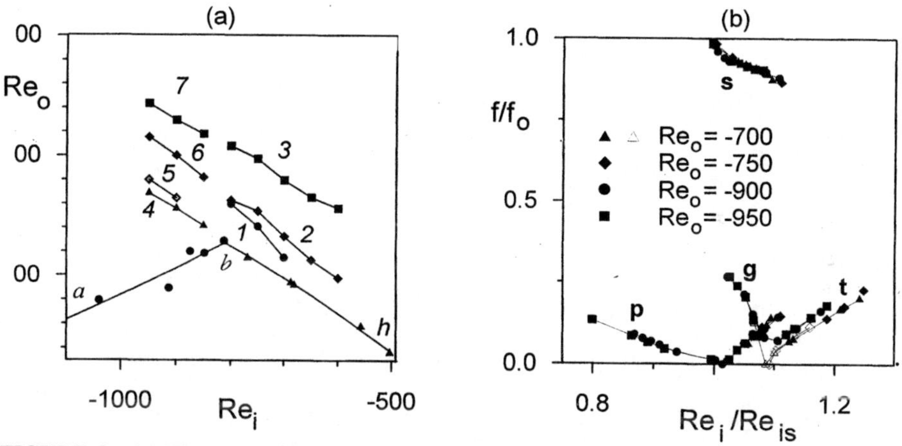

FIGURE 3. (a) Thresholds for the onset of the supercritical flow regimes at the segment *a* – *h*. (b) Relative frequencies for supercritical flow regimes.

Fig.3a also shows curves for the onset of supercritical flow regimes at the left of the point *b*: the quasi-periodical (*4*), three frequency (5), localized-vortex (*6*) and stochastic (*7*). It is easy to see that curve *4*, corresponding to the onset of quasi-periodical regime, continues the segment *b* – *h* of the stability curve. Moreover, the relative frequency f_s/f_2 has no breaks at point *b*. Due to this fact, we have attempted to find a unified description of the route to chaos in all considered region $-950 < Re_o < -700$.

It turns out that, at any Re_o, the relative frequencies of each supercritical regime depend only on the ratio Re_i/Re_{is}, where Re_{is} corresponds to the formation of spiral waves (Fig.3a, curves *4* and *b* – *h*). Fig.3b shows the unified dependencies off the relative frequencies on the ratio Re_i/Re_{is}. The relative frequencies of the spiral waves regime are described by curve *s*, which originate at the point $Re_i/Re_{is} = 1$. The difference lies in the fact that at $Re_o > -810$ this regime is secondary, i.e., arising at the stability threshold and at $Re_o < -810$ this regime is combined with the already existing circumpolar motion. Curve *p* describes circumpolar motion for all analyzed values of Re_o. For $Re_o < -810$ the onset of the circumpolar motion occurs always when $Re_i/Re_{is} < 1$, at the descending branch of the curve *p*, while for $Re_o > -810$ at the ascending one ($Re_i/Re_{is} > 1$). It is noteworthy that the frequency f_g/f_o, existing only at $Re_o < -900$, is close to frequency f_l/f_o in the hysteresis region of the localized-vortex regime corresponding to the segment *b* – *h* (Fig.3b).

Thus in the vicinity of the local maximum in point *b* frequency characteristics of all supercritical flow regimes are formed from the combinations of the frequencies which correspond to the following three spatial flow structures: spiral waves, circumpolar motion, and localized-vortex. The transition to stochastic behavior always occurs from the single-frequency localized-vortex regime with a structure symmetric with respect to the equator plane. At the same time, all bifurcation points preceding this regime arise according to the Ruelle-Takens scenario [11].

REFERENCES

1. Yavorskaya, I.M., Belyaev, Yu.N., Monakhov, A.A., *et al*, "Stability, Non-uniqueness and Transition to Turbulence in the Flow between Two Rotating Spheres" in *Theoretical and Applied Mechanics*, edited by F. P. J. Rimrott and B. Tabarrok, Proceedings of the XVth International Congress of Theoretical and Applied Mechanics, University of Toronto, Canada, 1980, pp. 431-443.
1. Munson, B. R., and Menguturk, M., *J. Fluid. Mech.* **69**, 705-719 (1975).
3. Nakabayashi, K., and Tsuchida, Y., *J. Fluid. Mech.* **195**, 495-522 (1988).
4. Nakabayashi, K., and Tsuchida, Y., *J. Fluid. Mech.* **295**, 43-60 (1995).
5. Yavorskaya, I.M., and Belyaev, Yu.N., *Eur. J Mech., B/Fluids* **10**, N 2 – Suppl., 267-274 (1991).
6. Wulf, P., Egbers, C., and Rath, H.J., *Physics of Fluids* **11**, 1359-1372 (1999).
7. Anderesk, C. D., Liu, S. S., and Swinney H. L., *J. Fluid. Mech.* **164**, 155-183 (1986).
8. Gertsenshtein, S. Ya., Zhilenko, D. Yu., Krivonosova, O. E., and Monakhov, A. A., *Fluid Dynamics* **34**, 500-504 (2000).
9. Gertsenshtein, S. Ya., Zhilenko, D. Yu., and Krivonosova, O. E *Doklady Physics* **45**, 697-700 (2000).
10. Belyaev, Yu.N., Yavorskaya, I.M., and Monakhov, A.A., *Izv.Akad. Nauk SSSR, Mekh. Zhidk. Gasa* N 2, 9-15 (1978).
11. Ruelle, D., and Takens, F., *Common Math. Phys.* **20**, 167-192 (1971).

CONDENSED MATTER

On-off Intermittency and Stochastically Driven Convection in Nematic Liquid Crystals

Ulrich Behn*, Thomas John* and Ralf Stannarius†

*Institute for Theoretical Physics, University of Leipzig, POB 920, 04009 Leipzig, Germany
†Institute for Experimental Physics I, University of Leipzig, POB 920, 04009 Leipzig, Germany

Abstract. Stochastically driven electrohydrodynamic convection in nematics, a standard example of dissipative pattern formation, has been found recently to exhibit on-off intermittency. Near threshold, one observes laminar phases interrupted by bursts of spatially regular stripes. Several statistical characteristics of the stochastic signal, such as the distribution of laminar periods, the distribution of amplitudes, and the power spectrum are determined by laser scattering techniques and compared with theoretical results and simulations. At the stochastic stability threshold, universal scaling laws are found over several orders of magnitude.

INTRODUCTION

Electrohydrodynamic convection in a liquid crystals is a standard system to describe pattern formation in spatially extended dissipative systems, for a recent review, see [1]. Sandwiching a nematic liquid crystal between two transparent electrodes and applying an electric field across the cell one observes above a given threshold voltage an instability of the homogeneous 'planar' state, where the director field is oriented parallel to the electrodes, against the formation of convective rolls, cf. Fig. 1.

The physical mechanism of the instability is closely connected with the anisotropic conductivity of the nematics [2, 3], cf. Fig. 2. Due to the anisotropic optical properties of the material, deflections of the director field can be observed by orthoscopic microscopy or laser scattering techniques. For instance, the convection rolls appear through the microscope as a stripe pattern.

The parameters are easy to control and the characteristic times of the phenomenon

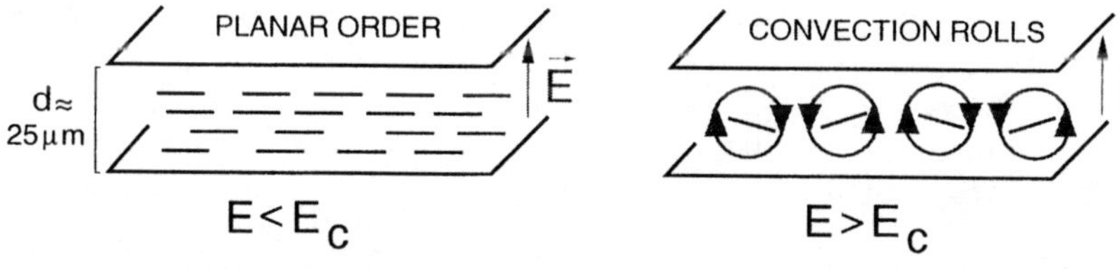

FIGURE 1. Formation of convection rolls in a sandwich cell. If the strength of the applied field is below a critical strength the director field is homogeneously aligned parallel to the electrodes (left). Above the critical strength the homogeneous planar order is unstable and hydrodynamic convection forms a roll pattern which leads to a periodically distorted director field amenable to optical observation (right).

CP622, *Experimental Chaos: 6th Experimental Chaos Conference*, edited by S. Boccaletti et al.

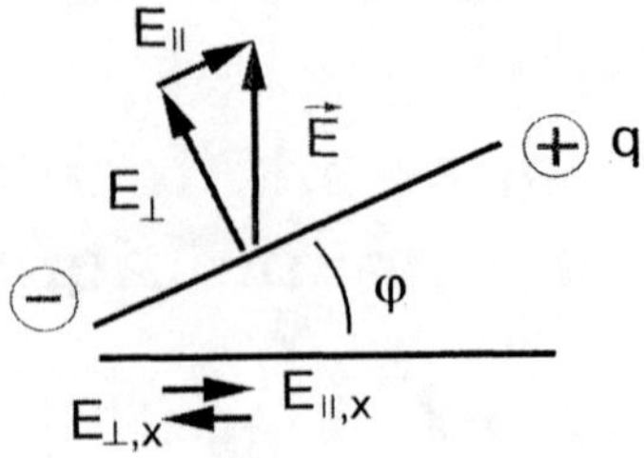

FIGURE 2. Carr-Helfrich mechanism. Shown is a distortion of the director from the parallel alignment due to thermal fluctuations which includes an angle φ with the electrode plane. The electric field perpendicular to the plane has components parallel und perpendicular to the director orentiation. They have x-components of same magnitude and opposite orientation. Due to the anisotropic conductivity a net current $j_x = (\sigma_{\parallel} - \sigma_{\perp})E_x$ is induced which leads to formation of space charges. Now a competition of destabilizing and stabilizing torques takes place. If the electric field is strong enough the homogeneous planar order becomes unstable against a periodic convection pattern. The relevant variables are the space charge density q and the director deflection field φ.

are convenient for experiments. The threshold voltage and the wave number of the pattern depend on the frequency of the driving field. There are two different regimes, the conductive regime and the dielectric regime, where the wavelength of the pattern is drastically different, separated by a cutoff frequency (see below). Increasing the voltage beyond the first threshold, a hierarchy of further instabilities is observed.

In the last two decades the system was investigated also for the case that the driving electric field has a stochastic, time dependent, component. A number of noise induced phenomena has been experimentally observed [4, 5, 6, 7, 8, 9, 10, 11] which has stimulated theoretical work [12, 13, 14, 15, 16]. The electric field couples in a multiplicative way to the variables of the system, we thus have an example of multiplicative noise.

The theoretical foundation describing the stability of the stochastically driven system was given in [16]. It was predicted that if the characteristic times of the driving process is at the threshold of the order of a characteristic time of the system, fluctuation phenomena may occur. The appearance of on-off intermittency in this system was experimentally observed by orthoscopic microscopy [17]. Near threshold the homogeneous planar state is interrupted by irregular bursts of convection rolls ('on'-states). At the threshold the distribution of quiescent periods ('off'-states) was determined as a power law with exponent $-3/2$ over several orders of magnitudes. Experiment and theory were in good quantitative agreement. With these results, a first example of the appearance of on-off intermittency in a spatially extended dissipative system with many degrees of freedom was found and theoretically explained.

On-off intermittency was first observed in theoretical investigations of nonlinear dynamical systems with few degrees of freedom driven by multiplicative noise [18, 19, 20, 21, 22, 23]. Subsequently also model systems with more degrees of freedom were considered [24, 25, 26, 27]. Very recently, a Swift-Hohenberg model with ad hoc assumed multiplicative noise was considered in [28].

In this paper we report on investigating the phenomenon by the more refined technique of laser diffraction. Using this method a number of other statistical characteristics

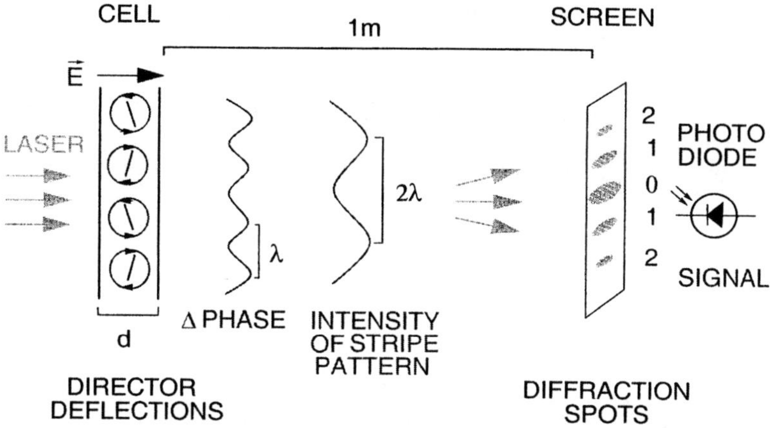

FIGURE 3. Experimental setup. The periodic convection pattern in the sandwich cell induces a perodic field of director deflections. Due to the anisotropic optical properties of the liquid crystal this leads to a phase grating of wavelength λ which is in the conductive regime of the order of the cell gap d. The intensity of the stripe pattern visible by orthoscopic observation with a microscope at one focus plane has a wavelength 2λ. A laser beam through the cell leads to a diffraction pattern on the screen. Either the whole pattern is monitored with a video camera or a photodiode is placed at a given diffraction spot.

of the stochastic signal are determined. Besides the distribution of laminar periods, the distribution of amplitudes, and the power spectral density is investigated. A good quantitative agreement with theory and simulations is observed in general.

EXPERIMENTAL SETUP

Commercial sandwich cells (*LINKAM*) with an electrode area of 5 mm×5 mm and a cell gap of 25 μm are filled with the nematic mixture *Mischung 5* (Halle) which consists of four disubstituted alkyl(oxy)phenyl-alkyloxybenzoates. Material characteristics of this substance can be found in [29]. Planar surface alignment is achieved by polyimide coating and antiparallel rubbing of the plates. The sample temperature is adjusted to 32 °C with an accuracy of 0.1 K.

The driving voltage is generated by a PC with a sampling rate of 7 kHz and subsequently amplified. The maximum peak-peak voltage available is 220 V.

We used a He-Ne laser with wavelength 632.8 nm at normal incidence on the cell. The laser spot is about 1 mm in diameter. The scattered light is monitored on a screen in a distance of about 1 m from the cell, cf. Fig. 3. From the two-dimensional scattering image the wave lengths and orientations of the patterns can be continuously extracted. We employ two different methods recording the scattering data.

A commercial video camera is used with an acquisition rate of 25 fps. This enables to study the evolution of the mode spectrum and to determine the full wave vector information, although with a limited time resolution. Using this setup it is visible that within the conductive and the dielectric regime the pattern is dominated by only a single mode. The wave vector remains fairly constant and only intensity modulations occur. In the parameter range near the cutoff separating the conductive from the dielectric regime a band of modes can become unstable in the stochastic case.

Within the conductive and the dielectric regime it suffices to measure the intensity of a given diffraction spot. There, a photodiode with an aperture of 7 mm^2 is positioned which has a linear output over the whole intensity range. The photo voltage is sampled with a shortest time resolution of 0.2 ms, digitized by an AD converter with a resolution of 12 bit, and stored as a time series.

OPTICS

The periodic director deflection field $\varphi(t)\sin(k_x x)sin(k_z z)$ associated with convection rolls leads to a modulation of the effective refractive index n_{eff}. For a transmitted laser beam a phase and amplitude grating is formed and a diffraction pattern appears, cf., eg., [30, 31, 32, 33, 34, 35].

The modulation of n_{eff} may cause different optical lengths for paths of the same geometric length which leads to a phase grating. Assuming straight paths through the cell, the phase differences can be calculated for small amplitudes of the deflection field $\varphi(t)$, it varies as $\varphi^2(t)\cos(2k_x x)$. Thus the phase grating has a wavelength $\lambda = \pi/k_x$. The light intensity $A(t)$ at the first scattering spot in the far-field diffraction pattern is related to the square of the Bessel function of first kind with the maximum phase difference in the argument [30]. Hence, for small φ one has $A(t) \sim \varphi^4(t)$ which enables to relate the measured intensities to a system variable of the theoretical description.

Numerical integration of the nonlinear Euler-Lagrange equation following from the Fermat principle $\int n_{\text{eff}} ds =$ Extr. allows to describe also deflections of the light path (cf., e.g., [36]). Together with the phase grating an amplitude grating appears which has a wavelength 2λ but the resulting diffraction pattern is of much weaker intensity than that of the phase grating. The numerical results show the validity of the relation $A(t) \sim \varphi^4(t)$ for values of φ up to the order of one [37].

NEMATO-ELECTROHYDRODYNAMICS

The basic equations of nemato-electrohydrodynamics are the Navier-Stokes equations, the Maxwell equations and the torque balance equations, a system of nonlinear partial differential equations. Since we are interested in phenomena near the threshold, first these equations are linearized. Assuming a simplified geometry (infinite extension of the probe parallel to the rolls) and inserting a test mode characterized by wave numbers $(k_x, k_z = \pi/d)$, and using so-called stress-free boundary conditions leads after adiabatic elimination of the velocity field to a system of two ordinary differential equations [38]. This system describes the dynamics of the space charge q and the spatial variation of the angle φ between the director and electrode plates, $\psi = \partial_x \varphi$. Introducing $\vec{z} = (q, \psi)^{\mathrm{T}}$ it reads

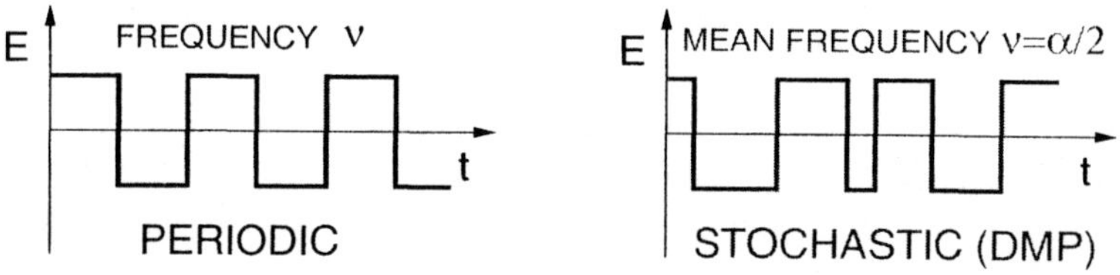

FIGURE 4. Square wave driving. Shown are a periodic square wave of frequency ν (left) and a realization of a dichotomous Markov process of mean frequency $\nu = \alpha/2$ (right) both jumping between $\pm E$. α is the mean number of jumps per unit time.

$$\dot{\vec{z}} = \mathbf{C}(t)\vec{z}\,, \quad \text{where} \quad \mathbf{C}(t) = -\begin{pmatrix} 1/T_q & \sigma_H E_t \\ aE_t & \Lambda_1 - \Lambda_2 E_t^2 \end{pmatrix}. \tag{1}$$

The coefficients in $\mathbf{C}$ depend on material properties and on the wave numbers k_x and k_z; explicit expressions are given in [1, 16].

For piecewise constant electric fields E_t, jumping e.g. between two values $E_\pm$ at times $t_\nu > t_{\nu-1}, \nu = 1, \ldots, n$ (cf. Fig.4) an explicit solution of (1) can be obtained [16]. We first consider a time interval between two jumps where $\mathbf{C}(t)$ is a constant $\mathbf{C}_\pm$ and diagonalize it by a unitary transformation $\mathbf{U}_\sigma \mathbf{C}_\sigma \mathbf{U}_\sigma^{-1} = \text{diag}\,(\lambda_i^\sigma)$, $\sigma = \pm$. In the diagonal representation the equation of motion is solved and the inverse transformation leads to $\vec{z}(t) = \mathbf{T}_\sigma(t-t')\vec{z}(t')$ where the time evolution matrix is $\mathbf{T}_\sigma(\Delta t) = \mathbf{U}_\sigma^{-1}\text{diag}[\exp(\lambda_i^\sigma \Delta t)]\mathbf{U}_\sigma$. For a given realization of the field with jumps at t_ν iteration leads to

$$\vec{z}(t_n) = \mathbf{T}_{\sigma_n}(\Delta t_n) \cdots \mathbf{T}_{\sigma_2}(\Delta t_2)\mathbf{T}_{\sigma_1}(\Delta t_1)\vec{z}(t_0). \tag{2}$$

In the periodic case, all $\Delta t_\nu = t_\nu - t_{\nu-1}$ are equal and the stability is determined by the eigenvalues of $\mathbf{T}_+(\Delta t)\mathbf{T}_-(\Delta t)$ reproducing the result of Floquet stability analysis.

In the stochastic case, the stability is determined by the largest Lyapunov exponent λ_1 of the product of random matrices in (2) in the long time limit. If $\lambda_1 > 0$ the trajectory diverges, if $\lambda_1 < 0$ it converges to zero. $\lambda_1 = 0$ defines the neutral curve, the mechanism of mode selection is explained in Fig. 5.

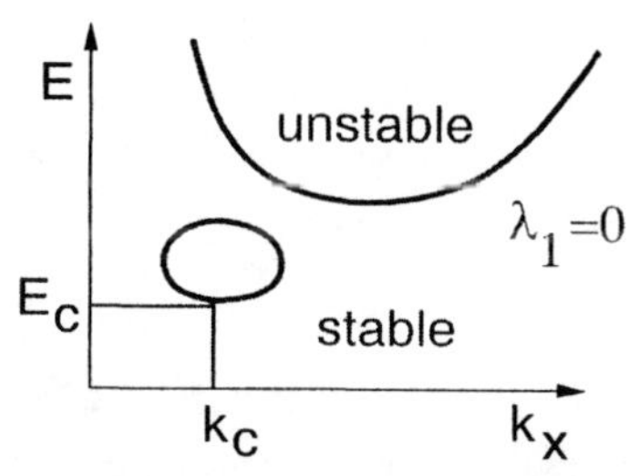

FIGURE 5. Mode selection. Shown are the nullclines of the largest Lyapunov exponent λ_1 in the k_x-E plane. Inside the island and in the upper region $\lambda_1 > 0$, there the homogeneous order is unstable against formation of rolls. Increasing the the magnitude of E up to E_c the wavenumber k_c is selected for which the unstable region is reached first.

STOCHASTIC STABILITY

The difficulty of evaluating an infinite product of random matrices can be circumvented [39] by a transformation from (q, ψ) to polar coordinates (r, ϕ), which leads to

$$\dot{r} = g(E_t, \phi) r, \quad (3)$$

$$\dot{\phi} = h(E_t, \phi), \quad (4)$$

explicit expressions for g and h are given in [16].

Equation (4) depends only on ϕ and it is possible to find the stationary probability density $P_\sigma(\phi)$ of the joint process (E_t, ϕ). Equation (3) is linear in r and can be solved for a given trajectory of the driving process. Thus the leading Lyapunov exponent is obtained as

$$\lambda_1 = \lim_{t \to \infty} \frac{1}{t} \int_0^t ds\, g(E_s, \phi) = \langle g(E_\sigma, \phi) \rangle, \quad (5)$$

where the brackets mean the average with respect to $P_\sigma(\phi)$. The second equality holds for almost all trajectories [40]. For the case which is of interest here, λ_1 can be evaluated explicitly in terms of generalized hypergeometric functions [14, 16].

The two Lyapunov exponents of the system are related by $\lambda_1 + \lambda_2 = \mathrm{Sp}(\mathbf{C}_- + \mathbf{C}_+)/2$. The threshold is defined by $\lambda_1 = 0$; the corresponding characteristic time $\tau_1 = |1/\lambda_1|$ diverges and is thus well separated from $\tau_{\mathrm{stoch}} = 1/(2\alpha)$. At threshold the second characteristic time $\tau_2 = |1/\lambda_2|$ can be obtained from the above relation. If $\tau_2 \sim \tau_{\mathrm{stoch}}$ which is here the case, one observes in experiment intermittent bursts of a regular spatial stripe pattern interrupting the quiescent state. As described below, at the sample stability threshold a number of universal power laws characterize the statistical properties of the trajectories.

Finally in this Section we transform the system (3,4) into a form similar to many models investigated in the context of on-off intermittency. For a given realization of the driving process, equation (4) can be integrated which gives $\phi_t[E_s]$ as a functional of the driving process E_s for times $s < t$. For our dichotomous model process explicit expressions can be given. Inserting this in (3) we obtain the single equation

$$\dot{r} = g(E_t, \phi_t[E_s]) r, \quad (6)$$

where the time dependent stochastic coefficient $g(E_t, \phi_t[E_s])$ depends on the prehistory of the driving process thus having complex statistical properties. Again, for the dichotomous process explicit expressions can be given.

ON-OFF INTERMITTENCY

Dynamical systems driven by multiplicative noise may exhibit at the threshold of stability a new type of intermittency characterized by specific statistical properties of the intermittent signal. Quiescent or 'laminar' periods ('off'-states) are interrupted in an irregular way by bursts of large activity ('on'-states). Therefore the phenomenon was named on-off intermittency [21]. At the threshold, a number of power laws hold, universal over a broad class of different systems. The early studies considered models with only few degrees of freedom, described by differential equations or mappings [18, 19, 20, 21, 22, 23]. Experimental realizations are found for instance in nonlinear circuits [41, 42, 43, 44], cf. also [45, 46, 47]. More recently, the phenomenon was investigated theoretically also in systems with many degrees of freedom [24, 25, 26, 27, 28].

The universality of the laws is closely related to statistical characteristics of persistent random walks which hold under quite general conditions for a broad class of stochastic processes. Various statistical laws characterizing on-off intermittency can be obtained, e.g., from the simple model [48, 49]

$$\dot{r} = [\lambda + \xi(t)]r + \zeta(t) \tag{7}$$

with reflecting boundary condition at $|r| = S$. Multiplicative and additive noise are independent, both are Gaussian white processes with zero mean and strength D_{mult} and D_{add}, respectively. For example, the autocorrelation of the additive process is $\langle \zeta(t)\zeta(t') \rangle = 2D_{\mathrm{add}}\delta(t-t')$.

At the threshold, $\lambda = 0$, the distribution of laminar periods scales like $P(\tau) \sim \tau^{-3/2}$. The result is related to the distribution of return times of a persistent random walk to an arbitrary level. In the vicinity of the threshold there are exponential corrections [22]. Further corrections are due to the additive noise [50].

Another scaling law holds for the distribution of amplitudes in the trajectory [18, 19, 49, 51]. In the vicinity of the threshold within a range $s < r < S$ the distribution scales like $P(r) \sim r^{-1+\eta}$. At the threshold, η vanishes and a r^{-1} power law holds. Due to the additive noise the validity region is bounded from below by $s = (D_{\mathrm{add}}/D_{\mathrm{mult}})^{1/2}$. For $r < s$ the additive noise dominates and the distribution is flat. S is an upper bound above which nonlinearities become important and the behaviour is not universal. This is a general result for systems driven by multiplicative noise in contrast to systems with only additive noise.

A further characteristics is the power spectral density, i.e. the Fourier transform of the autocorrelation function, $S(\omega) = \mathrm{FT}\langle r(t)r(t') \rangle$ [20, 48, 52, 53, 54, 55]. For the simple model (7) it has been shown that $S(\omega)$ scales over a range of frequencies $w < \omega < D_{\mathrm{mult}}$ like $S(\omega) \sim \omega^{-1/2}$. This scaling behaviour is universal for a broad class of systems. For $\omega < w$ there are no correlations and the power spectral density is flat. For $D_{\mathrm{mult}} < \omega$ one expects again a nonuniversal behaviour. For the model without additive noise, only off threshold there is a lower bound $w = \lambda^2/D_{\mathrm{mult}}$. With additive noise the lower bound is $\max\{\lambda^2/D_{\mathrm{mult}}, D_{\mathrm{mult}}/\log^2(D_{\mathrm{mult}}/\tilde{D}_{\mathrm{add}})\}$, where the tilde denotes the normalization of D_{add} by the amplitude of a typical burst. It should be noted here that the influence of amplitude truncation on the power spectral density is a complicatied issue for itself and a field of active research [56, 57, 58].

The results are rather robust against changing the nature of the noise (they are also found for chaotic driving), against replacing the additive noise by a lower bound, replacing the reflecting boundaries by nonlinear terms, considering discrete versions of this equation, etc.

A number of further characteristics have been considered, such as the scaling behaviour of higher moments and the fractal dimensions of level sets.

EXPERIMENT VS THEORY

Threshold voltages and selected wave numbers were experimentally determined driving the system with square wave voltages for a broad range of (mean) frequencies ν. In the deterministic case, the threshold towards the convective pattern is easily determined in a naive way: Below threshold there is no pattern, above there is. The results are compared with the thresholds and wave numbers of the theory where the known material parameters [29] are used, a fit of the remaining parameters yields an excellent agreement with the experimental data, cf. Fig. 6.

In the stochastic case (with the same set of material parameters) the thresholds and wave numbers are determined evaluating the threshold condition for stochastic stability and selecting the wave number where the threshold is minimal, cf. Fig. 5. These results are in good quantitative agreement with the experimental values, too. The threshold is experimentally defined looking at the distribution of amplitudes which is easier to determine than the distribution of laminar periods.

The intensity of a first order diffraction spot of the phase grating is observed over periods of about 2 hours and statistical properties of the stochastic signal are computed. A typical trajectory of the photo voltage is shown in Fig. 7 where also the definition of a laminar period, and the lower and upper bounds of the signal are explained.

In the following figures the distribution of laminar periods (Fig. 8), the distribution of amplitudes and the exponents of the power law as a function of the voltages (Fig. 9), and the power spectral density (Fig. 10) are shown for voltage amplitudes at the threshold, and about 5% below and above threshold. The driving process has a mean frequency $\nu = 80$ Hz, the system is well within the conductive regime.

The experimental results are compared with simulations which are performed on the base of Eq. (2) using lower and upper saturating bounds for the curvature $\psi_{\min} < \psi < \psi_{\max}$. The bounds are introduced as caricatures of the bounds appearing in the experimental systems, caused by thermal noise and the saturation due to nonlinearity. We have chosen $\psi_{\max}/\psi_{\min} \approx 10^2$ which is a reasonable dynamical range for the director

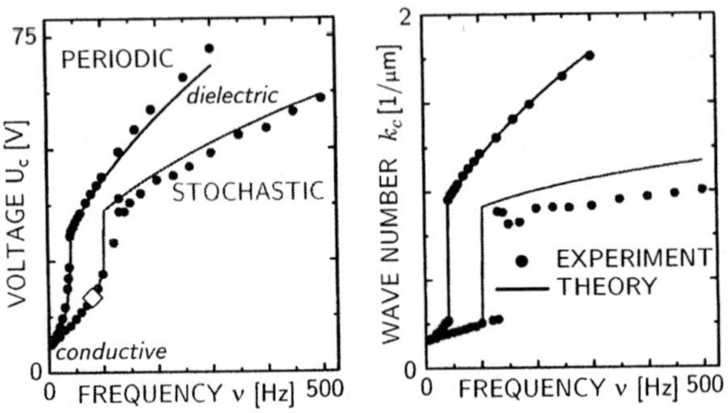

FIGURE 6. Threshold voltages (left) and selected wave numbers (right) for driving with periodic and stochastic square waves of (mean) frequency ν. Shown are the experimental data (full dots) in comparison with the theoretical results (solid lines). The diamond in the left figure indicates the parameter setting where the statistical characteristics of the pattern presented in the following have been determined.

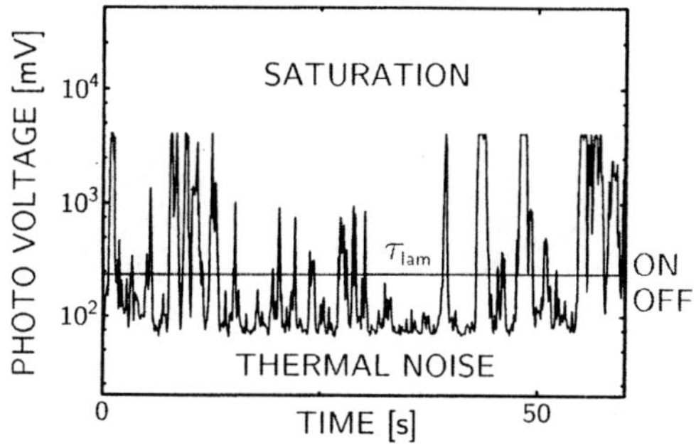

FIGURE 7. Stochastic signal from the photo diode. Shown is the signal on a semi-logarithmic scale obtained for driving the system with a dichotmous stochastic voltage of amplitude U=16 V and mean frequency $\nu = 80$ Hz which is below the threshold. Periods where the trajectory is above the horizontal line are considered as 'on', below this line as 'off'. The duration of a laminar (off) period is denoted by τ_{lam}. The trajectory is bounded from below by the noise produced by thermal fluctuations of the liquid crystal and from above by the limited range of the AD converter digitizing the signal from the photo diode. The trajectory shown here is well within the linear range of the theory.

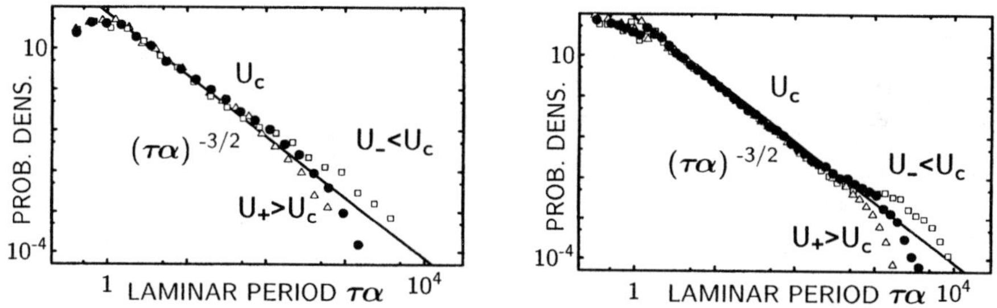

FIGURE 8. Distribution of laminar periods (in units of $1/\alpha$) in the conductive regime. Experiment (left) vs. theory (right). Shown are the results for $\nu = \alpha/2 = 80$ Hz at the threshold voltage $U_c = 17.2$ V and at voltages 6% below and above the threshold, U_- and U_+. The solid line indicates a $\tau^{-3/2}$ power law. At the threshold the power law holds over more than three orders of magnitude.

deflection. The theoretical thresholds and wave numbers are determined evaluating the analytic expression for $\lambda_1 = 0$ as described above.

CONCLUSION

We found for a system with truly many degrees of freedom, a spatially extended dissipative system driven by a multiplicative external noise, many characteristics of on-off intermittency. The on-state is a spatially ordered state of periodic convection rolls. At the threshold of stability, statistical characteristics of the stochastic signal obtained by laser diffraction experiments show power laws with the known characteristic exponents. The results are in good quantitative agreement with the analytical theory and the simulations of the basic equations of the theoretical model.

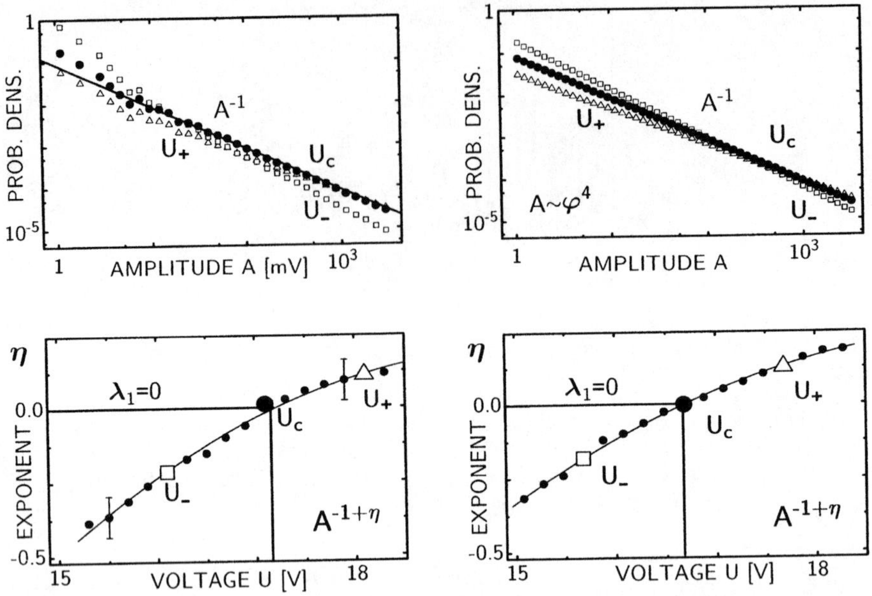

FIGURE 9. Distribution of amplitudes at the threshold U_c and at $U_\pm$ (top row). Experiment (left) vs. theory (right). To obtain the distribution of amplitudes from simulations, the relation $A \sim \varphi^4$ is adopted. The solid line is the A^{-1} power law which holds at the threshold over several magnitudes. In the vicinity of the threshold power laws are found with exponents $-1+\eta$. The bottom row shows the exponent η indicating the deviations from the A^{-1} law. At the threshold η and λ_1 equals zero. Again, experiment (left) vs. theory (right).

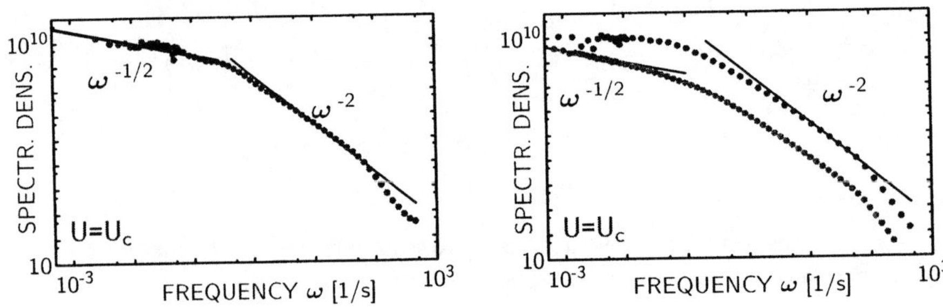

FIGURE 10. Power spectral density obtained from the stochastic signal of the photo diode (left) and from the simulation (right). The experimental result shows a regime where a $\omega^{-1/2}$ law seems to hold. For higher frequencies a ω^{-2} law can be seen. Including lower bounds for ψ as in the other cases the simulation shows only the ω^{-2} law, for low frequencies the power spectrum is flat (upper curve). Without lower bounds, thus increasing the dynamical range, the $\omega^{-1/2}$ behaviour is obtained (lower curve). The mean frequency of the driving process is 80 Hz which appears as upper bound for the ω^{-2} law.

In our case, the driving noise has a finite correlation time which is at the threshold of the order of a characteristic time of the physical system. Nevertheless the behaviour described here is very similar to that of simpler systems driven by Gaussian white noise. Moreover, the multiplicative driving process which appears in the formulation (6), which is the 'normal form' of equations describing on-off intermittency, has complicated statistical properties.

Further research is directed to describing the dielectric regime in more detail, the regime near the cutoff, where noise induced mode selection [59] takes place, and to the superposition of the stochastic field with a deterministic component which may change the ordering of the characteristic times of noise and system. The analytic derivation of the scaling laws and a systematic derivation of the nonlinear amplitude equation from the basic nemato-electrohydrodynamic equations for the case of stochastic driving is a challenge for theory.

A more detailed account and further results will be given elsewhere [60].

ACKNOWLEDGMENTS

Thanks is due to Hirokazu Fujisaka for stimulating discussions and for providing Ref. [28] prior to publication. Financial support by the DFG (grant Be 1417/4) is gratefully acknowledged.

We dedicate this paper to Professor Adolf Kühnel on the occasion of his 65th birthday with sincere thanks for his continuous interest in this subject and for many valuable discussions.

REFERENCES

1. Pesch, W., and Behn, U., "Electrohydrodynamic convection in nematics", in *Evolution of Spontaneous Structures in Dissipative Continuous Systems*, edited by F. H. Busse and S. C. Müller, Springer, Heidelberg, 1998, p. 335.
2. Carr, E. F., *J. Chem. Phys.*, **38**, 1536 (1963).
3. Helfrich, W., *J. Chem. Phys.*, **51**, 4092 (1969).
4. Kai, S., Kai, T., Takata, M., and Hirakawa, K., *J. Phys. Soc. Jpn.*, **47**, 1379 (1979).
5. Kawakubo, T., Yanagita, A., and Kabashima, S., *J. Phys. Soc. Jpn.*, **50**, 1451 (1981).
6. Brand, H. R., Kai, S., and Wakabayashi, S., *Phys. Rev. Lett.*, **54**, 555 (1985).
7. Kai, S., Tamura, T., S. Wakabayashi, M. I., and Brand, H. R., *IEEE-IAS Conf. Records*, **85CH**, 1555 (1985).
8. Kai, S., Fukunaga, H., and Brand, H. R., *J. Phys. Soc. Jpn.*, **56**, 3759 (1987).
9. Kai, S., Fukunaga, H., and Brand, H. R., *J. Stat. Phys.*, **54**, 1133 (1989).
10. Kai, S., "Electrohydrodynamic instability of nematic liquid crystals: growth process and influenc of noise", in *Noise in nonlinear dynamical systems*, edited by F. Moss and P. V. E. McClintock, Cambridge University Press, Cambridge, 1989, p. 22.
11. Brand, H. R., "Suppression of electrohydrodynamic instabilities by external noise", in *Noise in nonlinear dynamical systems*, edited by F. Moss and P. V. E. McClintock, Cambridge University Press, Cambridge, 1989, p. 77.
12. Behn, U., and Müller, R., *Phys. Lett.*, **113A**, 85 (1985).
13. Müller, R., and Behn, U., *Z. Phys. B*, **69**, 185 (1987).
14. Müller, R., and Behn, U., *Z. Phys. B*, **78**, 229 (1990).
15. Lange, A., Müller, R., and Behn, U., *Z. Phys. B*, **100**, 447 (1996).
16. Behn, U., Lange, A., and John, T., *Phys. Rev. E*, **58**, 2047 (1998).
17. John, T., Stannarius, R., and Behn, U., *Phys. Rev. Lett.*, **83**, 749 (1999).

18. Fujisaka, H., and Yamada, T., *Progr. Theor. Phys*, **74**, 918 (1985).
19. Fujisaka, H., and Yamada, T., *Progr. Theor. Phys*, **75**, 1087 (1985).
20. Yamada, T., and Fujisaka, H., *Progr. Theor. Phys*, **76**, 582 (1986).
21. Platt, N., Spiegel, E. A., and Tresser, C., *Phys. Rev. Lett.*, **70**, 279 (1993).
22. Heagy, J. F., Platt, N., and Hammel, S. M., *Phys. Rev. E*, **49**, 1140 (1994).
23. Fujisaka, H., Matsushita, S., Yamada, T., and Tominaga, H., *Physics Reports*, **290**, 27 (1997).
24. Yang, H. L., and Ding, E. J., *Phys. Rev. E*, **50**, R3295 (1994).
25. Platt, N., and Hammel, S. M., *Physica A*, **239**, 296 (1997).
26. Fujisaka, H., Ouchi, K., Hata, H., Masaoka, B., and Miyazaki, S., *Physica D*, **114**, 237 (1998).
27. Fukushima, K., and Yamada, T., *Phys. Lett. A*, **237**, 141 (1998).
28. Fujisaka, H., Ouchi, K., and Ohara, H., On-off convection: Noise induced intermittency near the convection threshold (2001), to appear in Phys. Rev. E.
29. Amm, H., Stannarius, R., and Rossberg, A. G., *Physica D*, **126**, 171 (1999).
30. Carroll, T. O., *J. Appl. Phys.*, **43**, 767 (1972).
31. Akahoshi, S., Miyakawa, K., and Takase, A., *Jpn. J. Appl. Phys.*, **15**, 1839 (1976).
32. Kondo, K., Arakawa, M., Fukuda, A., and Kuze, E., *Jpn. J. Appl. Phys.*, **22**, 394 (1983).
33. Miike, H., Kuriyama, Y., Hashimoto, H., and Ebina, Y., *J. Phys. Soc. Jpn*, **53**, 3280 (1984).
34. Vistin, L. K., and Yakovenko, S. S., *Sov. Phys. - Crystallogr.*, **28**, 587 (1983).
35. Bohatsch, H., and Stannarius, R., *Phys. Rev. E*, **60**, 5591 (1999).
36. Rasenat, S., Hartung, G., Winkler, B. L., and Rehberg, I., *Exp. in Fluids*, **7**, 412 (1989).
37. John, T. (2001), unpublished.
38. Goosens, W. J. A., "Electrohydrodynamic instabilities in nematic liquid crystals", in *Advances in liquid crystals Vol. 3*, edited by G. H. Brown, Academic Press, New York, 1973, p. 1.
39. Arnold, L., and Kloeden, P., *SIAM J. Appl. Math.*, **49**, 1242 (1989).
40. Oseledec, V. I., *Trans. Moscow Math. Soc.*, **19**, 197–231 (1968).
41. Hammer, P. W., Platt, N., Hammel, S. M., Heagy, J. F., and Lee, B. D., *Phys. Rev. Lett.*, **73**, 1095 (1994).
42. Yu, Y. H., Kwak, K., and Lim, T. K., *Phys. Lett. A*, **198**, 34 (1995).
43. Čenys, A., Namajuñas, A., Tamiševičius, A., and Schneider, T., *Phys. Lett. A*, **213**, 259 (1996).
44. Hun, Y. Y., Kim, D. C., Kwak, K., Lim, T. K., and Jung, W., *Phys. Lett. A*, **247**, 70 (1998).
45. Rödelsperger, F., Čenys, A., and Benner, H., *Phys. Rev. Lett.*, **75**, 2594 (1995).
46. Sauer, M., and Kaiser, F., *Phys. Rev. E*, **54**, 2468 (1996).
47. Feng, D. L., Yu, C. X., Xie, J. L., and Ding, W. X., *Phys. Rev. E*, **58**, 3678 (1998).
48. Venkataramani, S. C., Antonsen, T. M., Ott, E., and Sommerer, J. C., *Physica D*, **96**, 66 (1996).
49. Nakao, H., *Phys. Rev. E*, **58**, 1591 (1998).
50. Platt, N., Hammel, S. M., and Heagy, J. F., *Phys. Rev. Lett.*, **72**, 3498 (1994).
51. Fujisaka, H., Suetani, H., and Watanabe, T., *Progr. Theor. Phys. Suppl.*, **139**, 70 (2000).
52. Fujisaka, H., Ishii, H., Inoue, M., and Yamada, T., *Pogr. Theor. Phys.*, **76**, 1198 (1986).
53. Venkataramani, S. C., Antonsen, T. M., Ott, E., and Sommerer, J. C., *Physics Letters A*, **207**, 173 (1995).
54. Fujisaka, H., and Yamada, T., *Progr. Theor. Phys*, **90**, 529 (1993).
55. Miyazaki, S., and Hata, H., *Phys. Rev. E*, **58**, 7172 (1998).
56. Ishioka, S., Gingl, Z., Choi, D., and Fuchikami, N., *Phys. Letters A*, **269**, 7 (2000).
57. Gingl, Z., Ishioka, S., Choi, D., and Fuchikami, N., "Theoretical and experimental results concerning the amplitude truncation of Gaussian $1/f^{\alpha}$ Noises", in *Unsolved Problems of Noise and Fluctuations*, edited by D. Abbott and L. B. Kish, AIP, USA, 2000, p. 136.
58. Choi, D., and Fuchikami, N., *J. Phys. Soc. Japan*, **70**, 297 (2001).
59. Kurtze, D. A., *Phys. Rev. Lett.*, **77**, 63 (1996).
60. John, T., Stannarius, R., and Behn, U. (2001), in preparation.

First-Order Fréedericksz transition in a Liquid-Crystal Light Valve

M.G. Clerc*, T. Nagaya†, S. Residori** and C. S. Riera‡

*Departamento de Física, Facultad de Ciencias Físicas y Matemáticas, Universidad de Chile, casilla 487-3, Santiago, Chile
†Faculty of Education, Okayama University, Okayama, 700-8530, Japan
**Institut Non Linéaire de Nice, UMR 6618 CNRS-UNSA, 1361Route des lucioles, F-06560 Valbonne-Sophia Antipolis, France
‡Department of Mechanical Engineering, Massachussetts Institute of Technology,77 Massachussetts Avenue, Cambridge MA 02139, USA

Abstract. Light feedback in a Liquid-Crystal-light-Valve can render the Fréedericksz transition a first order one. A theoretical and experimental study of this first order transition is presented. An amplitude equation is derived, valid close to the transition. Depending on the values of the feedback parameters, both theory and experiment exhibit bistability, propagation of fronts and a Maxwell point.

INTRODUCTION

One of the most well-studied phenomena in the physics of liquid crystals is the field-induced distortion of a nematic liquid-crystal film, called the Fréedericksz transition (FT) [1]. For a nematic film, FT is usually a second-order transition [2]. The possibility of modifying FT into a first-order one has been considered, either through the simultaneous application of electric and magnetic field [3] or through the action of an optical field [4], for liquid crystals possessing a very large optical anisotropy. In both cases, an experimental verification is difficult to attain, especially over a large area of the liquid crystal film. Another approach, is to realize a global feedback by means of spatially integrated light intensity [5]. In this case, the FT experimentally displays clear features of a first-order transition. However, spatial dynamics are lost as a consequence of integration.

By means of a a Liquid Crystal Light Valve (LCLV), we have realized an experiment in which the local feedback is high enough to render FT first-order over a large size ($\simeq cm^2$) of the nematic film [6]. Thus, all the spatial phenomena that we expect to be associated with a first-order phase transition can be observed and investigated. In particular, the classical nucleation theory applies, and we show that a local perturbation leads to a front propagation which, depending on the distance from the Maxwell point, is associated to shrinking or expanding droplets.

Experimental setup. A LCLV consists essentially in a nematic liquid-crystal film sandwiched between a glass and a photoconductive plate over which a dielectric mirror is deposed. Coating of the two bounding plates induces a planar anchoring (nematic di-

CP622, *Experimental Chaos: 6th Experimental Chaos Conference*, edited by S. Boccaletti et al.

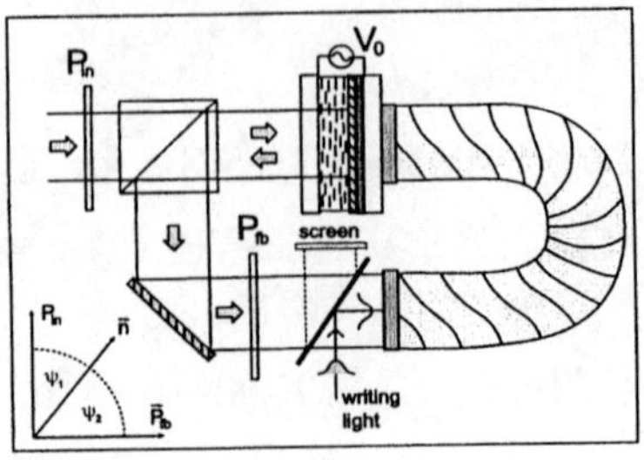

FIGURE 1. Experimental setup. Two confocal lenses, (not displayed in the scheme), provide a 1 : 1 image-forming system from the front side of the LCLV to its rear side. The optical feedback loop is closed by a fiber bundle, which is aligned in order to avoid any rotation or shift. P_{in} and P_{fb} are, respectively, the input and feedback polarizer. Their orientation with respect to the liquid crystal director is indicated in the left bottom of the figure.

rector $\vec{n}$ parallel to the walls) of the liquid crystal film. Moreover, transparent electrodes covering the two plates permit the application of an electric field across the liquid-crystal layer. The photoconductor behaves like a variable resistor, which resistance decreases for increasing illumination. The light-driven feedback is obtained by sending back onto the photoconductor the light which has passed through the liquid-crystal layer and has been reflected by the dielectric mirror. The voltage V_0 applied to the LCLV is sinusoidal and of frequency $f = 20\ KHz$, which is much larger than the liquid crystal response time.

The experimental setup, displayed in Fig.1, consists of a LCLV with optical feedback, [7]. Here, the optical loop is designed in such a way that the light beam experiences no diffraction as well as no geometrical transformation, so that no transverse structures are formed close to the FT. The light intensity I_w reaching the photoconductor is given by [8]

$$I_w = |\cos(\psi_1)\cos\psi_2 + \sin\psi_1 \sin\psi_2 e^{-i\beta cos^2\theta})|^2 I_{in} \tag{1}$$

where θ is the liquid crystal re-orientation angle, $\beta \equiv 2kd\Delta n cos^2\theta$ is the overall phase shift experienced by the light traveling forth and back through the liquid crystal layer, I_{in} is the input intensity, ψ_1 and ψ_2 are the angles formed by the input and feedback light polarization with the liquid crystal director $\vec{n}$, respectively, $k = 2\pi/\lambda$ is the optical wavenumber, d is the thickness of the liquid crystal layer and Δn is the difference between the extraordinary ($\parallel$ to $\vec{n}$) and ordinary ($\perp$ to $\vec{n}$) index of refraction of the liquid crystal. In our experiment, $\beta \simeq 120$ since $\lambda = 633\ nm$, $\Delta n = 0.2$, $d = 30\ \mu m$. Input light intensity is typically around $I_{in} \simeq 0.9\ mW/cm^2$.

THEORETICAL DESCRIPTION

Derivation of the normal form. For a nematic liquid crystal layer on which an electric field is applied, the dynamical equation for the director reads [2]

$$\gamma \vec{n} \wedge \partial_t \vec{n} = -\vec{n} \wedge \frac{\delta \mathcal{F}}{\delta \vec{n}}, \quad \vec{n}.\vec{n} = 1 \tag{2}$$

where γ is the rotational viscosity, and $\mathcal{F}$ the Frank free energy which is expressed by

$$\begin{aligned} \mathcal{F} &= \frac{1}{2} \int K_1 (\vec{\nabla}.\vec{n})^2 + K_2 (\vec{n}.(\vec{\nabla} \wedge \vec{n}))^2 + K_3 (\vec{n}.(\vec{\nabla} \wedge \vec{n}))^2 \\ &\quad -\varepsilon_\perp \vec{E}^2(\vec{n}) - \varepsilon_a (\vec{n}.\vec{E}(\vec{n}))^2 d\vec{r}. \end{aligned} \tag{3}$$

K_i are the elastic constants of the liquid crystal, ε_a the dielectric anisotropy and $\varepsilon_\perp$ the perpendicular dielectric permeability. For the sake of simplicity, we assume $K_1 = K_2 = K_3 = K$, leading to

$$\begin{aligned} \gamma \partial_t \vec{n} &= K(\nabla^2 \vec{n} - \vec{n}(\vec{n}.\nabla^2 \vec{n})) + \varepsilon_a (\vec{n}.\vec{E})(\vec{E} - \vec{n}(\vec{n}.\vec{E})) + \tfrac{\varepsilon_\perp}{2} \tfrac{\partial \vec{E}^2}{\partial \vec{n}} \\ &\quad -\tfrac{\varepsilon_\perp}{2} (\vec{n}.\tfrac{\partial}{\partial \vec{n}}) \vec{E}^2 + \tfrac{\varepsilon_a}{2} (\vec{n}.\vec{E}) \vec{n}.((\vec{n}.\tfrac{\partial}{\partial \vec{n}}) \vec{E}). \end{aligned} \tag{4}$$

The total electric field applied to the liquid crystal layer depends on the response of the photoconductor to the write intensity I_w and to the voltage applied to the liquid crystal layer ($V_0 = E_0/d$). As the light intensity is sufficiently small, the response of the photoconductor is linear. Thus, the total electric field reads $E(n) = E_0 + \alpha I_w(\vec{n})$, where $\alpha \simeq 4$ is a phenomenological parameter that we can evaluate from the experimental characteristics of the LCLV, (measuring I_w in mW/cm^2).

After substituting (1) in the expression for the electric field, it becomes $E(n) = E_0 + A + B\cos(\beta \cos^2\theta)$, where $A = \frac{1}{4}(\cos 2(\psi_1 - \psi_2) + \cos 2(\psi_1 + \psi_2) + 2)\alpha I_{in}$ and $B = \frac{1}{4}(\cos 2(\psi_1 - \psi_2) - \cos 2(\psi_1 + \psi_2))\alpha I_{in}$. Since the liquid crystals are aligned parallel to the walls (planar anchoring) and are submitted to a perpendicular electric field, the electric field is $\vec{E} = (0,0,E)$ and $\vec{n} = (n_x, 0, n_z)$ with $n_x^2 + n_z^2 = 1$. By inserting $E(n)$ in (4) and by means of standard bifurcation theory, we derive an amplitude equation for the first unstable Fourier mode, $n_z = u(x,y)\sin(\pi z/d)$ and $n_x = 1 - u^2 \sin^2(\pi z/d)/2$, which describes the director orientation at the onset of the instability. The equation reads

$$\partial_t u = c_1 u + c_3 u^3 + c_5 u^5 + K\nabla_\perp^2 u \tag{5}$$

where the coefficients c_1, c_3 and c_5 are given by

$$\begin{aligned} \gamma c_1 &= -\tfrac{K\pi^2}{d^2} + (A + E_0 + B\cos(\beta))(\varepsilon_a(A + E_0) + B(\varepsilon_a \cos(\beta) + 2\varepsilon_\perp \beta \sin(\beta))) \\ \gamma c_3 &= \tfrac{1}{3}\tfrac{K\pi^2}{d^2} - \tfrac{2}{3}\varepsilon_a (E_0 + A)^2 + \tfrac{2}{\pi^2}\varepsilon_\perp (2B\beta \sin(\beta))^2 - \tfrac{2}{3\pi^2}(12\beta^2 \varepsilon_\perp + \pi^2 \varepsilon_a)(B\cos(\beta))^2 \\ &\quad + \tfrac{\beta}{\pi^2}(\varepsilon_a (8 + \pi^2) - \varepsilon_\perp \pi^2)(A + E_0 + B\cos(\beta)) B \sin(\beta) \\ &\quad - \tfrac{4}{3\pi^2}(6\beta^2 \varepsilon \perp + \pi^2 \varepsilon_a)(E_0 + A) B\cos(\beta)) \\ \gamma c_5 &= (\tfrac{2B\beta}{\pi^2})^2 ((4 + \pi^2)\varepsilon_a - \pi^2 \varepsilon_\perp)(\sin(\beta)^2 - \cos(\beta)^2) \\ &\quad - \tfrac{B\beta}{12\pi^4}(192\beta^2 \varepsilon_\perp + (9\pi^2 + 64)\pi^2 \varepsilon_a)(E_0 + A)\sin(\beta) - (\tfrac{2\beta}{\pi^2})^2((4 + \pi^2)\varepsilon_a - \pi^2 \varepsilon_\perp)(E_0 \\ &\quad + A) B\cos(\beta) - \tfrac{B^2}{12\pi^4}(9\pi^2 + 64)\varepsilon_a + 768\beta^2 \varepsilon_\perp)\beta \sin(\beta)\cos(\beta). \end{aligned} \tag{6}$$

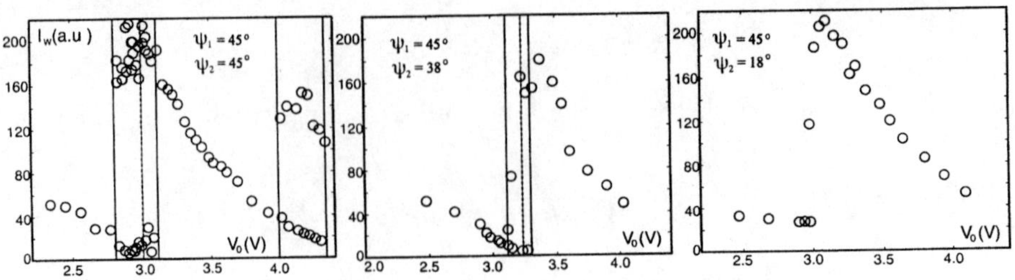

FIGURE 2. I_w as a function of the r.m.s. values of the applied voltage V_0, for different angles, ψ_1 and ψ_2, between the polarizers and the liquid crystal director.

Discussion. When c_3 is negative and of order one, Eq. (5) describes a second order FT. The FT becomes first order when c_1 and c_3 are positive (and small), with c_5 negative. Qualitatively, this situation well describes the experimental observations. c_3 is a function of the polarizer angles, of the input intensity and of the voltage applied to the LCLV. In particular, we have verified that changes in the polarizer angles produce effects with the same sign in the theory and in the experiment. However, it is important to underline that the theoretical description of the transition is valid only in the asymptotic limit $c_3 \sim (c_1)^{1/2}$. Instead, a quantitative description of the experiment would require $c_3 \sim 1$, which cannot be obtained from an asymptotic analysis [9]. Henceforth we consider $c_3 > 0$ and $c_5 < 0$.

RESULTS AND COMPARISON WITH THE MODEL

In Fig.2 we show I_w as a function of V_0 and in the presence of feedback. Three bifurcation diagrams are reported, for three different values of the polarizer angles ψ_1 and ψ_2. For $\psi_1 = \psi_2 = 45^0$, which maximizes the birefringence of the LCLV, the diagram displays an abrupt change at the transition point and an hysteresis region. For this value of ψ_1 and ψ_2, the extension of the hysteresis region is maximum, in agreement with the model. Notice that I_w is measured by a small area photodiode, i.e., it is a local measurement taken at the center of the feedback beam. By looking at the entire image of the feedback beam with a CCD camera, we see that the transition point is characterized by a white spot developing over a dark background. In the course of time, the front of the white spot expands and the white state overcomes the dark one. For the amplitude equation, this represents a front solution between the unstable state $u_0 = 0$ and the stable state $u_\pm^2 = (-c_3 \pm \sqrt{c_3^2 - 4c_1c_5})/(2c_5)$, a Kolmogorov-type front [10]. By further increasing the voltage, the LCLV birefringence changes and the white state becomes "grey" until the dark value is reached again. Successive transitions to the white state are present for larger values of the applied voltage. These states correspond to parameter values far from the FT, and hence they are not considered here.

By decreasing the voltage, we observe a hysteresis cycle. In order to determine the

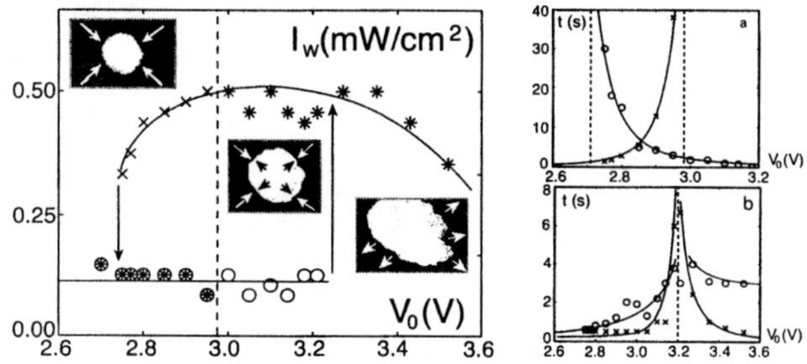

FIGURE 3. Bistability and front propagation. On the left, I_w *vs* V_0 plot; circles: dark state with writing light off; stars: white state with writing light off; cross: white state with writing light on. On the right, a) τ_r (circles) and τ_f (cross); b) τ_{on} (cross) and τ_{off} (circles) as a function of V_0.

size of the bistable region, we inject an additional light spot (low power He-Ne laser) into the feedback loop. This acts as a small perturbation, triggering the transition from the dark state to the white one. The white state persists when we block the additional writing light, while it switches to the dark state if we perturb the feedback. In Fig.2, the arrows delimit the region over which this writing-erasing procedure is robust. Notice that spatial inhomogeneities and other noise sources can influence the stability of the two states. Furthermore, just before the transition, there is an appreciable decrease of the intensity I_w. This is indeed a signature of the fact that we are approaching the transition point, where fluctuations become very large and light is diffused in all directions, in the same way as it occurs at the critical point for a liquid-vapor transition. This sort of critical opalescence [11], is responsible for the diminished efficiency of the light reflection from the LCLV.

For $\psi_1 = \psi_2 = 45^0$, a more detailed set of measurements is reported in Fig.3, left side, for a V_0 range centered around the transition point and together with three representative images of the feedback field. These images show the direction of the front propagation, depending on the mutual stability of the white and the dark state. The dashed line marks the point at which the front propagation is zero, usually called the Maxwell point. Below this point the white state is less stable than the dark one and the white spot, once created by the writing light, contracts to zero. Above the FT point, the white spot nucleates spontaneously and its front expands until the white state covers all the background. In between, the front of the spot expands or retracts depending on the size of the perturbation.

The three crucial points, i.e., the beginning of the bistability, the Maxwell point and the FT point, are identified by the divergence of the response times. We show in Fig.3, right side, the response times associated with the bifurcation diagram. In a, we show the rise time τ_r needed for a white spot to develop once the writing light is switched on and the fall time τ_f taken by the white spot to disappear when the writing light is switched off. The divergence of the first one identifies the beginning of the bistable region whereas the divergence of the second one corresponds to the Maxwell point. In b, we show the rise τ_{on} and fall τ_{off} time when the feedback is switched respectively on and off. The divergence of these two times corresponds to the

FT point. Correspondingly, in the amplitude equation (5), the beginning of the bistability is characterized by $c_3^2 - 4c_1c_5 \geq 0$, $c_1 < 0$, the Maxwell point by $c_3^2 = 16c_5c_1/3$ ($c_1 < 0$) and the FT by $c_1 = 0$.

CONCLUSION

In conclusion, we have derived an amplitude equation valid close to the FT for a nematic liquid crystal layer in the presence of a light-driven feedback. We have shown that, as a consequence of feedback, FT becomes first order. Our theoretical description is in a fair qualitative agreement with the experimental observations. The feedback is provided by the conversion of optical intensity into electric field via a photoconductive transducer (LCLV). Bistability is very robust here and is present for a relatively wide range of the experimental parameters. Spatial dynamics of front propagation is also reported.

ACKNOWLEDGMENTS

This work has been supported by the ACI Blanche of the French Ministry of Research. M. Clerc has benefited from the financial support from the Centro de Fisica No Lineal y Sistemas de Santiago. C. Riera was supported as a postdoctoral fellow by an ONR- NYI grant awarded to L. Mahadevan at the Massachusetts Institute of Technology (2000-03).

REFERENCES

1. V. Fréedericksz and V. Zolina, Trans. Faraday Soc. **29**, 919 (1933).
2. P. G. de Gennes and J. Prost, *The Physics of Liquid Crystals*, (Oxford Science Publications, Clarendon Press, second edition, 1993).
3. B.J. Frisken and P. Palffy-Muhoray, Phys. Rev. A **39**, 1513 (1989), Phys. Rev. A. **40**, 6099 (1989); S. Garg, S. Saeed and U.D. Kini, Phys. Rev. E **51**, 5846 (1995).
4. S.D. Durbin, S. M. Arakelian and Y.R. Shen, Phys. Rev. Lett. **47**, 1411 (1981); A. J. Karn, S.M. Arakelian, Y.R. Shen and H. L. Ong, Phys. Rev. Lett. **57**, 448 (1986).
5. Peng-Ye Wang, Hong-ju Zhang and Jian-hua Dai, Opt. Lett. **12**, 654 (1987).
6. M.G. Clerc, S. Residori and C.S. Riera, Phys. Rev. E **63**, 060701(R) (2001).
7. S.A. AKhmanov, M.A. Vorontsov and V.Yu. Ivanov, JETP Lett. **47**, 707 (1988).
8. see e.g. F.T. Arecchi, S. Boccaletti and P.L. Ramazza, Physics Reports **318**, 1 (1999) and references therein.
9. P. Manneville, *Dissipative structures and weak turbulence*, (Academic Press, San Diego 1990).
10. M. Cross and P. Hohenberg, Rev. Modern Phys. **65**, 581 (1993).
11. L. Mandel and E. Wolf, *Optical Coherence and quantum Optics*,(Cambridge University Press, New York, 1995).

Peculiarities of Stochastic Resonance in Ferroelectric Triglycine Sulfate

Martin Diestelhorst and Konstantin Drozhdin

Martin-Luther-University Halle-Wittenberg, Department of Physics/Experimental Physics II, Friedemann-Bach-Platz 6, D-06108 Halle, Germany

Abstract. Triglycine sulfate (TGS) is the first ferroelectric material that has been reported to exhibit the effect of stochastic resonance. We observed the effect experimentally, varying the noise strength added to a periodic signal, that was too weak to cause switching between the two possible states of polarization in ferroelectric TGS. Despite the fact that most of the theoretical predictions could be confirmed in our experiments there occured some peculiarities that have not been observed in other systems with stochastic resonance. The most obvious effect is the dependence of the spectral amplification and the signal-to-noise-ratio on the direction of noise variation. It can be explained taking into account the mechanism of domain switching in ferroelectric materials.

INTRODUCTION

Triglycine sulfate, usually abbreviated TGS is one of the most thoroughly investigated ferroelectric materials [1]. In the absence of an electric field it has two orientational states of electric polarization which are separated by an energetic barrier, if the temperature is below the phase transition temperature. The crystal may be switched from one state of polarization into the opposite state by an sufficiently high electric field. That means the electric field strength has to be higher than the coercive field. Applying an electric ac field lower than the coercive field to the ferroelectric, no polarization reversal occurs. Nevertheless already two of the three basic conditions for the occurence of the effect of stochastic resonance [2] are fulfilled:

(i) There exists an energetic barrier between two possible states.
(ii) A weak periodic input is applied to the material.

It is an easy task to fulfill the third condition by adding a noisy electric field to the weak periodic input signal. Consequently we were able to observe stochastic resonance in ferroelectric TGS recently [3].

Stochastic resonance has been demonstrated in a lot of systems [2]. These systems include Schmitt trigger, a bistable ring laser, magnetoelastic ribbons, bi-substituted ferrite-garnet films, neuronal systems and others.

CP622, *Experimental Chaos: 6th Experimental Chaos Conference*, edited by S. Boccaletti et al.

Here we present detailed investigations of stochastic resonance in this new system.

BASIC PROPERTIES OF TGS

Triglycine sulfate (TGS) is an organic ferroelectric with the compound formula $(NH_2CH_2COOH)_3H_2SO_4$. It consists of three glycine molecules and one H_2SO_4 group in the unit cell. In spite of its rather complicated structure TGS has become a model ferroelectric for different reasons. Firstly it is one of very few ferroelectrics that exhibit a second-order phase transition. So it offered the possibility to observe critical (i. e. fluctuation-dominated) phenomena very close to the phase transition temperature $T_C = 49° C$ [1]. Secondly it is an uniaxial ferroelectric. Thirdly large crystals of very high quality may be easily grown from water solution. And last but not least it is one of the best pyroelectric materials what made it very interesting for sensor applications.

The macroscopic behaviour of TGS may be understood on the basis of simple Landau theory of second order phase transitions. Especially the static dielectric properties close to the phase transition temperature can be derived quantitatively in terms of a Landau free-energy expansion in powers of an order parameter which is identic with the dielectric displacement D [1, 4]:

$$G(T,D) = \tfrac{\alpha}{2}(T - T_C)D^2 + \tfrac{\beta}{4}D^4 . \tag{1}$$

Using the well known thermodynamic formalism, it is an easy task to show that the dielectric susceptibility yields the Curie-Weiss law $\varepsilon = C/(T-T_C)$ and the spontaneous polarization is $P_S = 0$ above T_C (in the paraelectric phase) and $P_S = \pm\sqrt{\tfrac{\alpha}{\beta}(T - T_C)}$ in the ferroelectric phase below T_C.

From Equation (1) it can be seen that the free energy becomes a double well potential below T_C. That was the main motivation to look for the effect of stochastic resonance in this class of materials. On the other hand it is well known that in the ferroelectric phase the dielectric behaviour is mainly dominated by domain switching processes [1]. Because switching in ferroelectrics is connected with changes in the crystal lattice, it is very sensitive to the real structure of the crystal, for instance to the defect structure, nucleation processes etc.

After we have found the effect of stochastic resonance in TGS itself, it was interesting to investigate the underlying switching process in more detail. We used the same experimental setup as described earlier [3]. The signals of two generators are added electronically. One of them supplies a periodic sinusoidal voltage which may be varied in the range from $1\mu Hz$ to $21 MHz$. The other one is a noise generator. The amplified sum signal stimulates a TGS crystal plate which is electroded perpendicular to the ferroelectric axis. Along this axis switching can occur, if the periodic input is higher than the coercive field. In order to investigate stochastic resonance we fixed the amplitude of the periodic part of the sum signal to values from $\frac{1}{150} E_C$ to $\frac{1}{2} E_C$.

RESULTS AND DISCUSSION

In our experiments we observed different characteristics of stochstic resonance both at increasing and decreasing the noise strength at constant periodic voltage. Using a signal analyzer Agilent 89410A it was possible to determine both the spectral amplification and the signal-to-noise ratio. Simultaneously the time series of the charge at the ferroelectric capacitor and the periodic component of the input signal were measured to get the opportunity to inspect the dependence of the polarization of the ferroelectric on the periodic driving. The idea behind these measurements was to compare the behaviour of the ferroelectric in the regime of stochastic resonance with traditional hysteresis which can be observed using periodic fields above the coercive field strength E_C.

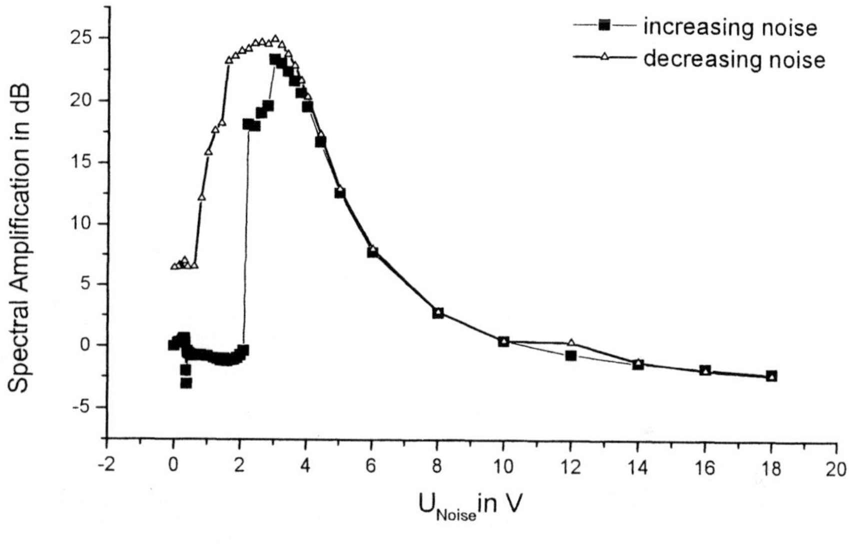

Figure 1. Spectral amplification at increasing and decreasing noise. Modulation voltage of 600mV corresponds to a field strength of about 1/5 of the coercive field. Modulation frequency: 20Hz

Fig. 1 presents as an example the spectral amplification according to the definition [2]

$$\eta = \frac{p_1}{p_A}. \tag{2}$$

Here p_1 is the power in the peak at the modulation frequency $f_{\mathrm{mod}} = 20Hz$, and p_A is the total power of the periodic part of the input signal. It is evident that there exists a threshold at about $U_{Noise} = 2.1V$ in the characteristic. Below that threshold the response of the ferroelectric is very weak. In terms of ferrolectricity that means that almost no polarization reversal occurs. At the threshold the power of the signal suddenly increases by a factor of about 100. Than it follows the route known from other systems exhibiting stochastic resonance. The maximum is reached at $U_{Noise} = 3.0V$. Further increase of the noise reduces the response of the ferroelectric, that means the part of polarization that is switched. At decreasing the noise the ferroelectric does not offer the threshold value. Instead it behaves like other systems with stochastic resonance.

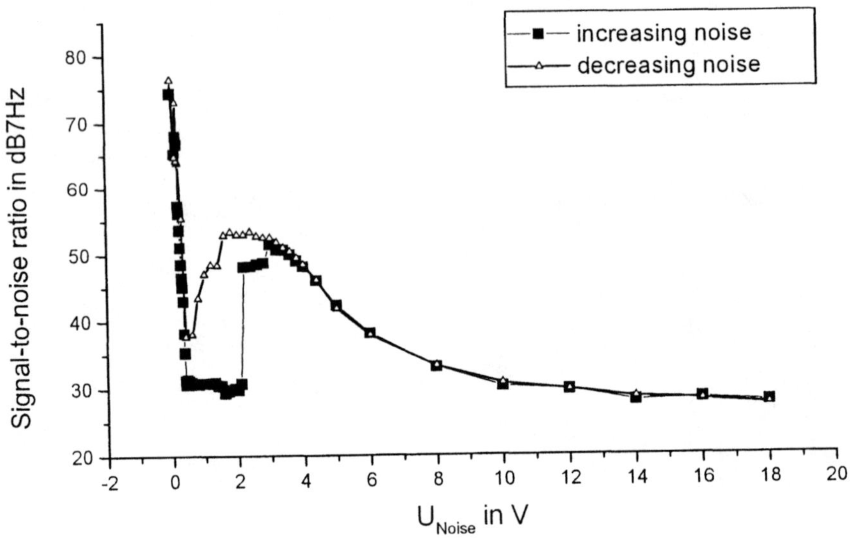

Figure 2. Signal-to-noise-ratio at increasing and decreasing noise. Parameters are the same as in Fig. 1

In Fig. 2 we present the signal-to-noise ratio as another characteristic of stochastic resonance. At increasing noise we find once more the same threshold value as in the spectral amplification. At low noise strengths the signal-to-noise ratio decreases. The amount of switched polar regions does not increase until the threshold is reached at increasing noise. So the power in the first harmonic does not increase whereas the noise in the sum output grows, resulting in a decreasing signal-to-noise ratio. This behaviour can be understood taking into account the intrawell motion of the polar regions in the ferroelectric. The material is not far from the equilibrium at low periodic driving. The additional noise does not increase the probability of switching. But after reaching the threshold of noise the ferroelectric begins switching in fast growing regions. So it was possible to observe the maximum in the signal-to-noise ratio nearly at the same noise strength as the maximum of the spectral amplification. Once more we found differences in the TGS-behaviour at increasing and decreasing noise. As can be seen from Fig. 2 the behaviour of the ferroelectric follows the route of other systems at decreasing noise. The threshold has been vanished. But the minimum according to the intrawell motion can be furthermore observed.

It is a standard method to investigate the hysteresis of the ferroelectric materials. From this point of view it is an interesting task to compare the behaviour of TGS in the regime of stochastic resonance with the hysteresis of complete switching, which can be observed using field strengths greater than the coercive field E_C. To do that we have measured long time series with exact 100 points per period of the sinusoidal part of the input signal. Averaging over 650 periods we were able to inspect the dependence of polarization on the periodic part of the input that gives some hint about the rate of switching in stochastic resonance compared with the fully developed hysteresis.

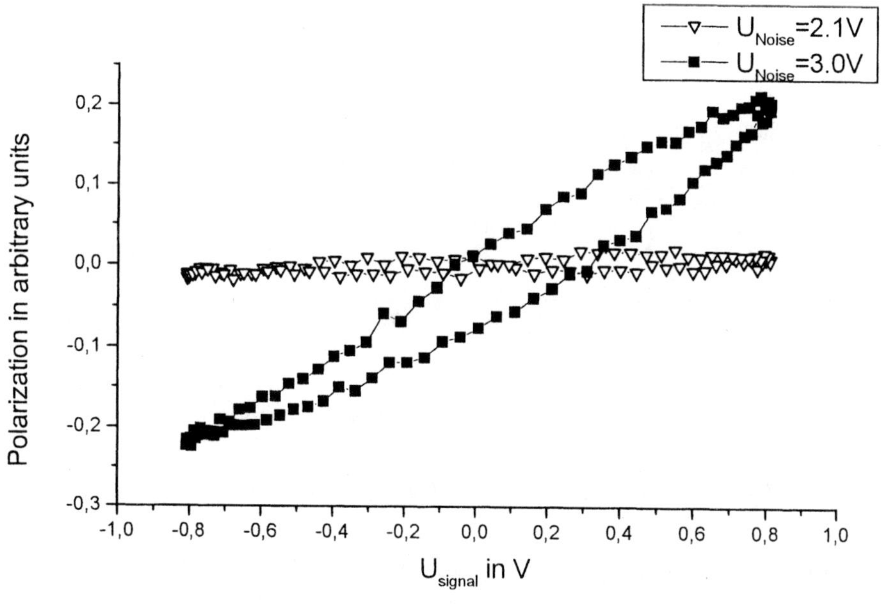

Figure 3. "Hysteresis" at increasing noise for two different noises: directly below the threshold and at the maximum of spectral amplification . Parameters are the same as in Fig. 1

Fig. 3 presents two "hysteresis" loops. The smaller one is recorded directly below the threshold noise strength $U_{Noise} = 2.1V$. It can be seen, that nearly no switching occurs. The larger curve represents the behaviour at the maximum of spectral amplification at increasing noise. Even in that case the TGS does not reach a saturation polarization as in the case of fully developed hysteresis. Comparing the value of spontaneous polarization of TGS at 45°C [5] with the change of polarization in the case of stochastic resonance it may be estimated, that the rate of switched polarization is about one fourth of the spontaneous polarization. In the case of decreasing noise this rate increases a little to one third of the spontaneous polarization. Another interesting fact follows from Fig. 3. There is a small asymmetry in the "hysteresis" which is visible especially at the maximum of the spectral amplification. This asymmetry gives rise to even harmonics in the power spectrum of the response as reported earlier [3]. The reason can be found in the defect structure of the real crystal [1]. Polar defects cause internal bias fields in the material causing asymmetries even in the hysteresis loops at large field strengths.

CONCLUSION

From the above presented results we can draw the following conclusions:

Ferroelectrics, especially ferroelectric TGS, are new interesting systems offering stochastic resonance. The process of stochastic resonance in ferroelectric TGS is dominated by domain switching processes. At increasing noise stochastic resonance is hindered up to a threshold noise. At decreasing noise the crystal behaves like other systems exhibiting stochastic resonance. These differences in the behaviour of the TGS at increasing and decreasing noise are reproducible. From the observation of hysteresis like dependences it is evident that the switching is incomplete. That can be concluded from comparison of spontaneous polarization with the switched charges at maximum spectral amplification. Asymmetric behaviour in the dependence of the polarization on the periodic driving is caused by polar defects.

REFERENCES

1. Lines, M. E. and Glass, A. M., *Principles and applications of ferroelectrics and related materials*, Clarendon Press, Oxford, 1977
2. Gammaitoni, L., Hänggi, P., Jung, P., and Marchesoni, F., *Rev. Mod. Phys.* **70**, 223 (1998)
3. Diestelhorst, M. and Drozhdin, K., *Ferroelectrics* **238**, 25 (2000)
4. Jona, F. and Shirane, G., *Ferroelectric crystals*, Macmillan, New York, 1962
5. Landolt-Börnstein, part III/3, *Ferro- and Antiferroelectric Substances*, edited by K.-H. Hellwege and A. M. Hellwege, Springer, Berlin, Heidelberg, New York, 1969

SYNCHRONY

Generalized Synchronization of Chaos in a Laser

D. Y. Tang

School of Electrical and Electronic Engineering,
Nanyang Technological University, Singapore 639798

Abstract. The optically pumped $^{15}NH_3$ single mode ring laser can exhibit various types of chaotic dynamics. These include the period-doubling chaos, intermittency and the Lorenz-like spiral chaos. Based on the chaotic dynamics of the laser we have designed experiments to study the interaction between coupled chaotic systems. In this paper we report on results of our investigations. We demonstrate experimental evidence of frequency entrainment and generalized synchronization of chaos between unidirectionally coupled chaotic lasers, and the antiphase frequency entrainment, antiphase chaotic synchronization and perfect chaotic synchronization between mutually coupled chaotic laser modes. Analyzing properties of the various forms of chaotic synchronization observed, we speculate that the physical origin of synchronization between coupled chaotic systems is the phase interaction between the coupled chaotic evolutions. In particular, the different forms of chaotic synchronization observed are results of different stages of chaotic phase locking.

INTRODUCTION

Due to the theoretical significance and potential practical applications in secure communication, synchronization of chaos between coupled chaotic systems has been intensively investigated recently [1-16]. Initially, synchronization of chaos was understood as a perfect coincidence of trajectories in phase space of coupled chaotic systems. This kind of synchronized chaos has actually been observed experimentally between coupled identical chaotic systems [4-7]. However, with the development of studies it turned out that synchronization of chaos in the form of identical chaotic oscillations is only a particular case of synchronized chaos. In fact other more complex forms of synchronized chaotic oscillations exist. Based on a drive-response system, Rulkov *et al.* have proposed a concept of generalized synchronization of chaos [10-11]. According to them, the dynamics of two coupled chaotic systems are synchronized if a functional relationship between their dynamical variables exists. Kocarev and Parlitz have studied the conditions for the occurrence of generalized synchronization [9]. They have shown that in the generalized synchronization the response is predictable, and the chaotic dynamics of the response system is conditionally equivalent to that of the driving system. Generalized synchronization of

CP622, *Experimental Chaos: 6th Experimental Chaos Conference,* edited by S. Boccaletti et al.

chaos can exist between any pair of coupled chaotic systems and has been experimentally observed in coupled nonlinear electronic circuits [10-11] and a chaotically driven chaotic laser system [16].

Depending on situations discussed, other forms of synchronization of chaos have also been proposed to describe particular dynamical relationships between the coupled chaotic systems. Rosenblum *et al.* have suggested a form of "phase synchronization" between two coupled self-sustained chaotic oscillators [12]. They show that in the regime of phase synchronization the phases of coupled chaotic oscillations are locked, while their amplitudes vary chaotically and are practically uncorrelated. Furthermore, the same authors have also proposed a case of "lag synchronization" [13]. They demonstrated numerically that starting from the phase synchronization, as the coupling strength is further increased, the phase synchronized state can evolve to a lag synchronization of chaos characterized by that the states of the coupled chaotic oscillators are nearly identical, but one system lags the other in time. The time lag reduces as the coupling strength is increased.

Synchronization between coupled periodic systems is a well-known phenomenon. When the oscillation frequencies of two coupled self-sustained periodic oscillators are within a certain range called the locking range, the frequencies of them will automatically lock together, and consequently the two systems will oscillate with the same frequency. Synchronization between two coupled periodic oscillators is well studied. Since the oscillation of a periodic oscillator is intrinsically regular, the effect of synchronization between coupled periodic systems is clear and unique. An essential difference between a chaotic system and a periodic system is that the dynamics of a chaotic system is intrinsically unstable. Instability of system dynamics results in that the evolution of trajectories in the phase space sensitively depends upon the initial conditions, and properties of the system are long–term unpredictable.

Despite the fact that the dynamics of a chaotic system is intrinsically unstable, synchronizations between coupled chaotic systems still exist. An intuitive question would be what's the physical origin of synchronization between coupled chaotic systems? Is the origin the same as that of synchronization between coupled periodic systems? Why do different forms of chaotic synchronization exist for coupled chaotic systems? And what is the relationship between them? To address all these questions we have designed and conducted various experiments to study the interaction between coupled chaotic systems. In this paper we will present our experimental results. In particular, we will try to explain the relationship between various forms of chaotic synchronization observed.

The paper is organized as follows: in the second section we will briefly describe the chaotic dynamics of an optically pumped $^{15}NH_3$ far infrared (FIR) single mode ring laser. As all our experiments were conducted on the laser, it is worthwhile to explain the configuration of the laser and recall some of its chaotic dynamics here again. In the third and fourth section we will describe the experimental arrangements and evidences of various forms of chaotic synchronization observed either under unidirectionally or mutually coupled chaotic systems. In particular, we will show experimentally that under the unidirectional coupling the coupled chaotic systems can have effects such as the frequency entrainment, the generalized synchronization of chaos, phase and lag synchronization. Under the mutual coupling we have observed the anti-phase frequency entrainment, anti-phase chaotic synchronization, and the perfect chaotic synchronization of various chaotic laser dynamics in our system. In the fifth section the physical mechanisms of various forms of chaotic synchronization are analyzed. We speculate that the fundamental physical origin of synchronization between coupled chaotic systems is essentially the same as that of the periodic systems. The various forms of chaotic synchronization observed are results of different stages of phase interaction between the coupled chaotic systems. The last section is the conclusion of the paper.

CHAOTIC DYNAMICS OF THE OPTICALLY PUMPED $^{15}NH_3$ SINGLE MODE RING LASER.

The optically pumped $^{15}NH_3$ single mode FIR ring laser is so far the only laser system that has exhibited the so-called intrinsic single mode chaotic laser dynamics. Here by "intrinsic" we mean that the chaotic dynamics of the laser is a result of the single mode laser equations describing interaction of a traveling wave laser beam with the homogeneously broadened two-level atoms. As firstly recognized by Haken in 1975 [17], the single mode laser equations are isomorphic to the Lorenz equations, one

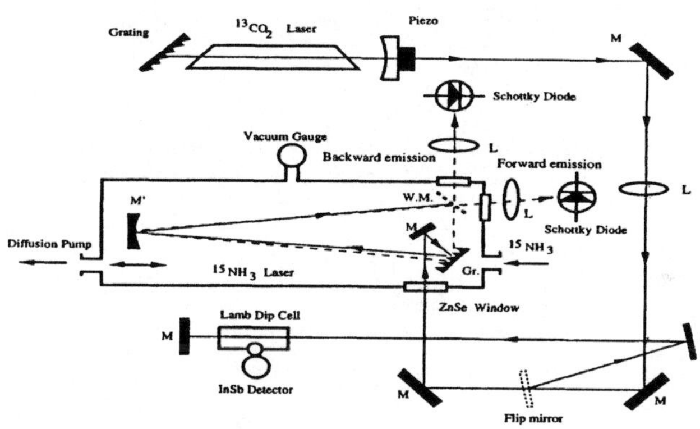

Figure 1. Experimental setup of the optically pumped $^{15}NH_3$ FIR ring laser. M: gold mirror; W.M.: wire mesh; Gr: grating; L: lens.

of the paradigms of the chaotic dynamic systems. The chaotic dynamics of the optically pumped $^{15}NH_3$ single mode ring lasers were extensively studied both theoretically and experimentally previously [18-20]. Here to better understand the experiments on the chaotic synchronization demonstrated later, we briefly recall some of the basic properties of the laser and its chaotic dynamics.

A typical experimental configuration of the optically pumped $^{15}NH_3$ ring laser is shown in Fig. 1. The laser gain medium is the $^{15}NH_3$ gas. The laser cavity is a triangular ring of perimeter of about 2m, which consists of a concave mirror of radius of curvature of 2m, an optical grating for coupling the pump beam into the laser cavity and an output coupler made of a gold mesh. The laser cavity has a semi-confocal structure. The FIR laser emission of the laser has a wavelength of 153 μm. The pump laser is an isotopic $^{13}CO_2$ laser of wavelength 10.76 μm. The pump laser frequency is finely tunable with a PZT mirror mount and precisely controlled by the Doppler free spectroscopy technique. The pump laser beam is mode matched to the TEM_{00} mode of the FIR laser cavity by use of an optical lens, and the pump beam power is controlled with wire-mesh-mirror attenuators. A Schottky-barrier laser diode is used to detect the laser output.

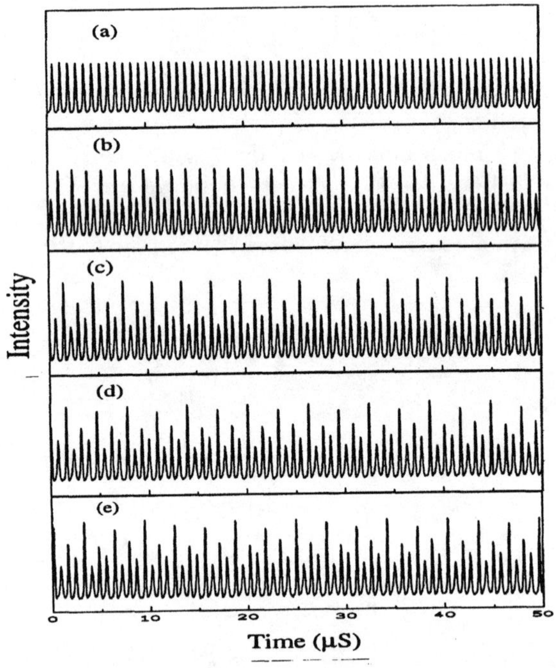

Figure 2. Typical period-doubling route to chaos observed in the laser. Gas pressure: 4 Pa. Pump intensity 4W/cm^2.

Because of the absorption line profile of the $^{15}NH_3$ molecules is Doppler broadened and the optical pumping excites selectively molecules that have the same axial velocity, by detuning the pump laser frequency in relation to the absorption line center, the gain for the forward and the backward laser modes has different central frequencies. Through tuning the pump frequency possibly far away from the absorption line center, and correspondingly shifting the laser cavity frequency to the forward or backward gain position, one can achieve the unidirectional lasing either in the forward or backward direction in the laser. While by tuning the pump laser frequency close to the absorption line center, as the frequency separation between the forward and the backward mode gain becomes small, bidirectional operation of the laser occurs. Under the bidirectional operation of the laser, the two laser modes share the same population inversion, so they are mutually coupled. The closer the pump frequency to the $^{15}NH_3$ absorption line center, the stronger is their coupling.

Due to its connection to the dynamics of the Lorenz equations, the chaotic dynamics of the optically pumped $^{15}NH_3$ single mode ring lasers were experimentally intensively investigated. Typical chaotic dynamics observed in the laser are the period-doubling route to chaos [21], intermittency [22] and the Lorenz-like spiral chaos [23]. To illustrate properties of the various types of chaotic dynamics of the laser, we have shown in Fig. 2 to 4 typical chaotic laser intensity evolutions of each of them observed. Fig. 2 shows a case of the period-doubling route to chaos. The period-doubling route to chaos is observed in the laser as the laser cavity detuning is decreased. With a large separation between the laser cavity mode frequency and the central frequency of the gain, the output of the laser is constant. However, as the frequency separation is decreased, the laser output intensity will first become pulsed, then experiences a series of period-doubling bifurcations, and eventually evolves into a chaotic state. Studying properties of the period-doubling chaos reveals that it has all the universal properties such as the existence of different periodic windows in the chaotic regime and the period-doubling route to chaos of some of the periodic windows etc.

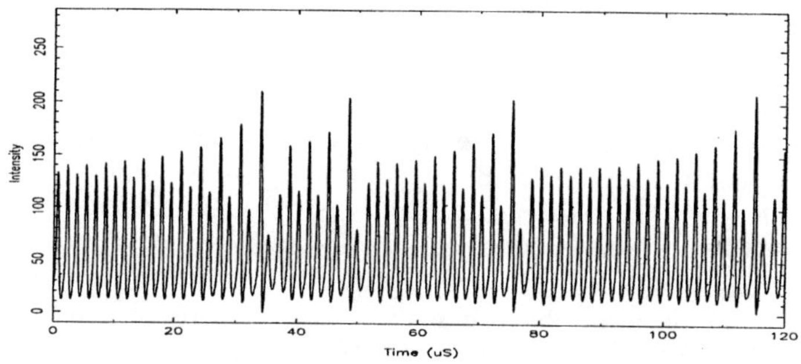

Figure 3. Type-III intermittency of the laser. Gap pressure: 4 Pa. Pump intensity: 7W/cm^2.

Fig. 3 shows a typical laser intensity evolution of the intermittent chaos observed. This kind of intermittent chaos had also been observed in a thermodynamic experiment, and was termed as the type-III intermittent chaos [24]. The characteristic of the intermittency is that the intensity pulsation peaks fall into two components. During one "laminar" phase the magnitude of one component grows up, while the other one decreases until a chaotic burst occurs. The frequency of the chaotic bursts depends on the control parameter. As the control parameter approaches a certain critical value, the laminar phases become short and shorter.

The most striking property of the chaotic dynamics of the laser is that under a certain condition its intensity output exhibits a spiral-kind chaos. Fig. 4 shows one of the chaos observed. The chaotic laser intensity evolution is characterized by that the peaks of the laser intensity pulsations increase monotonically within one spiral length, and the lengths of the spirals vary irregularly. This laser intensity evolution resembles closely those calculated from the Lorenz equations. Further experimental investigations have shown that the intensity variation corresponds to the spiral motion of the laser state vector on the two leaves of the Lorenz attractor, and the irregular lengths of the spirals are a result of the random jumps between the two leaves [25].

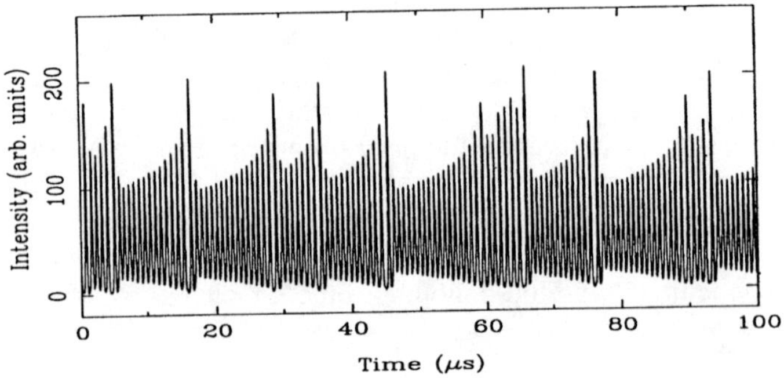

Figure 4. Lorenz-like spiral chaos of the laser. Gas pressure: 10 Pa. Pump intensity: 5W/cm^2.

Fig. 5 shows the Fourier power spectra of the laser intensity evolution under various chaotic laser dynamics. It is to see that for all kinds of chaotic laser dynamics, there exists a distinct sharp spectral component and its harmonics in the broadband spectral background. The existence of the sharp spectral component indicates that there is a predominant frequency in the chaotic oscillation. Chaotic attractors of this property are also called "coherent attractor" [26]. Our experimental results demonstrated that even for the Lorenz-kind spiral chaos, a predominant average chaotic pulsation frequency could still be identified.

A laser mode is in fact an optical field that consists of both the field amplitude and phase part. For an optical field as its oscillation frequency is very high, no detector can follow this change, when measured by a detector with the conventional method, only the field intensity (amplitude) dynamics can be detected, while the field phase dynamics will be lost. In order to get an insight into the field phase dynamics of a chaotic laser mode, we have used the optical heterodyne measurement technique to study its phase dynamics [25]. We find that when a laser mode is chaotic, both its amplitude and phase dynamics are chaotic, and in particular, the chaotic mode phase and amplitude dynamics exhibit closed relationship. Our experimental results confirm that the mode phase and amplitude are two independent dynamical variables of the dynamical system. They equally describe the chaotic dynamics of the system. Physically, the closed relationship between the mode phase and amplitude dynamics can be understood based on the Kramers-Krönig relations. As for the laser emission the intensity of a laser mode is determined by the gain property of the laser medium, the phase by the dispersion property, and the gain and dispersion of an active medium is related through the Kramers-Krönig relations.

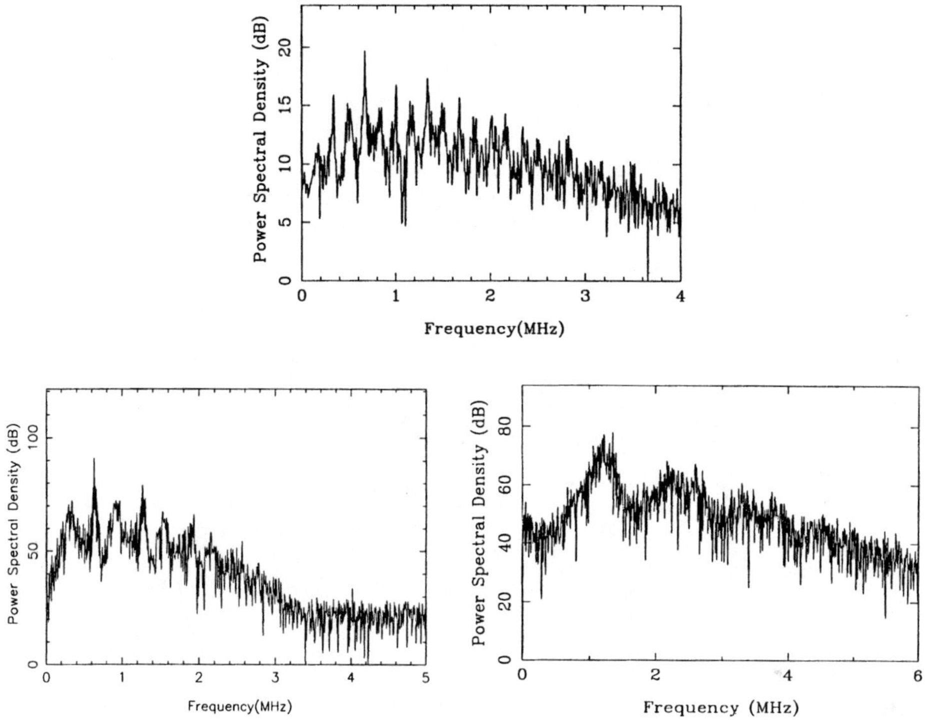

Figure 5. Power spectra of laser intensity evolution under various chaotic laser dynamics. Upper: period-doubling chaos. Lower left: type-III intermittency. Lower right: Lorenz-kind spiral chaos.

CHAOTIC SYNCHRONIZATION BETWEEN UNIDIRECTIONALLY COUPLED CHAOTIC LASERS

Based on the special experimental configuration of the optically pumped $^{15}NH_3$ laser a chaotically driven chaotic laser system can be easily constructed. Fig. 6 shows a schematic block diagram of the experiment setup we used. An acousto-optic modulator (AOM) is used to modulate the pump intensity of the $^{15}NH_3$ laser. Through chaotically modulating the pump intensity, we create an artificial chaotic driving system. In order to obtain a controllable chaotic driving signal, we first either record a chaotic laser intensity waveform without chaotic driving or calculate a chaotic waveform from the Lorenz equations. We then store this chaotic waveform in the memory of an arbitrary function generator (AFG). Based on the stored chaotic waveform, the AFG produces an analog signal that has exactly the same waveform as that of the stored one, but both its amplitude and average chaotic pulsation frequency can be continuously tuned. This analog signal is used to modulate the amplitude of the rf driving signal of the AOM, which in return transfers the modulation signal into the intensity change of the pump laser beam. The undiffracted beam of the AOM is used as the pump of the $^{15}NH_3$ laser. Effectively, the obtained pump intensity consists of a chaotically varying intensity part that acts as the chaotic driving, and a dc intensity part that ensures that without the chaotic driving the laser is itself chaotic. The chaotic intensity dynamics of the laser is compared with those of the driving signal. The advantage of the system is that the chaotic modulation of the pump laser intensity can be accurately controlled. By repeatedly driving the laser with the same chaotic signal under exactly the same experimental conditions, and comparing the chaotic laser outputs under the repeated driving, we can detect the generalized synchronization of chaos in our system. A rf spectrum analyzer is also used to monitor the real-time frequency distribution of the chaotic laser output. Based on this information the frequency relationship between the chaotic driving signal and the chaotic laser dynamics is controlled through adjusting the driving signal frequency.

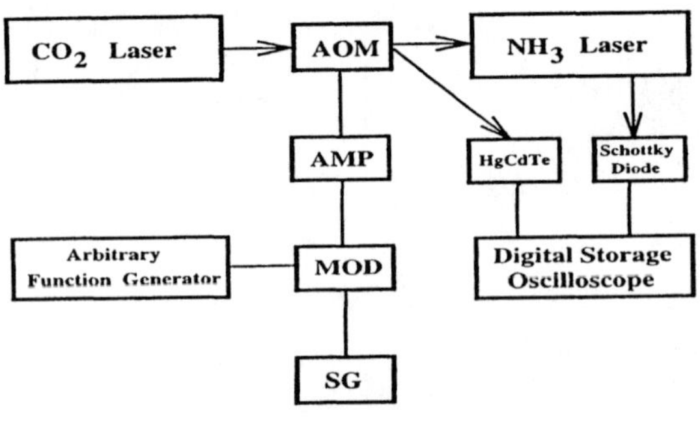

Figure 6. Block diagram of an unidirectionally coupled chaotic laser system. SG: signal generator. MOD: modulator; AMP: amplifier; AOM: acousto-optic modulator.

Synchronization of chaos of the laser under different driving signal strength and different frequency relationships was experimentally investigated. With a fixed driving signal strength, we found that the separation between the average pulsation frequencies of the laser dynamics and the driving signal plays an important role in achieving any forms of chaotic synchronization in our system. When the average pulsation frequencies are far away separated, no synchronization of chaos is observed.

In this case the existence of the chaotic driving makes the chaotic intensity evolution of the laser more complicated than its original single mode chaotic dynamics. Fig. 7 shows a typical experimental result observed in this situation. Comparing the laser output under the repeated driving, it is found that the chaotic dynamics of the laser is not synchronized even in the sense of generalized synchronization.

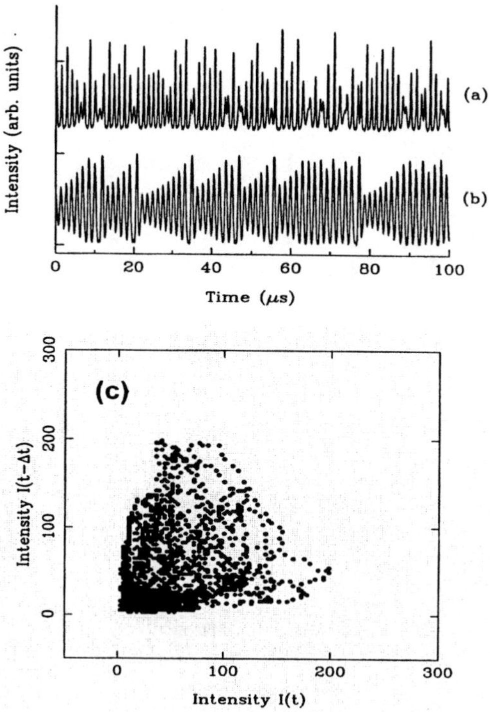

Figure 7. Chaotic dynamics of the laser without synchronization to the driving signal. a): chaotic laser intensity evolution. b): chaotic driving signal evolution. c): relationship plot of the laser dynamics under repeated driving.

When the separation between the average chaotic pulsation frequencies is small, within a certain range determined solely by the strength of the driving signal, the average pulsation frequencies can become locked together, and consequently the chaotic laser intensity pulsations become frequency entrained to those of the driving signal. In a frequency-entrained state, the chaotic dynamics of the laser is not necessarily synchronized to that of the driving signal, even in the sense of generalized synchronization of chaos. Fig. 8 shows one of these states observed. We note that a frequency-entrained state is in fact the state of "phase synchronization" called by Rosenblum et al [12].

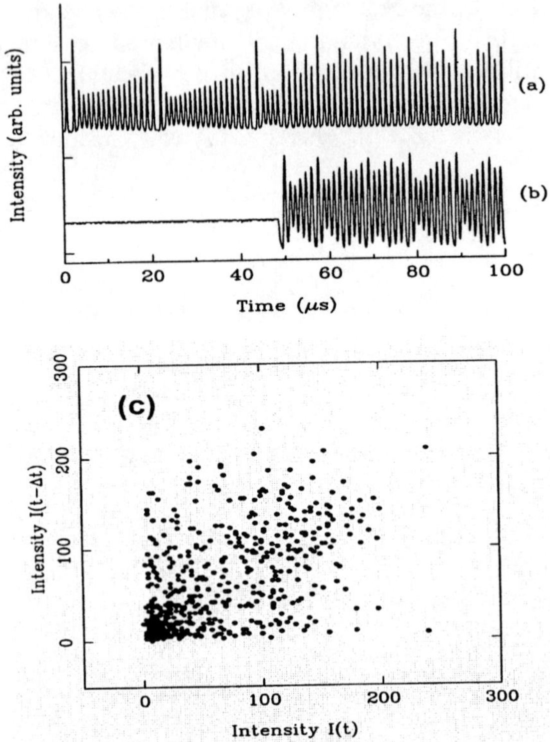

Figure 8. Chaotic dynamics of the laser under frequency entrainment. a): chaotic laser intensity evolution. b): chaotic driving signal evolution. c): relationship plot of the laser dynamics under repeated driving.

In our experiments the generalized synchronization of chaos is only observed when, based on a frequency-entrained state, the strength of the driving signal is further increased beyond a certain threshold. Fig. 9 shows one of the states. Under the generalized synchronization of chaos, although the laser intensity output still looks "chaotic", it is insensitive to the initial conditions and the system noise, in fact it is fully controlled by the driving signal. Our experimental results also confirm that under a generalized synchronization of chaos, the dynamics of the response system is functionally related to that of the driving system. With increase of the driving strength, we observed that the control of the driving signal on the dynamics of the response system becomes stronger and stronger. Fig. 10 shows that under even stronger driving strength, the dynamics of the response system approaches to that of the driving system. Comparing the peak heights of the laser intensity pulsation with those of the driving signals, there exists almost a linear relation. We believe that the Fig. 10 shows a case of the so called "lag synchronization of chaos" in our experiment.

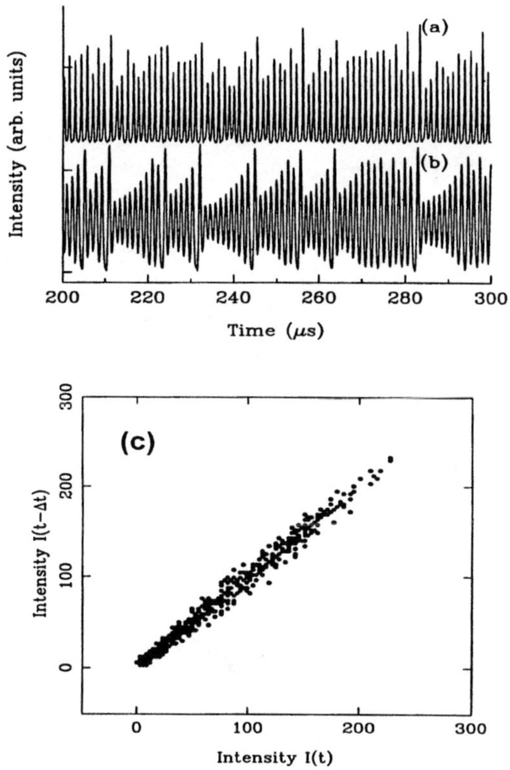

Figure 9. generalized synchronization of chaos of the laser. a): time evolution of the chaotic laser intensity. b): time evolution of the driving signal. c): relationship plot of the laser dynamics under repeated driving.

CHAOTIC SYNCHRONIZATION BETWEEN MUTUALLY COUPLED LASER MODES

By operating the laser bidirectionally, we obtain a situation where two counter-propagating chaotic laser modes coexist and are mutually coupled in the laser cavity through sharing the same population inversion and possibly the backscattering of the cavity mirrors. The two counter-propagating laser modes separate themselves spatially after leaving the laser cavity, which allows one to measure their individual chaotic intensity dynamics simultaneously. The interaction between these two chaotic laser modes represents a most general case of mutually coupled chaotic systems.

We have studied interaction between the chaotic modes under various experimental conditions. Generally it is observed that when they are coupled, the intensity evolution of each individual mode becomes very complicated. The chaotic dynamics of each mode under coupling is very different from that of the single mode chaotic dynamics of the laser. Normally no obvious synchronization between them can be identified as shown in Fig. 11. However, under certain conditions, even though the coupling

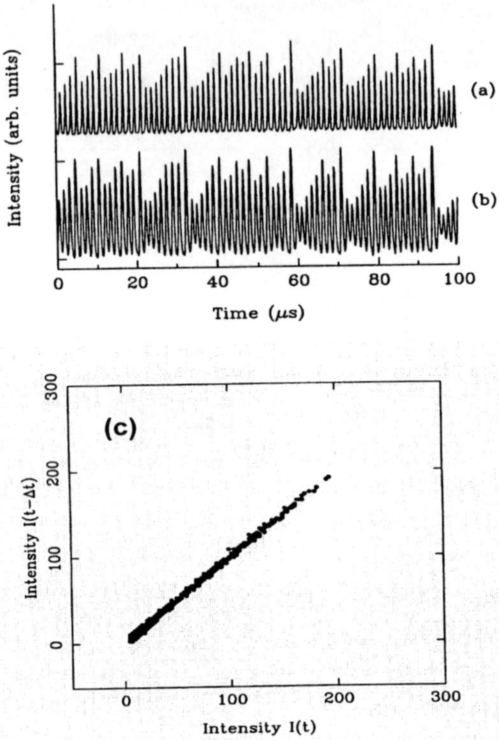

Figure 10. Generalized synchronization of chaos of the laser under stronger driving. a): chaotic laser intensity evolution. b): chaotic driving signal evolution. c): relationship plot of the laser dynamics under repeated driving.

strength between the modes is not strong, the intensity dynamics of the coupled modes can spontaneously become synchronized. Two kinds of synchronization were observed. Fig. 12 shows a case where the average mode beat frequency of the coupled modes is close to that of the predominant chaotic oscillation frequencies of the modes. Under the influence of the average beat frequency the average chaotic intensity pulsation frequencies of both the modes become locked, and consequently the chaotic intensity dynamics of the modes becomes "phase synchronized". We note that the chaotic intensity evolutions of the modes are exactly antiphase, and the chaotic dynamics of each mode is obviously different. The out-of-step pulsations of coupled periodic systems have been observed and intensively investigated previously [27-28]. It was identified as a cooperative self-organization of coupled systems. Our experimental results demonstrate that even in the case of coupled chaotic systems, their chaotic dynamics can be cooperatively self-organized.

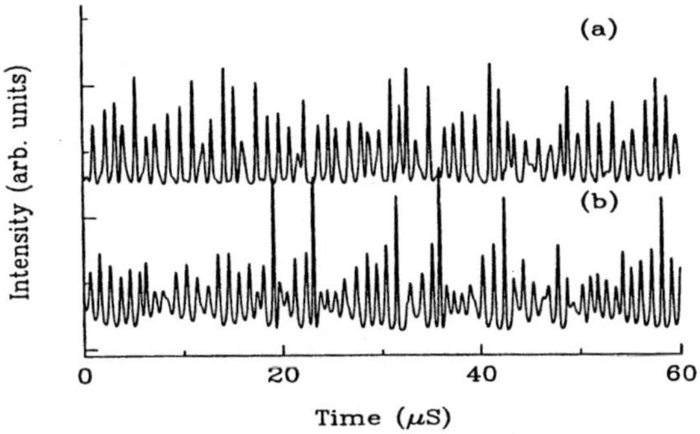

Figure 11. Typical mode intensity evolution of each of the two modes without synchronization.

Fig. 13 shows another case of chaotic synchronization observed, where the average mode frequencies are locked to each other. In our experiment the separation between the average mode frequencies reflects the interaction strength of them. In this situation the interaction between the modes is quite strong. As expected, under strong couplings the chaotic intensity dynamics of the modes approach to each other, as can be seen, the chaotic intensity dynamics of the modes exhibit close similarities. We emphasize that Fig. 13 is not a state of perfect chaotic synchronization, as their chaotic evolution is antiphase and not exactly the same.

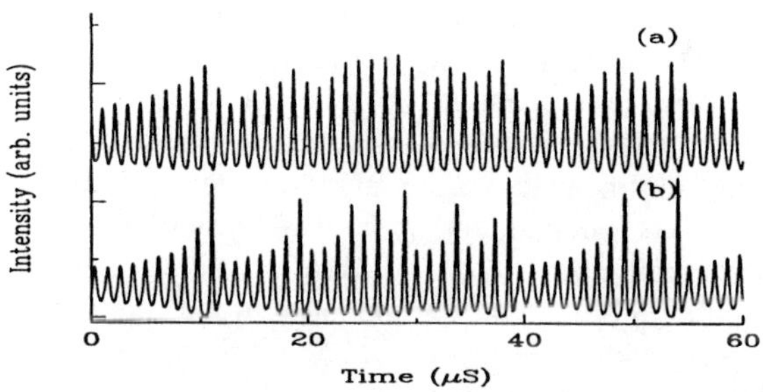

Figure 12. Mode intensity evolution of each of the two modes under the phase synchronization. The average beat frequency of the modes is locked to the chaotic oscillation frequencies of the modes.

With even stronger mode coupling phase locking between the modes is obtained, and consequently perfect chaotic synchronization is observed in our laser. Fig. 14 shows as an example one of the cases. Under very strong coupling the two modes behave synchronized like a single one, despite the fact that they propagate in opposite directions. Another interesting experimental result observed is that under the perfect synchronization of chaos the chaotic dynamics of the overall system is the same as that of the single mode laser. Under strong mode coupling we have experimentally observed the synchronized period-doubling route to chaos and the Lorenz-kind spiral chaos [14].

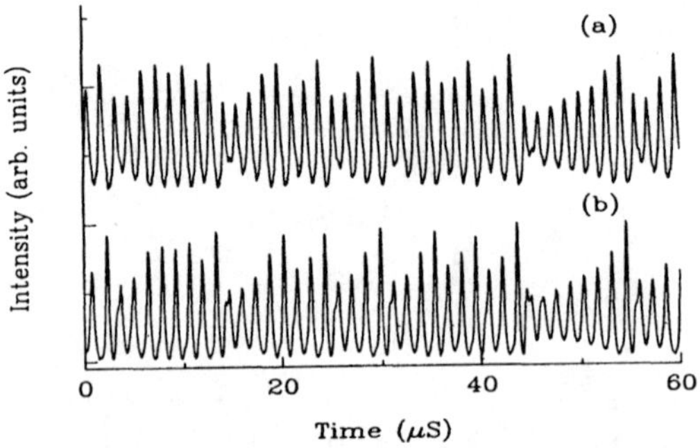

Figure 13. Mode intensity evolution of each of the modes under the antiphase chaotic synchronization. The average optical frequencies of the modes are locked.

PHYSICAL ORIGIN OF THE VARIOUS FORMS OF CHAOTIC SYNCHRONIZATION

To understand the formation of various forms of chaotic synchronization observed, we first analyze the interaction between the two mutually coupled chaotic laser modes. As we have pointed out previously, physically a laser mode is an oscillator. It consists of the mode amplitude and phase part. When a laser becomes chaotic, the mode dynamics actually consists of the chaotic amplitude dynamics and chaotic phase dynamics. From the nonlinear dynamics point of view, the amplitude and phase of a laser mode are just two independent dynamical variables of a chaotic laser system, they in fact carry equal amount of information about the system dynamics. Our previous experimental studies on the chaotic mode field dynamics have already confirmed the equivalence of the chaotic mode amplitude and phase dynamics [21].

It is a well known effect that when the frequencies of two counter-propagating modes of a bidirectional ring laser are close to each other, due to the mode phase interaction they will automatically lock together to a common value. This phenomenon is called mode locking. Phase interaction is a generic nature of coupled stable laser modes. Our experimental results on the mutually coupled chaotic laser modes firstly demonstrate that even when the interacting modes are intrinsically chaotic, their phases could still be locked, indicating that phase interaction between them still exist. However, different from the case of conventional laser mode locking, as the phase of a chaotic laser mode is intrinsically unstable, the mode locking between coupled chaotic laser modes will now occur in two steps: first only the average frequencies of the modes will be locked; when the locking force is further increased, the instantaneous frequencies of them will then become locked.

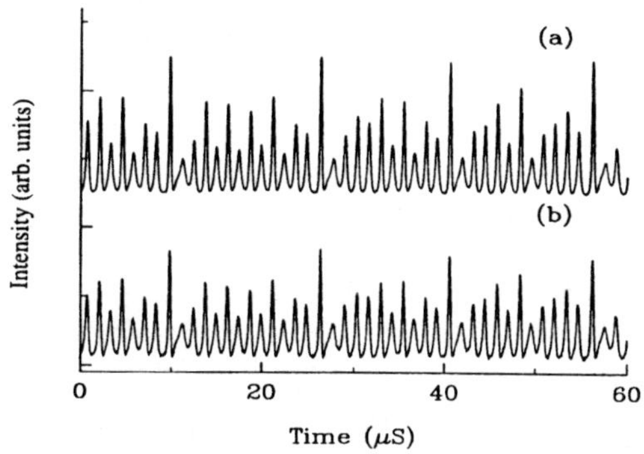

Figure 14. Mode intensity evolution of each of the modes under the perfect synchronization of chaos. The chaotic optical phases of the modes are locked in this situation.

As for a chaotic laser mode its chaotic amplitude dynamics and phase dynamics are closely related, the phase correlation between the modes results in that their amplitude dynamics also exhibit a certain correlation. Under the average frequency locking between the modes, only the carrier frequencies and the average chaotic phase drifts of the modes are constrained, the chaotic nature of the mode phase variations still exists. This has the consequence that the mode amplitude dynamics only show a certain relation, while details of them are different. Only when the phases of the modes are locked, their chaotic amplitude dynamics will then become exactly the same. Therefore, under the phase locking of the modes, their chaotic dynamics are perfectly synchronized.

Based on the phase interaction between the coupled laser modes the synchronization of their chaotic mode dynamics could be well explained, in particular, the different forms of the chaotic synchronization are natural result of their phase interaction. Although the coupled chaotic laser modes are a special case of coupled chaotic systems in the sense that a laser mode is a natural oscillator, whose phase and amplitude are clearly defined, while for a general chaotic system no such definition exists, considering that the dynamics of a chaotic system is mathematically fully described by the evolutions of trajectories in the phase space, and trajectories moving on a chaotic attractor have the property of recurrence, which physically resembles an oscillation evolution, one could principally extend the concept of phase interaction into coupled general chaotic systems. In fact we found that for an autonomous chaotic system the time evolution of its dynamical variables actually contains already the phase and amplitude properties of a chaotic oscillator. To demonstrate this, we have constructed the peak-to-peak intensity return map and the peak-interval-to-peak-interval return map based on the intensity evolution of the Lorenz kind spiral chaos shown in Fig. 4. The peak intensity return map actually represents the intensity dynamics of the signal, and the peak interval to peak interval return map represents the phase dynamics of the signal [26,29]. Comparing both of the return maps obtained we found that they exhibit similar cusp form as shown in Fig. 15, indicating that they equally represent the chaotic dynamics of the signal. This experimental evidence suggests that one could indeed describe the dynamics of coupled chaotic systems in terms of coupled chaotic oscillators.

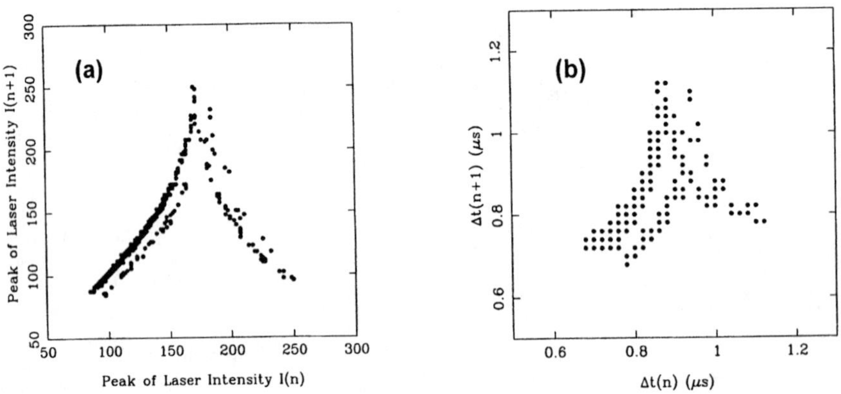

Figure 15. The peak-to-peak intensity return map and the peak interval to peak interval return map constructed from the Lorenz kind spiral chaos shown in Fig. 4.

In light of the phase interaction between the coupled chaotic oscillators we found that the various forms of synchronization between coupled chaotic systems could be uniquely explained. The phases associated with a chaotic evolution have recently been intensively investigated [30-31]. Using the Hilbert-transformation technique one can even quantitatively calculate the phase dynamics of a chaotic system [12]. We have calculated the phase evolutions associated with the intensity evolutions of the two coupled laser modes under the average pulsation frequency locking and perfect chaotic synchronization, respectively. Fig 16 shows the results. As expected, under the phase synchronization the average chaotic oscillation frequencies of the two modes are locked (they have the same average slope in the figure), while details of their chaotic phase variations are still random. Under the perfect chaotic synchronization, within the experimental measurement accuracy their instantaneous phases are locked.

Similarly the various forms of chaotic synchronization under the unidirectional coupling could be uniquely understood as well. The chaotically modulated signal carries both the amplitude and phase information of the driving system. Under the phase interaction between the driving and response systems, their average oscillation frequencies and instantaneous phases will be locked, respectively. Under the average frequency locking the frequency entrained states are obtained. As mentioned previously, under this synchronization the chaotic nature of the driven system is not fully suppressed, therefore, the driven system is still chaotic. When the phase interaction become strong enough, the phase of the response system will be locked to that of the driving system, and under the phase locking the chaotic dynamics of the

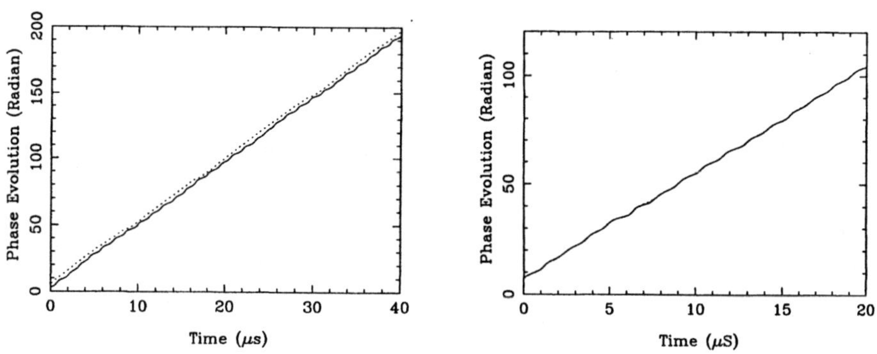

Figure 16. Phase evolution associated with the intensity dynamics of each of the coupled laser modes. Left: under the phase synchronization shown in Fig. 12. Right: under the perfect chaotic synchronization shown in Fig. 14.

response system is completely suppressed. However, depending on the dynamics of the driving and the response systems, and what parameter of the response system is driven etc., even under the phase locking their amplitude dynamics could still be different. If the dynamics of the response system is exactly the same, phase locking will result in exactly the same amplitude dynamics, the perfect chaotic synchronization will be obtained. While if the dynamics of the response system is different from that of the driving system, different amplitude dynamics will be obtained, resulting in the generalized synchronization of chaos. We point out that although for coupled chaotic systems there exist both amplitude and phase interaction between them, like the coupled periodic oscillators, the phase interaction between them is more sensitive than the amplitude interaction. To demonstrate this property we have shown in Fig. 17 an experimental result. We artificially constructed a modulation signal with pure phase chaos but constant amplitude. Under the driving of the signal, the phase variations of the laser chaos are locked to the driving signal, while the amplitudes are not. In this case the amplitude variations exhibit chaotic dynamics as determined by the phase-amplitude relation of the laser dynamics. This property of coupled chaotic systems is also obviously to see in fig. 9, 10. Under the generalized synchronization of chaos the chaotic dynamics of the response system is still mainly the intrinsic chaotic dynamics of its own rather than that of the driving system.

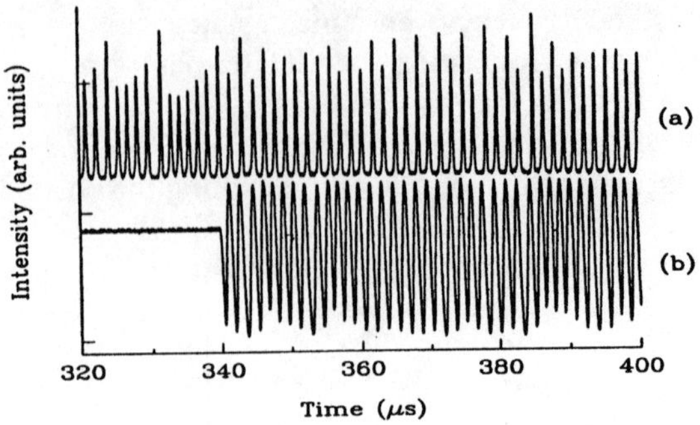

Figure 17. Chaotic laser intensity evolution under the driving of an artificial signal whose phase is chaotic while amplitude is constant.

As phase locking is a threshold effect, depending on the stability of the response system, the locking force required to achieve the synchronization of chaos is different. If the response system is at a fixed point, phase locking is immediately obtained. However, if the response system is on a limited cycle or a chaotic attractor in the phase space, larger locking forces are required. All these threshold properties of the chaotic synchronization have been confirmed in our experiments.

CONCLUSIONS

Although a chaotic oscillation is more complicated than a periodic oscillation, our experimental results demonstrate that the tendency toward synchronization is a generic property of coupled chaotic systems, either they are unidirectionally coupled or mutually coupled. Like the synchronization of coupled periodic oscillations, the physical origin of synchronization between coupled chaotic systems is still the phase interaction between them. Under the phase interaction phases of coupled chaotic oscillations will approach phase locking. However, in the cases of coupled chaotic oscillations, as their phases are intrinsically unstable, phase locking occurs in two steps: first under the weak coupling strength the statistical average of their chaotic oscillation frequencies will be locked. When the coupling strength is further increased, eventually their instantaneous frequencies will be locked together. Due to the existence of these two different stages of the chaotic phase locking, there exist different forms of chaotic synchronization. The so-called "phase synchronization" of chaos is in fact a state of the average frequency locking of the coupled chaotic systems, and the state of instantaneous frequency locking of them correspond to either the "perfect synchronization of chaos" or the "generalized synchronization of chaos", depending on whether the coupled chaotic systems have exactly the same chaotic dynamics or not. The perfect synchronization of chaos only occurs between coupled chaotic systems of exactly the same dynamics. If the dynamics of the coupled systems are different, even under the phase locking their amplitude dynamics will still be different, which results in the generalized synchronization of chaos. Under the average frequency locking, as the chaotic nature of the coupled system is not suppressed, the chaotic dynamics of them are still independent. The threshold coupling strength required for the phase locking of coupled chaotic systems depends upon the stability property of the coupled chaotic systems. In a chaotically driven stable system, if the dynamics of the driven system in the phase space is a fixed point, then the threshold is zero, otherwise it is not zero. As phase interaction between coupled chaotic systems is more sensitive than amplitude interaction, one could use this property for the secure communications, e.g., one could encode information in the amplitude dynamics of a chaotic signal while retain its phase dynamics unchanged. The phase locking between the carrier signal and the receiver will guarantee the generalized synchronization of chaos, while to achieve the perfect synchronization of chaos the chaotic dynamics of the receiver must be exactly the same as that of the chaotic carrier signal. To accurately retrieve the encoded information the perfect synchronization of chaos is necessary.

REFERENCES

1. Pecora, L. M., and Carroll, T. L., *Phys. Rev. Lett.* **64**, 821(1990).
2. Carroll, T. L., and Pecora, L. M., *Int. J. Bifurcation Chaos,* **5**, 563(1995).
3. Anishchenko, V. S., Vadivasova, T. E., Postnov, D. E., and Safonova, M. A., *Int. J. Bifurcation Chaos* **2**, 633(1992).
4. Murali, K., and Lakshmanan, M., *Phys. Rev. E.* **49,** 4882(1994).
5. Roy, R, and Thornburg, Jr., K. S., *Phys. Rev. Lett.* **72,** 2009(1994).
6. Sugawara, T, Tachikawa, M, Tsnkamoto, T, and Shimizu, T, *Phys. Rev. Lett.* **72,** 3502(1994).
7. Winful, H. G., and Rahman, L, *Phys. Rev. Lett.* **65,** 1575(1990).
8. Pecora, L, and Carroll, T. L., *Phys. Rev. A.* **44,** 2374(1991).
9 Kocarev, L, and Parlitz, U, *Phys. Rev. Lett.* **76,** 1816(1996).
10. Rulkov, N. F., Sushchik, M. M., Tsimring, L. S., and Abarbanel, H. D. I., *Phys. Rev. E.* **51,** 980(1995).
11. Rulkov, N. F., *Chaos,* **6,** 262(1996).
12. Rosenblum, M. G., Pikovsky, A. S., and Kurths, J., *Phys. Rev. Lett.* **76,** 1804(1997).
13. Rosenblum, M. G., Pikovsky, A. S., and Kurths, J., *Phys. Rev. Lett.* **78,** 4193(1997).
14. Tang, D. Y., Dykstra, R., and Heckenberg, N. R., *Phys. Rev. A,* **54,** 5317(1996).
15. Tang, D. Y., and Hechenberg, N. R., *Phys. Rev. E,* **55,** 6618(1997).
16. Tang, D. Y., Dykstra, R., Hamilton, M. W., and Heckenberg, N. R., *Phys. Rev. E.* **57,** 3649(1998).
17. Haken, H., *Phys. Lett.* **53A,** 77(1975).
18. Roldan, E., Valcarcel, G. J. de, Vilaseca, R., Corbalan, R., Martinez, V. J., and Gilmore, R., *Quantum Semiclass. Opt.* **9,** R1(1997).
19. Weiss, C. O., Hübner, U., Abraham, N. B., and Tang, D. Y., *Infrared Phys. Technol.* **36,** 489(1995).
20. Weiss, C. O., Vilaseca, R., Abraham, N. B., Corbalan, R., Roldan, E., Valcarcel, G. J. de, Pujol, J, Hübner, U., Tang, D. Y., *Appl. Phys. B* **61,** 223(1995).
21. Tang, D. Y., Li, M. Y., Malos, J. T., Heckenberg, N. R., and Weiss, C. O., *Phys. Rev. A,* **52,** 717(1995).
22. Tang, D. Y., Pujol, J., and Weiss, C. O., *Phys. Rev. A* , **44,** R35(1991).
23. Weiss, C. O., and Brock, J., *Phys. Rev. Lett.* **57,** 2804(1986).

24. Dubois, M., Ubio, M. A., and Berge, P., *Phys. Rev. Lett.* **51,** 1446(1983).
25. Tang, D. Y., Li, M. Y., and Weiss, C. O., *Phys. Rev. A* , **44,** 7597(1991).
26. Farmer, J. D., *Phys. Rev. Lett.* **47,** 179(1981).
27. Wiesenfeld, K., Bracikowski, C., James, G., and Roy, R., *Phys. Rev. Lett.* **65,** 1749(1990).
28. Otsuka, K, Mandel, P., Bielawski, S., Derozier, D., and Glorieux, P., *Phys. Rev. A,* **46,** 1692(1992).
29. Candaten, M., and Rinaldi, S., *Int. J. Bifurcation Chaos,* **10,** 1805(2000).
30. Pikovsky, A. S., Rosenblum, M. G., Osipov, G. V., and Kurths, J., *Physica (Amsterdam)* **104D,** 219(1997).
31. Yalcinkaya, T., and Lai, Y. C., *Phys. Rev. Lett.* **20,** 3885(1997).

Experimental Observation of Coherence and Stochastic Resonances in an Electronic Chua Circuit

Oscar Calvo*, Iacyel Gomes*, Claudio R. Mirasso* and Raúl Toral*,†

* *Departament de Física, Universitat de les Illes Balears, 07071-Palma de Mallorca, Spain*
† *Instituto Mediterráneo de Estudios Avanzados , CSIC-UIB, 07071-Palma de Mallorca, Spain*

Abstract. Stochastic and coherence resonances appear in nonlinear systems subjected to an external source of noise and are characterized by a maximum response at the optimal value of the noise intensity. This paper shows experimentally that it is possible to observe them in a *chaotic* system. To this end we have analysed an electronic Chua circuit running in the chaotic regime and added noise to its dynamics. In the case of coherence resonance, we observe an optimal periodicity for the jumps between chaotic attractors, whereas in the case of stochastic resonance we observe a maximum in the signal–to–noise ratio at the frequency of an external sinusoidal perturbation.

INTRODUCTION

Contrary to the intuitive role of noise as a disordering agent, it has been shown that the response of a nonlinear dynamical system can improve in some occasions by the addition of noise. The paradigmatic and pioneering example is that of *stochastic resonance* in bistable systems[1, 2] which had been originally proposed as a mechanism for the observed periodicity of the Earth's ice ages. In this case, a dynamical system with two stable fixed points is subjected to the action of a periodic forcing and noise. The periodic forcing is sufficiently weak that, by itself, can not overcome the separation barrier between the stable fixed points and the system oscillates around one of them, chosen by the initial conditions. The presence of noise, however, allows the system to make random excursions and visit both fixed points. The jumps between the fixed points are easier to make when the effect of the external forcing is to lower enough the separation barrier. If the average time for the stochastic walk to go from one fixed point to the other coincides with the periodicity of the external signal, a maximum coincidence of the jumps with the external period is observed. The coincidence is optimal for the right value of the noise intensity. This is, basically, the mechanism for stochastic resonance. The original example has been extended to many other cases of physical as well as biological interest and there exists now a well established number of results concerning theoretical, numerical and experimental studies of stochastic resonance under a large variety of conditions[3, 4, 5].

A related but different phenomenon is that of *coherence resonance* by which a quasi-periodic signal is generated under the effect of the fluctuations in a system not subjected to a periodic forcing[6, 7, 8]. Again, the periodicity is optimal (resonance) for a given value of the noise intensity. The simplest example is an excitable system. This is a

CP622, *Experimental Chaos: 6th Experimental Chaos Conference*, edited by S. Boccaletti et al.

system with a fixed point which is stable for low amplitude perturbations. However, perturbations larger than a threshold value result in the system returning to the fixed point after a large excursion, a pulse, in phase space. In the case of coherence resonance, the perturbations are induced by noise and, if the noise has the appropriate intensity, the system is excited at the end of every pulse in such a way that a quasi-periodic train of pulses is generated. Coherence resonance has also been observed in bistable and oscillatory systems[9, 10] or close to the onset of a bifurcation[11]. Experimental evidence has been shown for excitable electronic circuits[12, 13] and a laser system[14].

In this paper we show experimental evidence for both resonances in electronic Chua circuits operating in the chaotic regime. The possibility of having stochastic resonance in a chaotic circuit with two attractors under the presence of noise had been shown theoretically and numerically by Anishchenko and coworkers[15, 5]. This is different from the so-called *deterministic stochastic resonance*[16] which appears in deterministic chaotic systems close to the onset of intermittency and that has been shown experimentally in ferromagnetic[17] and laser[18, 19] systems.

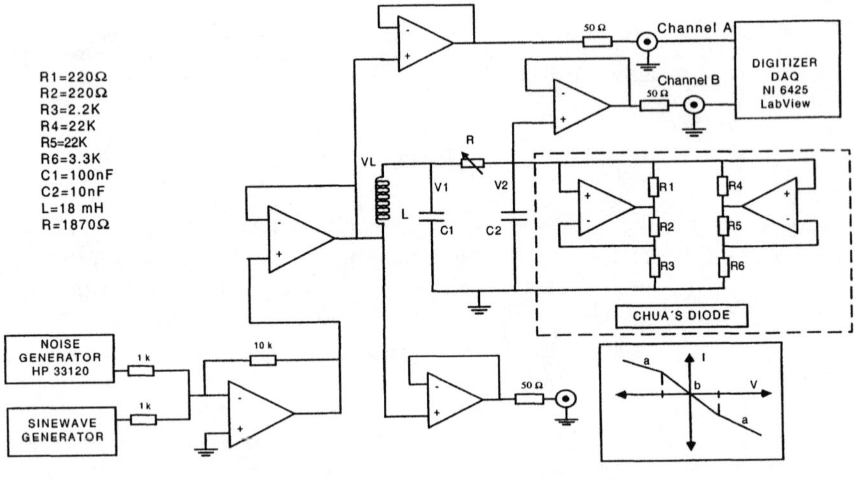

FIGURE 1 Experimental setup.

Our Chua circuit is shown in figure 1. A digital acquisition board from National Instruments NI-DAQ plugged into a laptop computer was used to digitize the signal at a sampling rate of 10KHz,. A Labview program controlled the board in a continuous acquisition mode and an HP 33120 function generator was used to provide the noise signals from 0 to 5 V[rms]. A Chua circuit can be described in terms of three non-linear first order differential equations[20]. These equations predict that the system has, for some range of parameters, three unstable fixed points. In this case, the output voltages V_1 and V_2 oscillate chaotically in time around different dynamical attractors. In our experiments, we choose the main parameters $C_1 = 10$ nF, $C_2 = 100$ nF, $R = 1673\ \Omega$, $a = -8/7$, $b = -5/7$, where a and b characterize the slope of the non-linear Chua resistance (see inset fig. 1). For this set of parameters, there is a *single scroll* attractor (as shown in the fig. 2(a), and its mirror image (not shown in the figure). In the absence of any external perturbation, V_1 and V_2 chaotically oscillate around one of the unstable fixed points with a mean oscillation frequency of ~ 2.3 KHz and without any possibility to jump to its mirror image attractor. Depending on the initial conditions, the system moves within this attractor or on its mirror image. The time evolution of V_2 is shown in the figure 2(b). We can see that V_2 (and similarly for V_1) oscillates chaotically with a rather large fluctuations in the minima and maxima values. This fact will be important when studying the stochastic and coherence resonance phenomena, as will be discussed later.

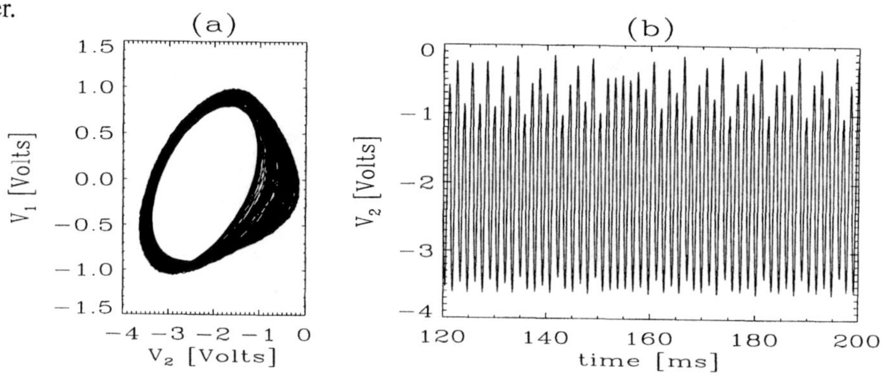

FIGURE 2 (a) Phase space representation of the attractor; (b) Time series of V_2

To study the stochastic resonance effect we start by adding a small periodic signal (a sinusoidal one) to the circuit in the branch of the impedance L as shown in figure 1. The amplitude of the signal is sufficiently small (~ 20 mV) that the system is not able to jump to its mirror image attractor. The frequency of the signal is ~ 2.1 KHz, close to the mean frequency of the Chua attractor. A white noise source is also added to the system through the same branch of the periodic signal source. We have checked that in the absence of the periodic signal, the system remains in the same attractor, and does not jump to its mirror image, for small noise levels. Once the noise is increased the system starts to jump from one attractor to the other. This situation is depicted in Figure 3(a), where we plot the time evolution of V_2 for three different noise levels, together with the sinusoidal signal. In a typical stochastic resonance behavior, e.g. when a system jumps between two potential wells with stable minima, its is expected that an optimum noise level exists for which the system jumps at every external signal period. In our case, however, we do not move between two stable minima but between two chaotic attractors, each with its own internal dynamics. When looking in more detail at the jumping process between the attractors, we observe that after a jump the system starts to move close to the center of the attractor and not in the outer orbits. In this case it is unlikely that the system would jump to the other attractor until it reaches an outer orbit. Therefore, for the sinusoidal frequency we have chosen, we do not expect a jump from one attractor to the other at every sinusoidal period. This fact is also reflected in figure 3(a). The center panel corresponds to the time trace of V_2 with the optimum noise level. Although the resonance between the noise and the external periodic signal is not clear, it becomes evident when plotting the Fourier transform of the signal, $| S(\nu) |$ (power spectrum). In figure 3(b) it can be observed a maximum of the Signal-to-Noise Ratio (SNR) at an intermediate noise level, in this case a value of ~ 1.5 V [rms]. This fact is confirmed when plotting the SNR vs. the noise level (see figure 4). A clear maximum appears indicating an optimum response of the system to this intermediate noise intensity.

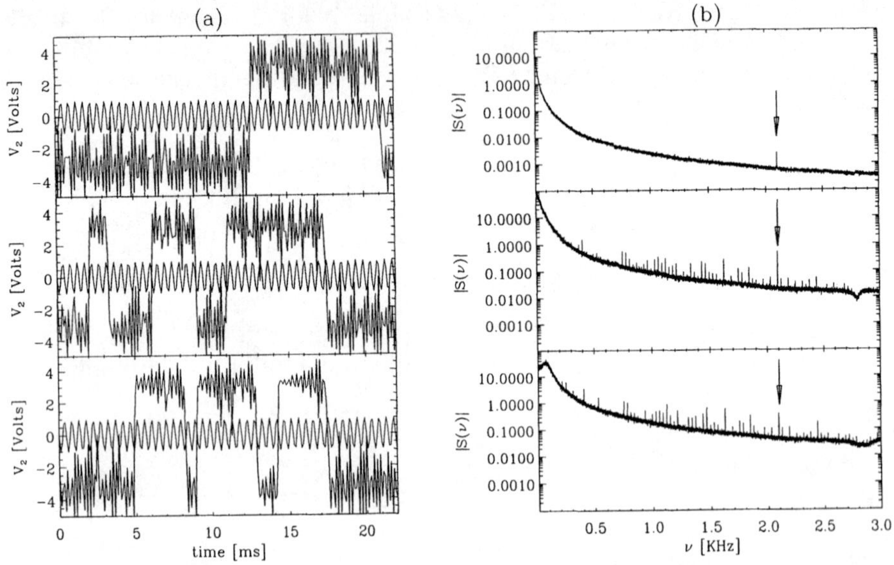

FIGURE 3 (a) Time evolution of V_2 for three different noise levels. From top to bottom: 1V [rms], 1.5 V [rms] and 2.25 V [rms]. The sinusoidal signal is added in the three panels with an arbitrary amplitude; (b) Power spectrum for the same noise intensity as in panel (a). The arrow indicates the frequency of the sinusoidal signal.

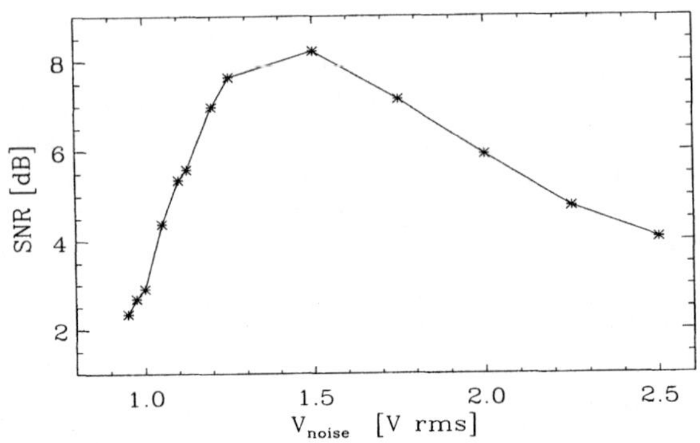

FIGURE 4 Signal-to-noise ratio as a function of the noise level.

The phenomenon of coherence resonance is characterized by an optimum response of a system for an intermediate noise level without any other forcing. In this sense it is not very different to the stochastic resonance except for the absence of the sinusoidal source, and for this reason it is sometimes called *stochastic resonance without external forcing*. We start our experiment in the same situation as we started the previous one. Again for small noise levels the system is not able to jump to its mirror image attractor. As we do not have a periodic external forcing we do not expect any resonance in the SNR. This means that we have to look for other indicators. To quantify the situation we calculate the variance σ of the residence time in the attractors normalized to the mean value $\langle t \rangle$ and the normalized autocorrelation function of the time series[8]. Three typical time traces are shown in the figure 5(a) for the voltage V_2 at low, intermediate and large noise levels. It can be seen that for the intermediate noise level (the optimum one), the regularity of the jumping process becomes evident. This is the first indication that noise is inducing a more regular behavior in this chaotic system. The autocorrelation functions depicted in figure 5(b) present the deepest minimum value at the same noise level, indicating maximum anticorrelation between the two states. These results reveal the existence of a constructive effect of the noise that is capable to yield a maximum regularity in the process of transition from one chaotic attractor to the other.

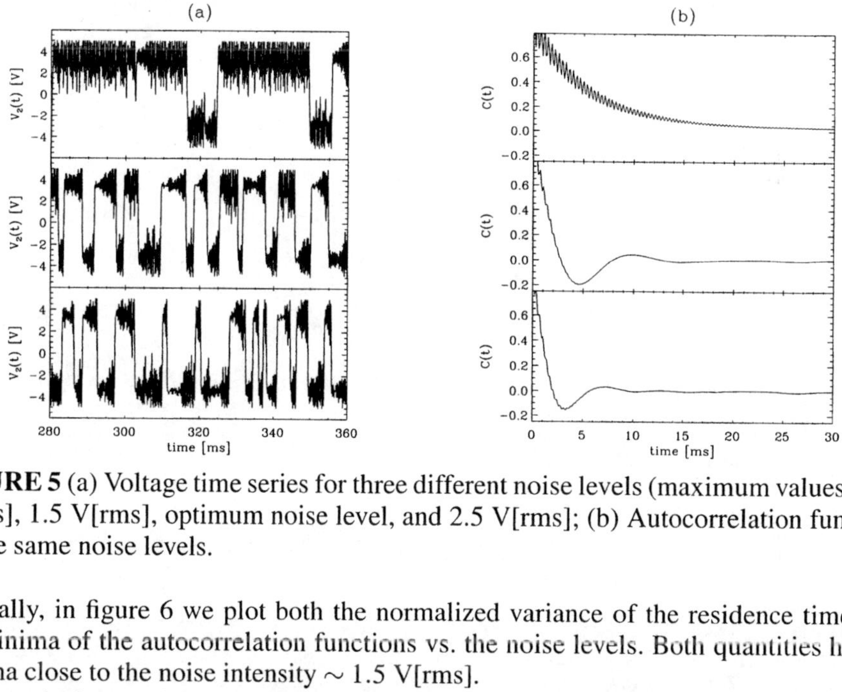

FIGURE 5 (a) Voltage time series for three different noise levels (maximum values): 0.5 V[rms], 1.5 V[rms], optimum noise level, and 2.5 V[rms]; (b) Autocorrelation function for the same noise levels.

Finally, in figure 6 we plot both the normalized variance of the residence time and the minima of the autocorrelation functions vs. the noise levels. Both quantities have a minima close to the noise intensity $\sim$ 1.5 V[rms].

In conclusion, we have experimentally shown that the main features of both stochastic and coherence resonance are present in some chaotic systems. We have done the demonstration using a chaotic Chua circuit. In the case of the stochastic resonance we observed a clear maximum for an intermediate noise level of the signal-to-noise ratio at the external frequency. For the coherence resonance, a quasi-periodic response can be induced by just applying an appropriate amount of noise to a circuit which, otherwise, exhibits a non-regular behavior. The quality of the periodicity was quantified by the variance of the residence time and the minimum of the time correlation function.

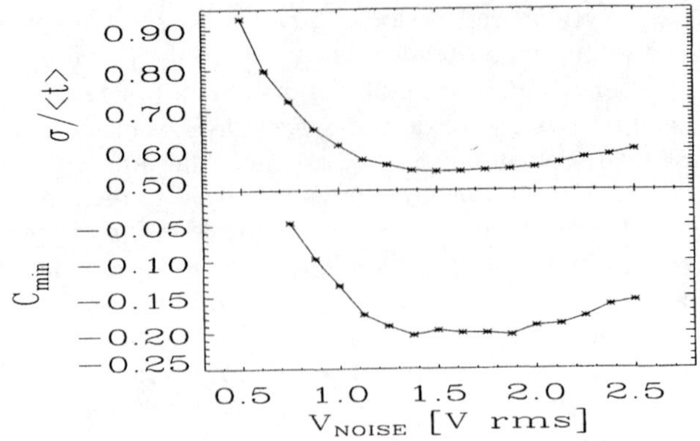

FIGURE 6 (a) Variance of the residence time normalized by the mean time $\sigma/\langle T\rangle$ and (b) minimum of the correlation function C_{min} as a function of the noise level V_{NOISE}.

ACKNOWLEDGMENTS

We thank financial support from DGESIC (Spain) projects numbers PB97-0141-C02-01 and BFM2000-1108.

REFERENCES

1. R. Benzi, A. Sutera and A. Vulpiani, J. Phys. **A14**, 453 (1981).
2. C. Nicolis and G. Nicolis, Tellus **33**, 225 (1981).
3. Proceedings of the NATO Advanced Research Workshop: Stochastic Resonance in Physics and Biology. F. Moss, A. Bulsara and M.F. Shlesinger, eds. J. Stat. Phys. **70** (1993).
4. L. Gammaitoni, P. Hänggi, P. Jung and F. Marchesoni, Rev. Mod. Phys. **70**, 223 (1998).
5. V.S. Anishchenko, A.B. Neiman, F. Moss and L. Schimansky-Geier, Physics-Uspekhi **42**, 7 (1999).
6. H. Gang, T. Ditzinger, C.Z. Ning and H. Haken, Phys. Rev. Lett. **71**, 807 (1993).
7. W. Rappel and S. Strogatz, Phys. Rev. **E50**, 3249 (1994).
8. A.S. Pikovsky and J. Kurths, Phys. Rev. Lett. **78**, 775 (1997).
9. B. Lindner and L. Schimanwky-Geier, Phys. Rev. E **61**, 6103 (2000).
10. J.R. Pradines, G.V. Osipov and J.J. Collins, Phys. Rev. E **60**, 6407 (1999).
11. A. Neiman, P. Saparin and L. Stone, Phys. Rev. E **56**, 270 (1997).
12. D.E. Postnov, S.K. Han, T.G. Yim and O.V. Sosnovtseva, Phys. Rev. **E59**, R3791 (1999).
13. S.K. Han, T.G. Yim, D.E. Postnov and O.V. Sosnovtseva, Phys. Rev. Lett. **83**, 1771 (1999).
14. G. Giacomelli, M. Giudici, S. Balle and J.R. Tredicce, Phys. Rev. Lett. **84**, 3298 (2000).
15. V.S. Anishchenko, M.A. Safanova and L.O. Chua, Int.J. Bif. Chaos **2**, 397 (1992).
16. V.S. Anishchenko, A.B. Neimann and M.A. Safanova, J. Stat.Phys. **70**, 183 (1993).
17. E. Ribold, W. Just, J. Becker and H. Benner, Phys. Rev. Lett. **78**, 3101 (1997).
18. A. N. Pisarchik and R. Corbalán, Phys. Rev. **E58**, R2697 (1998).
19. V.N. Chizhevsky, R. Vilaseca and R. Corbalán, Phys. Rev. **E61**, 6500 (2000).
20. *Chua's Circuit: A Paradigm for Chaos*, R.N. Madan, ed. World Scientific Publishing (1993).

CHAOS-SYNCHRONIZATION IN SEMICONDUCTOR LASER SYSTEMS: AN OPTICAL PHASE DEPENDENT SCENARIO

Michael Peil, Tilmann Heil, Ingo Fischer and Wolfgang Elsäßer

Institute of Applied Physics, Darmstadt University of Technology
Schloßgartenstr. 7, D-64289 Darmstadt, Germany

Abstract. Synchronization of chaotic oscillators is of high current interest in various areas of science. Semiconductor laser systems offer a great potential for experimental studies of synchronization phenomena, because of well-controllable parameters, well-studied nonlinear dynamical behavior and their broad spectrum of applications. We investigate chaos-synchronization of two unidirectionally coupled semiconductor lasers with delayed optical feedback. We present a characteristic synchronization scenario in dependence of the relative optical cavity phase of the subsystems. For adjusted phase, we find excellent synchronization of the intensity dynamics in combination with coherence among the emitted fields of the lasers, despite of the fast chaotic wavelength-fluctuations. Variation of the phase leads to conspicuous changes in the intensity-dynamics associated with drastically reduced correlation and loss of coherence among the lasers. Our results provide insight into the consequences of vectorial coupling for the synchronization scenario and open the perspective for innovative concepts for encrypted GHz data communication.

INTRODUCTION

Synchronization phenomena of coupled nonlinear oscillators are of fundamental interest, as they are encountered in various areas of science [1–3]. Recently, chaos synchronization phenomena of coupled chaotic semiconductor laser (SL) systems have attracted much attention, since their well-controllable parameters and their well-studied nonlinear behavior [4,5] makes them ideally suited for studies on the fundamental synchronization phenomena of coupled nonlinear systems. Furthermore, they enable the realization of innovative applications in the field of telecommunication, using chaotic carriers for encrypted high-bandwidth data transmission [6–10]. The understanding of the synchronization scenario related to variations of the system parameters is of high relevance for both aspects.

CP622, *Experimental Chaos: 6th Experimental Chaos Conference*, edited by S. Boccaletti et al.

A particularity of SL systems is the possibility of vectorial coupling via amplitude and phase of the optical fields. We experimentally study the importance of this kind of coupling on the chaos synchronization scenario of two unidirectionally coupled SL systems, where each system is subject to delayed optical feedback, thus emitting chaotically [4,11,12]. We concentrate on the chaos synchronization scenario of these coupled systems in dependence of the well-controllable relative cavity phase and show that the vectorial nature of the optical coupling is of particular importance. For matched cavity-phase-conditions, we find excellent synchronization of the intensity dynamics in combination with constructively interfering fields of the lasers, hence coherence among the lasers. Varying the relative cavity phase, we find a characteristic scenario. We observe gradual loss of synchronization, conspicuous changes in the intensity dynamics associated with drastically reduced cross correlation coefficients, and loss of the coherence among the lasers. We demonstrate the stability of the synchronization manifold and provide perspectives for the functional use of these properties in innovative concepts for encrypted high-bit data communication.

EXPERIMENTAL RESULTS

A sketch of the experimental setup is depicted in Fig. 1. Our system consists of two SLs, subjected to delayed optical feedback ($\tau_d = 2.9\,\mathrm{ns}$) from high reflecting mirrors, thus emitting chaotically. Each laser is pumped by a DC low noise current source at 1.01 times its solitary threshold current I_{th}^{sol}, and is temperature stabilized to better than 0.01 K. We have selected two device-identical SLs, as coinciding laser parameters are essential for chaos synchronization. In order to avoid asymmetries with respect to detuning effects, we monitor the optical spectra of both lasers with an optical spectrum analyzer (OSA) with 0.1 nm resolution, such that their frequencies match within an accuracy to better than 1 GHz. By changing the length of the cavities on sub-wavelength scale via piezo translators (PZT), we vary the

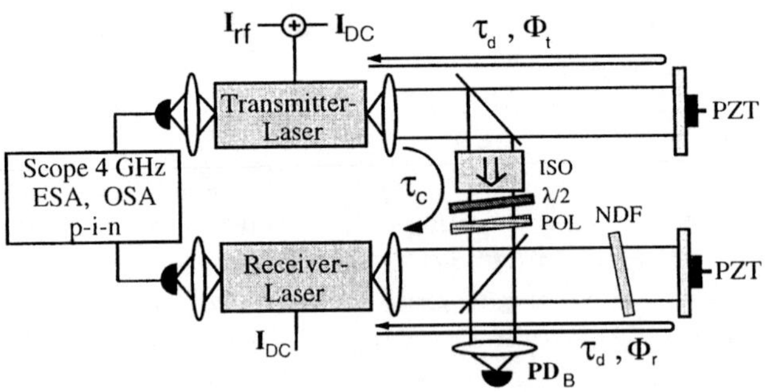

FIGURE 1. Experimental setup of coupled transmitter and receiver external cavity lasers.

feedback phase of the reflected light (cavity phase) $\Phi_{t,r}$. In the following, we will call the driving laser *transmitter* and the driven laser *receiver*. The unidirectional coupling is realized optically by injecting a well-defined fraction of the optical field of the transmitter via an optical isolator (ISO) into the receiver. The polarizer (POL) and the $\frac{\lambda}{2}$-plate guarantee a coupling via the dominant TE component of the optical field. We detect the intensity dynamics of both lasers simultaneously with photodetectors and an electrical spectrum analyzer (ESA). A 2-channel oscilloscope of 4 GHz analog bandwidth resolves the fast intensity fluctuations on the relevant sub-ns time scales and the low frequency fluctuations, which are the typical dynamical effects for these conditions (LFF) (see e.g. [11]). We couple an intensity equal to 40 % of the transmitters feedback intensity into the receiver, whose feedback intensity is reduced by a neutral density filter (NDF) to 70 % compared to the transmitters feedback intensity. For this condition, we achieve excellent synchronization of the intensity time series of the lasers [13]. We find that the relative cavity phase $\Phi_{rel} = \Phi_r - \Phi_t$ is a key parameter determining the intensity dynamics of the receiver. In this contribution we concentrate on the influence of Φ_{rel} by varying Φ_r. In Figure 2 we present intensity time series of the coupled systems for three dynamical regimes corresponding to different relative cavity phase conditions. In Fig. 2 a) $\Phi_{rel} = 0\,\pi$, in c) $0.7\,\pi$, and $1.4\,\pi$ in d). The time series of the transmitter

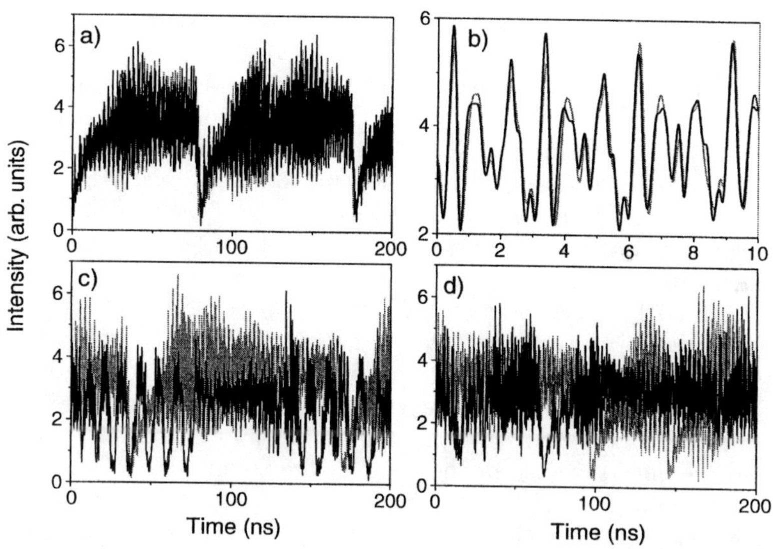

FIGURE 2. Intensity time series of the transmitter (grey solid line) and the receiver (black solid line) for various relative cavity phases: $\Phi_{rel} = 0\,\pi$ rad a), $0.7\,\pi$ rad c), and $1.4\,\pi$ rad d). For ease of comparison the time lag of the receiver time series $\tau_c = 2.9\,ns$ has been compensated for. Figure b) depicts a 10 ns zoom of a). The parameters for the transmitter are kept constant for the different phase conditions.

and the receiver are represented by grey solid and black solid lines, respectively. Fig. 2 a) depicts the time series for optimized phase condition $\Phi_{rel} = 0\,\pi$. For this regime we observe excellent synchronization of the intensity dynamics of the transmitter and the receiver. We obtain a maximum cross correlation coefficient of 0.90 of the time series, if the time series of the receiver is shifted forward in time by the coupling time $\tau_c = 4.6$ ns [8,13]. For ease of comparison, the lag of the time series of the receiver has been compensated for in Fig. 2. We monitor the rf and the optical spectra of the transmitter and the receiver and find remarkable correspondence to each other, thus confirming the synchronization. Comparison of Fig. 2 a) and its 10 ns zoom Fig. 2 b), reveals that the time series are almost identical not only with respect to the intensity dropouts, but even more on the dynamically relevant sub-ns time scales. We find that the synchronization is independent of the coupling time τ_c and robust against small variations of the injection current and the coupling strength. Next we vary the relative cavity phase from $\Phi_{rel} = 0$ to $\Phi_{rel} = 2\,\pi$ and monitor the influences on the the emission dynamics of the receiver. We observe intermittent loss of synchronization with still high correlated intervals of the time series, leading to slowly decreasing correlation coefficients. Reaching $\Phi_{rel} = 0.7\,\pi$, we find drastic changes in the intensity dynamics of the receiver. Figure 2 c) depicts the intensity time series for this striking regime. The still highly correlated intervals have vanished and we observe well pronounced intensity oscillations in combination with a strong decrease of the correlation coefficient down to 0.2. For this regime, the optical spectra of the transmitter and the receiver show deviations in their relative intensities of the longitudinal modes. Increasing Φ_{rel} further to $\Phi_{rel} = 1.4\,\pi$, the strong oscillations in the intensity dynamics of the receiver vanish, and the transmitter and receiver seem to run independently. The corresponding time series are depicted in Figure 2 d). The correlation coefficient for the time series has decreased to 0.1. Further increasing Φ_{rel} up to $2\,\pi$, we regain synchronization passing through a steep increase of the correlation coefficients. We find that Φ_{rel} is a $2\,\pi$ periodic parameter for variations of the cavity length within a range of several wavelengths. In order to get more detailed insight into the properties of the coupling conditions, we measure the intensity at the photodetector PD_B. We find low intensities for the synchronization regime, due to constructive interference of the coupling field and the feedback field of the receiver, hence the lasers couple coherently, despite of their fast wavelength-fluctuations. For low correlated states, corresponding to $1.2\,\pi < \Phi_{rel} < 1.6\,\pi$, we neither find constructive nor destructive interference effects, hence in these states the coupling field and the receiver feedback field are incoherent. An overview over the the synchronization scenario depending on Φ_{rel} represented by the cross correlation coefficient for the intensity time series of the transmitter and receiver is depicted in Fig. 3. We note that the figure exhibits asymmetries for the correlation coefficients in dependence of Φ_{rel}, since the coefficients for Φ_{rel} differ significantly from those for $\Phi'_{rel} = 2\,\pi - \Phi_{rel}$.

In order to verify the existence of a stable synchronization manifold and to distinguish the observed behaviour from linear amplification, we apply a perturbation to the DC pump current of the transmitter by adding a small sinosoidal AC mod-

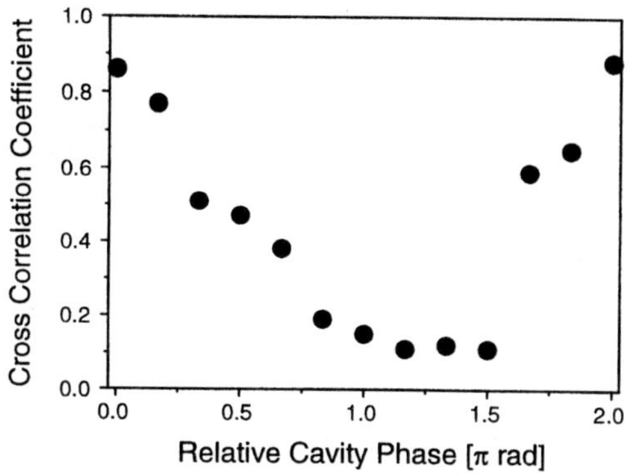

FIGURE 3. Cross correlation coefficients of the intensity time series of the transmitter and receiver with respect to the relative cavity phase Φ_{rel}.

ulation of approximately 1 % of I_{th}^{sol}. We analyze the behaviour by monitoring the rf spectra of both lasers. For a stable synchronization manifold, the system should exhibit chaos pass filtering properties [8], which means that a small perturbation superimposed onto the driving system is suppressed by the driven system. Figure 4 depicts the rf spectra of the transmitter and the receiver, represented by the grey and the black solid line, respectively. For clarity the spectrum of the transmitter is shifted vertically. The broadened peaks in the spectra correspond to the chaotic emission of the SL systems and are remarkably similar for the transmitter and the receiver. In this experiment the frequency of modulation has been chosen to 300 MHz. The peak, corresponding to the modulation frequency of the transmitter, is suppressed by 10 dB in the receiver. We could achieve up to 20 dB signal suppression for frequencies up to 2 GHz. We note that the suppression depends on the

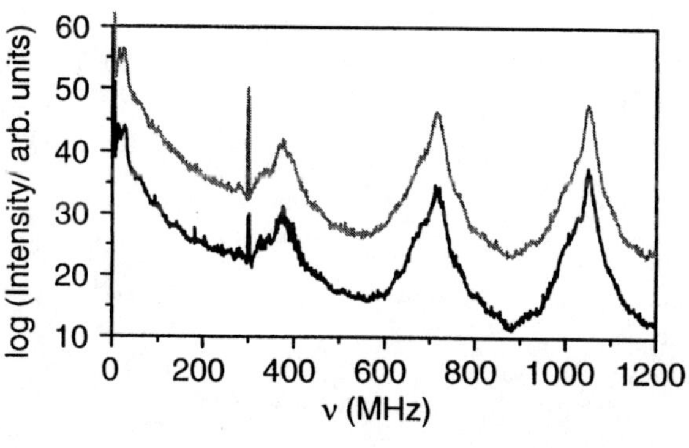

FIGURE 4. rf spectra of the intensity time series of the transmitter (grey line) and the receiver (black line). Frequency of pump current modulation for the transmitter is 300 MHz.

frequency of modulation. These results show that the synchronization manifold is stable. For Φ_{rel} deviating significantly from zero, we did not observe signal suppression, as synchronization of the lasers brakes down. This combination of excellent signal suppression at high frequencies, plus the high dimensional chaotic dynamics are very interesting for utilizing them in a high-bit rate encrypted communication system based on chaotic carriers.

SUMMARY

We have reported chaos synchronization for two unidirectionally optically coupled external cavity semiconductor lasers. We have classified four regimes of receiver's intensity dynamics depending on the relative cavity phases, i.e. synchronization, intermittent synchronization, large intensity oscillations, and uncorrelated intensity dynamics of the lasers. We find coherent emission of the optical fields of the two lasers for the synchronized state, despite of the fast wavelength-fluctuations. In contrast, for low correlated states the lasers are incoherent. Finally, we have demonstrated highly selective chaos pass filtering properties for our experiment, which make our receiver system attractive for the realization of innovative concepts for encrypted high-bit data communication utilizing chaos synchronization phenomena.

REFERENCES

1. Glass, L., *Nature* **410**, 277 (2001).
2. Coffman, K., McCormick, W.D., and Swinney, H.L., *Phys. Rev. Lett.* **56**, 999 (1986).
3. Schäfer, C., Rosenblum, M.G., Kurths, J., and Abel, H.H., *Nature* **392**, 239 (1998).
4. Fischer, I., Heil, T., and Elsäßer, W., in *Fundamental Issues of Nonlinear Laser Dynamics*, edited by B. Krauskopf, and D. Lenstra, AIP Conference Proceedings **548**, 218, Melville, New York (2000).
5. Ahlers, V., Parlitz, U., and Lauterborn, W., *Phys. Rev. E* **58**, 7208 (1998).
6. Van Wiggeren, G.D., and Roy, R., *Phys. Rev. Lett.* **81**, 3547 (1998).
7. Mirasso, C.R., Colet, P., and García-Fernández, P., *IEEE Photonics Technology Letters* **8**, 299 (1996).
8. Fischer, I., Liu, Y., and Davis, P., *Phys. Rev. A* **62**, 011801(R) (2000).
9. Pecora, L., *Physics World* **4**, 25 (1998).
10. Larger, L., Goedgebuer, J.-P., and Delorme, F., *Phys. Rev. E* **57**, 6618 (1998).
11. van Tartwijk, G.H.M., Levine, A.M., and Lenstra, D., *IEEE J. Sel. Top. Quantum Electron.* **Vol. 1, No. 2**, 466 (1995).
12. Mørk, J., Tromborg, B., Mark, J., *IEEE J. Quantum Electron.* **QE-28**, 93 (1992).
13. H. Fujino and J. Ohtsubo, *Opt. Lett.* **25**, 625 (2000).

Noise Robust Synchronized Chaotic Circuits

Thomas L. Carroll

US Naval Research Lab, Wsshington, DC 20375 USA

Abstract. The synchronization of chaotic circuits is greatly degraded by noise, so that for noise as large or larger than the synchronizing signal, no useful synchronization is possible. I demonstrate here a set of synchronized chaotic circuits that is robust to additive white noise. For any noise level, it is possible to achieve arbitrarily small synchronization error by properly changing one circuit parameter.

INTRODUCTION

While chaotic synchronization has been proposed as a method of spread spectrum communication [1-11], self-synchronizing chaotic systems suffer from a problem common to all self-synchronizing communication systems: since the synchronizing signal is also the carrier signal, any noise present during transmission contaminates the synchronizing signal and degrades synchronization [12].

The circuit described in this paper reduces the noise added to the synchronizing signal by averaging the chaotic signal over a long time scale. The circuit actually has dynamics at two distinct time scales. The synchronizing signal drives the faster of the two time scales while the slower time scale acts to average out the noise.

THE CIRCUIT

The basic idea behind this circuit was suggested by experimental work involving ferromagnetic resonance in yttrium iron garnet [21-24]. When driven with the proper microwave signal, a sample of yttrium iron garnet displays dynamics at two different time scales. The driving signal is in the gigahertz range, but the yttrium iron garnet may display chaotic dynamics in the kilohertz range. If a circuit with two different time scales was constructed, then noise contaminating the high frequency signals might be averaged out in the low frequency part of the circuit.

Such a circuit has been built. This circuit is based on coupling together two Piecewise Linear Rossler circuits [25] with different frequencies. Mathematically, the circuit may be described by

CP622, *Experimental Chaos: 6th Experimental Chaos Conference,* edited by S. Boccaletti et al.
2002 American Institute of Physics 0-7354-0071-7

$$
\begin{aligned}
\frac{dx_1}{dt} &= -\frac{1}{RC_1}\left(\gamma_1 x_1 + 0.5x_2 + x_3 + 0.5|x_4|\right) \\
\frac{dx_2}{dt} &= -\frac{1}{RC_1}\left(-x_1 + \gamma_2 x_2 + x_6\right) \\
\frac{dx_3}{dt} &= -\frac{1}{RC_1}\left(-g(x_1) + x_3\right) \\
\frac{dx_4}{dt} &= -\frac{1}{RC_2}\left(x_1 + 0.05x_4 + 0.5x_5 + x_6\right) \\
\frac{dx_5}{dt} &= -\frac{1}{RC_2}\left(-x_4 + 0.11x_5\right) \\
\frac{dx_6}{dt} &= -\frac{1}{RC_2}\left(-g(x_4) + x_6\right) \\
g(x) &= \begin{Bmatrix} 0 & x < 3 \\ 15(x-3) & x \geq 3 \end{Bmatrix}
\end{aligned} \tag{1}
$$

where γ_1 and γ_2 may be varied, $R = 10^5\ \Omega$, and C_1 and C_2 may be varied to alter the relative time scales of the two parts of the circuit. The signals x_1 through x_3 form the low frequency part of the circuit, while x_4 through x_6 form the high frequency part.

Figure 1 shows different projections of the attractor for the circuit when $C_1 = 0.01$ μF , $C_2 = 0.001$ μF, $\gamma_1 = 0.05$, and $\gamma_2 = 0.02$, so that the time scales are separated by a factor of 10. Figure 1(a) is a plot of x_2 vs. x_1 , while 1(b) shows x_5 vs. x_4 . Figure 2 shows power spectra of two different signals from the circuit. Figure 2(a) shows the power spectrum of x_1 , while 2(b) shows the power spectrum of x_4 . The higher frequency peak in the power spectrum, most prominent in Fig. 2(b), is at 1110 Hz. The lower frequency peak in the power spectrum, largest in Fig. 2(a), is at 111 Hz.

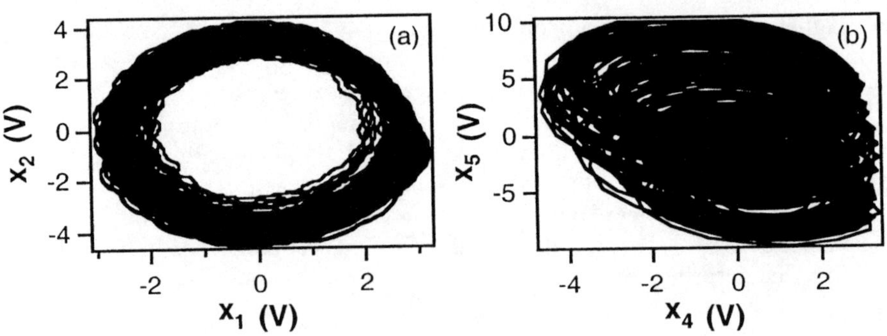

FIGURE 1. Two projections of the attractor for the circuit described by eq. (1) when $C_1 = 0.01$ μF and $C_2 = 0.001$ μF (data from the circuit). (a) x_2 vs. x_1 (lower frequency part) (b) x_5 vs. x_4 (higher frequency part).

Choosing different capacitors altered the relative frequencies of the main peaks in the power spectrum of the circuit. If $C_1 = 0.1$ μF, $C_2 = 0.001$ μF, $\gamma_1 = 0.02$, and $\gamma_2 = 0.02$, the high frequency peak was in the same location, but the low frequency peak was lower by a factor of 10 than the low frequency peak in Fig. 2.. The peak frequency in x_4 is still at 1110 Hz, but the peak frequency in x_1 is now at 11.1 Hz.

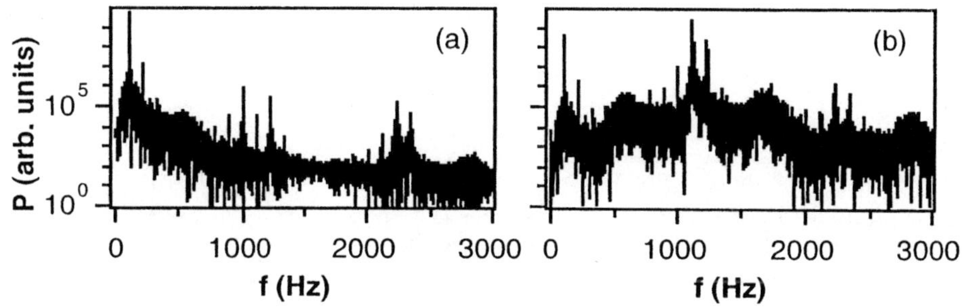

FIGURE 2. (a) Power spectrum P for the x_1 signal from the circuit when $C_1 = 0.01$ μF and $C_2 = 0.001$ μF. (b) Power spectrum P for the x_4 signal from the circuit when $C_1 = 0.01$ μF and $C_2 = 0.001$ μF.

SYNCHRONIZATION

A second circuit was built that closely matched the drive circuit described by eq. (1). A linear combination of signals from the drive circuit was used to synchronize the response circuit. The response system was described by

$$x_t = \sum_{i=1}^{6} k_i x_i^d \qquad x_r = \sum_{i=1}^{6} k_i x_i^r$$

$$\begin{aligned}
\frac{dx^r{}_1}{dt} &= -\frac{1}{RC_1}\left(\gamma_1 x^r{}_1 + 0.5x^r{}_2 + x^r{}_3 + 0.5\left|x^r{}_4\right|\right) \\
\frac{dx^r{}_2}{dt} &= -\frac{1}{RC_1}\left(-x^r{}_1 + \gamma_2 x^r{}_2 + x^r{}_6\right) \\
\frac{dx^r{}_3}{dt} &= -\frac{1}{RC_1}\left(-g\left(x^r{}_1\right) + x^r{}_3\right) \\
\frac{dx^r{}_4}{dt} &= -\frac{1}{RC_2}\left(x^r{}_1 + 0.05x^r{}_4 + 0.5x^r{}_5 + x^r{}_6 + b_4\left(x_t - x_r\right)\right) \\
\frac{dx^r{}_5}{dt} &= -\frac{1}{RC_2}\left(-x^r{}_4 + 0.11x^r{}_5 + b_5\left(x_t - x_r\right)\right) \\
\frac{dx^r{}_6}{dt} &= -\frac{1}{RC_2}\left(-g\left(x^r{}_4\right) + x^r{}_6 + b_6\left(x_t - x_r\right)\right)
\end{aligned} \tag{2}$$

where the superscript d refers to the drive circuit and r refers to the response circuit. Note that the error signal $x_t - x_r$ is fed back only into the high frequency part of the circuit, x_4 to x_6 .

Table 1

i	k_i	b_i
1	-4.59	
2	5.61	
3	3.16	
4	-0.79	1.11
5	-0.21	-0.38
6	0.36	0.33

The parameters k_i and b_i are set to minimize the largest Lyapunov exponent for the response circuit corresponding to eq. (2) [26,27]. The k_i 's and b_i 's are varied by a linear optimization routine in order to minimize the largest Lyapunov exponent for the response circuit. For the parameters listed in Table 1, the largest Lyapunov exponent for the response circuit was -1160 s^{-1} (when C_1 = 0.01 μF and C_2 = 0.001 μF). There are many other possible combinations of the k's and b's that give similar Lyapunov exponents.

NOISE EFFECTS

The most interesting feature of this circuit is that by varying the ratio between C_1 and C_2 (varying the ratio between the frequencies of the two parts of the circuit), the amount of synchronization error caused by noise could be varied. To test this feature, white noise was added to the transmitted signal x_t (defined in eq. (2)). The signal to noise ratio (S/N) was calculated by dividing the RMS amplitude of x_t by the RMS amplitude of the white noise. The synchronization error for the low frequency part of the circuit, $\delta(x_1)$, was calculated by dividing the RMS amplitude of $x^d_1 - x^r_1$ by the RMS amplitude of x^d_1 , where the superscript d refers to the drive circuit and r refers to the response circuit.

Figure 3 shows the synchronization error as a function of S/N for two different ratios of C_1 and C_2 . The filled-in circles show the synchronization error when C_1 = 0.01 μF and C_2 = 0.001 μF, so that the fast frequency in the circuit is 1110 Hz and the slow frequency is 111 Hz (corresponding to the power spectra of Fig. 2). The synchronization error at a S/N of 1 (0 dB) is about 0.07, climbing to about 0.14 at a S/N of 0.33 (-4.8 dB). The minimum synchronization error at high S/N is about 0.02 because of mismatch between the circuits.

The open squares in Fig 3 show the synchronization error when C_1 = 0.1 μF and C_2 = 0.001 μF, so that the fast frequency in the circuit is 1110 Hz and the slow frequency is 11.1 Hz. For this larger frequency ratio, the synchronization error at a S/N of 1 is about 0.035, while at a S/N of 0.33 the error is about 0.07.

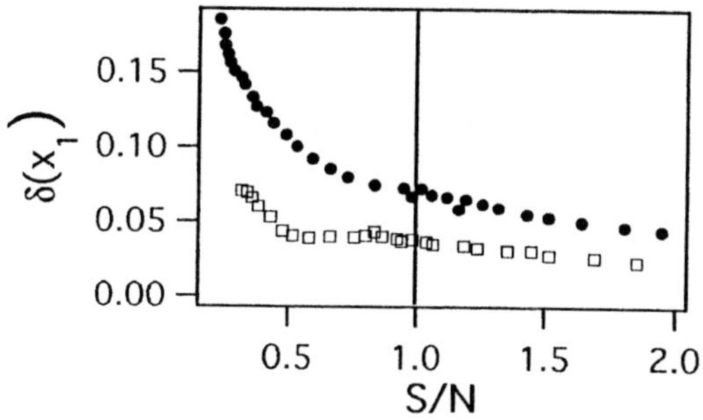

FIGURE 3. RMS synchronization error $\delta(x_1)$ from the circuit as a function of RMS signal to noise ratio S/N. (a) $C_1 = 0.01\ \mu F$ and $C_2 = 0.001\ \mu F$. (b) $C_1 = 0.1\ \mu F$ and $C_2 = 0.001\ \mu F$

Clearly, when the frequency difference between the main frequencies in the circuit is larger, the synchronization error is smaller (the synchronization error for the higher frequency x_4 signals in the circuit was larger than the synchronization for the low frequency signals). The circuit acts as its own filter, so that the noise which contaminates the transmitted signal x_t is averaged out by the x_1 to x_3 part of the circuit. The relative improvement in synchronization error will depend on the specific circuit as well as the ratio of high to low frequencies, but by increasing the frequency ratio it should be possible to make the synchronization error smaller.

CONCLUSIONS

In previous synchronous chaotic circuits, there was a minimum S/N below which useful synchronization was not possible. In this circuit, the synchronization error can always be made smaller by making the ratio between fast and slow frequencies larger. Increasing the ratio of high to low frequencies is equivalent to averaging the received signal over a longer time in order to average out noise.

This frequency separation puts the chaotic system on a similar footing with other simple linear communications systems; signal averaging to remove a signal from noise is now a possibility for self-synchronizing chaotic systems as well as linear periodic systems. While this development in itself does not necessarily make chaotic systems practical for communications, it does give us access to a common noise reduction tool.

REFERENCES

[1]L. M. Pecora and Thomas L. Carroll, "Synchronization in Chaotic Systems," Physical Review Letters **64** (Feb.), 821-824 (1990).

[2]T. L. Carroll and L. M. Pecora, "Synchronizing Chaotic Circuits," IEEE Transactions on Circuits and Systems **38** (April), 453-456 (1991).

[3]K. M. Cuomo, A. V. Oppenheim, and S. H. Strogatz, "Synchronization of Lorenz-Based Chaotic Circuits with Applications to Communications," IEEE Transactions on Circuits and Systems **40** (Oct.), 626-633 (1993).
[4]H. Dedieu, M. P. Kennedy, and M. Hasler, "Chaos Shift Keying: Modulation and Demodulation of a Chaotic Carrier Using self Synchronizing Chua's Circuits," IEEE Transactions on Circuits and Systems **40** (Oct.), 634-642 (1993).
[5]T. L. Carroll and L. M. Pecora, "Cascading Synchronized Chaotic Systems," Physica D **67** (Aug.), 126-140 (1993).
[6]L. Kocarev, K. S. Halle, K. Eckert et al., "Experimental Demonstration of Secure Communications via Chaotic Synchronization," International Journal of Bifurcations and Chaos in Applied Science and Technology **2** (Sept.), 709-713 (1992).
[7]G. Kolumban, M. P. Kennedy, and L. O. Chua, "The Role of Synchronization in Digital Communications Using Chaos-Part I: Fundamentals of Digital Communications," IEEE Transactions on Circuits and Systems-Part I: Fundamental Theory and Applications **44** (10), 927-935 (1997).
[8]K. Murali and M. Lakshmanan, "Transmission of signals by synchronization in a chaotic Van der Pol-Duffing oscillator," Physical Review E **48** (Sept.), 1624-1626 (1993).
[9]U. Parlitz, L. Kocarev, T. Stojanovski et al., "Encoding messages using chaotic synchronization," Physical Review E **53** (May), 4351-4361 (1996).
[10]N. F. Rul'kov, A. R. Volkovski, A. Rodriguez-Lozano et al., "Synchronous chaotic behavior of a response oscillator with chaotic driving," Chaos, Solitons and Fractals **4** (#2), 201-211 (1994).
[11]L. S. Tsimring and M. M. Sushchick, "Multiplexing chaotic signals using synchronization," Physics Letters A **213** (3-4), 155-166 (1996).
[12]R. Brown, N. F. Rulkov, and N. B. Tufillaro, "Synchronization of Chaotic Systems: The Effects of Additive Noise and Drift in the Dynamics of the Driving," Physical Review E **50** (6), 4488-4508 (1994).
[13]E. J. Kostelich and J. A. Yorke, "Noise Reduction in Dynamical Systems," Physical Review A **38** (3), 1649-1652 (1988).
[14]R. Cawley and G. H. Hsu, "Local Geometric projection Method for Noise Reduction in Chaotic Maps and Flows," Physical Review A **46** (6), 3057-3082 (1992).
[15]E. Rosa, S. Hayes, and C. Grebogi, "Noise Filtering in Communication with Chaos," Physical Review Letters **78** (7), 1247-1250 (1997).
[16]T. Sauer, "A Noise Reduction Method for Signals from Nonlinear Systems," Physica D **58** (1-4), 193-201 (1992).
[17]T. L. Carroll, "Approximating chaotic time series through unstable periodic orbits," Physical Review E **59** (2), 1615-1621 (1999).
[18]T. L. Carroll, "Synchronizing Chaotic Systems Using Filtered Signals," Phys Rev. E **50** (Oct.), 2580-2587 (1994).
[19]T. L. Carroll, "Communicating using Filtered Synchronized Chaotic Signals," IEEE Transactions on Circuits and Systems **42** (Feb.), 105-110 (1995).
[20]N. F. Rulkov and L. Tsimring, "Communication with Chaos over Band-Limited Channels," International Journal of Circuit Theory and Applications **27** (6), 555-567 (1999).
[21]G. Gibson and C. Jeffries, Physical Review A **29**, 811 (1984).
[22]X. Y. Zhang and H. Suhl, Physical Review A **32**, 2530 (1985).
[23]S. M. Rezende, O. F. de Alcantra Bonfim, and F. M. de Aguiar, Physical Review B **33**, 5153 (1986).
[24]T. L. Carroll, L. M. Pecora, and F. J. Rachford, "Chaos and Chaotic Transients in an Yttrium Iron Garnet Sphere," Physical Review A **40** (1), 377-386 (1989).
[25]T. L. Carroll, "A simple circuit for demonstrating regular and synchronized chaos," American Journal of Physics **63** (April), 377-379 (1995).
[26]J. H. Peng, E. J. Ding, M. Ding et al., "Synchronizing Hyperchaos with a Scalar Transmitted Signal," Physical Review Letters **76** (Feb), 904-907 (1996).
[27]G. A. Johnson, D. J. Mar, T. L. Carroll et al., "Synchronization and imposed bifurcations in the presence of large parameter mismatch," Physical Review Letters **80** (18), 3956-3959 (1998).

DYNAMO

Liquid Sodium Experiments: The Effect of Turbulence and Lorentz Forces

Daniel R. Sisan, Woodrow L. Shew, and Daniel P. Lathrop

Dept. of Physics, University of Maryland, College Park
College Park, MD 20742 USA

Abstract. We are pursuing several liquid sodium experiments in order to understand magnetic field generation in astrophysical bodies. In one experiment, pulse decay techniques using small externally applied magnetic fields quantify the system's approach toward a dynamo. Using the same apparatus, a large constant external magnetic field was applied modifying the flow through the Lorentz force. The induced field fluctuations indicate four distinct regimes that were uniquely quantified by the Elasasser number, the ratio of the Lorentz force to the advective force. Two future experiments are also described briefly.

INTRODUCTION

There is a large body of theoretical and numerical research suggesting that Earth and other astrophysical bodies with magnetic fields are dynamos [1]. Indeed, the equations governing flowing liquid conductors in the presence of magnetic fields (specifically, the Navier-Stokes equation and the Maxwell equations) have been known for years, and many researchers have demonstrated analytical and numerical self-generating solutions [2,3,4]. However, the parameter values used to obtain these self-generating solutions differ substantially from the parameter values of astrophysical dynamos. Approximations are necessary in theoretical and numerical models because, at relevant parameter values, the fluid motion is highly turbulent, making the solutions to these equations very difficult to resolve to appropriate length scales. Experimental models, on the other hand, do not have this resolution restriction—"calculations" can be performed in seconds with highly turbulent experiments that would take decades, or even centuries, to perform with current computer technology.

So liquid metal experiments are an excellent complement to theoretical and numerical work, and they have the potential to provide a better understanding of magnetic field generation. Nonetheless, only a few research groups have attempted, or are attempting, this type of experiment. (A significant proportion of the groups are, in fact, represented at this conference.)

In this paper we report results of, and plans for, several liquid sodium experiments at the University of Maryland. The results come from a 30 cm diam. mechanically forced experiment. In this system, measurements were taken using pulsed external magnetic fields (100 G) and constant external magnetic fields (up to 2000 G). Our

CP622, *Experimental Chaos: 6th Experimental Chaos Conference*, edited by S. Boccaletti et al.

future plans include a 60 cm dia. rotating convection experiment, and a 3 m dia. experiment that will be mechanically forced and rotating.

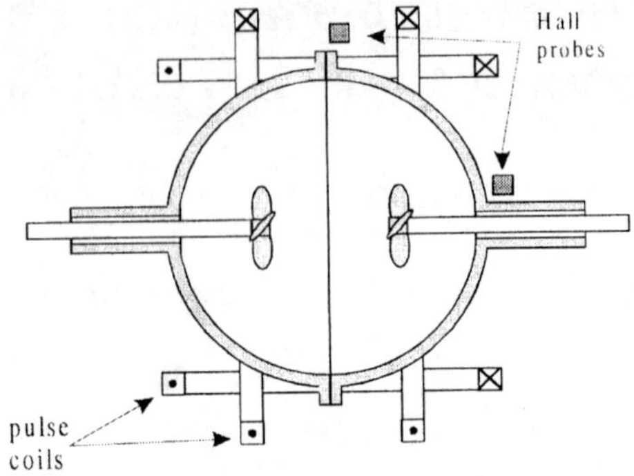

FIGURE 1. Cross-section of the 30 cm dia. experiment. Rotating shafts attached to propellors drive the flow. Hemholtz coil pairs supply external magnetic fields along, or perpendicular to, the rotation axis.

THE APPROACH TO SELF-GENERATION

The device of the 30 cm dia. experiment (Fig. 1) is a hollow stainless steel sphere filled with liquid sodium [5]. Two propellors, powered independently by 7.5 kW motors, drive the flow. Hall effect probes outside the sphere measure the magnetic field.

This system does not self-generate, but, using a pulse decay technique, a shift toward self-generation is observed. In this technique, an external magnetic field is applied and then rapidly turned off. The magnetic field decays exponentially due to induced currents in the liquid sodium (see Fig. 2). With no fluid motion, the decay constant was within 3% of the theoretical value for field diffusion in a conducting sphere. With the propellors spinning, however, there may be fluid motion that enhances field growth, in which case the decay constant becomes less negative. If the fluid motion enhances the field strongly enough to overcome resistive damping, then the system would self-generate. The competition between resistive decay and the stretching properties of the fluid motion is quantified by the magnetic Reynolds number, $\mathrm{Re}_m = UL/\eta$, where U is a characteristic velocity, L is a characteristic length, and η is the magnetic diffusivity, the inverse of the product of conductivity and magnetic permeability. We define U as the tip speed of the propellor, and L as the radius of the sphere. A typical applied magnetic pulse is shown in Fig. 2(a). The decay constant, measured as in Fig. 2(b), depends on the magnetic Reynolds number. For external fields aligned with the axis of rotation, the decay constant shifts toward self-generation, as seen in Fig. 2(d). Available power limits the magnetic Reynolds number that can be obtained practically.

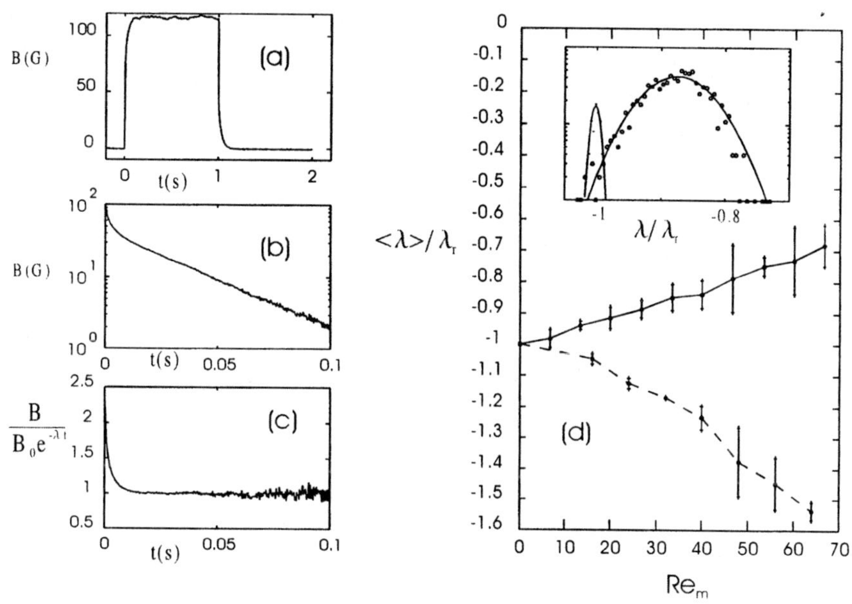

FIGURE 2. A pulse decay measurement estimates the largest decay rate of the system and shows a trend toward self-generation for external fields aligned with the shafts. A Hall probe signal shows the external magnetic field pulse (a) used to estimate the decay rate (the slope of the semilog plot in (b)). (c) shows the same data as in (b) divided by the exponential function. The mean decay rate is plotted in (d). The solid line, showing shifts toward self-generation, is for fields aligned with the shafts and the dotted line, showing shifts away from self-generation, is for fields perpendicular to the shafts. The error bars indicate shot-to-shot fluctuations in the decay constants. The distribution of decay constants is shown in the inset of (d); the narrow distribution is for no flow and the broad distribution is for $Re_m=42$. The decay rates have been normalized to yield –1 for the purely resistive case (no flow).

The distribution of decay constants shown in the inset plot of Fig. 2(d) leads to a prediction that the transition to self-generation will exhibit intermittent bursts in magnetic fields. This would occur as the mean decay rate nears zero (at larger Re_m)—flow states associated with the positive side of this distribution should lead to growth of the magnetic field. As the field grows the Lorentz force will cause a back-reaction on the flow, suppressing the growth, if the system is non-hysteretic. Thus, just below transition the system will have short bursts of growing fields followed by periods of decaying fields [6].

FLUCTUATIONS

While the external field is applied, fluctuations are observed in the induced field (Fig. 3). This external field (100 G) is too small to cause a back reaction on the flow, and the magnetic field is a passive vector modified by the fluid. Thus, the fluctuations in the induced field reflect the fluctuations of the velocity in the underlying fluid turbulence. Using Taylor's hypothesis, Fig 3(b) shows that, indeed, the system contains Kolmogorov turbulence—not surprising for our system which has a (hydrodynamic) Reynolds number of order 10^7.

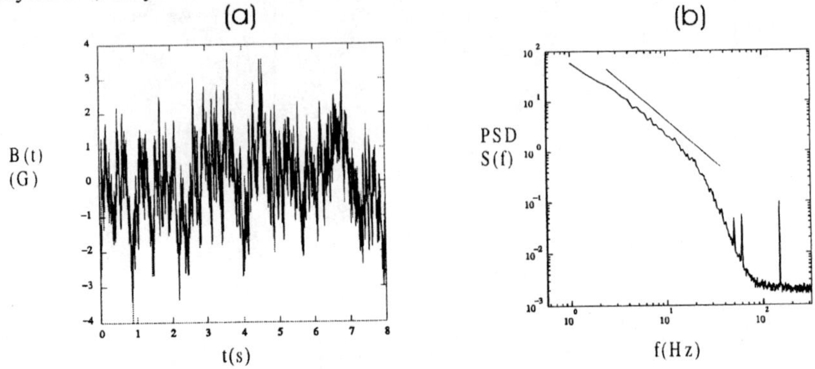

FIGURE 3. Typical field fluctuations. A time series is shown in (a) with the mean applied field (120 G) and the mean induced field (-17 G) subtracted. The power spectra (b) shows features of broadband turbulence including a power law decay at low frequency. The offset line represents a best fit exponent, $f^{-1.69\pm0.03}$, consistent with Kolmogorov scaling.

THE EFFECT OF THE LORENTZ FORCE

The field produced by the coils used in our pulse decay measurements is too small to have a significant back reaction on the fluid motion. In order to better understand the saturation process we installed a magnet system that produces fields (2 kG) with magnetic energy comparable to the kinetic energy of the flow, allowing the Lorentz force to influence the fluid motion. As we increased the field strength, we observed qualitatively distinct changes in the system's response, as shown in Fig. 4.

For the Lorentz force to affect the system it needs to overcome the inertia of the flow, thus a logical parameter quantifying the relative strength of the field is the Elsasser number, the ratio of the Lorentz force to the advective force (sometimes called the Interaction parameter). We define the Elsasser number as:

$$\Lambda = \frac{B_0^2 b}{\rho \mu_0 \eta \Omega a} \tag{1}$$

where B_0 is the externally applied field, Ω is the rotation rate, b is the sphere radius, a is the propellor radius, μ_0 is the permeability of free space (and sodium), ρ the density of sodium, and η is the magnetic diffusivity of sodium. The Elsasser number for Earth is thought to be order one [4]. The Elsasser number for a general self-generating system depends on the effect the Lorentz forces have on the dynamics, which, with this experiment, we are examining.

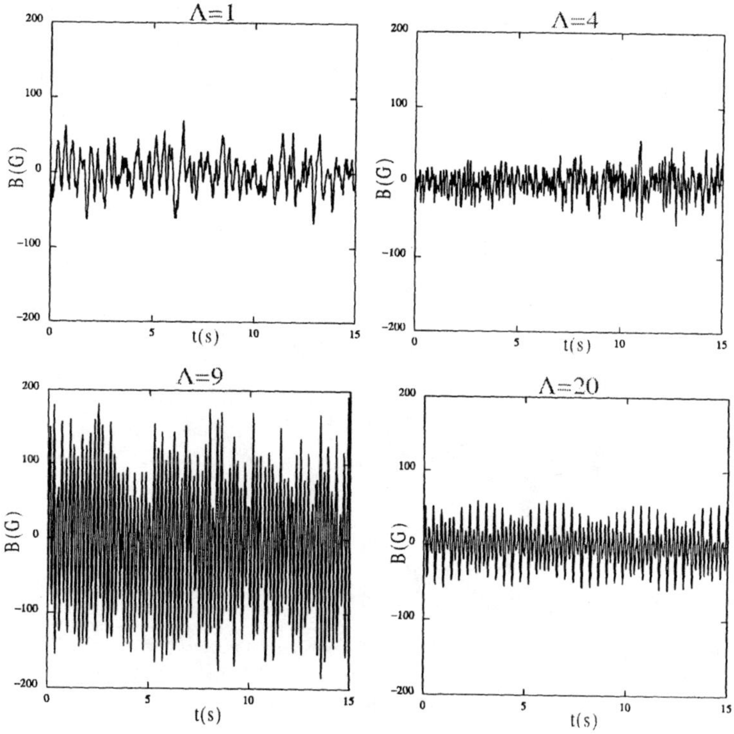

FIGURE 4. The induced field passes through four distinct regimes as the externally applied magnetic field is increased. For Λ=1, the external magnetic field is too weak to alter the fluid motion and the magnetic field is a passive vector; the induced field fluctuations reflect the underlying fluid turbulence. For Λ=4, the magnetic field suppresses turbulence in the system, seen in the decreased magnitude of the induced field fluctuations. For Λ=9, large oscillations emerge; the frequency of oscillation increases linearly with applied field strength. Finally, for Λ=20 the oscillations decrease again in magnitude and become cleaner; the oscillations consist of three sharp frequencies—the dominant frequency at ½ the propellor rotation rate and two sidebands near ¼ and ¾ the rotation rate—that are independent of applied field strength.

FUTURE EXPERIMENTS

We are close to completing construction of a new 60 cm dia. rotating convection experiment, shown in Fig. 5. This experiment has a spherical geometry that is similar to the geometry in the Earth's outer core. Sodium fills the region between the smaller inner sphere, which is cooled with hexane, and the larger spherical shell, which is heated from the outside with halogen heat lamps. Centrifugal forces and the radial temperature difference drive the flow. We are collaborating with Gary Glatzmaier at the University of California, Santa Cruz, who is numerically simulating the experiment. This comparison between simulation and experiment will serve as a benchmark for the simulation, which, as with simulations of dynamos in general, includes necessary approximations.

Another, much larger experiment (3 m diameter), is in the planning stages. The linear trend of Fig. 2(d) suggests that self-generation may occur at larger values of Re_m for a similar geometry. For this planned experiment, Re_m up to 400 can be obtained.

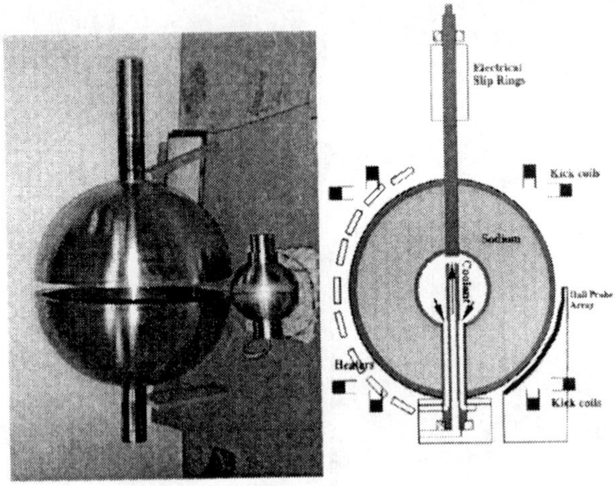

FIGURE 5. Basic structures of the 60 cm diam. rotating convection experiment. The smaller sphere in the photo fits inside the larger shell to create the geometry in the cross-sectional drawing on the left. The system will have a maximum rotation rate of 100 RPS and a maximum heat flux of 10 kW.

ACKNOWLEDGMENTS

This work was funded by the NSF Earth Sciences-Geophysics and the NSF Instrumentation Development programs, the Research Corporation, and the University of Maryland. We gratefully acknowledge assistance from Thomas Antonsen, A.B. Cawthorne, James Drake, Khurrum Ghulani, Adil Hassam, Donald Martin, Edward Ott, and Nicholas L. Peffley.

REFERENCES

1. P.H. Roberts, "Fundamentals of Dynamo Theory" In Lectures on Solar and Planetary Dynamos, ed. M.R.E. Proctor, A.D. Gilbert. 1-58. Cambridge, UK: Cambridge Univ. Press, 1994.
2. F. H. Busse, "Homogeneous Dynamos in Planetary Cores and in the Laboratory," Annu. Rev. Fluid Mech. 2000, **32**:383-408.
3. M.L. Dudley and R.W. James, "Time-dependent kinematic dynamos with stationary flows," Proc. R. Soc. Lond. A **425**, 407 (1989).
4. E. Dormy, J.-P. Valet, and V. Courtillot, "Numerical models of the geodynamo and observation constraints," Geochem. Geophys. Geosys., vol. 1 (2000).
5. N.L. Peffley, A.B. Cawthorne, and D.P. Lathop, "Toward a Self-generating Magnetic Dynamo: the Role of Turbulence," Phys. Rev. E **61**, 5287 (2000).
6. D. Sweet, E. Ott, J.M. Finn, T.M. Antonsen, Jr. and D.P. Lathrop, "Blowout Bifurcations and the Onset of Magnetic Activity in Turbulent Dynamos," Phys. Rev. E **63**, 6621 (2001).

Water Experiments Related To The "Von Karman Sodium" Dynamo Project

L. Marié[1], M. Bourgoin[2], F. Pétrélis[3], J. Roy[1], J. Burguete[1], A. Chiffaudel[1], F. Daviaud[1], S. Fauve[3], P. Odier[2], J.-F. Pinton[2]

[1]*SPEC, CEA / SACLAY, 91191 Gif-sur-Yvette Cedex,*
[2]*ENS Lyon, 46 allée d'Italie, 69364 Lyon*
[3]*ENS, 25 rue Lhomond, 75005 Paris*

Abstract. The purpose of the "Von Karman Sodium" (V.K.S.) experiment is to study the "Dynamo Effect", namely the spontaneous generation of magnetic field in a flow of electrically conducting fluid. The device has been built at CEA / Cadarache, in collaboration with CEA / Saclay, Ecole Normale Supérieure de Lyon and Ecole Normale Supérieure de Paris. It consists of a cylindrical vessel, filled with liquid Sodium, in which two coaxial rotating disks induce a Von-Karman type flow. Several experimental runs have taken place since June 2000. In order to optimize the V.K.S. set-up, a half-scale water prototype has also been built. It has allowed us to measure mean velocity profiles, as well as pressure fluctuations and mechanical power dissipation. We have observed that under certain circumstances the mean component of the turbulent flow can undergo a global bifurcation.

INTRODUCTION

The generic term "Von Karman flows" designates the class of flows induced by the rotation of two coaxial impellers located at both ends of a cylindrical vessel. These flows have been studied in the past both experimentally[6,7,9] and numerically. They are supposed to be good candidates for the realization of an experimental homogeneous fluid dynamo. In particular, kinematic dynamo simulations in a sphere[2] and direct numerical simulations of the Taylor-Green geometry[3] have shown that these flows can lead to self-excitation for accessible values of the magnetic Reynolds number $R_m = \mu\sigma LV$.

The "Von Karman Sodium" (V.K.S.) experiment uses liquid sodium as a working fluid in a cylindrical vessel with diameter 40 cm. Two co-axial counter-rotating impellers driven by two 75 kW motors generate the flow. The magnetic Prandtl number of liquid metals being extremely small ($P_m = \mu\sigma\nu \approx 10^{-5}$), one can expect any flow of liquid metal having a strong magnetic Reynolds number to also have a very large kinetic Reynolds number. In order to characterize the flow, extensive preliminary studies have been performed on a half-scale water prototype. Mean velocity profiles have been obtained and introduced in kinematic dynamo simulations of the induction equation[4,5]. In the following, we present torque and pressure fluctuation measure-

CP622, *Experimental Chaos: 6th Experimental Chaos Conference*, edited by S. Boccaletti et al.

ments obtained in the water experiment. We have observed in particular a global bifurcation of the flow topology in experimental conditions where the integral kinetic Reynolds number is of the order of 10^6.

WATER EXPERIMENTS

Experimental Setup

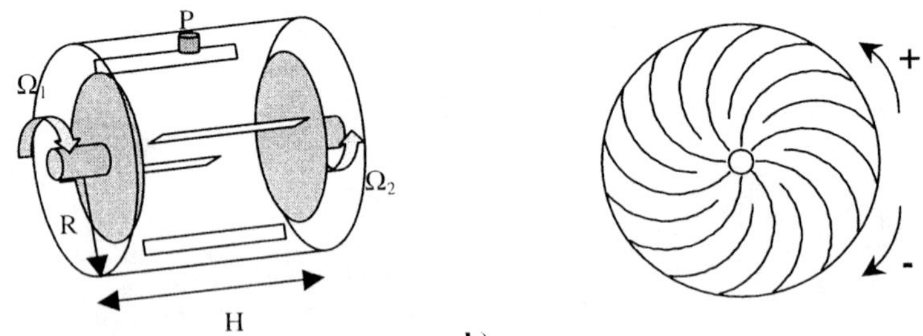

a) b)

FIGURE 1. Sketch of a) the Von Karman Apparatus b) the impellers. When the motor rotation rates are set to positive values, the impellers are rotated counter-clockwise.

The experimental setup is shown in figure 1. It consists of a cylindrical container, whose radius and height are R_c = 100 mm, H_c=200 mm. Four 10 mm high baffles can be mounted on the cylinder wall, parallel to the axis. The impellers, located at the ends, consist of 185 mm diameter plates, each fitted with 16 blades. Eight of the blades stop at r=30 mm from the plate center, while the others stop at r=10 mm. All the blades have a 50 mm curvature radius. A central hole, 15 mm in diameter, allows for securing to the motor shaft. Blades 10 mm high and 20 mm high have been used. The distance between the plates has been kept equal to H_d = 180 mm. The impellers are driven by two 2 kW ac motors. The angular velocities can be varied independently between –25 Hz and 25 Hz. The variable frequency drives are used to monitor torque and angular velocity values. A Kistler 6031 acceleration-compensated piezoelectric pressure sensor, mounted in the mid-plane of the apparatus, measures the fluctuations of the dynamical pressure. The signal is low-pass filtered at 1 kHz, then recorded at a sampling rate of 3 kHz on a PC, using a Data Translation DT3001 board.

Torque Measurements

Dimensional analysis indicates that the torque exerted by the fluid on the impellers can scale either as $T \propto \rho L^3 \nu \Omega$ or, in the limit of very large kinetic Reynolds numbers, as $T \propto \rho L^5 \Omega^2$. In these expressions, $\rho \approx 1000 \mathrm{kg \cdot m^{-3}}$ is the density of the fluid, $\nu \approx 10^{-6} \mathrm{m^2 \cdot s^{-1}}$ is its kinematic viscosity, $L = R_c = 0.1\mathrm{m}$ is a typical length-scale of the flow, and Ω is a typical frequency of the flow, which we have chosen to be equal to $(\Omega_1^2 + \Omega_2^2)^{1/2}$. Using ΩR_c as a typical value of the velocity, one can define the integral kinetic Reynolds number as $R_e = \Omega L^2 / \nu$. Even for a moderate rotation rate of 2 Hz, this yields a value of R_e well in excess of 10^5. Figure 2 shows a typical example of the variation of the torque exerted by motor 1 with Ω. One observes a very good agreement with the $T \propto \rho L^5 \Omega^2$ scaling. This has prompted us to study the dimensionless coefficients $C_1 = T_1 / \rho L^5 \Omega^2$ and $C_2 = T_2 / \rho L^5 \Omega^2$ rather than the actual torques exerted by the motors, T_1 and T_2.

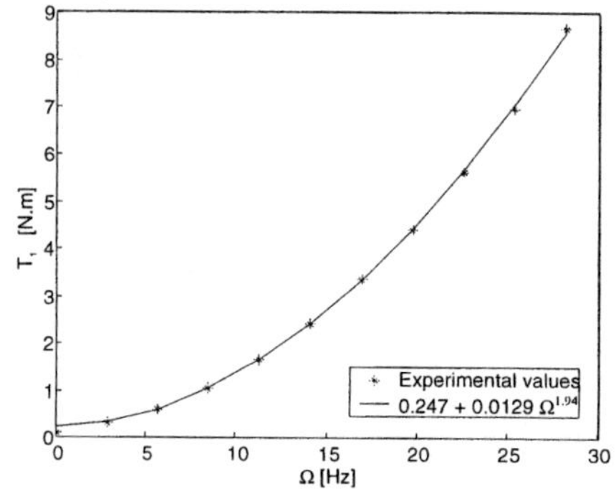

FIGURE 2. Variation of the torque exerted by the motor on the fluid, as a function of $\Omega = (\Omega_1^2 + \Omega_2^2)^{1/2}$. Here both impellers were rotated in the counter-clockwise (positive) direction, at the same rotation rate. The solid line represents the best fit obtained for a function with shape $\alpha + \beta\Omega^\gamma$. The value of the exponent, 1.94, is in good agreement with the expected value of 2. Friction torque in the bearings and sealings accounts for the 0.25 N.m offset. This value will be substracted from the torque data in all the subsequent graphs.

On Figure 3, we have represented the values of C_1 and C_2, as a function of the polar coordinate angle θ such that $(\Omega_1, \Omega_2) = (\Omega\cos(\theta), \Omega\sin(\theta))$. The torque exerted by a motor is counted positive if its natural effect would be to rotate the impeller in the positive direction.

A striking feature of the graphs of figure 3 is the good collapse on a single curve of values obtained at very different values of Ω, in the range 4 Hz to 30 Hz. This proves that the scaling shown on figure 2, which was obtained in the case $\Omega_1 = \Omega_2$, is indeed quite general. Another interesting point is that, while changing the blade height only seems to introduce a (non-trivial) scaling factor in the y-direction, mounting the baffles in the vessel changes vastly the shape of the curves. One can actually see that in the range of angles included between 160° and 225°, mounting the baffles *reduces* the mechanical power dissipated by the flow. A common feature of all the curves is that the strongest torque values are measured when both impellers are rotated in the negative direction. A last point that we find noteworthy is that on figures 3a, 3b and to some extent 3d, the two curves do not intersect at $\theta = 225°$. This means that, though the two motors operate at the same speed, the torque they exert on the fluid is significantly different. This has prompted us to perform further measurements in that region, which are presented in the next paragraph.

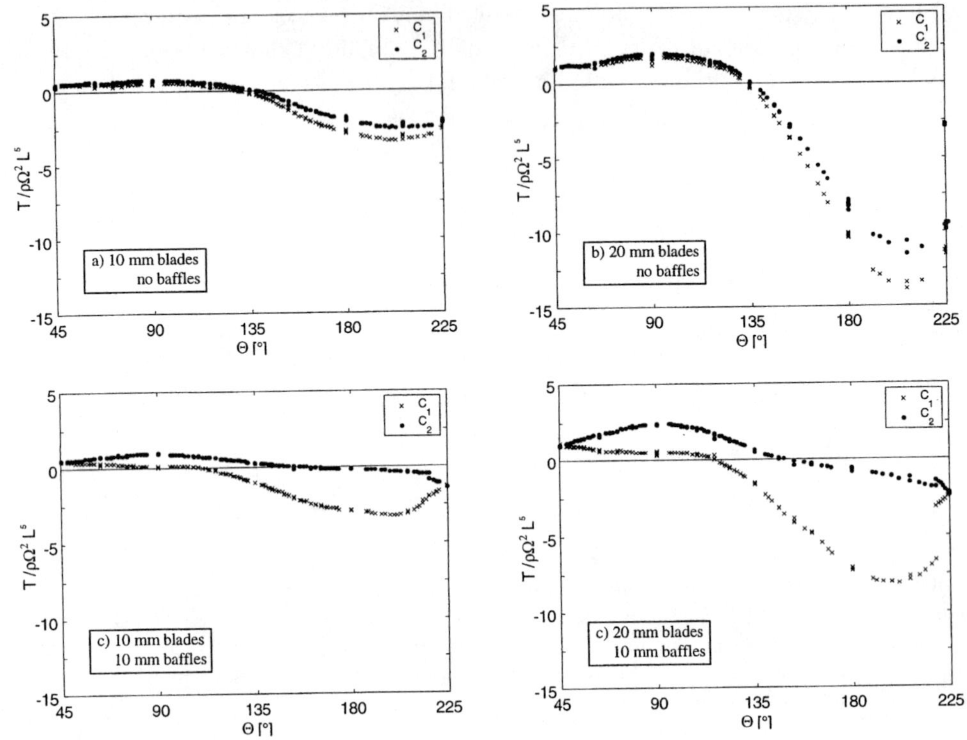

FIGURE 3. Variations of the two dimensionless coefficients c_1 and c_2 as functions of the angle θ, for the different flow configurations. θ =45° corresponds to $\Omega_1 = \Omega_2 > 0$. For values of the angle between 45° and 90°, the two impellers move in the positive (counter-clockwise) direction. For values between 90° and 180°, impeller 1 moves in the negative (clockwise) direction and impeller 2 moves in the positive direction. For θ =135°, we have $-\Omega_1 = \Omega_2$, $\Omega_2 > 0$. Values between 180° and 225° correspond to the case where both impellers move in the negative direction. θ =225° means $\Omega_1 = \Omega_2 < 0$.

Global Bifurcation

We have seen on figure 3a and 3b that when the two impellers are rotated at the same speed in the negative direction, they do not exert the same torque on the fluid. More measurements have been performed using the 20 mm high blades, without baffles. The results are shown on figure 4, together with sketches of the various flow configurations observed. Seeding the flow with air bubbles, we have been able to visualize three distinct flow configurations for θ =225°. If one first starts motor 1, then brings motor 2 to the same (negative) rotation rate, one gets to the value θ =225° from below. The measured torques follow the two curves on the left part of figure 4. The flow observed in this case has only one poloidal recirculation cell. The central column of fluid flows from impeller 2 to impeller 1.

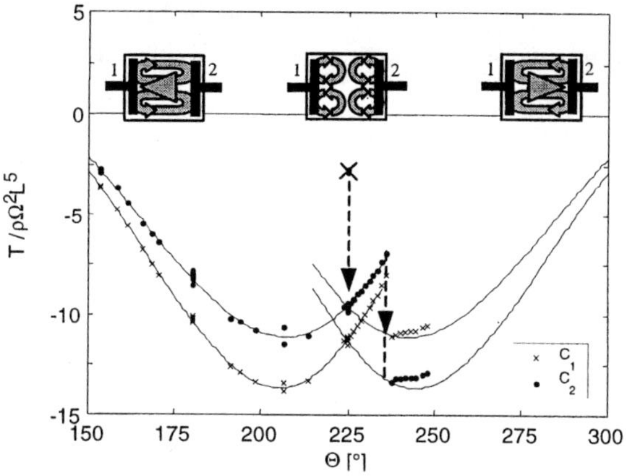

FIGURE 4. Values of C_1 and C_2 as functions of the angle θ, in the vicinity of $\theta = 225°(\Omega_1 = \Omega_2 < 0)$. The drawings represent the various flow configurations that can be observed.

If one then goes on increasing the value of θ, either by decreasing the rotation rate of motor 1 or by increasing that of motor 2, the flow relaxes, after a stretch of time of the order of one minute, to the mirror-symetric situation, where the central column of fluid flows from impeller 1 to impeller 2. The torques start following the curves on the right part of figure 4. If one then brings the value of θ back under 225°, the flow responds in the symmetric manner, relaxing after a while to its original state. This hysteretic behaviour has also been observed with the 10 mm high impeller blades, and is apparent on the curves of figure 3d, obtained with baffles. The flow in that last situation being less easy to visualize with air bubbles, we are not yet able to describe the different configurations.

The third configuration is obtained when the rotation rates of the two motors are kept equal all the way from standstill to their final value. The flow obtained in that case has two poloidal recirculation cells of roughly equal size. The torques exerted on the fluid by the two motors are equal, and their value is much smaller than the ones observed in the previous cases. This situation corresponds to the data points situated immediately below the sketch in the middle of figure 4. This flow configuration is meta-stable, in that it spontaneously relaxes, after a stretch of time that can be as long as one or two hours, in one of the other two configurations. We have never observed the converse evolution.

Pressure fluctuation measurements

Pressure root-mean-squared values

The typical value of pressure fluctuations one would expect in a turbulent flow is

$\rho V^2 \approx \rho\Omega^2 R^2$. Figure 5 shows the variations of the root-mean-squared value of the pressure fluctuations divided by $\rho\Omega^2 R^2$, in the case of 20 mm high blades and no baffles. Once again, the collapse of the points on the curve is quite good, except near the two end points, where the large values obtained for P_{rms} suggest that statistical convergence may not have been achieved.

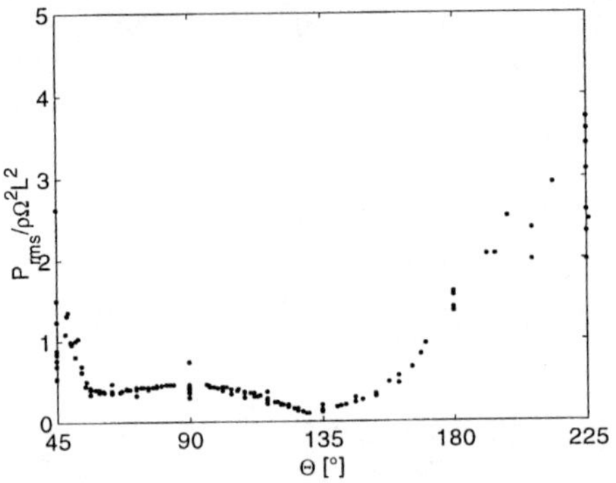

FIGURE 5. Variation of $P_{rms}/\rho\Omega^2 R^2$ as a function of θ, in the case of 20 mm high impeller blades, without baffles.

One can see that the pressure fluctuations amplitude depends strongly on the value of θ. Seeding with air bubbles allowed us to connect modifications of the flow structure to some of the features of figure 5. At θ =45°, which is the case of balanced counter-rotation ($\Omega_1 = \Omega_2 > 0$), we have observed two poloidal recirculation cells, separated by a vigorous mixing layer, in which very powerful vortices evolve. Between 45° and 60°, this mixing layer is gradually shifted towards the slower impeller, while the pressure fluctuations measured in the mid-plane diminish. Between 60° and 90°, the mixing layer stays confined between the impeller blades, where it remains quite active. The pressure fluctuations remain at a nearly constant level. After 90°, the two impellers start moving in the same direction in the laboratory frame. The pressure fluctuations gradually get smaller, until θ reaches 135° ($-\Omega_1 = \Omega_2, \Omega_2 > 0$). Here the two impellers move at the same speed in the laboratory frame, and the fluid is nearly in solid body rotation. The flow is very quiet, as can be seen from the pressure fluctuations level. For values of θ between 135° and 225°, the flow seems to have a single poloidal recirculation cell, in which a "bubble packet" rotates in a periodic fashion. Here the pressure fluctuations gradually increase, and reach their most important values at θ =225°.

Pressure time series and power spectral density

We have displayed time series and power spectral densities of three examples of pressure signals on figure 6. All of them have been obtained with 20 mm high impeller blades, without baffles.

Figure 6a and 6b have been obtained in the case of counter-rotation with both impellers rotating at 20 Hz (θ =45°). As we have mentioned above, the flow has two distinct poloidal recirculation cells, separated by a mixing layer. The time series exhibits strong downward peaks. Such events have been studied in the past [6,7], they are usually associated with regions of very concentrated vorticity. The power spectral density exhibits two distinct scaling regions. At low frequencies, the spectrum decays algebraically with a slope –1.2. At high frequencies, the spectrum slope is –3.1.

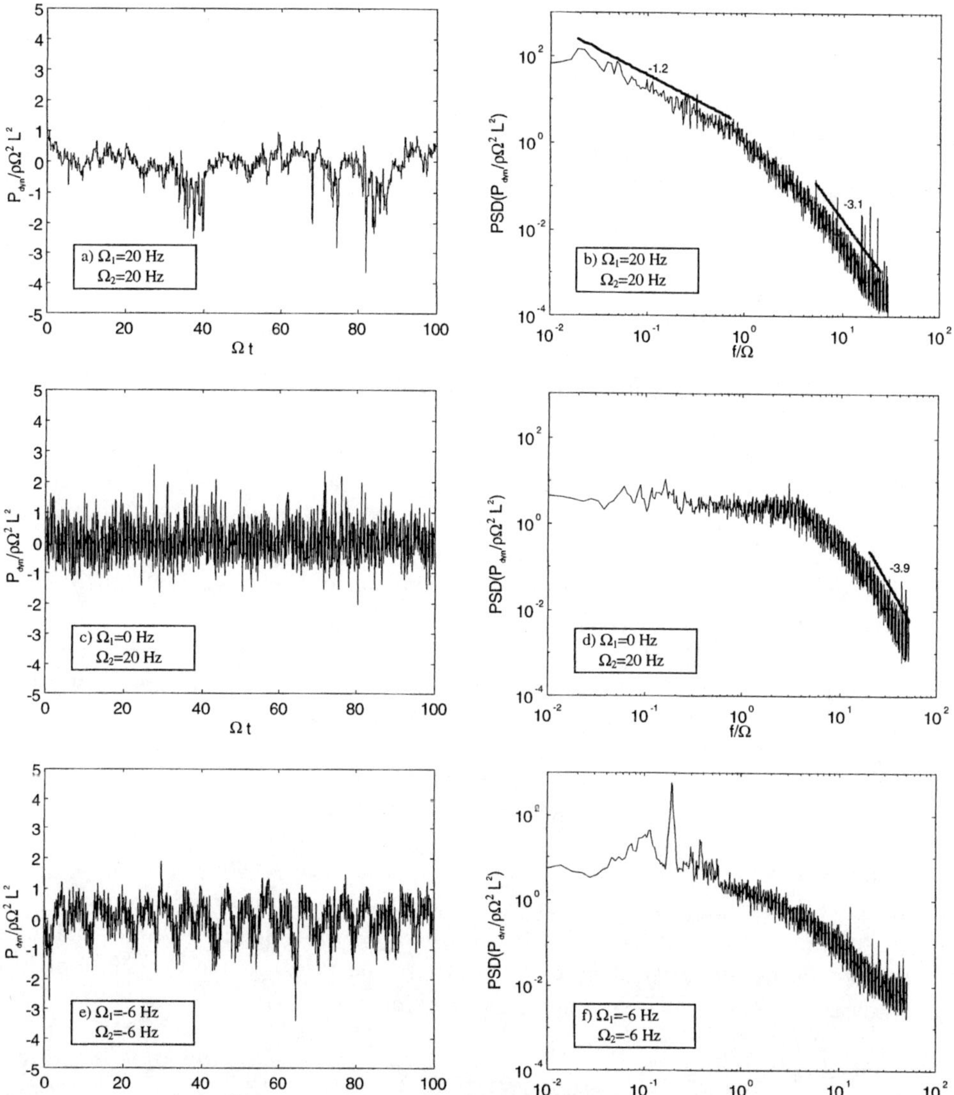

FIGURE 6. Time series and power spectral densities of typical pressure signals. The time has been scaled by Ω, and the pressure has been scaled by $1/\rho\Omega^2 L^2$ on all graphs

Figure 6c and 6d have been obtained with only one impeller rotating at 20 Hz (θ =90°). The time series has a very different ("whiter") texture. The power spectrum is flat at low frequencies for two and a half decades, roughly until $f \approx 4\Omega$. At high frequencies, it displays an algebraic decay, with exponent –3.9. This value is reminiscent of the −11/3 exponent predicted by [8] for the spatial spectrum of pressure fluctuations. However, it is not clear that the very high amplitude of the velocity fluctuations we observe in the flow permits the use of a Taylor hypothesis to translate from the temporal to the spatial domain.

Last, figure 6e and 6f have been obtained with both impellers rotating at –6 Hz (θ =225°). This quite low rotation rate is actually the largest our motors can sustain, since the torques values involved are already as high as 15 N.m. The flow studied in that particular run is one of the flows that have only one poloidal recirculation cell. Both the time series and the power spectral density show that the flow has a very strong periodic component at a frequency $f \approx 0.2\Omega$. Seeding the flow with air bubbles, we have seen them concentrate in a diffuse "packet" located near the cylinder wall, roughly in the midplane, which precesses in the azimuthal direction. No such things can be observed in the case of the symetric, meta-stable flow.

CONCLUDING REMARKS

We have performed systematic torque and pressure fluctuations measurements on a fully turbulent Von Karman flow induced in a cylindrical container by two impellers. We have studied in particular the influence of the height of the impellers' blades, as well as the influence of baffles mounted on the cylinder walls. This allowed us to observe and characterize the strongly hysteretic behaviour of the mean flow structure, as well as a global bifurcation of its topology.

Quantitative Laser Doppler Velocimetry measurements are underway to study the different flow structures. Particular attention is paid to the cases where the perturbations induced by the baffles preclude good visualization using air bubbles.

ACKNOWLEDGMENTS

We thank Vincent Padilla for actually building the experimental apparatus, and Cécile Gasquet for helping us in setting up the various instruments and the programs associated with them. We thank D. Ericher, P. Meininger, D. Bonamy and B. Dubrulle for fruitful discussion. J. B. thanks the Ministerio de Educación y Ciencia (spanish government) for a post-doctoral grant when working at the CEA.

REFERENCES

1. Zandbergen P.J., Dijkstra D., *Ann. Rev. Fluid Mech.* **19**, 465 (1987).
2. Dudley N.L., James R.W., *Proc. Roy. Soc. Lond.* **A425**, 407 (1989).

3. Nore C., Brachet M.-E., Politano H., Pouquet A., *Phys. Plasmas* **4**, 1 (1997).
4. Marié L.,Burguete J., Chiffaudel A., Daviaud F., Ericher D., Gasquet C., Petrelis F., Fauve S., Bourgoin M., Moulin M., Odier P., Pinton J.-F., Guigon A., Luciani J.-B., Namer F., Léorat J., "M.H.D. in Von Karman swirling flow, development and first run of the Sodium experiment" in *Lecture Notes in Physics. Dynamo and Dynamics, A Mathematical Challenge. Cargèse (France), 2000* edited by P. Chossat, D. Armbruster, I. Oprea, NATO ASI series, Kluwer Academic Publ., 2001.
5. Burguete J., Daviaud F., Léorat J., Marié L., "Homogeneous dynamo: numerical analysis of experimental von Kármán type flows", preprint, (2001).
6. Fauve, S., Laroche C., Castaing B., *J. Phys. II,* **3**, 271, (1993).
7. Cadot, O., Douady, S., Couder, Y., *Pys. Fluids* **A7**, 630-646 (1995).
8. George W.K., Beuther P.D., Arndt R.E.A., *J. Fluid Mech.* **148**, 155-191 (1984).
9. Mordant N., Pinton J.-F., Chillà F., *J. Phys. II,* **7**, 1 (1997).

ECOLOGY

Self Organization of Two-dimensional Insect Neural Networks

Amir Ayali [a], Orit Shefi [a,b] and Eshel Ben-Jacob [b]

[a] *Department of Zoology, and* [b] *School of Physics and Astronomy*
Tel-Aviv University, Tel-Aviv 69978, Israel

Abstract. How single neurons self-organize to form a complex functional system, the neural network, is a fundamental question in developmental neuroscience, computation science, and pattern formation. Two-dimensional *in vitro* invertebrate preparations offer an attractive model system to tackle this question due to the large size of the neurons, and their ability to grow in relative isolation as well as to develop elaborate networks. Using cultured locust neurons, we monitor and analyze their morphology and growth process under various density conditions. Neurons are affected by their neuronal vicinity; they actively target neighbor cells, and their overall structure in the 2D plane is oriented or directional in the presence of nearby cells. Connections formed between neurons lead to a process of simplification of the network members morphology. As the network matures the tendency for structural reduction becomes apparent in the neuronal as well as the whole network morphology. The preparation we have developed has vast potential for the study of form-function relation issues.

INTRODUCTION

How does a functioning network evolve from single entities? What is the role of pre-determined factors *vs.* the effects of the network environment? These questions have been a major focus of many scientists in recent years, as the study of networks in general and neural networks in particular have crossed traditional fields and disciplines of science, from computer science through physics to biology (for recent review, see [17] and references within).

The neural network is a bridging link between the neuron and that marvel of complexity, the brain [11,12,18,19,20,24]. However, even the simplest neural networks in the nervous system show surprisingly complex wiring diagrams [1]. Furthermore the single entities, the neurons themselves, come in a bewildering

CP622, *Experimental Chaos: 6th Experimental Chaos Conference*, edited by S. Boccaletti et al.

variety of shapes and forms. In order to gain some insight into the rules governing the self-organization of neural networks, one needs both to choose a system that is simple (relative to any *in-vivo* network), and to be able to control for as many of its variables as possible.

We are currently studying the morphological development and organization of two-dimensional neural networks in cultures of neurons dissociated from locust ganglia. Invertebrate cells offer an attractive model because of their large size and the facility with which they can be cultured in various conditions (e.g. extremely low densities) [2,6,7,9,10,13,16,23]. Most importantly, the *in vitro* two-dimensional culture enables easy access for non-invasive optical observations, allowing us to follow the dynamics of neural growth and network organization. It also opens the way to simultaneous recordings of electrical activity from many neurons (utilizing Multi Electrode Array technology). Furthermore, many invertebrate neurons are identifiable neurons [14], which may allow identifying neurons in situ and comparative characterization of neuronal structure and network circuitry *in-vitro* (in culture) and *in-vivo*.

In our work we focused on two aspects of network formation. First, we characterized single neuron morphology, and studied the effects of neuronal vicinity on the neuron's growth pattern. Second, we investigated the dynamics of the neuron growth. We followed the timing and sequence of events leading from single isolated cultured neurons to an interconnected network.

METHODS

Our culturing method followed Smith and Howes [22] with slight modifications. The neurons were dissociated from the frontal ganglion (FG) of adult desert locusts, *Schistocerca gregaria*. After dissection, enzymatic treatment and mechanical dissociation, the neurons were plated on Petri dishes pre-coated with Concanavalin A (Sigma). The number of ganglia per dish varied determining the density of the culture and thus the average distance between cells. Cultures were maintained in 70% relative humidity, at a temperature of 29 degrees Celsius, and complete darkness.

Plated neurons were fully differentiated adult neurons that have lost their original dendrites and axon, leaving only the soma. We monitored the neurons for their entire growth period, during which neurites regenerated, outgrew and connected to form morphologically stable networks. We first examined the cells 24 h after dissociation

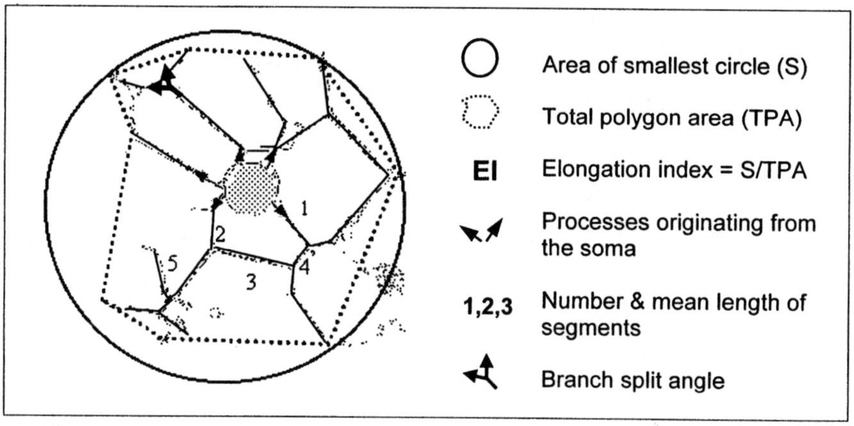

FIGURE 1. A processed image of a single neuron and some of the morphometric parameters that were analyzed: The smallest circle that encompasses all the cell neurites, a polygon that represents the area of the cell, number of processes that originate from the soma, total segments of the neurite tree.

and then once a day. A CCD camera mounted on a phase contrast microscope was used to acquire single images into the PC for analysis.

In order to observe the growing process continuously, some culture dishes were maintained in a special controlled chamber mounted on the phase contrast microscope. Time lapse observations were performed with the CCD camera connected and controlled by the PC. Due to the sensitivity of the cultured cells to light, an optical shutter was momentarily opened by an external trigger, simultaneously with the computer controlled frame grabber.

We measured descriptive morphometric parameters of single neurons, which relate to the neurite outgrowth, the arborization of neurites, and to the spatial distribution of neurites in the 2D-growth plane (Fig. 1). These included the number of processes originating from the neuronal soma, the number of neurite segments per cell, the length of the segments, the soma area, general cell orientation etc. [3, 4, 8; see Fig. 1]. We divided the developing neurons into different categories according to their cellular vicinity (e.g. isolated neurons *vs.* neurons with close neighbors), as well as the cell's stage of development (days in culture and connectivity to other neurons). Mean values were calculated, and compared for neurons of the various categories.

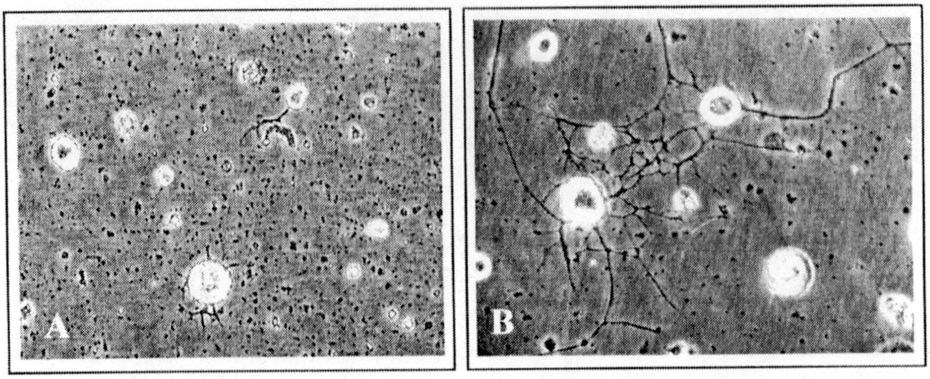

FIGURE 2. **A.** After 24 h 50% of the plated neurons had already started developing neurites. **B.** A 3-day-old culture showing elaborate branching pattern and interconnections of the neurons.

RESULTS

The Frontal Ganglion Culture

The FG is a small ganglion in the locust head consisting of approximately 100 neurons. The FG neurons *in situ* are typical invertebrate ganglion cells. They are located in the outer skirts of the ganglion, and send their processes to the dense neuropil region in the core of the ganglion. Based on descriptions from other invertebrates (e.g. [21]), most cells are known to be uni-polar, that is they grow one process that branches distally to give rise to an elaborate dendritic tree and axon.

Neurons within the intact locust FG form neural circuits that generate rhythmic motor outputs controlling foregut movements (Ayali and Zilberstein, in preparation).

Our cultured neurons varied in size between 10 μm to 50 μm. After 24 h in culture about 50% of the cells survived, out of which 25% had already started developing neurites (Fig 2A). Most neurons demonstrated multi-polar growth.

We maintained the culture for up to a month. The most intense stage of development and process growing was between day 1 and day 4 (Fig 2B). After this rapid growth stage there was a pronounced decrease in growth rate, few new neurites were formed and there were relatively few changes in connectivity.

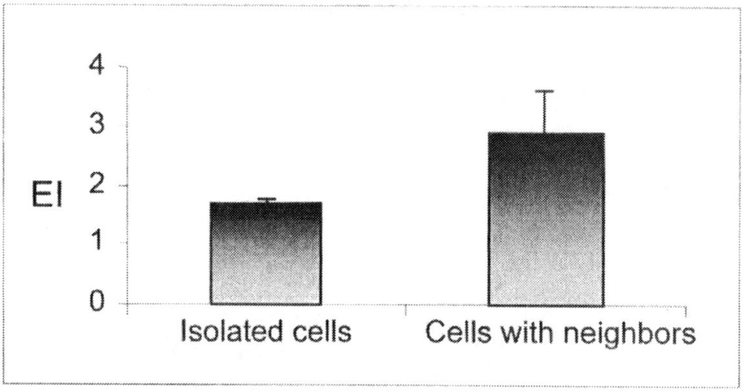

Figure 3. The Elongation index (see Fig. 1) calculated for practically isolated neurons (n=8) *vs*. neurons with close neighbors (n=29).

Neuronal Vicinity Affects Growth Morphology

Neurons could be plated at various densities, generating very low density cultures, even to the extent that the neurons were practically isolated, or high density cultures, where most of the neurons had close neighboring cells. The average distance between cells ranged between 20 μm, in a typical high-density dish, to 350 μm, in a low-density dish.

To characterize possible effects of the neuron's cellular vicinity on its growth pattern, an elongation index (EI) was introduced. This index measures the ratio between the smallest circle that encompasses the neuron and the area of the polygon that connects the neurite tips (Fig. 1). It reflects whether the growth of a single neuron is homogenous or symmetrical *vs*. oriented or directed. As expected, we found the EI to be higher in neurons with close neighboring cells, as compared to neurons with no neighbors. As shown in Figure 3, the latter demonstrated EI closer to 1, i.e. more symmetrical growth pattern.

We followed single neurons by time lapse observations for up to 90 hours after plating, during which they interconnected to generate intricate networks. Close examination of the neurite outgrowth process suggested that some neurons actively targeted neighbor cells (Fig. 4). The neurites elongated toward specific neighbors, split and connected to the target neurons. Cell bodies that had not yet began

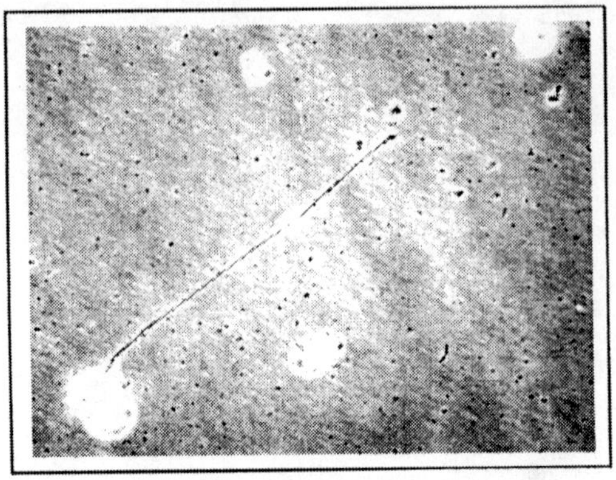

FIGURE 4. A neurite outgrowth toward a neighboring cell demonstrating a selective directional growth pattern.

regenerating processes were seemed to be awaken or triggered by the contact with the approaching neurite, and started growing their own neurites.

As the neurons regenerated, the processes elongated and then branched to give rise to new, second order ones. Generally a mother branch would split into two daughters branches (Fig. 5). We were interested in determining whether a predominant, possibly pre-determined, splitting angle exists. We were also interested in the effect of neuronal vicinity on the branching angle. Throughout the growth period there was constant remolding of the cells neurite pattern, and after the new segments started to grow the original splitting angles would change. Therefore, the angles were measured immediately after the new branches have formed (Fig. 5). We found that neurons have an innate preferred branching angle of around 80° (Fig 6A). In the presence of stimuli from the environment, presumably coming from nearby neurons and their growth cones, this preferred angle may be masked to give rise to a much wider, environment-dependent range of angles (Fig 6B).

Neurons and Networks Go Through Structural Growth Followed by Reduction

As the neurons grew and started connecting to each other the first outburst of neurite outgrowth, rich branching and elaborate pattern formation gave rise to a

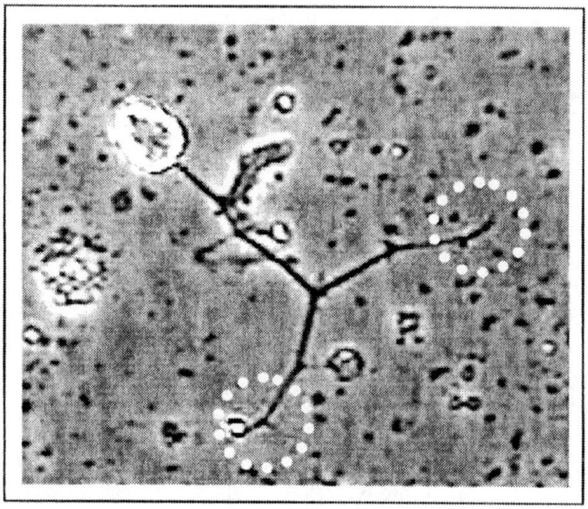

FIGURE 5. An image of a neuron 2 days after plating demonstrating branching points (marked with dotted circles) at the time the splitting angles were measured (see text for details).

second very different process of reduction and simplification of the pattern. This was clearly demonstrated by the differences in growth pattern of neurons connected *vs.* not connected to neighbor neurons. As long as the neurons remained isolated, the number of processes originating from their soma significantly increased with time, culminating in a 50% increase between days 2 and 3 after plating ($p<0.05$). In marked contrast, neurons that were already connected to neighbor cells, demonstrated a 50% reduction in the number of processes originating from the soma between days 2 and 3 ($p<0.005$). Thus, by day 3 there was a highly significant difference between connected and not yet connected neurons in the number of processes originating from the soma ($p<0.005$). Connected neurons also had a significantly lower average number of neurite segments per cell compared to not connected neurons ($p<0.005$).

As mentioned earlier, the rapid growth period of the networks was limited to the first 4 to 5 days. As the networks reached days 6 or 7 in culture, two other phenomena were observed that fall under the reduction and simplification process heading. Both phenomena involved re-arrangements and migration. First, the previously very complicated web of neurites started forming much thicker bundles. In time-lapse observation this process comprised two events; absorption of "unneeded" branches and even whole neurites, together with re-arrangements of neurites and what

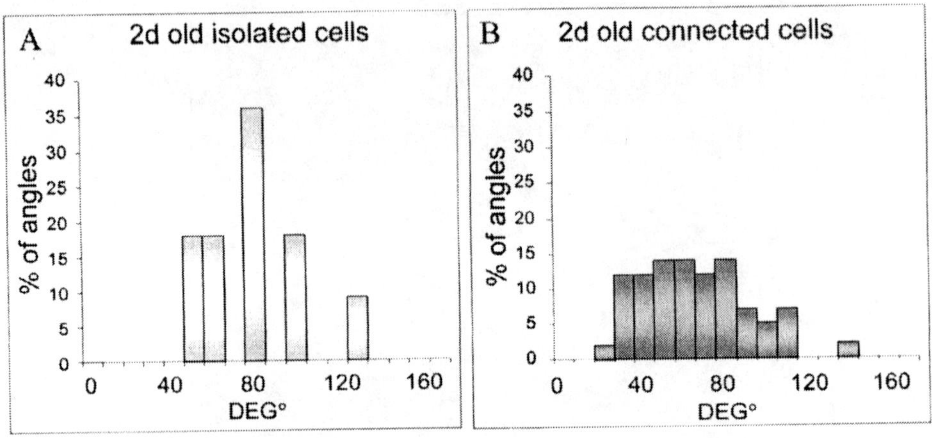

FIGURE 6. Branching split angles measured just after the neurites split (see Fig.4) in practically isolated neurons (**A**, n=11) and in neurons with close neighbors that had already formed connections to other cells (**B**, n=43). Data clearly suggest that the neurons have an innate preferred branching angle that is masked by the effects of the neuronal vicinity, or the environment.

appeared to be fusion of parallel ones. Second, the neuronal cell bodies started to gather into packed clusters. Again in time-lapse observation, the somata were observed to migrate along the newly formed bundles toward one another. Thus relatively homogenous cultures in which single neurons were scattered, evolved into cultures organized into a few centers comprising clusters of neurons and thick nerve-like bundles connecting them (Fig. 7).

SUMMARY

One of the most profound open questions in science is how a functioning brain emerges from a collection of individual neurons [15, and references within]. The study of this question translates to investigation of how a collection of initially simple elements, namely neurons, self-organize to form a new and extremely complex system: the neural network [5,18,19,20,24].

As in networks in general, there is a strong relation between the network's wiring diagram or structure, and its function. Individual neurons' branching pattern, and the

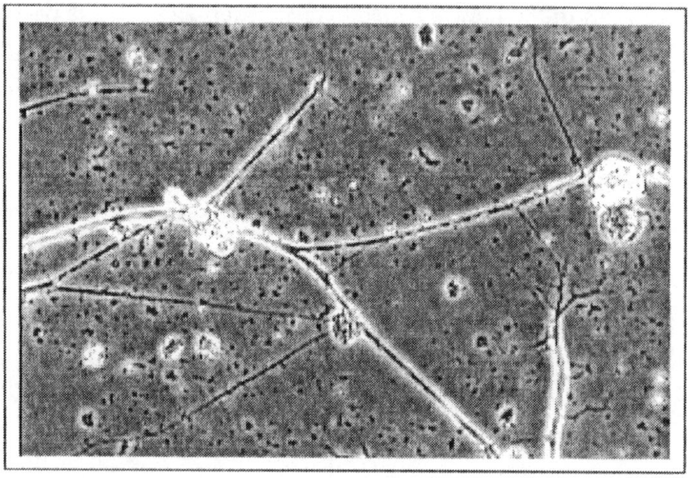

FIGURE 7. The neural network in a 7-day-old culture appears much less complex than after 3 or 4 days following plating. Many of the neurites now form thick bundles. The cell bodies have completed a process of migration along these bundles to aggregate in few clusters of neurons.

formation of distinct interconnection between neurons, are dominant factors, instrumental in the future output of the neural circuit, behaviour.

The overall objective of our research was to study the adaptive self-organizing rules that govern the formation of neural networks. *In vivo* nervous systems are largely inaccessible for such a task. We therefore investigated two-dimensional *in vitro* preparations, in which insect neurons are grown in a controlled environment. These cultures offer an attractive model system due to the large size of the neurons, which enables manipulation, and the ability of the cultured neurons to grow in relative isolation as well as to develop elaborate networks [6,7,9,10,13,16,23].

In our work we concentrated on the changing morphology of single neurons as they regenerate and outgrow processes to become members of the evolving neural network. By controlling the culture density, we tested the effect of neurons on each other's structure and growth pattern. By following the culture continuously throughout the generation of the neural network we could identify and characterize landmarks or milestones in neural network formation.

Our results demonstrate that neurons are affected by their neuronal vicinity: they actively target neighbor cells, and their overall growth morphology is affected by the presence of close neighbors. Connections formed between neurons lead to a simplification process of the network members' morphology. This is part of a general simplification and reduction process that characterizes the mature neural network and culminates in the generation of cell clusters that are connected by nerve-like structures.

These results will serve as a basis for a study of the interrelations between morphology and activity of evolving networks.

ACKNOWLEDGMENTS

We sincerely thank Dr. Netta Cohen for her helpful comments on an early version of this manuscript.

REFERENCES

1. Achacoso, T.B., and Yamamoto, W. S., *Neuroanatomy of C. elegans for computation,* CRC Press, Boca Raton, FL, (1992).
2. Acklin, S. E., and Nichlls, J. G., *J. Neurosci.,* **10,** 1082-1090, (1990).
3. Cannon, R. C., Wheal H. V., and Turner D. A., *J. Comp. Neurol.,* **413,** 619-633, (1999).
4. Devaud, J. M., Quenet, B., Gascuel, J., and Masson, C., *Neuroreport,* **6,** 214-218, (1994).
5. Goodhill, G. J., Mathematical guidance for axons. *TINS,* **21,** 226-231, (1998).
6. Hayashi, J. H., and Hildebrand, J. G., *J. Neurosci.,* **10,** 848-859, (1990).
7. Howes, E. A., Cheek, T. R., and Smith, P. J. S., *J. Exp. Biol.,* **156,** 591-605, (1991).
8. Kawa, A., Stahlhut, M., Berezin, A., Bock, E., and Berezin, V., *J. Neurosci, Meth.,* **79,** 53-64, (1998).
9. Kirchhof, B. and Bicker, G., *J. Comp. Neurol.,* **323,** 411-422, (1992).
10. Kloppenburg, P. and Horner, M., *J. Exp. Biol.,* **201,** 2529-2541, (1998).
11. Koch, C., And Laurent, G., *Science,* **284,** 96-98, (1999).
12. Mpitsos, G. J., *Brain Behav. Evol.,* **55,** 256-277, (2000).
13. Lapied, B., Tribut, F., Sinakevitch, I., Hue, B., And Beadle, D. J., *Tiss. Cell,* **25,** 893-906, (1993). 7.
14. Leonard, J. L., *Brain Behav. Evol.* **55,** 233-235, (2000).
15. Quartz, S. R., and Sejnowski, T. J., *Behav. Brain Sci.* **20,** 537-596 (1997).
16. Rossler, W., and Bickmeyer, U., *J. Exp. Biol.,* **183,** 323-339, (1993).
17. Strogatz, S. H., *Nature,* **410,** 268-276, (2001)
18. Segev, R., and Ben-Jacob, E., *J. Comp. Sys.,* **1,** 67-78, (1998).
19. Segev, R., and Ben-Jacob, E., *Neural. Net.* **13,** 185-199, (2000).
20. Segev, R., and Ben-Jacob, E., Self-wiring of neural networks. *Physics Letters A* 237: 307-313, (1998).

21. Selverston, A. I., Russel, D. F., Miller, J. P. and King, D. G., *Prog. Neurobiol.,* **7,** 215-290, (1976).
22. Smith, P. J. S., and Howes, E. A., *J. Neurosci. Meth.,* **69,** 113-122, (1996).
23. Whitington, P. M., *Pharma. Therap.,* **58,** 263-299, (1993).
24. Wilkinson, C., and Curtis, A., *Phys. World,* **12,** 45-48, (1999).

Complex Synchronization Phenomena in Ecological Systems

Lewi Stone*, Ronen Olinky*, Bernd Blasius†, Amit Huppert* and Bernard Cazelles**

*Biomathematical Unit, Life Sciences, Tel Aviv University, Ramat Aviv, 69978, Israel
†Institute of Physics, University of Potsdam, Am Neuen Palais 10, D-14469 Potsdam, Germany
**CNRS UMR 7625 - Universite Pierre et Marie Curie, 7 quai Saint Bernard, case 737, 75252 Paris, France.

Abstract.
Ecological and biological systems provide us with many striking examples of synchronization phenomena. Here we discuss a number of intriguing cases and attempt to explain them taking advantage of a modelling framework. One main focus will concern synchronized ecological end epidemiological cycles which have Uniform Phase growth associated with their regular recurrence, and Chaotic Amplitudes – a feature we term UPCA. Examples come from different areas and include decadal cycles of small mammals, recurrent viral epidemics such as childhood infections (eg., measles), and seasonally driven phytoplankton blooms observed in lakes and the oceans. A more detailed theoretical analysis of seasonally synchronized chaotic population cycles is presented.

INTRODUCTION

Synchronization is a phenomenon that is as important as it is widespread in ecological systems. Consider Australia's Great Barrier Reef. Here many coral species have the remarkable ability to tune their reproductive cycle with astonishing precision. On a single day each year, just after the November full moon, corals synchronize by collectively releasing millions of gametes in a remarkable large-scale mass-spawning event, whose purpose is to enhance the survival of the next generation. Similar large-scale synchronization phenomena may be found in the insect world. Certain Asian firefly species are famous for their ability to gather in dense swarms where they rhythmically flash on and off in perfect unison, in a synchronized display of courtship. Equally curious is the manner in which many animal populations oscillate in a chaotic fashion but are nevertheless able to collectively synchronize over enormous spatial scales, sometimes over continents. For the classic ten-year Canadian hare-lynx cycle, populations from different regions have been tightly synchronized for hundreds of years now, creating one of Ecology's greatest enigmas. Major human epidemics are also synchronized both spatially and temporally, as the disease spreads in waves through a network of interconnected suburbs. Here we review various aspects of ecological synchronization and methods that have been adopted to model the many aspects of this phenomenon.

CP622, *Experimental Chaos: 6th Experimental Chaos Conference*, edited by S. Boccaletti et al.

THE EVOLUTION OF SYNCHRONIZATION STRATEGIES

A perplexing question that did not escape the notice of some of the greatest thinkers working in evolutionary theory can be stated very simply: "Why do synchronized behavioral strategies evolve over evolutionary time?" Unfortunately there is no straightforward answer to this innocent looking question. Many bird species, for example, prefer to breed together in a synchronized mating season at much the same time each year (for reasons not necessarily connected to climatic conditions), while other related species breed continuously throughout the year and show no such signs of synchronized behavior. Why does one species synchronize and not the other? Many similar questions are still a great puzzle for biologists. Nevertheless, evolutionary arguments have made some important successes, and we briefly summarize an interesting model attributed to Maynard Smith [1].

In the last few decades, reproductive synchrony has often been considered a sexual strategy adopted by the female sex and forced on the entire population to decrease the likelihood of the male deserting his partner. Should a male attempt to cheat, he will probably find the mating season too short to allow him time to find and mate with a new partner. He will inevitably lose by running from, rather than investing in, his offspring. Thus it is to the females advantage to keep the mating season synchronized and short. In other words, synchronization has evolved to prevent males cheating on their partners. Maynard Smith's evolutionary model is simple and elegant, and begins by assuming that if both parents are involved in raising their young, the number of surviving offspring is v_2. However, if the male deserts on his partner, only v_1 young will survive (with $v_1 < v_2$). Now suppose that there is a probability p that a male will desert his partner after she has laid eggs, and then succeeds to mate with another female. The fitness of such a deserting male is determined by the number of offspring he produces, which is on the average $v_1 + pv_2$. Thus desertion will be favored (or is an evolutionary stable strategy) only if $v_1 + pv_2 > v_2$ i.e., if

$$p > \frac{v_2 - v_1}{v_2}. \tag{1}$$

Hence desertion is favored only if there is a good chance (p) of the male finding and mating with a second partner. Note that should the female force a short synchronized breeding season on the male, this acts to reduce p, and increase her chances that the male will stay and help raise their young. So it is a worthwhile strategy for females to synchronize their breeding season. In recent years behavioural ecologists have far more sophisticated arguments for understanding parental care and reproductive synchrony - our aim here was to provide a glimpse into the sometimes unusual approaches used in evolutionary models.

There are many other types of synchronized reproductive strategies, and they are often context dependent. As we have mentioned, some corals, spawn on a single day every year. It is believed that this is either to 'fool' predators or to minimize their impact through 'predator satiation.' Similarly, different coral species spawn in different months (i.e., with phase lags between events) possibly to prevent hybridization of different species and/or as a mechanism to release larvae from interspecific competition. Cicadas in America periodically emerge with regular synchronized cycles of prime number intervals of 7 or 13 years. Evolutionary theorists try to explain the vexing question: "Why prime?"

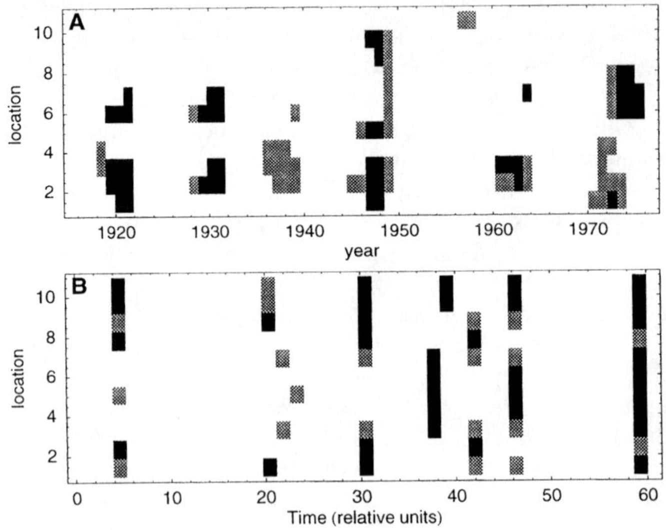

FIGURE 1. a) Occurence of outbreaks of Lepidoptera (Douglas-fir tussok moth) in British Columbia (Canada) from 1916 - 1976, at 11 locations. Data from [2]. b) Simulation of a CML of 11 subpopulations modelled by the Ricker model. The synchronized state ($x_i = x_j$ for all i, j) is unstable but one can observe imperfect synchrony. Black color: severe outbreak ($x > 0.70\,\max(x_i)$); grey color: moderate outbreak ($x > 0.40\,\max(x_i)$); white color: no outbreak.

SYNCHRONIZED POPULATION OSCILLATIONS AND EXTINCTIONS

Physicists often study the dynamics and synchronization of Coupled Map Lattices (CML's, see eg., [3]). But this also proves to be highly relevant for studying ecological systems [4]. Fig.1a shows the occurence of outbreaks of Lepidoptera (tussock moths) at 11 locations spread over $100'$s of kilometers across Canada (from [2]). Note the erratic synchronized patterns which probably originate from factors other than environmental cues. The dataset raises several problems which CML models may help in exploring. For example: Why are there apparent population cycles? Why are the populations synchronized? And why is the synchronization erratic?

To gain insight into this 'imperfect' synchronization, imagine a spatial landscape of interconnected patches, each with its own subpopulation subject to the usual localized birth-death growth process (eg., the simple logistic equation is a useful model here). The following CML model may then be set up:

$$x_i(t+1) = \sum_{j=1}^{N} m_{ij} F(x_j(t)). \tag{2}$$

Here $x_i(t)$ represents the population of the i'th patch, and m_{ij} describes the migration from patch-j to patch-i. The term $F(x_j)$ models the birth death process of the populations, and might typically be the logistic model $F(x) = rx(1-x)$ or the Ricker model $F(x) = x \cdot e^{r(1-x)}$.

The criterion for the fully synchronized state ($x_i = x_j$) for all i, j is well known (eg., [3, 5]) and depends on the subdominant eigenvalue λ of the dispersal matrix M, and the Lyapunov exponent μ of a single (uncoupled) one-dimensional map F. Full (or 'perfect') synchronization is assured if the transverse Lyapunov exponent ($\mu_\perp$) of system 2 satisfies:

$$\mu_\perp = \mu + \ln(|\lambda|) < 0. \tag{3}$$

For parameters that violate this criterion it is possible to arrive at states of 'imperfect synchrony.' However the above criterion is based on long-term average quantities (eg., $(\mu, \mu_\perp)$). In the case of complex dynamics (chaotic or stochastic) these quantities may fluctuate and their distributions may explain unexpected dynamics (eg., riddled basins, on-off intermittency). Hence even when the invariant synchronous manifold is stable $\mu_\perp \approx< 0$, there can still be initial conditions for which the transverse instantaneous Lyapunov exponent becomes weakly positive and trajectories can jump to another attractor. This effect can produce dynamics with riddled basins: depending on initial conditions one can observe both perfect and imperfect synchrony. Conversely, when the synchronous manifold is weakly unstable ($\mu_\perp >\approx 0$), it is still possible to observe synchronized dynamics although this behaviour may be interrupted by on-off intermittency [6].

Fig.1b displays simulations of a metapopulation of $N = 11$ subpopulations (CML) connected via nearest neighbour coupling with the Ricker model taken for F. Parameters values used are $r = 6.5$ and the dispersal coefficient is $m = 0.35$. In this particular case, the synchronized state ($x_i = x_j$ for all i, j) is unstable but one can observe 'imperfect synchrony' due to the development of synchronized chaotic population clusters. Very similar erratic synchronization can emerge for other reasons. For example, with globally coupled populations, when $r = 4$, $m = 0.35$ the synchronized state ($x_i = x_j$ for all i, j) is weakly unstable and the On-Off intermittency produces synchronized patterns very similar to the observed data.

Spatial synchronization is a phenomenon that was studied by ecologists in the late 1970's well before CML's became popular in the physics literature. Den Boer, for example, from the Netherlands realized from his field work that migration between patches tends to enhance synchronization of the larger metapopulation. As such, synchronization could create conditions that were sometimes uniformly bad for all local populations. Synchronized patch populations that periodically attained dangerously low levels, could more easily become globally extinct. On the other hand, asynchrony between patches acts to prevent global extinction. At any point in time some local populations increase and act as source populations that (through migration) prevent the disappearance of declining sink populations. The conditions for synchrony, asynchrony and chaotic dynamics in CML's are thus of interest for ecologists who use these models in an attempt to understand the dynamics of population extinctions and its relation to asynchrony.

PHASE SYNCHRONIZATION OF CHAOTIC POPULATION CYCLES

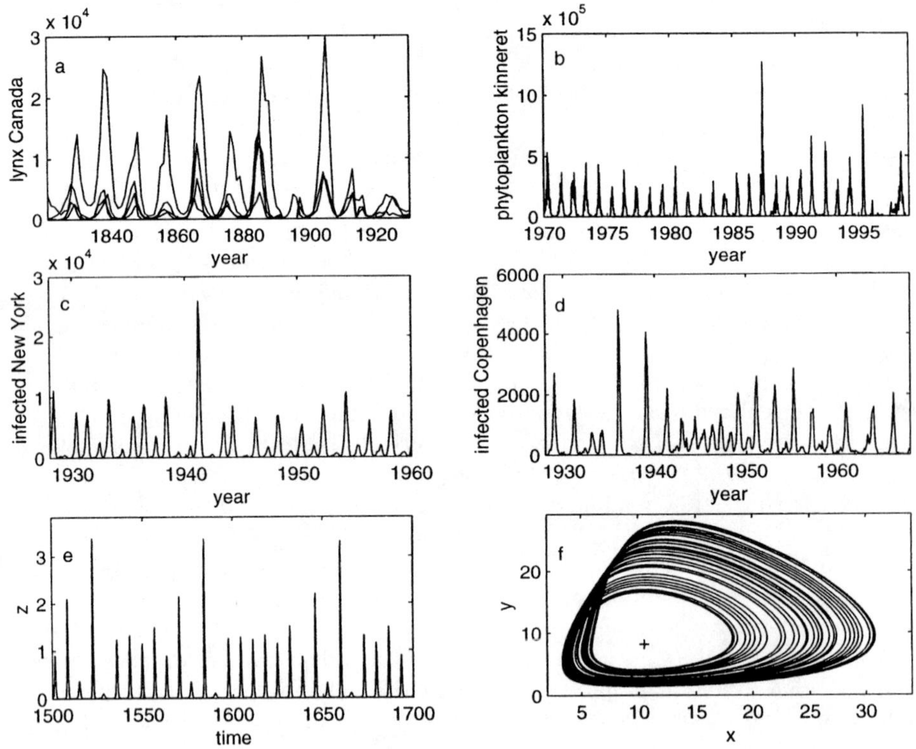

FIGURE 2. a) 10 year cycle in the Candian lynx record 1820-1930, plotted for seven separated districts in Canada; b) phytoplankton bloom in Lake Kinneret, Israel (with kind permission of Prof. Utza Pollingher); c) Number of cases of measles in New York City 1920-1960 and d) Copenhagen 1920-1970; e) simulated UPCA dynamics from (4) and f) $x-y$ phase plane with parameters: $a = b = 1, c = 10, k = 0.1, e = 0.6, x_0 = 1.5, y_0 = 0, z_0 = 0.01$.

We are also studying synchronized population oscillations, such as those observed in Canada's unusual 10-year hare-lynx cycle (Fig.2a). These populations have the intriguing feature of highly erratic or chaotic abundances but remarkably constant period length – a behaviour we refer to as UPCA (Uniform Phase evolution with Chaotic Amplitudes). As yet, conventional ecological models have failed to reproduce this behaviour, despite the fact that UPCA is an important and commonly observed ecological and biological dynamic. It has been noted in many systems, ranging from small mammal populations to neural network oscillations. Recently, we designed a simple deterministic foodweb model that generates chaotic UPCA oscillations [7, 8] (Fig.2e,f & Legend):

$$\begin{aligned}\dot{x} &= a(x-x_0)-kxy\\ \dot{y} &= -b(y-y_0)+kxy-eyz\\ \dot{z} &= -c(z-z_0)+eyz.\end{aligned} \tag{4}$$

The UPCA dynamics of this model may be explained in simple terms as follows. Firstly, note that the model has three unstable equilibrium points. The relevant equilibrium in the positive orthant may be approximated for $y_0, z_0 \ll 1$ as

$$x^* = b/k, \quad y^* = a/k - ax_0/b, \quad z^* = cz_0/(c-ey^*). \tag{5}$$

For the given parameters, $(x^*,y^*,z^*) \approx (10,8.5,0.1)$ and this forms the centre of an unstable oscillation which spirals with increasing radius about the point (x^*,y^*) in the $x-y$ plane ($z \approx 0.1$, see Fig.2f). When y increases beyond the critical threshold level, $\tilde{y} = c/e$, the z-variable 'triggers' and the trajectory shoots rapidly out of the plane $z \approx 0$ only to be reinjected back, soon after, to return close to the fixed point (x^*,y^*,z^*). These dynamical features have similarities to the Rössler equations, the main difference being that now all populations are always positive - a realistic feature that is nontrivial to implement [8].

Linearizing the equations about the equilibrium and taking eigenvalues of the resulting Jacobian makes it possible to approximate the frequency of the system as

$$\Omega = \sqrt{ab - akx_0 - (akx_0/2b)^2}. \tag{6}$$

Ω quantifies the 'frequency' of the trajectory as it spirals about the unstable equilibrium in the $x-y$ plane, and the estimate is of good accuracy according to numerical simulations. The result is useful for synchronization studies where fine control of oscillator frequency is often required. Note that for $x_0 = 0$, the result simplifies to $\Omega = \sqrt{ab}$ as it should for the classical Lotka-Volterra equation.

Examining more closely the Canadian hare-lynx cycle (Fig.2a), one notes that although the lynx populations of many different regions phase-lock to the same extraordinary collective rhythm, their peak population abundances are nevertheless chaotic and largely uncorrelated to one another. What makes this so peculiar is that there is synchronization in phase or rhythm, but not in amplitude; in other words, this is a clear case of 'phase-synchronization.' The ecological UPCA model provides us with a tangible system for exploring this form of complex chaotic synchronization and, in fact, phase-synchronization immediately emerges upon coupling several UPCA foodweb models of different frequencies Ω_i via diffusive migration. We have investigated spatially extended systems by examining a spatial lattice of patches, where each patch is a single UPCA foodweb model. Such a 'metacommunity' could, for example, represent the Canadian hare-lynx system when viewed over a large spatial scale. Nonidentical model foodwebs with random natural frequencies that act as spatial 'inhomogeneities' over the lattice induce interesting dynamics. In the regime of phase-synchronization, synchronized patch populations are typically separated by a phase lag which can give rise to complex spatio-temporal patterns [7, 8]. Often a remarkable chaotic travelling wave structures are produced, where population abundances remain chaotic, but unusual circular waves form and spread in time across the spatial landscape. Ecological field studies of synchronized populations (eg plant, plankton, vole, hare-lynx populations) have reported similar travelling wave structures.

SEASONALLY DRIVEN UPCA POPULATION DYNAMICS

Another important aspect of synchronization is the manner in which ecological communities entrain to periodic environmental or climatic driving variables [9, 10]. In Lake Kinneret (Sea of Gallilee), Israel, for example, seasonal stratification and nutrient availability patterns ensure that almost every March a large-scale algae bloom appears (see Fig.2b). As in many other freshwater lakes, the presence of a phytoplankton bloom signals dangerous nutrient loading (eutrophication) and water quality problems. The seasonal forcing adds a complicating factor to the study of bloom dynamics since it can induce irregular chaotic oscillations synchronized to the seasonal cycle. This is a form of synchronization through unidirectional coupling or one-way driving. Including seasonal forcing such a model in dimensionless variables would read as [11]:

$$\begin{aligned} \dot{N} &= I - \beta(t)NP - qN \\ \dot{P} &= \beta(t)NP - P, \end{aligned} \tag{7}$$

where N represents nutrients and P represents phytoplankton in the water column. The two dimensionless parameters I and q are an effective influx and nutrient loss rate respectively. The seasonal forcing is given by the yearly periodic function $\beta(t)$ and could be represented by

$$\beta(t) = 1 - \alpha\cos(\omega t). \tag{8}$$

Seasonal forcing induces UPCA dynamics that are generally phase locked to the seasonal cycle. Although we have succeeded in obtaining the desired goal of generating UPCA dynamics from a periodically forced model, Eqns. (6,7) lead to behaviour that is to some degree unrealistic for describing phytoplankton dynamics. In our simulations it is often the case that many years go by without algae blooms occurring. These ‘skippings,’ which we discuss further below, are not desirable but appear to be intrinsic to the forced model (7). Nevertheless, with subtle modifications [11], we have succeeded in constructing a seasonally forced model that generates the desired UPCA behaviour producing simulations similar to Fig.2e,f.

EPIDEMIOLOGY

In the same spirit but in the context of epidemiology synchronization has great control over the (often chaotic or irregular) dynamics of diseases as they spread through complex networks of suburbs and cities. Here we concentrate on the dynamics of childhood

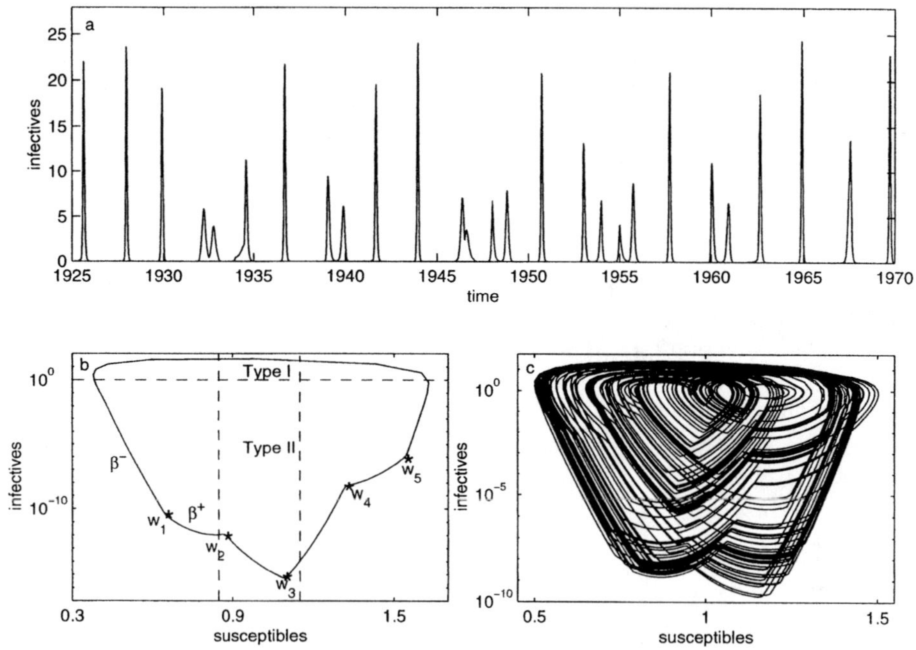

FIGURE 3. Simulation results of the epidemic model (9) in the UPCA region. a) Chaotic time series exhibiting years of fade-outs or 'skips' when the epidemics fail to trigger. b) Illustratory phase plane diagram with logarithmic scale for infectives. Type I solutions (major epidemics) occur in upper portion of phase plane above the i-nullcline (dashed horizontal line). Type II solutions appear below i-nullcline. The vertical dashed lines represent the s-nullclines for the two different seasons. The w_i demark the points in the phaseplane when the seasons change and the s-nullcline jumps from β^- to β^+ or vice-versa. c) The phase plane diagram of the time series in (a).

diseases such as measles which are also entrained by an annual seasonal cycle - the epidemics usually coincide with the beginning of the school year. The oscillations are irregular (see measles in New York City and Copenhagen displayed in Figs.2c,d) being characterized by inherent UPCA-like dynamics. Yet conventional epidemiological models are incapable of reproducing this behavior. The UPCA models we are developing provide a technical-fix and form the basis of a theoretical study of the spread of epidemics.

Our focus here is on the synchronization of measles to the seasonal cycle and address the interesting problem of 'fade-out' dynamics. Namely, in some years the epidemic fails to take off at some locations (as in Copenhagen), while at other locations fade-outs are less frequently or never occur (as in New York).

We start with the standard SIR model for epidemics

$$\begin{aligned}\dot{S} &= \mu - \mu S - \beta_0\beta(t)SI \\ \dot{I} &= \beta_0\beta(t)SI - \gamma I \\ \dot{R} &= \gamma I - \mu R. \end{aligned} \tag{9}$$

Here, S, I, R represent the density of susceptible, infected and recovered individuals, respectively, in a constant population of size $N = S+I+R = 1$, scaled to unity without loss of generality. The average life span of an individual and the infectious period are given by μ^{-1} and γ^{-1}, respectively. The infection contact rate is given by $\beta_0\beta(t)$ which is a periodic function of time that takes into account seasonality. Here, we use a square wave forcing with amplitude δ, so that

$$\beta(t) = (1 \pm \delta). \tag{10}$$

The year is divided into two seasons, each of time T. In winter $\beta(t) = \beta^+ = (1+\delta)$ while in summer $\beta(t) = \beta^- = (1-\delta)$. Typical parameter values for measles are $\mu = 0.02$, $\gamma = 75$. For the infection rate we set $\beta_0 = 1700$ and δ varies from $\delta = 0.15$ to $\delta = 0.2$ [5].

After a change of variables, the equations read

$$\begin{aligned} \dot{s} &= m(1-\beta(t)si) - \mu s \\ \dot{i} &= \gamma i\,(\beta(t)s - 1). \end{aligned} \tag{11}$$

where we have introduced

$$i = \frac{\gamma}{\mu} I, \quad s = \frac{\beta_0}{\gamma} S, \quad m = \frac{\mu\beta_0}{\gamma}. \tag{12}$$

Note the structural similarity between the epidemic model eqns. (10) and the phytoplankton model (6), hinting that the dynamics of epidemics and phytoplankton blooms might not be unrelated. Fig.3a provides a time-series of the SIR equations in which the dynamics are chaotic. Since the numbers of infected can change many orders of magnitudes during an epidemic outbreak it is useful to describe the infected numbers by their logarithm, $w = \log i$. Fig.3c shows the solution in (s,w) phase space. Note, that epidemics are strongly phase locked and occur at roughly the same time every year. For some years, however, epidemics do not appear to trigger at all. These are the above mentioned 'skips' or fadeouts that are characteristic of the observed data.

Analytical solutions

In the following analysis we will reconstruct the phase plane by explicit expressions in order to understand the dynamical foundations of epidemic fadeouts. In the process we sketch a method for constructing a first return map that approximates the dynamics of system (9). We make use of the fact that the number of infected individuals change many orders of magnitude during a typical epidemic (see also Schartz et al. [12] who use a similar approach to describe a forced laser systems). Note, that the scaled variable s in (1) is of the order of 1, so that the susceptible death term $-\mu s$ in the first equation of (11) can be neglected with no major effect on the dynamics in the parameter regime of interest.

When $i \gg 1$ we approximate (11) as

$$\begin{aligned} \dot{s} &= -m\beta(t)si \\ \dot{i} &= \gamma i\,(\beta(t)s-1), \end{aligned} \tag{13}$$

and when $i \ll 1$ (11) may be approximated as

$$\begin{aligned} \dot{s} &= m \\ \dot{i} &= \gamma i\,(\beta(t)s-1). \end{aligned} \tag{14}$$

These approximations distinguish two solution types in the phase plane. A solution of the first system would be referred as Type I solution. This solution approximates trajectories located above the s nullcline which lies on the horizontal $i = \frac{1}{\beta s} \approx 1$. Type II solutions approximate trajectories below this nullcline and are solutions of (14).

In Type I solutions, during an epidemic the infected population i changes by many orders of magnitude, while the susceptible population s is always decreasing. Type II solutions are characterized by small numbers of infected individuals, while the susceptibles s are always increasing. The rate of change of s is characterised by different time scales and is slow for Type II solutions but very fast for Type I. Therefore, the system spends most of its time in solution Type II, characterised by minimal infected numbers and a slow build up of susceptibles.

In the following analysis we will make use of the property that the i -nullcline is a straight line located at $s = 1/\beta^{\pm}$. These nullclines are marked as vertical dotted lines in Fig.3b. The effect of seasonality can be regarded as a periodical jump in the horizontal position of this nullcline from $s = 1/\beta^{-}$ to $s = 1/\beta^{+}$ (or vice-versa).

Type I

Eqns. (13) is the well known SIR epidemic model with solution

$$\frac{m}{\gamma}(i(t) - i_0) = s_0 - s(t) + \frac{1}{\beta^+}\ln\frac{s(t)}{s_0}. \tag{15}$$

Note that Type I solutions are initiated when the trajectory crosses the s-nullcline and enters the upper half of the phase plane (Fig.3b). This entry point is the end point of solutions of Type II which we specify shortly. For now asume that the forcing function β^+ remained constant over the entire epidemic period. In this scenario the maximum number of individuals i_m is reached when $s(t) = \frac{1}{\beta^+}$, which gives

$$i_m = \frac{m}{\gamma}(s_0 - \frac{1}{\beta^+} + \frac{1}{\beta^+}\ln\frac{1}{(\beta^+)s_0}) + i_0. \tag{16}$$

In the scenario where the forcing jumps during the epidemic build up it is possible to derive a suitable formulation for i_m.

The epidemic ends when the infective population crosses the s-nullcline (approximately when $i = 1$ or $w = \log i = 0$). At this point the trajectory passes to Type II solutions with initial conditions that can now be determined easily from (15).

Type II

Recall that a year is taken to have only two seasons. It starts in summer at times $t_n,\quad n = 0,2,4,..$ and a low infection rate $\beta^- = 1-\delta$, followed by winter at times $t_n,\quad n = 1,3,5,..$ with high infection rate $\beta^+ = 1+\delta$. Each season is half a year long ($T = \frac{1}{2}$). The overall strategy for solving the system (14) is to calculate the solution in the time intervals between t_n and t_{n+1}. Each time interval will have a constant infection rate (i.e. either β^+ or β^-) and inital values s_n and w_n. For the case of Type II solutions we assume that at time t_0, $\beta(t) = \beta^-$ and the inital conditions are s_0 and $w_0 = 0$.

First, observe that the susceptibles are always increasing with constant rate $s(t) = mt + c_0$ where $c_0 = s_0 - mt_0$. From the phase plane (Fig.3b) we see that the point (s_1, w_1) is a sharp point that results from the change of $\beta(t)$ as it jumps from β^- to β^+. All sharp points (s_n, w_n) are produced by this mechanism. By integrating system (14) between two neighboring sharp points we get

$$\frac{w_1 - w_0}{\gamma} = \frac{w_1}{\gamma} = \beta^- \int_{t_o}^{T} s(t)dt - T + t_0 = \beta^- \int_{t_o}^{T} mt + c_0 dt - T + t_0 \tag{17}$$

$$\frac{w_{n+1} - w_n}{\gamma} = \beta^{\pm} \int_{nT}^{(n+1)T} s(t)dt - T \tag{18}$$

which enable us to express w_n in recursive form to ultimately obtain:

$$\frac{w_n}{\gamma} = \frac{1}{8} m[n^2 + (-1)^n n\delta] + \frac{1}{2}[n(c_0 - 1) + \frac{\delta}{2} c_0((-1)^n - 1)] - c_1 \tag{19}$$

where $c_1 = \frac{1}{2} t_0^2 (1-\delta) + t_0((1-\delta)c_0 - 1)$ and $T = 1/2$ This recursive scheme stops whenever $w_n > 0$. At this point we return back to Type I solutions.

A point where the infected population function reaches a local maximum w_m and $w_m < 0$ is referred as a skipping point. This is because in the case $w_m > 0$ an epidemic could have otherwise been triggered. The difference between a skipping point and an epidemic peak is not just a question of magnitude. During a skipping point the susceptible population continues to build up, while at an epidemic peak the susceptible population declines. From the real data we know that different cities manifest different skipping dynamics. By this analysis of the Type II solution we can deduce the maximum number of skips from the system parameters. Let us define N to be the smallest n where $w_n > 0$, then the overall number of skips is given by $N/2$.

We have extended the argument to show that the number of skips

$$N \sim \frac{1}{m} = \frac{\gamma}{\mu\beta_0}. \tag{20}$$

In order to determine the initial condition to enter again a Type I solution for calculating the next cycle ones need to keep track of time. By defining $w^* = 0$ to be the entrance point for Type I solutions we can calculate the time t^* it would take w_n to reach $w^* = 0$.

$$\frac{w^* - w_n}{\gamma} = -\frac{w_n}{\gamma} = \beta^+ \int_{nT}^{nT+t^*} (mt + c_0)dt + t^*, \tag{21}$$

Rearranging (21), we find $t^* = \frac{-b+\sqrt{(}b-2m(1+\delta)\frac{w_n}{\gamma})}{(1+\delta)m}$ and the initial conditions are given in the form of $(s^*,0)$ where $s^* = m(nT + t^*) + c_0$.

Putting all this together, we have arrived at a method that allows us in principal to deduce a first return map of the chaotic SIR equations. In so doing we have arrived at an explicit expression for the number of fadeouts that can be expected, and the conditions which control this number. Thus for the first time fadeouts can be understood and predicted in terms of explicit epidemiological parameters. Finally, we are able to see more directly the mechanism of synchronization through unidirectional coupling or one-way driving, and we are planning to explore this in much more detail with this framework.

CONCLUDING REMARKS

Over the last ten years physicists have invested a great deal of effort in studying the theory of synchronization. Most of this effort has been directed towards understanding chaotic synchronization of model systems. Much less is known about synchronization phenomena in natural systems [7, 13]. In this paper we investigated four different forms of complex biological synchronization using a range of different theoretical approaches, in an attempt to cross the bridge between the fields of physics and ecology. We believe that with the help of tools from nonlinear dynamics, these examples should provide inspiration and motivation for furthering our understanding of the diverse but subtle aspects of synchronization.

ACKNOWLEDGMENTS

L.S., A.H. and R.O. were supported by the McDonnell Foundation and EU grant Phytoplankton On-Line; B.B. was supported by the German Volkswagen-Stiftung.

REFERENCES

1. Maynard-Smith, J., *Anim. Behav.*, **25**, 1–9 (1977).
2. Myers, J. H., *Advances in Ecol. Research*, **18**, 179–242 (1988).
3. Kaneko, K., *Physica D*, **41**, 137–149 (1990).
4. Earn, D. J. D., Rohani, P., Bolker, B. M., and Grenfell, B., *Science*, **287**, 667–670 (2000).
5. Earn, D. J. D., Levin, S. A., and Rohani, P., *Science*, **290**, 1360–1363 (2000).
6. Cazelles, B., *Phys Rev. E*, in press.
7. Blasius, B., Huppert, A., and Stone, L., *Nature*, **399**, 354–359 (1999).
8. Blasius, B., and Stone, L., *Int. J. Bif. and Chaos*, **10**, 2361–2380 (2000).
9. Cazelles, B., and Boudjema, G., *Am. Nat.*, **157**, 670–676 (2001).
10. Vandermeer, J., Stone, L., and Blasius, B., *Chaos, Solitons and Fractals*, **12**, 265–276 (2001).
11. Huppert, A., Blasius, B., and Stone, L., *American Naturalist*, in press.
12. Schwartz, I. B., and Erneux, T., *SIAM J. Appl. Math.*, **54**, 1083–1100 (1994).
13. Blasius, B., and Stone, L., *Nature*, **406**, 846–847 (2000).

AUTHOR INDEX

S

T

U

V

W

Y

Z